The Evolution of Human Populations in Arabia

Vertebrate Paleobiology and Paleoanthropology Series

Edited by

Eric Delson
Vertebrate Paleontology, American Museum of Natural History,
New York, NY 10024, USA
delson@amnh.org

Ross D. E. MacPhee
Vertebrate Zoology, American Museum of Natural History,
New York, NY 10024, USA
macphee@amnh.org

Eric J. Sargis
Anthropology, Yale University
New Haven, CT 06520, USA
eric.sargis@yale.edu

Focal topics for volumes in the series will include systematic paleontology of all vertebrates (from agnathans to humans), phylogeny reconstruction, functional morphology, Paleolithic archaeology, taphonomy, geochronology, historical biogeography, and biostratigraphy. Other fields (e.g., paleoclimatology, paleoecology, ancient DNA, total organismal community structure) may be considered if the volume theme emphasizes paleobiology (or archaeology). Fields such as modeling of physical processes, genetic methodology, nonvertebrates or neontology are out of our scope.

Volumes in the series may either be monographic treatments (including unpublished but fully revised dissertations) or edited collections, especially those focusing on problem-oriented issues, with multidisciplinary coverage where possible.

For other titles published in this series, go to
www.springer.com/series/6978

The Evolution of Human Populations in Arabia

Paleoenvironments, Prehistory and Genetics

Edited by

Michael D. Petraglia

Research Laboratory for Archaeology and the History of Art, School of Archaeology
University of Oxford, UK

Jeffrey I. Rose

Department of Anthropology and Geography,
Oxford Brookes University, UK

🐎 Springer

Editors
Michael D. Petraglia
Research Laboratory for Archaeology
  and the History of Art
School of Archaeology
University of Oxford
UK
michael.petraglia@rlaha.ox.ac.uk

Jeffrey I. Rose
Department of Anthropology and Geography
Oxford Brookes University
UK
jeffrey.i.rose@gmail.com

ISBN 978-90-481-2718-4   e-ISBN 978-90-481-2719-1
DOI 10.1007/978-90-481-2719-1
Springer Dordrecht Heidelberg London New York

Library of Congress Control Number: 2009928843

Cover illustration: Desert landscape in the Sharqiyah Sands, Oman, photo by Daniel Richter. Stone tool
from Wadi Qilfah 4, photo by Jeffrey Rose.

Printed on acid-free paper

Springer is part of Springer Science+Business Media (www.springer.com)

Preface

The romantic landscapes and exotic cultures of Arabia have long captured the interests of both academics and the general public alike. The wide array and incredible variety of environments found across the Arabian peninsula are truly dramatic; tropical coastal plains are found bordering up against barren sandy deserts, high mountain plateaus are deeply incised by ancient river courses. As the birthplace of Islam, the recent history of the region is well documented and thoroughly studied. However, legendary explorers such as T.E. Lawrence, Wilfred Thesiger, and St. John Philby discovered hints of a much deeper past during their travels across the subcontinent. Drawn to Arabia by the magnificent solitude of its vast sand seas, these intrepid adventurers learned from the Bedouin how to penetrate its deserts and returned with stirring accounts of lost civilizations among the wind-swept dunes.

We now know that, prior to recorded history, Arabia housed countless peoples living a variety of lifestyles, including some of the world's earliest pastoralists, communities of incipient farmers, fishermen dubbed the "Ichthyophagi" by ancient Greek geographers, and Paleolithic big-game hunters who were among the first humans to depart their ancestral homeland in Africa. In fact, some archaeological investigations indicate that Arabia was inhabited by early hominins extending far back into the Early Pleistocene, perhaps even into the Late Pliocene.

The extraordinarily rich cultural record of the region is set against a tapestry that can only be described as one of the hottest and most inhospitable deserts in the world. Yet, geologists have discovered that the bleak Arabian environment swung many times in the past between hyperarid and wet phases, which resulted in the formation of interior lakes, perennial rivers, coastal springs, mangrove swamps, and large estuaries. The archaeological record demonstrates that our ancestors thrived along these ancient waterways. This is why early explorers traversing the Empty Quarter consistently reported finding scatters of stone tools littering the surface, abundant evidence for prehistoric occupation. One wonders what happened to these populations at the onset of adverse arid periods; how rapid were these environmental changes across the peninsula; did they cause groups to contract, disperse, or die out? These are just some of the questions we have set out to answer in this book, given their bearing on past, contemporary, and future peoples in Arabia.

Both of the volume's editors were originally drawn to Arabia for its spectacular, albeit rather poorly known prehistoric record. Although we come from different academic backgrounds, we are both interested in addressing a similar set of questions. After having worked in India for many years, Petraglia was drawn to Arabia as a step towards tracing Out of Africa dispersals along the Indian Ocean rim. Concurrently and independently, Rose was conducting fieldwork in southern Arabia, aimed at placing the Paleolithic archaeological record there in a temporal and inter-regional context. The editors' paths first crossed in 2006 at the annual Seminar for Arabian Studies held at the British Museum in London. We began a dialogue at the conference that continued through to the next day in Cambridge, discussing the need for a fresh evaluation of the Arabian Paleolithic record in light of contemporary issues in human evolutionary studies. As a result of this fortuitous meeting, we organised a special session at the Arabian Seminar in 2007 entitled "Defining the Palaeolithic of Arabia." Following this session, we felt the need to further expand upon the papers by broadening our understanding of the archaeological, environmental, and genetic history of the region. In the process of putting together this book, we decided that more recent time periods should be represented as well, in order to include demographic and environmental changes that occurred in Arabia through the Early Holocene. It quickly became apparent to us that geographic variations across the peninsula are a critical facet of Arabian prehistory; therefore, we sought contributions from scholars who could address human occupation in different regions and microclimates of the subcontinent. To contextualize the chapters dealing with archaeology, we felt that including complementary research in paleoenvironmental studies and human population genetics was essential.

Although we have tried to fill out the temporal and spatial picture as much as possible, this book still has considerable gaps in coverage owing to a dearth of fieldwork and our fragmented body of knowledge. Indeed, as Arabia is 2.5 million km^2 in size, many areas are yet to be surveyed, and much evidence still awaits discovery. While there have been important archaeological studies in the recent past, comprehensive, interdisciplinary research is woefully lacking. Despite these shortcomings, we felt that this book was badly needed; too many global overviews depict Arabia simply as a blank spot on the map. This must be rectified if we are to understand historical events and evolutionary processes, such as the mechanisms and patterns that have governed demographic change over the course of the Pleistocene and Holocene. We should no longer accept a situation where arrows of hominin dispersal are drawn across Arabia, while no localities are cited as potential evidence. Indeed, we often read about the spread of domestication from elsewhere in the Near East, yet there is rarely discussion of the innovative cultural experiments that took place in Arabia, incorporating plant and animal domestication strategies with different resource management techniques such as water manipulation.

In reviewing this book, it will become apparent to the reader that there is myriad research to be done in Arabian prehistory. Many of the questions we pose, such as the impact of climate change on human populations, are not limited to the pursuit of academic knowledge. Indeed, obtaining information about human responses to previous episodes of climate change may produce insights into future resource management strategies and practices. As we find ourselves living on an increasingly warmer and drier earth, this information might come in handy some day.

This book was enthusiastically supported by Eric Delson, one of the Series Editors for Springer. Each individual chapter underwent formal peer-review, and we thank the following individuals for providing comments on draft papers: Abdullah Alsharekh, Geoff Bailey, Amanuel Beyin, Paolo Biagi, Roger Blench, Nicole Boivin, Ueli Brunner, Vicente Cabrera, Rob Carter, Remy Crassard, Harriet Crawford, Nick Drake, Sarah Elton, Dorian Fuller, Andy Garrard, Naama Goren-Inbar, Michael Haslam, Lamya Khalidi, Toomas Kivisild, Krista Lewis, Lisa Maher, Anthony Marks, Joy McCorriston, Maru Mormina, Adrian Parker, Dan Potts, Jakub Ridl, Garth Sampson, Julie Scott-Jackson, Margareta Tengberg, Alan Turner, Philip VanPeer, Pierre Vermeersch, Ghanim Wahida, and Tony Wilkinson. We thank Anthony Marks for his insightful concluding remarks in the closing section of this book. We hope that readers will find this compilation to be of interest to them, and that it will inspire others to join in future endeavors waiting to be carried out in Arabia.

Michael D. Petraglia
Jeffrey I. Rose
Oxford, UK

Contents

Contributors

Khaled K. Abu-Amero
Shafallah Genetics Medical Center, Lusail Street, P.O. Box 4251, Doha, Qatar
abuamero@gmail.com

Abdullah Alsharekh
Department of Archaeology and Museology, King Saud University, P.O. Box 2456, Riyadh, 11451, Kingdom of Saudi Arabia
alsharekh@hotmail.com

Geoff Bailey
Department of Archaeology, King's Manor, University of York, York, YO1 7EP, UK
gb502@york.ac.uk

Mark J. Beech
Historic Environment Department, Abu Dhabi Authority for Culture and Heritage, P.O. Box 2380, Abu Dhabi, United Arab Emirates
mark.beech@cultural.org.ae

Roger Blench
Kay Williamson Educational Foundation, 8 Guest Road, Cambridge, CB1 2AL, UK
R.Blench@odi.org.uk

Nicole Boivin
Research Laboratory for Archaeology and the History of Art, School of Archaeology, University of Oxford, Dyson Perrins Building, South Parks Road, Oxford, UK, OX1 3QY
nicole.boivin@rlaha.ox.ac.uk

Vicente M. Cabrera
Genética, Biología, Universidad de La Laguna, La Laguna, 38271 Tenerife, Spain
vcabrera@ull.es

Viktor Černý
Institute of Archaeology of Academy of Sciences of the Czech Republic, Prague, v.v.i., Letenská 4, Prague 1, CZ-11801, Czech Republic
cerny@arup.cas.cz

Rémy Crassard
Leverhulme Centre for Human Evolutionary Studies, University of Cambridge, The Henry Wellcome Building, Fitzwilliam Street, Cambridge, CB2 1QH, UK
rc461@cam.ac.uk

Nick Drake
Department of Geography, King's College London, Strand, London, WC2R 2LS, UK
n.drake@kcl.ac.uk

Christopher M. Edens
American Institute for Yemeni Studies, P.O. Box 2658, Sana'a, Republic of Yemen
aiysyem@y.net.ye

Francesco G. Fedele
Laboratory of Anthropology, University of Naples "Federico II", via Mezzocannone 8,
80134 Naples, Italy
ffedele01@yahoo.it

Carlos A. Fernandes
Centro de Biologia Ambiental, Universidade de Lisboa, C2 2.3.25, 1749-016 Campo
Grande, Lisboa, Portugal
CaFernandes@fc.ul.pt

Dorian Q. Fuller
Institute of Archaeology, University College London, 31-34 Gordon Square,
London, WC1H 0PY, UK
d.fuller@ucl.ac.uk

Ana M. González
Genética, Biología, Universidad de La Laguna, La Laguna, 38271, Tenerife, Spain
amglez@ull.es

Reto Jagher
Institute of Prehistory and Science in Archaeology, University of Basel,
Spalenring 145, CH-4055, Basel, Switzerland
Reto.Jagher@unibas.ch

Lamya Khalidi
Centre d'Études Préhistoire, Antiquité, Moyen Age, UMR 6130 – CNRS,
Université de Nice, Sophia Antipolis, 250, rue Albert Einstein, Sophia-Antipolis,
F.-06560 Valbonne, France
lamya.khalidi@gmail.com

José M. Larruga
Genética, Biología, Universidad de La Laguna, La Laguna, 38271, Tenerife, Spain
jlarruga@ull.es

Lisa A. Maher
Leverhulme Centre for Human Evolutionary Studies, University of Cambridge,
The Henry Wellcome Building, Fitzwilliam Street, Cambridge, CB2 1QH, UK
l.maher@human-evol.cam.ac.uk

Anthony E. Marks
Department of Anthropology, Southern Methodist University, 3225 Daniel Avenue,
Heroy Building 408, Dallas, TX, 75275, USA
amarks@mail.smu.edu

Louise Martin
Institute of Archaeology, University College London, 31-34 Gordon Square,
London, WC1H 0PY, UK
louise.martin@ucl.ac.uk

Joy McCorriston
Department of Anthropology, The Ohio State University, 244 Lord Hall,
124 W. 17th Ave., Columbus, OH, 43210, USA
mccorriston.1@osu.edu

Ali Al Meqbali
Historic Environment Department, Abu Dhabi Authority for Culture and Heritage
(ADACH), Al Ain National Museum, P.O. Box 15715, Al Ain,
United Arab Emirates
aly_elmegbali@yahoo.com

Adrian G. Parker
Department of Anthropology and Geography, Oxford Brookes University,
Oxford, OX3 0BP, UK
agparker@brookes.ac.uk

Michael D. Petraglia
Research Laboratory for Archaeology and the History of Art,
School of Archaeology, University of Oxford, Dyson Perrins Building,
South Parks Road, Oxford OX1 3QY, UK
michael.petraglia@rlaha.ox.ac.uk

Daniel T. Potts
Department of Archaeology, The University of Sydney, NSW 2006, Australia
dpot3385@usyd.edu.au

Jakub Rídl
Institute of Molecular Genetics of Academy of Sciences of the Czech Republic,
Prague, v.v.i., Vídeňská 1083, Prague 4, CZ-14220, Czech Republic
ridlj@img.cas.cz

Jeffrey I. Rose
Department of Anthropology and Geography, Oxford Brookes University,
Oxford, OX3 0BP, UK
jeffrey.i.rose@gmail.com

Julie Scott-Jackson
PADMAC Unit, University of Oxford, Institute of Archaeology, 36 Beaumont Street,
Oxford, OX1 2PG, UK
julie.scott-jackson@arch.ox.ac.uk

William Scott-Jackson
PADMAC Unit, University of Oxford, Institute of Archaeology, 36 Beaumont Street,
Oxford, OX1 2PG, UK
william.scott-jackson@arch.ox.ac.uk

Walid Yasin Al-Tikriti
Historic Environment Department, Abu Dhabi Authority for Culture
and Heritage (ADACH), Al Ain National Museum, P.O. Box 15715,
Al Ain, United Arab Emirates
wyasin11@yahoo.com

Hans-Peter Uerpmann
Institut für Ur- und Frühgeschichte und Archäologie des Mittelalters
Naturwissenschaftliche Archäologie, Eberhard-Karls-Universität Tübingen,
Rümelinstr., 19-23, D-72070, Tübingen, Germany
hans-peter.uerpmann@uni-tuebingen.de

Margarethe Uerpmann
Institut für Ur- und Frühgeschichte und Archäologie des Mittelalters,
Naturwissenschaftliche Archäologie, Eberhard-Karls-Universität Tübingen,
Rümelinstr., 19-23, D-72070, Tübingen, Germany
margarethe.uerpmann@uni-tuebingen.de

Vitaly I. Usik
Institute of Archaeology, National Academy of Science, B. Khmelnitsky Street 15,
01030, Kiev-30, Ukraine
vitaly_usik@yahoo.com

Ghanim Wahida
McDonald Institute for Archaeological Research, University of Cambridge,
Downing Street, Cambridge CB2 3ER, UK
ghanimwahida@hotmail.com

Tony J. Wilkinson
Department of Archaeology, University of Durham, South Road, Durham,
DH1 3LE, UK
t.j.wilkinson@durham.ac.uk

Chapter 1
Tracking the Origin and Evolution of Human Populations in Arabia

Jeffrey I. Rose and Michael D. Petraglia

Keywords Demography • Dispersals • Genetics • Holocene • Paleolithic • Quaternary Environments • Refugia

All this opens out a fascinating field for comprehensive research, involving several sciences, biological as well as physiographical, which it would be richly worth while for an active prehistorian to undertake.

(Caton-Thompson, 1957: 384)

Introduction

Take a glance at any world map and it is immediately apparent that Arabia occupies a critical geographic position, linking Africa, Europe, and Asia. This singular point echoes across every chapter, noted by nearly every author who has contributed to this volume. It is odd, then, that the prehistory of such a critical corner of global real estate has languished in such obscurity until now. As archaeologists begin to shed further light on this relatively unknown region, the emerging picture seems to underscore what is so cartographically obvious – that the Arabian peninsula has probably played a central role in the dispersal of our species and closely related ancestors.

The geographic designation 'Arabian peninsula' refers to the 2.5 million km² landmass fringed by the Red Sea to the west, Arabian Sea to the south, and Persian Gulf to the east. Politically, it encompasses the Kingdom of Saudi Arabia, the Hashemite Kingdom of Jordan, the Republic of Yemen, the Sultanate of Oman, the United Arab Emirates, the State of Qatar, the Kingdom of Bahrain, and the State of Kuwait.

J.I. Rose (✉)
Department of Anthropology and Geography,
Oxford Brookes University, Oxford, OX3 0BP, UK
e-mail: jeffrey.i.rose@gmail.com

M.D. Petraglia
Research Laboratory for Archaeology and the History of Art
School of Archaeology, University of Oxford, South Parks Road,
Oxford, OX1 3QY, UK
e-mail: michael.petraglia@rlaha.ox.ac.uk

Arabia's most evocative landscape features are the expansive dune fields that sprawl across much of the subcontinent, filling the huge interior basins with heaping deposits of rust-colored sand. Juxtaposed in and around these vast wastelands are lush sub-tropical forests, deflated gravel plains, jagged mountain ranges, and some 7,000 km of coastline.

History of Prehistoric Research in Arabia

Paleoenvironmental researchers have discovered that climatic conditions within Arabia's different ecological niches were far from stable over the course of the Quaternary, swinging between wet and dry extremes in the past 2 million years. During the Late Pliocene and Early Pleistocene, some Arabian river systems carried volumes of water equivalent to the Nile (Thompson, 2000). It is reasonable to suppose that these significant environmental fluctuations had a profound effect on the development of early and later human populations in the region.

Recent archaeological and genetic research suggests that human occupation in Arabia was as rich and varied over time as the landscapes upon which these early inhabitants dwelt. The peninsula was one of the first stops for our incipient human ancestors expanding out of Africa. It is a stone's throw from East Africa, where a wealth of hominin fossils have been unearthed (not to mention the oldest anatomically modern human remains), the earliest farming communities developed along its northern margin, and at one point it was surrounded by the world's first three complex civilizations. Despite these significant biological and cultural evolutionary milestones documented around Arabia, the peninsula itself has remained a virtual *terra incognita* in prehistoric studies.

That is not to say the region has been ignored. Scholars have long speculated as to the role of Arabia in the development of our species. Seventy-five years ago, Henry Field dubbed this part of the world "the cradle" of early humans and suggested that "southwestern Asia, including the African territory, may well have nurtured the development of Homo

Fig. 1 Photos of diverse
ecosystems found throughout the
Arabian peninsula

Longitudinal dune overlooking a drainage
channel in the Wahiba Sands desert

Top of Jebel Harim in the Musandam peninsula, looking west toward
the Persian Gulf

Cows and camels grazing along the grass-covered banks of Wadi Darbat
in the Dhofar Mountains, Oman

Groundwater fed spring at the southern end
of the Nejd Plateau, Dhofar Region, Oman

Terraced fields in the Yemen Highlands

sapiens" (Field, 1932: 426). For two decades following his presage remark, scientists and explorers such as St John Philby (1933), Bertram Thomas (1938), Gertrude Caton-Thompson (1939, 1954, 1957), Henry Field (1951, 1955, 1956, 1958, 1960a, 1960b, 1961), Frederick Zeuner (1954), and Wilfred Thesiger (1959) combed the surface of the sub-continent. In nearly every case, they reported finding stone tools associated with old river beds and dry lake basins, leading to the conclusion that the barren interior had once been significantly more conducive to supporting prehistoric hunter-gatherer communities. Even at this early stage, the question of Pleistocene connections between East Africa and South Arabia across the Bab al Mandab was under consideration. Given the conclusions reached by various contributors in this book, it is interesting to note that Caton-Thompson (1957) too observed minimal evidence for demographic exchange across the Red Sea.

A series of obstacles such as war, isolationism, and impenetrable geography significantly impeded research during the latter half of the twentieth century, at which time Paleolithic studies were more or less abandoned while rigorous field-work was conducted in nearby, more accessible parts of the Levant and East Africa. Notable exceptions to this were Whalen's surveys in Yemen and Oman (Whalen and Pease,

1991; Whalen and Schatte, 1997; Whalen et al., 2002), de Maigret's Italian Mission to North Yemen (de Maigret, 1981, 1984, 1985, 1986), the joint Yemeni-Soviet Expedition to South Yemen (Amirkhanov, 1987, 1991, 1994, 2006), the Danish Expedition to Qatar (Kapel, 1967), and the Comprehensive Survey Project in Saudi Arabia (e.g., Adams et al., 1977; Masry, 1977; Zarins et al., 1979, 1980, 1981, 1982; Whalen et al., 1981, 1983, 1984, 1988). Although few and infrequent, these expeditions surveyed huge tracks of land, recording a plethora of lithic scatters that underscore the scope of habitation throughout the region. By the close of the twentieth century, it was abundantly clear that prehistoric occupation in Arabia was extensive, yet stratified and datable Paleolithic sites continued to elude archaeologists (Petraglia and Alsharekh, 2003).

A seemingly unrelated scientific development in the late 1990s had a dramatic impact on prehistoric research in Arabia. While studying the phylogenetic structure of modern human groups distributed in the Horn of Africa, a team of geneticists discovered traces of one of our species' most ancient mitochondrial DNA lineages, a branch that was previously thought to have its roots in Asia (Quintana-Murci et al., 1999). The occurrence of haplogroup M markers in East Africa indicated to scientists that the 'Arabian Corridor' (i.e., Yemen, Oman, and the U.A.E.) served as a conduit for populations moving between Africa and Asia; thereby confirming the existence of a hypothesized southern dispersal route out of Africa (Tchernov, 1992; Lahr and Foley, 1994, 1998; Stringer, 2000).

The haplogroup M genetic discovery recalibrated research agendas and reinvigorated fieldwork activities by empirically demonstrating the Arabian peninsula's geographic prominence in early human expansion. As geneticists shone the spotlight on Arabia, the pace of discovery quickened accordingly. Within the past few years, datable Paleolithic sites have finally been discovered. New high-resolution systematic surface surveys provide greater understanding of stone tool distribution across the landscape. A rapidly growing body of paleoenvironmental data allow for reconstructions of variable climatic conditions over the course of the Pleistocene. Genetic samples obtained from modern Arabian populations enable us to assess their position on the human family tree.

The recent flurry of discovery has now reached a boiling point and permits the first comprehensive examination of Arabian prehistory across multiple disciplines. Hence, we have compiled this volume to present, synthesize, and discuss the state of research in Arabia with an emphasis on the dynamic relationship between human and landscape evolution. Many of the contributions are pioneering studies that force us to refine, or in some cases entirely re-evaluate fundamental issues in human prehistory. *The Evolution of Human Populations in Arabia* is intended to engender interest in the region, to serve as a foundation for future

research, to inform scholars in related disciplines, and to facilitate inter-disciplinary dialogue.

We have organized the volume into five themed parts that examine different facets and phases of Arabian prehistory. 'Part I: Quaternary Environments and Demographic Response' is concerned with evaluating signals of Pleistocene and Holocene climate change. These chapters describe the evolution of the landscape and the impact of climate change on human populations in Arabia. The ensuing part, 'Genetics and Migration' examines the phylogenetic structure of primate populations currently living in Arabia (both humans and baboons) in order to assess the relationship of these groups within their respective family trees. New data are used to address issues such as early human expansion through the Arabian Corridor and different migratory contributions to the Arabian gene pool. Part III 'Pleistocene Archaeology' and Part IV 'Holocene Prehistory' present new archaeological findings across the subcontinent, at sites ranging from the Lower Paleolithic to the Neolithic periods. These discoveries are considered within a broader regional context and in relation to the genetic and paleoenvironmental records. The final part discusses these new data from the perspective of a scholar who has spent nearly half a century conducting archaeological investigations in and around Arabia.

Quaternary Environments and Demographic Response

Since the underlying premise of this book is the inexorable link between humans and environments in Arabia, the opening section provides detailed descriptions of Arabian landscapes and the history of climate change across the subcontinent. These contributions provide the scenery upon which the drama of human evolution in Arabia was enacted. The authors weave together paleoenvironmental and archaeological evidence to assess the relationship between climate change and demographic response. In several later chapters, the predictive models set forth in Part I are used to frame the Pleistocene and Holocene archaeological records.

As the most probable starting point for hominins entering Arabia, we begin with an examination of coastal landscapes along the Red Sea and demographic movement across this waterway based on ongoing underwater research around the Farasan Islands (Bailey, 2009). Bailey reviews evidence indicating that there was no land bridge linking Africa and Arabia across the Bab al Mandab Strait since the Pliocene period. In contrast to recent arguments that have been made for the development of aquatic subsistence strategies by early humans, which facilitated their rapid expansion out of Africa along the rim of the Indian Ocean (e.g., Stringer, 2000; Mellars, 2006), Bailey concludes that "the case for marine resources as a primary factor in promoting a process of coastal colonization by early human populations whether anatomically modern or earlier remains weak." Furthermore, he writes "the belief that general aridity prevailed during glacial periods and would have deterred all but ephemeral settlement is certainly an oversimplification, and indeed substantially incorrect."

Parker (2009) summarizes the paleoenvironmental record of climate change over the last 350 ka, presenting an exhaustive database of proxy environmental signals used to calculate pluvial-arid oscillations. Until quite recently, we have had only a vague sketch of paleoenvironmental conditions during the late Middle and Upper Pleistocene periods. Parker presents substantial evidence for one or more wet pulses during MIS 3, a heretofore poorly understood environmental phase.

Additionally, there are indications for a series of pluvial episodes during MIS 6, an isotopic stage that was initially characterized by prolonged aridity (Anton, 1984). These and other findings have a direct bearing on the timing and nature of hominin movement onto the peninsula, as they push the feasibility of hunter-gatherer range expansions into Arabia as far back as 200 ka, if not earlier.

Wilkinson (2009) discusses paleoclimatic and archaeological evidence from the highlands of Yemen and adjacent geomorphic zones. The terrain of southwestern Arabia is predominantly comprised of mountains and upland plateaus between 2,000 and 3,600 m above sea level. These Tertiary volcanic and granite peaks trap moisture from the Indian Ocean monsoon, which is responsible for depositing between 200 and 700 mm of rainfall per annum. This favorable climatic regime played a key role in the development of Holocene agricultural communities, affecting not only the

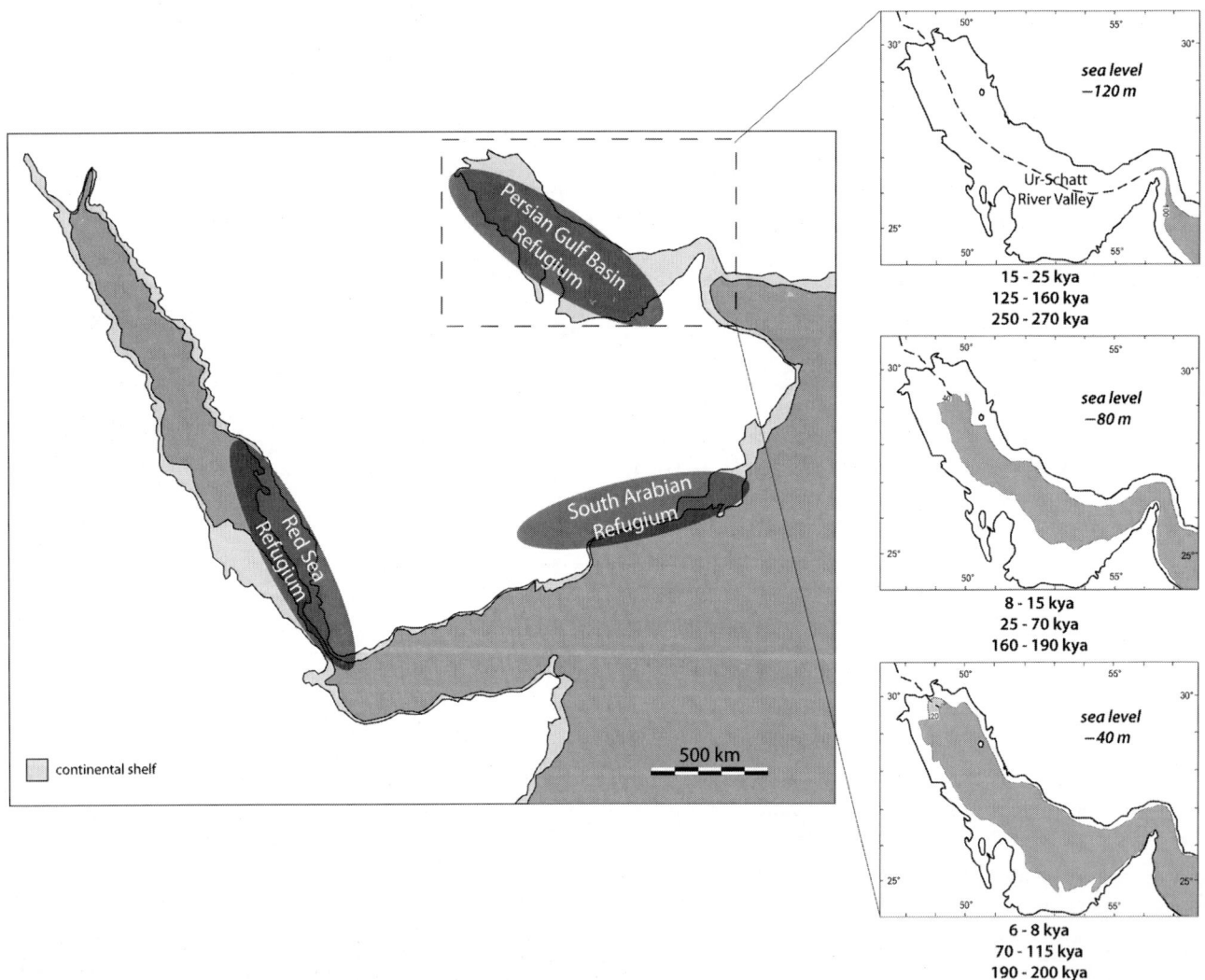

Fig. 2 Map of Arabian refugia showing extent of landscape during periods of reduced sea level. The Persian Gulf shorelines at various sea level increments have been adapted from Kennett and Kennett (2006; Fig. 2, pp. 72). Date ranges were calculated based on global eustatic sea level curves as well as the Late Pleistocene/Holocene record of sea level transgression into the Gulf basin (e.g., Lambeck, 1996)

highlands but surrounding Red Sea and Arabian Sea coastal plains that receive a large portion of the runoff.

Together, the chapters comprising Part I paint the picture of a mosaic landscape encompassing a multitude of varied micro-environments. From the authors' descriptions, we can begin to articulate at least three coherent core zones that served as population refugia during environmental downturns: the Red Sea coastal plain, the Dhofar Mountains and adjacent littoral zone in Yemen and Oman, and the emerged floodplain within the Persian Gulf basin (Fig. 2).

To understand the development of human populations in Arabia during the Pleistocene and Early Holocene, it is helpful to conceptualize populations tethered to these different refugia, expanding and contracting from such habitats during cycles of amelioration and desiccation. These zones represent the only abundant and predictable sources of freshwater anywhere to be had in Arabia during periods of aridity.

Genetics and Migration

New genetic research permits further examination of the various demographic scenarios set out in Part I. We include a description of the mtDNA genetic structure of modern Yemeni population (Rídl et al., 2009), a phylogeographic analysis of Arabian mtDNA lineages throughout the peninsula (Cabrera et al., 2009), and an examination of *Papio hamadryas* mtDNA variation on either side of the Red Sea (Fernandes, 2009).

Rídl and colleagues synthesize multiple datasets published over the last decade (i.e., Richards et al., 2000; Thomas et al., 2002; Richards et al., 2003; Kivisild et al., 2004; Rowold et al., 2007; Černý et al., 2008) to address whether populations in southern Arabia bear traces of an initial human migration out of East Africa, and to what extent subsequent demographic input has affected the Yemeni gene pool. Their analysis is concerned with the basal nodes upon the human phylogenetic tree. Rídl et al. observe that the modern Yemeni population exhibits a mix of sub-Saharan haplogroups (L-type derivatives) and West Eurasian haplogroups (M-type and N-type derivatives), leading them to conclude that "the overall composite nature of Yemeni gene pool also supports its probable role as a recipient of gene flows from different parts of Africa and Eurasia." Most African L-type haplogroups recorded in Yemen are attributed to relatively recent East African slave trade.

Cabrera et al. (2009) use the mtDNA evidence from Arabia to assess the relationship between haplogroups M and N, and whether their phylogeographic distribution indicates one or more dispersal events. The authors argue that the commonly held belief that haplogroup L3 split into haplogroups M and N within Africa and subsequently expanded outward does not adequately explain modern phylogeographic patterning. They support an exodus during MIS 5, in which early human groups expanded out of Africa during a favorable episode sometime around 100 ka.

Fernandes (2009) examines the timing of mtDNA coalescence between *Papio hamadryas* baboons living on either side of the Red Sea; this is particularly germane since every recorded Pleistocene expansion of sub-Saharan African species into southwest Asia is associated with a hominin dispersal. As the only living evidence (other than *Homo sapiens*) of a successful primate colonization out of Africa and into Arabia during the Middle/Late Pleistocene, Fernandes uses these data to examine Pleistocene faunal movement across the Bab al Mandab. He observes that the divergence ages indicating when the ancestral baboon population crossed the Red Sea into Arabia correspond with periods of *high* sea level rather than low, supporting the argument that there was no Pleistocene land bridge connecting Africa and Arabia.

All three genetic analyses are largely in agreement. The phylogenetic structures of both human and baboon populations suggest minimal demographic movement across the Bab al Mandab between Africa and Arabia in the Upper Pleistocene. In fact, these genetic data point to more significant genetic exchange with the Levant at this time.

Paleolithic Archaeology

Early/Middle Pleistocene (ca. 2.0 Ma–200 ka)

There is clear and abundant evidence for the presence of Acheulean hominins in Arabia (Petraglia, 2003). The identification of Acheulean sites is of course significant for understanding the earliest adaptations of hominins in Arabia and for assessing their landscape usage behaviors. Petraglia et al. (2009) address and discuss archaeological finds at Acheulean site complexes along the Wadi Fatimah, near the Red Sea in Saudi Arabia, and from the hillslopes near the modern town of Dawādmi in central Saudi Arabia.

In considering the findings at Dawādmi and Wadi Fatimah, Petraglia et al. note that sites within both complexes are positioned in carefully chosen locations. They are situated on elevated spots that afforded early hominins with high visibility vantage points, enabling them to spot plant resources, standing water, and animal movements across large distances. The sites were found near springs or stream channels, and in proximity to abundant lithic resources in the form of andesite and rhyolite dikes, where some stone tool quarries, with giant cores, manufacturing debris and finished tools were identified.

Given the spatial distribution of Acheulean sites in Arabia, Petraglia and colleagues postulate hominins initially spread into Arabia along the coasts. They travelled into the interior via river valleys during brief wet pulses, and contracted back into surrounding refugia at the onset of more arid conditions. Therefore, interior occurrences probably represent short term population expansions associated with episodic pluvial phases. As one of the largest systems draining into the Red Sea, Wadi Fatimah would have been a particularly plentiful source of water, as well as attracting diverse plant and animal species. The authors also discuss inter-regional patterning among Arabian Acheulean tools and types found elsewhere in the Levant, Africa, and India. Examining the morphology of the Large Cutting Tools, Petraglia and colleagues claim that the Arabian tools are similar to those identified as Acheulean in Africa.

Late Middle and Upper Pleistocene (ca. 200–12 ka)

Wahida et al. (2009) describe an assemblage of surface artifacts collected at Jebel Barakah in the Western Region of the Abu Dhabi Emirate, which they suggest belongs to an early phase of the Middle Paleolithic. The artifacts were manufactured on high quality chert derived from a deflated surface capping the hill. The primary reduction strategy observed within the assemblage is the centripetal Levallois technique, which grades into biconical radial and high-backed radial cores. While the authors note the occurrence of the bidirectional Levallois technique and Nubian Type I cores, they emphasize the predominance of radial cores and debitage from the Jebel Barakah findspots. In addition to an array of non-diagnostic tool types such as denticulates and notches, Wahida et al. report a cordiform bifacial handaxe and bifacially retouched sidescraper.

Scott-Jackson et al. (2009) collected a similar array of Paleolithic material in their surveys around the interior foot-hills of the Hajar mountain chain in Sharjah and Ras al Khaimah Emirates. Artifacts were recovered from large workshops situated on limestone ridges approximately 300 m in elevation with views of the Al Madam plain to the west and wadi channels directly below. The lithic assemblages described by Scott-Jackson and colleagues were categorized into Groups A1, A2, A3, and B1 based on the presence/absence of certain techno-typological features. All of the "A Group" artifacts contained some Levallois variant, ranging between centripetal and unipolar-convergent, as well as an assortment of biconical and high-backed radial cores. As for tools, there was a high frequency of sidescrapers and bifacial

implements including bifacially-worked sidescrapers with flat invasive retouch, backed bifacial knives, foliates, limandes, and, in the case of the A3 Group, a large elongated bifacial handaxe. The combination of Levallois cores associated with Micoquian-like bifacial forms hints at a Middle Paleolithic chronological attribution.

Archaeological survey results from investigations in central Oman are presented by Jagher (2009). In 2007 and 2008, Jagher and his team conducted a systematic survey of the Huqf region, recording over 350 archaeological sites. The Huqf is a series of low hills encompassing a diverse array of ecological habitats sandwiched between the Jidat Al Harasis plain to the west and Indian Ocean coastline to the east. Since the landscape consists predominantly of deflated surfaces within minimal aeolian cover, archaeological visibility was deemed exceptionally high. Jagher estimates some workshop localities in southern Huqf comprise up to 1 million artifacts, if not more. Jagher suggests the frequency and density of material is due to the extended usage of these sites over long periods of time.

Crassard (2009) systematically analyzes and categorizes a series of surface sites discovered in the eastern reaches of the Wadi Hadramaut drainage system in Yemen. The author recognizes three general categories of core reduction: centripetal Levallois, unipolar-convergent Levallois, and radial cores. Bifacial tools produced via façonnage reduction are notably absent among these assemblages. In considering inter-regional patterning based on techno-typological features observed within the Hardamaut material, Crassard observes: "relations, whose character remains to be defined, with the Levantine Mousterian would be then more probable than with an African Middle Stone Age (MSA) or Nubian Mousterian."

Rose and Usik (2009) report findings from a series of archaeological surveys and excavations carried out in and around the Dhofar Mountains, southern Oman, between 2002 and 2008. Multiple assemblage types were collected throughout this region; by far the most common technologies present are elongated blanks produced by simple unidirectional-parallel or unidirectional-convergent reduction strategies. The authors adopt the Eurasian term "Upper Paleolithic" (UP) to classify this and other seemingly related assemblages from southern Arabia. Although the material is technologically classified as UP, Rose and Usik caution against assigning a temporal range to this archaeological phase in Arabia at this stage in our understanding. They argue that the absence of UP connections with Africa is yet another indication that the modern human exodus from Africa occurred no later than MIS 5 (128–74 kya).

Maher (2009) reviews what is perhaps the least understood phase in Arabian prehistory, the Late Pleistocene, which she defines as the interval spanning MIS 3 and MIS 2. There is a paucity of evidence from this time period, restricted to a handful of deflated surface scatters. In most cases, sites were attributed to the Upper Paleolithic based on the presence of technological and typological features such as burins, perforators, blades, and endscrapers. Maher points out, however, that classic UP points, carinated pieces, and other blade tools that are common elsewhere are absent in Arabia. The author discusses potential UP artifacts in Yemen, in which two primary core reduction strategies were observed: flat cores with parallel flake and blade removals, and unidirectional or bidirectional prepared cores for the production of elongated points.

Regarding the question of Terminal Pleistocene microlithic assemblages in Arabia, Maher observes that no microlithic material (*sensu* Levantine Epipaleolithic) has yet been found in southern or central Arabia, and there are only a few isolated surface finds from the northernmost extent of the peninsula. On the whole, Maher concludes that "the Late Pleistocene cannot be assigned as 'Levantine' in character, but there are some interesting hints at Levantine connections in the north, and African connections in the south…It is not that the entire peninsula was abandoned during the Late Pleistocene. Rather, something else may have been going on here that we don't yet fully recognize…"

Crassard, Rose and Usik, and Maher describe Middle and Upper Paleolithic archaeological assemblages from Yemen, Oman, and Saudi Arabia. While the artifacts presented come from different areas of the Arabian peninsula and are presumably separated by large gaps of time, it is noteworthy that the contributions independently and concurrently arrive at similar conclusions. They all note some degree of affinity with archaeological evidence from the Levant, and minimal resemblance to material from East Africa.

Early Holocene Archaeology (ca. 12–8 ka BP)

Holocene demographic expansion onto the peninsula is one of the central themes in Part IV of this book. The 'Neolithic' peopling of Arabia is critical for deciphering the genetic composition of modern Arabian populations. In addition, understanding the dynamics of Early Holocene population expansion into Arabia is useful for setting up a frame of reference to understand the processes involved in prior Pleistocene range expansions. Contributions to Part IV by Uerpmann et al., Fedele, McCorriston and Martin, and Boivin et al. provide insights into the repopulation of Arabia from various perspectives around the peninsula.

Uerpmann et al. (2009) review the archaeological evidence for Early Holocene occupation in eastern Arabia. They address the issue of population discontinuity or continuity leading into the Holocene. According to the authors, the

available evidence "is still incapable of confirming or denying the hypothesis of population continuity or replacement between the Pleistocene and Holocene in eastern Arabia, it does raise a few question marks over another thesis that has been commonly enunciated in the archaeological literature of this area, namely the Levantine, PPNB-related origins of the earliest lithic industry and, by extension, Holocene inhabitants, found there." Uerpmann and colleagues offer three hypotheses exploring the dynamics of the Neolithic expansion into Arabia:

1. The peopling of eastern Arabia by PPNB-related settlers was the result of widespread climatic deterioration to the north of the Arabian peninsula around 6,200 BC.
2. The peopling of eastern Arabia by PPNB-related settlers was the result of widespread population dispersal during the Early Holocene.
3. The earliest settlement in southeastern Arabia reflects repopulation from South Arabia and/or northeastern Africa.

The authors lend greater credence to the second possibility, given the timing of the expansion amid a period of environmental amelioration. In their view, expansion into the desert was more likely triggered by a pulling, rather than pushing mechanism. While they do not discount the third hypothesis of indigenous and/or African origins, there is not yet enough evidence to properly assess this possibility.

Fedele (2009) examines Early Holocene occupation in the Yemeni Highlands, a region that little has been written about in discussions of the Neolithic peopling in Arabia. The chapter presents heretofore unpublished evidence from archaeological surveys in Wādī at-Tayyilah and Wadi Khamar attesting to an Early Holocene 'Pre-Neolithic' habitation throughout the eastern Yemen Plateau. Pre-Neolithic material excavated from site WTH3 in Wādī at-Tayyilah is described as a microblade technology with expediently utilized blanks. Based on the present evidence, Fedele views the Pre-Neolithic and Neolithic of this region as belonging to a single continuum, supported by the continuity of occupation at WTH3. The features of this highland industry display hints of similarities with East Africa rather than the Fertile Crescent, leading the author to adopt the African terminology 'LSA'.

McCorriston and Martin (2009) examine the evidence for the development of Early Holocene pastoralist societies along the desert margins of southern Arabia. The authors examine three specific issues related to the origins of animal domestication in this area: where and when do the earliest domesticates appear, from where were these animals brought, and what do these data imply about the transition from hunter-gatherer to fully pastoral societies?

Based on the available evidence, they conclude that domesticated taurine cattle could have arrived from the

Levant or possibly from Africa by the sixth millennium BC, if not earlier. Just as Uerpmann et al. (2009) have suggested there were multiple waves of expansion into Arabia, the data presented by McCorriston and Martin indicate a similar scenario in which cattle were introduced into Arabia "with differing human populations and population densities in very different adoption strategies at different times." Arguing against a single Levantine expansion, the authors note that the dates and lithic techno-typological features from the earliest known site of cattle domestication in southern Arabia, Manayzah, do not support an introduction of people culturally or temporally related to the Levantine PPNB. Instead, McCorriston and Martin suggest that the earliest herd animals were probably introduced as a pioneering strategy among local hunters.

The composite evidence presented in this section suggests that the Holocene peopling into Arabia emanated from multiple sources in and around Arabia at different times. Uerpmann et al. (2009) write: "the situation may be far more complex than previously assumed, and that it is wrong to speak of 'colonization' in the singular. At all events, it is unlikely to have been an 'event', and much likelier to have been a process which may initially have involved hunters and gatherers coming from the south, soon followed by aceramic herders from the northwest using some variant of PPNB-related lithic technology. How and where they met and mixed and how they sorted out their subsistence economies is a fascinating topic for future research."

While many domesticates undoubtedly arrived into Arabia overland, the unique sea-embedded context of the peninsula has also likely resulted in key maritime routes of dispersal for some of the region's domesticated species. Boivin et al. (2009) accordingly emphasize Arabia's maritime position, and the new corridors of dispersal, trade routes and connections that emerged in the region with the advent of seafaring technologies and know-how in the mid-Holocene. The chapter presents an important, and unprecedented, synthesis of the evidence for maritime activity in the Arabian subcontinent and surrounding areas from the mid-Holocene to the Classical period. It traces the emergence of the first Ichthyophagi, or 'Fish-Eaters', as they were known in the Classical texts, the development of maritime subsistence and seafaring capabilities, and the gradual growth of maritime trade and exchange activities into the Bronze Age and beyond. The chapter takes a multidisciplinary approach to the theme of Arabia's maritime past, drawing upon not just archaeology, but also paleoenvironmental, historical linguistic and textual sources of data. It argues for a broadly synchronous emergence of maritime subsistence and trade activities on both the eastern and western sides of the peninsula, but emphasizes the unique trajectories of development that subsequently characterize the Red Sea and the Gulf, and that by the Classical period

cause them to sometimes alternate as key routes of maritime trade within the wider Indian Ocean. The chapter by Boivin et al. explores the critical question of exactly who was responsible for the trade activities that brought zebu and other domesticates to Arabia, and also moved key crops between Africa and India from as early as the second millennium BC. One of its most important points is that these translocations were likely at least partly the result of the activities of small-scale Arabian societies, rather than solely the work of the Bronze Age states to whom they are often attributed.

The final chapter by Khalidi (2009) addresses these maritime issues through more detailed study of precisely these small-scale communities, and in doing so comes full circle back to the question concerning the movement of populations between Africa and Arabia. It is ironic that so much speculation surrounds the question of connections across the Red Sea during the Paleolithic – one of the central themes of this volume – yet the only definitive evidence comes from the Middle Holocene and onward. Khalidi examines obsidian exchange networks between Eritrea, Yemen, and Saudi Arabia. Based on data collected from both the African and Arabian sides of the Red Sea, Khalidi combines lithic analysis and obsidian sourcing to articulate regional trade networks. She recognizes two primary modes of obsidian circulation in Yemen: (1) limited exchange using local highland sources and (2) trade over long distances either 300 km away in Saudi Arabia or 100 km to the west in Eritrea. Khalidi cites the presence of geometric microliths and bipolar flaking technology on both coasts of the Red Sea as evidence in favor of Red Sea exchange.

Human Evolution in Arabia

The environmental, genetic, and archaeological pictures painted by the various contributions within this volume illustrate a region that was ever in flux, where stasis was the exception rather than the rule. Arabia was undoubtedly a place of genetic mixing, mirrored in the phylogenetic structure of the modern Arabian population. For the last 1 million years, hominin groups from all three continents surrounding the Arabian peninsula episodically expanded onto the subcontinent when conditions permitted: whether it was hunter-gatherers tracking the expansion of their geographic range, cattle-pastoralists exploiting the ameliorated grasslands of the Arabian interior, fishermen moving along the emerged continental shelf during glacial maxima, or displaced communities forced out of the Ur-Schatt River Valley by marine incursion into the basin.

If the Holocene frame of reference is any indication, then we can assume that demographic expansion into Arabia was not a straightforward process involving a single source population. It is likely that different groups moved into the region from multiple points of origin at different times and for variable reasons. The observed inter-regional patterning leads to the question of which species were responsible for the creation of Middle and Upper Pleistocene archaeological sites in Arabia. There are a number of possibilities for the identity of Arabia's inhabitants during the interval between 200 and 100 ka:

1. *Homo heidelbergensis* – Given the presence of Acheulean technologies in Arabia, and the possibility that early Middle Paleolithic technologies were derived from Large Cutting Tool assemblages, it is necessary to entertain the possibility that local, indigenous groups of *Homo heidelbergensis* or their descendents evolved and survived in the peninsula. If this were the case, it is therefore possible that these populations were coeval with other groups of hominins expanding into the peninsula in the late Middle and Upper Pleistocene.

2. *Homo neanderthalensis* – It is possible that there was some demographic input of populations in Arabia as groups travelled southward from the Levant and the Zagros mountains. Neanderthal fossils are geographically close to Arabia; their presence in Shanidar Cave, at the foothills of the Zagros mountains in Iraq, raises the possibility that these hominins ranged into parts of northern Arabia. Yet, there are some morphological and archaeological factors which indicate that Neanderthals may not have commonly inhabited the southern Arabian landscapes. Their robust morphological traits have been often cited as adaptations to colder environments than predicted in Arabia. With respect to their technology, consistent and systematic Levallois technology with distinct platform faceting, a defining characteristic of the Mousterian, is virtually absent in Arabia.

3. *Homo helmei* – *Homo helmei* is considered an archaic variant of early *Homo sapiens* postulated to have spread from Africa in a wave of dispersal sometime after 200ka and associated with a Mode 3 technology (Lahr and Foley, 2001). Though fossils of this taxon are thus far only found in Africa, the possibility exists that these early moderns reached the Arabian peninsula during an early dispersal event. From an archaeological perspective, Marks (2009) observes some overlapping features between Jebel Faya 1, Assemblage C and Sangoan-type assemblages in East Africa, and Rose (2007) has noted Sangoan/Lupembam-type features among Sibakhan assemblages. A Sangoan/Lupembam-type assemblage collected at Abu Hagar in Sudan (Lacaille, 1951) was found in proximity to and in the same geological context as the cranium from Singa which has been sometimes identified as *Homo helmei* (McBrearty and Brooks, 2000).

4. *Homo sapiens* – Most scholars posit that anatomically modern humans dispersed into the Arabian peninsula, although the chronology for this event(s) remains unknown. There is a possibility that early groups of *Homo sapiens* expanded into the peninsula during pluvial pulses in MIS 6, potentially bearing the macrohaplogroup L3 marker (Cabrera et al., 2009) or some other now extinct lineage. One possibility is that this emigrating group belonged to the ancestral population that gave rise to all subsequent M and N mtDNA branches. Anatomically modern human remains excavated at Skhul and Qafzeh Caves in the Levant, approximately 100 ka, provide evidence of this expansion out of Africa. If, indeed, early *Homo sapiens* were displaced from the Levant by expanding Neanderthal groups around MIS 4, it is possible that they contracted back into the Red Sea and South Arabian refugia, particularly given the preponderance of Levantine-related laminar technologies reported from the southern zone. As such, the identity of Arabian technologies between 200 and 100 ka may indicate the presence of early *Homo sapiens* populations. If, indeed, the Skhul/Qafzeh population represents a successful expansion and not an evolutionary dead-end, as many researchers have written, this provides a more parsimonious explanation as to why the early humans that reached Australia during early MIS 3 are morphologically similar to the Skhul/Qafzeh specimens (Schillaci, 2008). Inter-regional examination of the archaeological evidence from the Middle-Upper Paleolithic transition throughout Africa and Eurasia leads Marks (2005) to a similar conclusion.

Although we still do not have answers to many evolutionary questions, the contributions in this book clearly demonstrate Arabia's record of extreme climate change, diverse populations, and wide array of variables impacting the movement of human groups. Anthropological "systems analysis" sees human systems in a state of positive or negative feedback loops (e.g., Flannery, 1968; Renfrew, 1979; Lowe and Barth, 1980). Groups remain in stasis (positive feedback) until one or more variables within the system changes, forcing people to adapt some aspect of the system to compensate for this alteration. From this perspective, Arabia represents a massive negative feedback loop in the process of human development. Populations inhabiting Arabia faced a recurring succession of profound changes in their environment triggered by aridification, amelioration, and sea level fluctuation. These dramatic landscape transformations must have had greatly influenced the trajectory of human biological and cultural evolution.

It is an exciting time to be involved in the investigation of Arabian prehistory. The genetics revolution has enabled us to address Pleistocene and Early Holocene demographics in a manner no scholar would have dreamed just 25 years ago.

The sudden accessibility of archaeological sites in this once obscure region has filled in a major piece of the puzzle that, until now, was consistently relegated to *terra incognita*. We are at but the vanguard of discovery in a part of the world that promises rich rewards indeed.

References

Adams R, Parr P, Ibrahim M, al-Mughannum AS. Preliminary report on the first phase of the Comprehensive Survey Program. Atlal. 1977;1:21–40.

Amirkhanov HA. The Acheulean of southern Arabia. Sovetskaia Arkheologiia. 1987;4:11–23 (in Russian).

Amirkhanov HA. The Paleolithic of South Arabia. Moscow (in Russian): Nauka; 1991.

Amirkhanov HA. Research on the Paleolithic and Neolithic of Hadramaut and Mahra. Arabian Archaeology and Epigraphy. 1994;5:217–28.

Amirkhanov HA. Stone Age of South Arabia. Moscow (in Russian): Nauka; 2006.

Anton D. Aspects of geomorphological evolution: paleosols and dunes in Saudi Arabia. In: Jado AR, Zötl JG, editors. Quaternary period in Saudi Arabia. ii. Sedimentological, hydrogeological, hydrochemical, geomorphological, and climatological investigations of Western Saudi Arabia. Vienna: Springer; 1984. p. 275–96.

Bailey G. The Red Sea, coastal landscapes, and hominin dispersals. In: Petraglia MD, Rose JI, editors. The evolution of human populations in Arabia: paleoenvironments, prehistory and genetics. The Netherlands: Springer; 2009. p. 15–37.

Boivin N, Blench R, Fuller DQ. Archaeological, linguistic and historical sources on ancient seafaring: a multidisciplinary approach to the study of early maritime contact and exchange in the Arabian peninsula. In: Petraglia MD, Rose JI, editors. The evolution of human populations in Arabia: paleoenvironments prehistory and genetics. The Netherlands: Springer; 2009. p. 251–78.

Cabrera VM, Abu-Amero KK, Larruga JM, González AM. The Arabian peninsula: gate for human migrations out of Africa or cul-de-sac? A mitochondrial DNA phylogeographic perspective. In: Petraglia MD, Rose JI, editors. The evolution of human populations in Arabia: paleoenvironments, prehistory and genetics. The Netherlands: Springer; 2009. p. 78–87.

Caton-Thompson G. Climate, irrigation, and early man in the Hadramaut. Geographical Journal. 1939;93(1):18–35.

Caton-Thompson G. Some paleoliths from South Arabia. Proceedings of the Prehistoric Society. 1954;29:189–218.

Caton-Thompson G. The evidence of south Arabian paleoliths in the question of Pleistocene land connections with Africa. Pan-African Congress on Prehistory. 1957;3:380–4.

Černý V, Mulligan CJ, Rídl J, Žaloudková M, Edens C, Hájek M, et al. Regional differences in the distribution of the Sub-Saharan, West Eurasian and South Asian mtDNA lineages in Yemen. American Journal of Physical Anthropology. 2008;136:128–37.

Crassard R. The Middle Paleolithic of Arabia: the view from the Hadramawt Region, Yemen. In: Petraglia MD, Rose JI, editors. The evolution of human populations in Arabia: paleoenvironments, prehistory and genetics. The Netherlands: Springer; 2009. p. 151–68.

Fedele FG. Early Holocene in the highlands: data on the peopling of the eastern Yemen Plateau, with a note on the Pleistocene evidence. In: Petraglia MD, Rose JI, editors. The evolution of human populations in Arabia: paleoenvironments, prehistory and genetics. The Netherlands: Springer; 2009. p. 215–36.

Fernandes C. Bayesian coalescent inference from mitochondrial DNA variation of the colonization time of Arabia by the hamadryas baboon (*Papio hamadryas hamadryas*). In: Petraglia MD, Rose JI, editors. The evolution of human populations in Arabia: paleoenvironments, prehistory and genetics. The Netherlands: Springer; 2009. p. 89–100.

Field H. The cradle of Homo sapiens. American Journal of Archaeology. 1932;36(4):426–30.

Field H. Reconnaissance in Southwestern Asia. Southwestern Journal of Anthropology. 1951;7(1):86–102.

Field H. New Stone Age sites in the Arabian peninsula. Man. 1955;55:136–8.

Field H. Ancient and modern man in Southwestern Asia. Coral Gables: University of Miami Press; 1956.

Field H. Stone implements from the Rub' al Khali, Southern Arabia. Man. 1958;58:93–4.

Field H. Stone implements from the Rub' al Khali. Man. 1960a;60: 25–6.

Field H. Carbon-14 date for a 'Neolithic' site in the Rub' al Khali. Man. 1960b;60:172.

Field H. Palaeolithic implements from the Rub' al Khali. Man. 1961;61: 22–3.

Flannery KV. Archaeological systems theory and early Mesoamerica. In: Meggers BJ, editor. Anthropological archaeology in the Americas. Washington, DC: Anthropological Society of Washington; 1968. p. 67–87.

Jagher R. The Central Oman Palaeolithic Survey: recent research in Southern Arabia and reflection on the prehistoric evidence. In: Petraglia MD, Rose JI, editors. The evolution of human populations in Arabia: paleoenvironments, prehistory and genetics. The Netherlands: Springer; 2009. p. 139–50.

Kapel H. Atlas of the stone-age cultures of Qatar. Denmark: Aarhus University Press; 1967.

Kennett DJ, Kennett JP. Early state formation in southern Mesopotamia: sea levels, shorelines, and climate change. Journal of Island and Coastal Archaeology. 2006;1:67–99.

Khalidi L. Holocene obsidian exchange in the Red Sea region. In: Petraglia MD, Rose JI, editors. The evolution of human populations in Arabia: paleoenvironments, prehistory and genetics. The Netherlands: Springer; 2009. p. 279–91.

Kivisild T, Reidla M, Metspalu E, Rosa A, Brehm A, Pennarun E, et al. Ethiopian mitochondrial DNA heritage: tracking gene flow across and around the gate of tears. American Journal of Human Genetics. 2004;75:752–70.

Lambeck K. Shoreline reconstructions for the Persian Gulf since the Last Glacial Maximum. Earth and Planetary Science Letters. 1996; 142:43–57.

Lahr M, Foley RA. Multiple dispersals and modern human origins. Evolutionary Anthropology. 1994;3:48–60.

Lahr M, Foley RA. Towards a theory of modern human origins: geography, demography, and diversity in recent human evolution. Yearbook of Physical Anthropology. 1998;41:137–76.

Lahr MM, Foley RA. Mode 3, Homo helmei, and the pattern of human evolution in the Middle Pleistocene. In: Barham L, Robson Brown K, editors. Human roots: Africa and Asia in the Middle Pleistocene. Bristol: Western Academic & Specialist Press; 2001. p. 23–39.

Lowe JWG, Barth RJ. Systems in archaeology: a comment on Salmon. American Antiquity. 1980;45(3):568–74.

Maher L. The Late Pleistocene of Arabia in relation to the Levant. In: Petraglia MD, Rose JI, editors. The evolution of human populations in Arabia: paleoenvironments, prehistory and genetics. The Netherlands: Springer; 2009. p. 187–202.

de Maigret A. Ricerche archeologiche italiane nella Repubblica Araba Yemenita: notizia di una seconda ricignizione. Oriens Antiquus. 1981;21:237–53.

de Maigret A. Archaeological activities in the Yemen Arab Republic, 1984. East and West. 1984;34:423–54.

de Maigret A. Archaeological activities in the Yemen Arab Republic, 1985. East and West. 1985;35:337–97.

de Maigret A. Archaeological activities in the Yemen Arab Republic, 1986. East and West. 1986;36:376–470.

Marks A. Comments after four decades of research on the Middle to Upper Paleolithic transition. Mitteilungen der Gesellschaft für Urgeschichte. 2005;14:81–6.

Marks AE. The Paleolithic of Arabia in an inter-regional context. In: Petraglia MD, Rose JI, editors. The evolution of human populations in Arabia: paleoenvironments, prehistory and genetics. The Netherlands: Springer; 2009. p. 295–308.

Masry AH. Notes on the recent archaeological activities in the Kingdom of Saudi Arabia. Proceedings of the Seminar for Arabian Studies. 1977;7:112–9.

McBrearty S, Brooks A. The revolution that wasn't: a new interpretation of the origin of modern human behavior. Journal of Human Evolution. 2000;39:453–563.

McCorriston J, Martin L. Southern Arabia's early pastoral population history: some recent evidence. In: Petraglia MD, Rose JI, editors. The evolution of human populations in Arabia: paleoenvironments, prehistory and genetics. The Netherlands: Springer; 2009. p. 237–50.

Mellars P. Why did modern humans populations disperse from Africa ca. 60, 000 year ago? A new model. Proceedings of the National Academy of Sciences USA. 2006;103(25):9381–6.

Parker AG. Pleistocene climate change from Arabia: developing a framework for hominin dispersal over the last 350 ka. In: Petraglia MD, Rose JI, editors. The evolution of human populations in Arabia: paleoenvironments, prehistory and genetics. The Netherlands: Springer; 2009. p. 39–49.

Parker A, Rose J. Climate change and human origins in southern Arabia. Proceedings of the Seminar for Arabian Studies. 2008;38:25–42.

Petraglia MD. The Lower Paleolithic of the Arabian peninsula: occupations, adaptations, and dispersals. Journal of World Prehistory. 2003;17(2):141–79.

Petraglia MD, Alsharekh A. The Middle Palaeolithic of Arabia: implications for modern human origins, behaviour and dispersals. Antiquity. 2003;77(298):671–84.

Petraglia MD, Drake N, Alsharekh A. Acheulean landscapes and large cutting tool assemblages in the Arabian peninsula. In: Petraglia MD, Rose JI. The evolution of human populations in Arabia: paleoenvironments, prehistory and genetics. The Netherlands: Springer; 2009. p. 103–16.

Philby H. Rub al' Khali: an account of exploration in the Great South Desert of Arabia under the auspices and patronage of His Majesty 'Abdul' Aziz ibn Saud, King of the Hejaz and Nejd and its dependencies. Geographical Journal. 1933;81:1–21.

Quintana-Murci L, Semino O, Bandelt H, Passarino G, McElreavey K, Santachiara-Benerecetti AS. Genetic evidence of an early exit of Homo sapiens from Africa through eastern Africa. Nature Genetics. 1999;23:437–41.

Renfrew C. System collapse as social transformation. In: Renfrew C, Cooke KL, editors. Transformations: mathematical approaches to culture change. New York: Academic; 1979.

Richards M, Macaulay V, Hickey E, Vega E, Sykes B, Guida V, et al. Tracing European founder lineages in the Near Eastern mtDNA pool. American Journal of Human Genetics. 2000;67:1251–76.

Richards M, Rengo C, Cruciani F, Gratrix F, Wilson JF, Scozzari R, et al. Extensive female-mediated gene flow from sub-Saharan Africa into Near Eastern Arab populations. American Journal of Human Genetics. 2003;72:1058–64.

Rídl J, Edens CM, Černý V. Mitochondrial DNA structure of Yemeni population: regional differences and the implications for different migratory contributions. In: Petraglia MD, Rose JI, editors. The evolution of human populations in Arabia: paleoenvironments, prehistory and genetics. The Netherlands: Springer; 2009. p. 69–78.

Rose JI. The question of Upper Pleistocene connections between East Africa and South Arabia. Current Anthropology. 2004;45(4):551–5.

Rose JI. Among Arabian sands: defining the Paleolithic of southern Arabia. Ph.D. dissertation, Southern Methodist University, Dallas; 2006.

Rose JI. The Arabian corridor migration model: archaeological evidence for hominin dispersals into Oman during the Middle and Upper Pleistocene. Proceedings of the Seminar for Arabian Studies. 2007;37:219–37.

Rose JI. Modern human origins: the 'out of Arabia' hypothesis. In: Cleuziou S, Tosi M, editors. In the shadow of the ancestors. 2nd ed. Muscat (in Arabic): Ministry of Heritage and Culture; 2008.

Rose JI, Usik VI, as-Sabri B, Schwenninger J-L, Clark-Balzan, L, Parton A, et al. Archaeological evidence for indigenous human occupation in southern Arabia at the end of the Pleistocene. n.d. Manuscript in possession of authors.

Rose JI, Usik VI. The "Upper Paleolithic" of South Arabia. In: Petraglia MD, Rose JI, editors. The evolution of human populations in Arabia: paleoenvironments, prehistory and genetics. The Netherlands: Springer; 2009. p. 169–85.

Rowold D, Luis J, Terreros M, Herrera R. Mitochondrial DNA gene flow indicates preferred usage of the Levant corridor over the Horn of Africa passageway. Journal of Human Genetics. 2007;52:436–47.

Schillaci MA. Human cranial diversity and evidence for an ancient lineage of modern humans. Journal of Human Evolution. 2008;54:814–26.

Scott-Jackson J, Scott-Jackson W, Rose JI. Paleolithic stone tool assemblages from Sharjah and Ras al Khaimah in the United Arab Emirates. In: Petraglia MD, Rose JI, editors. The evolution of human populations in Arabia: paleoenvironments, prehistory and genetics. The Netherlands: Springer; 2009. p. 125–38.

Stringer C. Paleoanthropology: coasting out of Africa. Nature. 2000;405:24–6.

Tchernov E. Biochronology, paleoecology and dispersal events of hominids in the southern Levant. In: Akazawa T, Aoki K, Kimura T, editors. The evolution and dispersal of modern humans in Asia. Tokyo: Hokusen-Sha; 1992. p. 149–88.

Thesiger W. Arabian sands. New York: E.P. Dutton & Co.; 1959.

Thomas B. Arabia Felix: across the Empty Quarter of Arabia. Oxford: Readers' Union Limited; 1938.

Thomas MG, Weale ME, Jones AL, Richards M, Smith A, Redhead N, et al. Founding mothers of Jewish communities: geographically separated Jewish groups were independently founded by very few female ancestors. American Journal of Human Genetics. 2002;70:1411–20.

Thompson A. Origins of Arabia. London: Stacey International; 2000.

Uerpmann H-P, Potts DT, Uerpmann M. Holocene (re-)occupation of Eastern Arabia. In: Petraglia MD, Rose JI, editors. The evolution of human populations in Arabia: paleoenvironments, prehistory and genetics. The Netherlands: Springer; 2009. p. 205–14.

Wahida G, Yasin A-T, Beech MJ, Al Meqbali A. A Middle Paleolithic assemblage from Jebel Barakah, Coastal Abu Dhabi Emirate. In: Petraglia MD, Rose JI, editors. The evolution of human populations in Arabia: paleoenvironments, prehistory and genetics. The Netherlands: Springer; 2009. p. 117–24.

Whalen N, Pease DW. Archaeological survey in Southwest Yemen, 1990. Paléorient. 1991;17(2):127–31.

Whalen N, Schatte KE. Pleistocene sites in southern Yemen. Arabian Archaeology and Epigraphy. 1997;8:1–10.

Whalen N, Killick A, James N, Morsi G, Kamal M. Saudi Arabian archaeological reconnaissance 1980: B. Preliminary report on the Western Province survey. Atlal. 1981;5:43–58.

Whalen N, Sindi H, Wahida G, Siraj-Ali J. Excavation of Acheulean sites near Saffaqah in al-Dawadmi (1402/1982). Atlal. 1983;7:9–21.

Whalen N, Siraj-Ali J, Davis W. Excavation of Acheulean sites near Saffaqah, Saudi Arabia. Atlal. 1984;8:43–58.

Whalen N, Siraj-Ali J, Sindi H, Pease D, Badein M. A complex of sites in the Jeddah-Wadi Fatima Area. Atlal. 1988;11:77–87.

Whalen N, Zoboroski M, Schubert K. The Lower Palaeolithic in south-western Oman. Adumatu. 2002;5:27–34.

Wilkinson TJ. Environment and long-term population trends in southwest Arabia. In: Petraglia MD, Rose JI, editors. The evolution of human populations in Arabia: paleoenvironments, prehistory and genetics. The Netherlands: Springer; 2009. p. 51–66.

Zarins J, Ibrahim M, Potts D, Edens C. The preliminary report on the third phase of the Comprehensive Archaeological Survey Program – the central province. Atlal. 1979;3:9–42.

Zarins J, Whalen N, Ibrahim M, Mursi A, Khan M. Comprehensive Archaeological Survey Program: preliminary report on the Central and Southwestern Province Survey: 1979. Atlal. 1980;4:9–36.

Zarins J, Murad A, al-Yish K. Comprehensive Archaeological Survey Program: the second preliminary report on the Southwestern Province. Atlal. 1981;5:9–42.

Zarins J, Rahbini A, Kamal M. Preliminary report on the archaeological survey of the Riyadh area. Atlal. 1982;6:25–38.

Zeuner FE. Neolithic sites from the Rub Al-Khali, southern Arabia. Man. 1954;54:1–4.

Part I
Quaternary Environments and Demographic Response

Chapter 2
The Red Sea, Coastal Landscapes, and Hominin Dispersals

Geoff Bailey

Keywords Bab al Mandab • Coasts • Farasan Islands • Marine Resources • Paleoenvironment • Paleoclimate • Red Sea

Introduction

The Red Sea has typically been viewed as a barrier to early human movement between Africa and Asia over the past 5 million years, and one that could be circumvented only through narrow exit points at either end, vulnerable to blockage by physical or climatic barriers (Fig. 1). It is one of several significant obstacles cutting across 'savannahstan' (Dennell and Roebroeks, 2005), a broad swathe of herbivore-rich savannah and grassy plains that began to extend over a vast area stretching from West Africa to China with climatic cooling from at least 2.5 Ma, and a key macro-environmental context for early hominin dispersal[1]. However, this concept of the Red Sea Basin as a barrier should not obscure the fact that its coastal regions also hold considerable potential attractions for early human settlement, especially under climatic conditions wetter than today, including a complex tectonic and volcanic topography not unlike that of the African Rift, capable of providing localized fertility for plant and animal life, tactical opportunities for pursuit of herbivores and protection from predators (King and Bailey, 2006), along with inshore and intertidal marine resources.

G. Bailey (✉)
Department of Archaeology, King's Manor,
University of York, York, YO1 7EP UK
e-mail: gb502@york.ac.uk

[1] Other comparable barriers are the Sahara desert, which would have constrained movement between sub-Saharan Africa and North Africa, and the arc of the Taurus-Zagros mountains and the Iranian plateau, providing a series of physical or climatic barriers extending from Anatolia to the Indian subcontinent.

The modern climate of the region as a whole is generally arid or semi-arid, with spasmodic rainfall and limited supplies of surface water, and similar climates would have imposed a major limitation on past human settlement. However, there have clearly been periods of wetter climate in the past, and both marine and terrestrial environments have been subject to considerable change resulting from the climatic and sea level fluctuations of the glacial–interglacial cycle, as well as to longer-term factors of tectonic deformation associated with rifting, faulting and volcanic activity.

The Red Sea region therefore presents a rather complex valve controlling movement between Africa and Asia, with limited points of transit determined by physical and climatic barriers that are likely to have varied with long-term changes in paleogeography and climate. Although the possibility of direct human movement out of Africa across the Mediterranean has been raised, such routes would always have required sea crossings, even at the narrowest point of the Gibraltar Straits (ca. 11 km), and there is currently no decisive evidence in favor of such movements (cf. Derricourt, 2005; O'Regan, in press).[2] The Red Sea region remains the most obvious and most probably the only transit region for hominin movement between Africa and Eurasia throughout most of the Plio-Pleistocene. Therefore, an understanding of its long-term environmental history and potential for early human settlement is central to an understanding of the wider picture of hominin dispersal.

The case for an African origin of the *Homo* lineage and subsequent wider dispersal beyond Africa some time after about 1.8 Ma continues to command a wide consensus, so too the case for an African origin of anatomically modern *Homo sapiens* and their dispersal out of Africa sometime after 150 ka. However, this does not rule out a priori the possibility that earlier hominin crossings took place at 2.5 Ma or even earlier, that earliest hominin populations originated over a wider zone that encompassed Africa, Arabia and

[2] It would be fair to say that there is also no decisive evidence to refute the possibility of such movements.

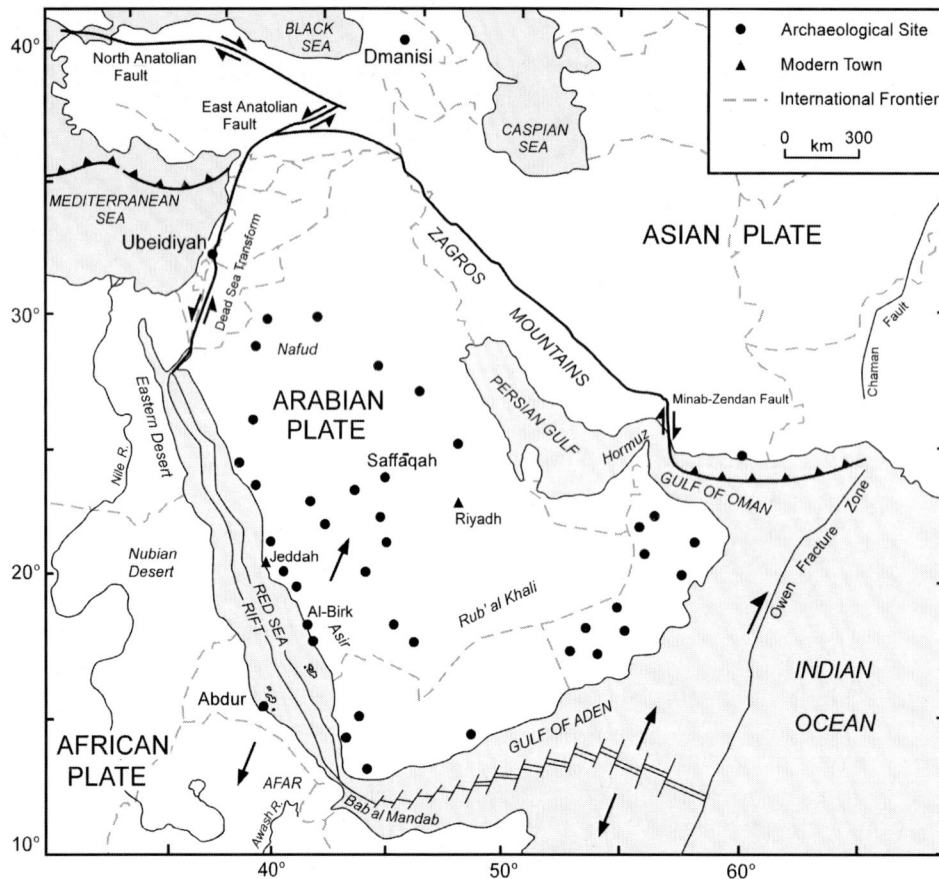

Fig. 1 General map of the Red Sea and adjacent regions, showing plate boundaries and major faults. *Arrows* indicate direction of plate motions. Also shown is a simplified distribution of Lower and Middle Paleolithic archaeological sites in the Arabian peninsula, together with sites elsewhere mentioned in the text (© G. Bailey and C. Vita-Finzi)

Asia, or that movements between Africa and Asia may have been in both directions and not just one-way out of Africa (see Dennell and Roebroeks, 2005). Studies of Plio-Pleistocene mammalian fossils show that there has been two-way traffic between Asia and Africa, albeit intermittent (Tchernov, 1992; Turner and O'Regan, 2007), and some genetic studies of human ancestry also suggest a pattern of repeated contact implying two-way movement (Templeton, 2002). Other mammals, of course, provide at best an imperfect analogy for human biogeography, not least because humans are omnivores who can feed on a wide range of resources including marine ones, and have some capacity for surmounting water barriers by swimming or rafting, both traits that could have extended far back into the earliest stages of human evolution. Whatever the outcome of these debates, the long-term history of the Red Sea Basin is likely to play an important role in their resolution, and the time span of investigation here is taken to be the past 5 million years, in order to encompass the widest range of possible scenarios for hominin dispersal. It is, however, inevitable that the most detailed reconstructions of environmental change and paleogeography are for the later stages of the

Pleistocene, and that uncertainties and gaps in knowledge increase as one goes further back in time.

The aim of this chapter, then, is to examine critically the archaeological and paleoenvironmental evidence pertaining to the Red Sea region both as a pathway of dispersal between Africa and Asia and as a zone of occupation in its own right that may have offered varied attractions for early human settlement regardless of the possibilities of onward dispersal to the north and the east – or to the south and west.

Environmental and Archaeological Context

Geographical Factors

The Red Sea extends for 2,000 km in a north–south direction through more than 17 degrees of latitude, from 12.5° N to 30° N. Over most of its length it is very wide with an average width of 280 km and a maximum width of 354 km, and an offshore topography that plunges quite steeply to reach a maximum depth along the central axis of 2,850 m (Head,

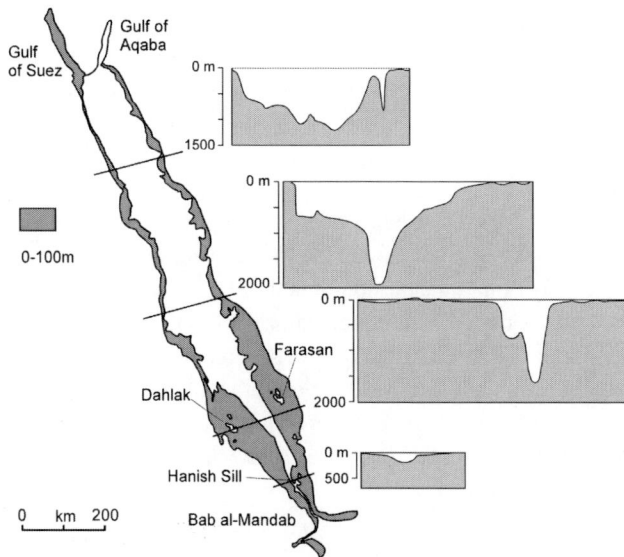

Fig. 2 The Red Sea, showing features mentioned in the text and the amount of land exposed at the −100 m bathymetric contour (information from Head 1987a, © G. Bailey)

1987a). It is therefore impassable without modern seafaring technology, even assuming lowered sea levels, except at the northern and southern extremities. In the north it divides into two branches on either side of the Sinai peninsula, the relatively shallow (50–70 m) Gulf of Suez to the west, and the narrower and deeper (250–1,800 m) Gulf of Aqaba to the east. In the south the basin is connected to the Indian Ocean by the Bab al Mandab Straits, which is 29 km wide. The shallowest part of this southern channel is −137 m at the Hanish Sill in the vicinity of the Hanish Islands, over 100 km to the north of the Straits (Fig. 2). The geographical configuration of both northern and southern extremities is sensitive to changes of relative sea level linked to the glacial–interglacial cycle, and especially at the southern end, where the sea-channel is shallow enough that it might have been closed or easily crossed at low sea-level stands.

Under present day conditions there is only one means of circumventing the Red Sea on dry land and that is in the north across a neck of low-lying land about 120 km wide between the Mediterranean coastline and the Gulf of Suez, extending eastwards from the region of the Nile Delta to the Sinai peninsula. This region would have become broader during periods of low sea level with the drying out of the Gulf of Suez, supplemented by a narrow extension of the Mediterranean coastal plain. There is no obvious physical barrier to human movement through this region. However, climatic conditions may have presented an obstacle during arid periods. Access from the vicinity of the Nile may also have been constrained by deep, steep-sided gorges during marine regressions as the Nile cut down to the lower sea level, as deep as 200 m during the Lower Pleistocene (Butzer, 1980), and by an extensive delta at high sea levels (cf. Tchernov, 1992). This northern

route by no means offered a permanently open or easy pathway of dispersal. Moreover, the Nile appears to have had a much reduced flow of water intermittently during the Pleistocene, notably during the earlier part of the Pleistocene and during the Last Glacial Maximum (LGM) (Lamb et al., 2007). Nevertheless, by convention this northern route, whether via the Nile and the Mediterranean coast, or more directly between the Red Sea coast and the Jordan Valley via the Gulf of Suez and the shores of the Gulf of Aqaba, has been assumed to be the principal artery of contact between Africa and western Asia, an assumption reinforced by the abundant finds of Paleolithic archaeology and African species of mammalian fauna in the Levant and early dates in the 1.8– 1.4 Ma range at sites such as Ubeidiyah (Tchernov, 1992; Ron and Levi, 2001), and further north, in the Caucasus, at Dmanisi at 1.7 Ma (Lordkipanidze et al., 2000).

More recently, considerable attention has focused on the 'southern corridor' across the southern end of the Red Sea and around the coastlines of the Indian Ocean into the Indian subcontinent (Lahr and Foley, 1994), reinforced by growing awareness of the substantial and widely distributed record of Paleolithic archaeological sites in the Arabian peninsula (Petraglia, 2003, 2007; Petraglia and Alsharekh, 2003; Rose and Usik, 2009). The popularity of this route has been further strengthened by genetic studies based on comparisons of DNA characteristics in modern populations, which seem to suggest a single rapid dispersal of modern humans out of Africa at about 70 ka (Oppenheimer, 2003; Forster and Matsumura, 2005; Macaulay et al., 2005; Thangaraj et al., 2005), an idea that has been coupled with a supposedly new emphasis on marine resources that attracted modern human populations to productive coastlines and propelled them eastwards around the rim of the Indian Ocean (Stringer, 2000; Walter et al., 2000; Mellars, 2006; Bulbeck, 2007). Similarities of early stone-tool industries between East Africa, Arabia and the Indian subcontinent have been discussed in relation to this hypothesis (e.g., Rose, 2004; Beyin, 2006; Mellars, 2006), but the technological and typological characteristics of the industries in question are variable and sites and industries are patchily distributed over large territories (James and Petraglia, 2005; Petraglia, 2007). The balance of independent convergence versus cultural or demic diffusion in the interpretation of such widely separated material is difficult to establish with any confidence, and there is little as yet in this material that would argue decisively in favor of the southern route. At any rate, an interest in the southern corridor has stimulated closer investigation of the possibilities for transit of the southern end of the Red Sea under different climatic, topographic and sea level conditions (Bailey et al., 2007a).

Models of human dispersal based on deductions from the variability of DNA in present-day populations have exercised a particularly powerful influence on the scientific and popular imagination (Oppenheimer, 2003; Cabrera et al., 2009; Rídl et al., 2009). The evidence they provide of a single African source for all anatomically modern populations, broadly supported by the dating of human fossils, now commands a strong consensus, perhaps also the deduction of a single dispersal event out of Africa, while their combination with the notion of developing maritime adaptations and coastal dispersal represents a compelling synthesis. However, the capacity of DNA models to specify the date of this dispersal event is questionable, even more so their ability to discriminate between alternative pathways of dispersal between Africa and southern Asia. In this regard, such models provide, at best, hypotheses in need of further exploration and testing against independent sources of evidence, and raise as many questions as they purport to answer.

The concept of a maritime dispersal out of Africa, linked to the developing behavioral adaptations of modern humans, has proved especially attractive (cf. Walter et al., 2000; Mannino and Thomas, 2002; Oppenheimer, 2003; Bulbeck, 2007; Marean et al., 2007; Turner and O'Regan, 2007), but the evidence in its support is at best weak (Bailey et al., 2007a; Bailey, in press), and raises a number of unresolved questions. Examination of potential routes between Northeast Africa and the Indian subcontinent (Field and Lahr, 2005; Field et al., 2007) suggests that there are a number of significant barriers along the coastal corridor that would have required long diversions inland, although the question of what constitutes a barrier, especially under paleogeographic and environmental conditions unlike those of today, remains to be explored in more detail. Crossing the southern end of the Red Sea during periods of high sea level as at present would certainly require seaworthy boats, but less obviously so during periods of low sea level. The extension of the coastal landscape during periods of low sea level also would have altered the potential of many coastlines to act as zones of settlement and dispersal. The likelihood of short sea crossings across the Red Sea without the aid of boats, the potential of marine resources in this region, and evidence of their early exploitation are all matters in need of further investigation and are discussed later.

If human groups took the southern route across the Red Sea at 70 ka, when sea levels were relatively low, why should not earlier migrants have taken the same route during earlier periods of low sea level, as has long been hypothesized by others (e.g., Whalen et al., 1989; Whalen and Fritz, 2004)? Sea level, of course, is not the only factor. Climatic changes are also relevant, both as 'pull' and 'push' factors. Improved (wetter) climatic conditions might make both sides of the southern channel more attractive and fertile regions for plant and animal life and human settlement ('pull' factors). More arid conditions might have compelled populations to disperse more widely and to cross previously unpenetrated barriers in the search for new territory ('push' factors). Moreover,

while it is true that the facility to cross the southern end of the Red Sea would have broadened the possibilities of contact and movement between Africa and Asia, such a dispersal route is not essential to populate Arabia or initiate movement thence further to the east. The whole of the Arabian peninsula could have been filled with human and other mammalian species derived from Africa, or western Asia, via the northern end of the Red Sea, more or less instantaneously within the chronological resolution of existing dating techniques. Thus the nature of the environments, landscapes and resources around the Red Sea, and especially on the Arabian side, and their potential attractiveness to human settlement under different climatic regimes, may be as critical a question to pose as the possibility of transit across the southern end.

Geology

The Red Sea Basin originated as a terrestrial depression, perhaps as early as the Jurassic period over 150 Ma, with crustal thinning and depression accompanied by slow uplift of the surrounding flanks (Braithwaite, 1987). Later on, from about 40 Ma onwards in the late Oligocene, volcanic activity accompanied by occasional marine incursions from the Mediterranean became more marked, the main outlines of the Basin took shape, the Gulf of Aden began to open as a result of rifting that originated in the Indian Ocean, and rifting and seafloor spreading accentuated the separation of the Arabian Plate from Africa and turned the Red Sea into a progressively wider and deeper marine basin linked to the Indian Ocean (Bonatti, 1985; Omar and Steckler, 1995). The timing of some of these processes is not well dated. According to Girdler and Styles (1974), a first phase of sea-floor spreading occurred between 41 and 34 Ma. During the Miocene, between about 25 and 5 Ma, there was little further rifting, and high rates of evaporation in a semi-enclosed basin resulted in the formation of thick evaporites (salt deposits), suggesting conditions of considerable aridity. The evaporites are interleaved with marine deposits indicating intermittent incursion of the sea from the Mediterranean but rather unstable conditions for marine life. Opening of the Gulf of Aden had begun by 13 Ma (Manighetti et al., 1997; Hubert-Ferrari et al., 2003), and a second phase of sea-floor spreading within the Red Sea Basin took place after about 4–5 Ma (Girdler and Styles, 1974), accompanied by uplift in the area of Suez cutting off any further connection with the Mediterranean, and the establishment of a permanent marine connection to the Indian Ocean.

Rifting is the result of thinning and separation of the Earth's crust and is accompanied by volcanism and faulting, subsidence of the rift floor, and progressive uplift of the rift

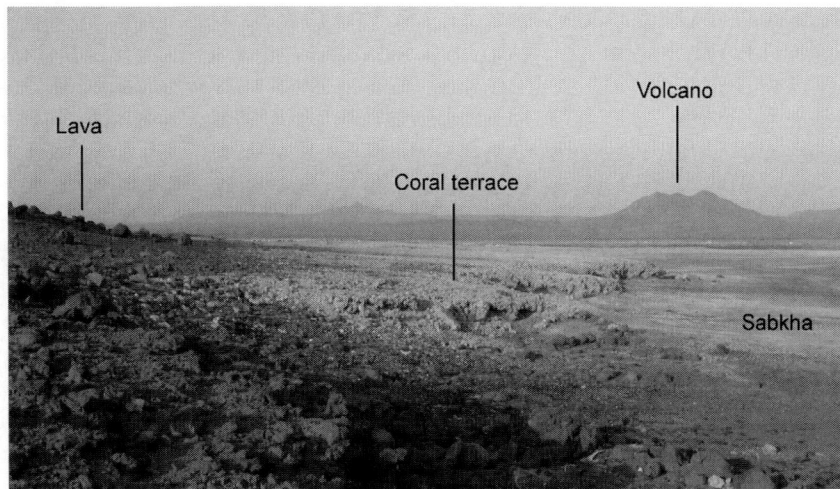

Fig. 3 Geological features in the vicinity of the site of Al Birk on the coast of Saudi Arabia, looking north. Extinct volcanoes are visible on the far horizon. The lava cone on the *left* is dated at 1.3 Ma and the sea is to the *left*
on the other side of the lava cone. Banked up against the lava cone is an elevated coral terrace believed to be of Last Interglacial age with Middle Stone Age artifacts on the surface (photo G. Bailey, March 2004)

flanks to form mountain escarpments. During the second phase of seafloor spreading after 5 Ma, rifting created a deep axial trough in the center of the Basin (the Rift or Graben), which cut through earlier deposits. The uplifted escarpments are best developed on the Arabian side and towards the southern end, with highest elevations of over 3,000 m in the Asir and Yemeni highlands in the Southwest corner of the Arabian peninsula and in the Ethiopian highlands.

Bordering the mountain escarpments is a well-developed coastal plain, known on the Arabian side in its southern sector as the Tihama, which has an average width of about 60 km. There is a similar feature on the Eritrean side, which widens out into the Afar Depression in the southwest corner. Further north, and especially in the Sudanese and Egyptian sectors, relief is less marked, the coastal plain narrower and the desert hinterland encroaches more closely on the coast.

Both sides of the Basin are composed of crystalline basement and sedimentary rocks, and these are capped locally by basalt flows, especially on the Arabian side, and in Ethiopia, where flood basalts are nearly 4,000 m thick. Sand and gravel outwash deposits of Pleistocene or Holocene age cover much of the coastal plain, which is bordered by evaporites and coral reefs on the seaward side, and, in places, salt flats (sabkha) subject to periodic tidal inundation along the shore edge (Fig. 3).

The Miocene evaporites are often of great thickness and occur mostly at depth beneath younger deposits, but because of their low density and high mobility they tend to push upwards locally beneath overlying sediments to form salt domes or diapirs, resulting in local crustal distortion. These are sometimes accompanied by deep offshore depressions associated with salt withdrawal. These features are especially well marked in the region of the Farasan Islands, but are present elsewhere also (Dabbagh et al., 1984; Bosence et al.,

1998; Plaziat et al., 1998; Warren, 1999). Coral reefs are forming at the present day, and older cemented coral terraces are associated with previous high sea levels at higher elevations than the present day shoreline (Butzer and Hansen, 1968; Gvirtzman, 1994; Taviani, 1998).

Offshore topography is highly variable. In the Gulf of Aqaba, the offshore gradient dips almost vertically as a result of tectonic controls, and there are many other coastal sectors where the submerged continental shelf is quite narrow (Fig. 2). Elsewhere, the seabed is shallower as in the Gulf of Suez, and in the southern sector of the Red Sea, where the most extensive areas of continental shelf would have been exposed during periods of low sea level, especially in the vicinity of the Farasan Islands and the Dahlak Archipelago.

The Red Sea Basin is thus a region of considerable geological instability, resulting from a combination of rifting and volcanism, more localized crustal movements caused by salt tectonics, and isostatic warping of coastal regions caused by the global effects of ice loading and alteration of water masses on the continental shelf with changes in sea level (Lambeck, 2004). All of these processes have important archaeological implications, both for reconstructing the changing paleogeography of coastline configuration and possibilities of sea crossings, and for understanding the detailed character of the landscape and its relative resource-richness and attractiveness for human settlement in different areas and at different time periods.

Climate and Environment

Climatic conditions throughout the Red Sea coastal regions are semi-arid today and relatively uniform apart from temperature gradients associated with latitude (Edwards, 1987). In the north, maximum daily temperatures range from a low of 20°C in January to a high of 35°C in July, and in the south the corresponding range is 29–40°C. Rainfall rarely exceeds 180 mm per year and is mostly concentrated in the winter months, resulting in semi-desert vegetation that supports a sparse fauna of rodents, antelopes and gazelle, and patchy areas of greater fertility around permanent water sources and at higher altitude where rainfall is higher. Rainfall increases at higher elevation and especially in the high mountains of the south, which receive summer rains from the Indian Ocean monsoon system. In the Southwest corner of the Arabian peninsula, in the highlands of Yemen and the Asir mountains of Saudi Arabia, annual rainfall ranges between about 300 and 1,000 mm. In the Ethiopian highlands, rainfall ranges between about 500 and over 2,000 mm, and nourishes permanent rivers such as the Blue Nile and the Awash, but these are far removed from the coastal regions of the Red Sea. The northern part of the Red Sea is influenced by Mediterranean cyclones, which bring winter rains, and this region is also sensitive to the effects of the North Atlantic oscillation, which leads to periods of greater or lesser aridity approximately every 6 years (Felis et al., 2000). The absence of perennial streams or rivers places a high premium on the availability of surface water in the form of springs, wells or oases, making the region as a whole sensitive to climatically-induced changes in precipitation resulting from shifts in the path of the main rain-bearing systems in the north and the southeast.

The assumption of prevailing aridity is one of the reasons why the Red Sea region and the Arabian peninsula more widely has been discounted both as a zone of early human habitation and as a pathway of dispersal. However, there is little to choose in climatic terms between the north and south ends of the Red Sea, at least under present-day conditions. If climatic aridity was a deterrent to human settlement and dispersal in the south, it would seem to have been equally so in the north. There is no doubt that supplies of fresh water are a major limiting factor on human settlement under present-day climatic conditions, and that conditions have been periodically wetter in the past. But it is generally true to say that water supplies in the region as a whole are quite sensitive to climate change, and the extent of territory available or suitable for human occupation has shown corresponding changes, with expansion into the desert regions to east and west during periods of more favorable climate.

The marine environment of the Red Sea is one of the saltiest in the world because of high rates of evaporation and limited interchange with the Indian Ocean. In the Bab al Mandab Straits salinity is close to the global ocean average of about 35%, and then increases steadily on a south–north axis to reach a maximum of 40.5% in the north (Edwards, 1987). In winter, warmer and fresher water flows into the

Red Sea near the surface from the Gulf of Aden, while cooler and saltier water flows outwards at depth. In summer the surface flow is reversed, and intermediate water from the Gulf of Aden flows into the Red Sea between the two outflowing layers (Siddall et al., 2002). Marine fertility and plankton production is highest in the south, as a result of the inflow of nutrients from the Indian Ocean and the extensive areas of submerged shelf that are shallow enough to facilitate recycling of nutrients from the seabed to the photosynthetic zone near the surface (Weikert, 1987). Fringing coral reefs and extensive beds of sea grass are also a significant source of nutrients for marine life in the form of plant detritus and organic matter. But in many areas nutrient productivity is low because of the deep water, the establishment of temperature and salinity gradients, and lack of disturbing currents, all of which inhibit recycling of nutrients, resulting in areas of marine 'desert'.

Coral reefs, sea grasses, and intermittent mangroves support a varied suite of reef fish and molluscs (Mastaller, 1987; Ormond and Edwards, 1987), and many inshore and intertidal organisms are adapted to conditions of high salinity (Jones et al., 1987). Marine food chains also support pelagic fish, turtles and sea mammals such as dugong, whales and dolphins (Frazier et al., 1987). The marine fauna is of Indo-Pacific origin (apart from a small number of species that have migrated recently from the Mediterranean through the Suez Canal). Organisms that inhabit the intertidal and shallow sublittoral zones are generally impoverished in numbers of species compared to the Indian Ocean because of extreme conditions of high temperature and salinity, and these conditions would have become more extreme during periods of low sea level, resulting in a degree of endemism, with species that are adapted to much higher salinities and temperatures than their Indian Ocean equivalents (Jones et al., 1987). The most abundant fisheries are in the south in the vicinity of the Farasan and Dahlak islands, though these do not apparently compare in potential abundance with the richer fisheries of the Persian Gulf (Head, 1987b; Ormond and Edwards, 1987).

As on land, marine productivity is likely to have been sensitive to climatically induced changes, especially the increased rates of evaporation and higher salinity associated with a drop in sea level and reduced inflow from the Indian Ocean.

Archaeological Context

The Afar region in the southwest corner of the broader region under discussion here offers one of the longest sequences and some of the earliest hominin fossils in Africa, including finds of *Australopithecus afarensis* and *Ardipithecus ramidus*, and earliest fossil remains of modern humans (Johanson and Taieb, 1976; Woldegabriel et al., 1994; White et al., 2003). Whether this reflects unusually favorable geological conditions for exposing early deposits compared to other regions or a genuine focus of early hominin evolution and settlement remains unclear. But, as the northernmost sector of the African Rift, it shares many of the dynamic landscape features associated more generally with the Rift and its attractiveness to early hominin settlement, including complex topography, mosaic environments, diverse resources, and abundant supplies of surface water. It is also an obvious bridgehead for movement into the coastal regions of the Red Sea. Substantial finds of material have also, of course, been recorded in the Nile Valley and the adjacent desert regions, notably the Fayum (Wendorf and Marks, 1968; Wendorf, 1976; Vermeersch, 2001), though early Pleistocene or earlier material is elusive, perhaps because of lack of suitably early geological exposures.

On the African side of the Red Sea in coastal regions proper, sites are more patchily distributed, reflecting amongst other factors the vagaries of exploration and geological visibility. Acheulean and Middle Stone Age material has been recovered, often in association with elevated coral terraces formed at previous periods of high sea level, notably in the north (Plaziat et al., 1998), in Djibouti (Faure and Roubet, 1968), and in Eritrea, where there is an important concentration of sites (Beyin and Shea, 2007) including the important find of Abdur dated at ca. 130 ka and associated with faunal remains and claims for the exploitation of marine resources (Walter et al., 2000). Sondheim in the hinterland of the Egyptian coastal sector represents a rare cave site with a Late Pleistocene archaeological sequence (Van Peer, 1998).

Similar material has been found on the Arabian side. Although the Arabian peninsula has often been discounted as an arid and inaccessible cul-de-sac cut off by the Red Sea, there are large numbers of Paleolithic sites widely distributed across the region, mostly discovered during a series of surveys organized in the 1970s and 1980s by the Comprehensive Archaeological Survey Program of Saudi Arabia (for the Red Sea zone and Arabian escarpment, see in particular Zarins et al., 1979, 1980, 1981; Ingraham et al., 1981; Killick et al., 1981; Gilmore et al., 1982; also Caton-Thompson, 1953), with additional surveys and excavation by Norman Whalen (Whalen et al., 1983, 1984, 1986, 1988), and by early expeditions in the Yemen (Amirkhanov, 1991), added to by more recent explorations (see Petraglia, 2003, 2007; Rose, 2004; Crassard, 2000, 2009; Rose and Usik, 2009; Scott-Jackson et al., 2009; Wahida et al., 2009). These include Acheulean and Middle Stone Age sites associated with coral terraces on the Red Sea shoreline showing general similarities with the material on the African side of the Red Sea. Sites are also widely distributed in other landscape settings, including coastal settings in the broad sense, that is

sites on or close to the present-day shoreline and in the coastal hinterland, the mountain escarpment, and inland basins associated with paleo-lakes, springs and drainage channels, many of which are dry under present-day climatic conditions (Petraglia, 2003, 2007; Petraglia and Alsharekh, 2003; Jagher, 2009; Petraglia et al., 2009; Rose and Usik, 2009). There are particularly important concentrations of sites on the Red Sea side of the Arabian escarpment in the region of Jeddah, the southern coastal sector between Al Birk and Jizan, the Asir Highlands, and the wadis draining to the east (Fig. 1).

Most of these sites are surface sites, and dates are for the most part lacking. Some material has been described as Oldowan, but whether this material is genuinely as early as that label suggests, or simply the result of poor quality local raw material, expedient tool use, or incomplete sampling remains unclear in the absence of any geological or radiometric dates. Uranium series dating of calcite concretions on artifacts from Saffāqah, near Dawadmi, indicates a minimum age of ca. 100 or ca. 200 ka (Whalen et al., 1984). At Al Birk, on the coast, the maximum age for the material is 1.3 Ma (the date of the lava cone which provided the raw materials for artifact manufacture in the vicinity, Bailey et al., 2007a, b), while other material at that site and elsewhere along the Red Sea coastline is associated with an elevated coral terrace of presumed Last Interglacial age (Bailey et al., 2007a).

Paleoenvironment and Resources

Sea-Level Change and the Southern Pathway

The general pattern of eustatic sea-level change over the Last Glacial–interglacial cycle according to a variety of sources of information is shown in Fig. 4. The different sea level curves are derived from different sources of information, subject to varying margins of uncertainty, and show differences of detail but broad agreement in general trends and a maximum amplitude between interglacial and glacial-maximum sea levels of about 115–130 m. The 100 m bathymetric contour in the Red Sea provides a useful approximation of coastline configuration at the glacial maximum, and highlights the extensive areas of new land exposed in the southern basin and the narrowness of the channel over the Hanish Sill and through the Bab al Mandab Straits (Fig. 2).

A more detailed analysis of changing coastline configuration in the southern channel at different sea level positions and dates is shown in Fig. 5, based on bathymetric data and modeling of isostatic distortion (Bailey et al., 2007a, in prep). Margins of error in this method of reconstruction make it impossible to be certain whether or not there was a land connection at extreme low sea level, but any such connection would have been only a few meters in elevation and unlikely to have formed an effective barrier to the movement of sea water between the Red Sea and the Gulf of Aden. Interpretations of the Red Sea deep-sea isotope record confirm the absence of an enduring barrier at any time during

the past 400 ka, based on the absence of extreme isotope values that would be expected had the Basin become cut off from the Indian Ocean and subjected to very high salinity (Siddall et al., 2003). Fernandes et al. (2006) claim that the sea channel would never have been less than about 4 km wide and 15 m deep. However, the deduction of channel geometry from the isotope record is subject to its own margins of uncertainty, and while the evidence suggests uninterrupted flow of water between the Red Sea and the Gulf of Aden even at the lowest sea level, the detailed modeling of coastline configuration shown in simplified form in Fig. 5d suggests that the channel might equally well have comprised a series of narrow braided channels of varying depth rather than a single broad one.

How far we should regard such a crossing as a barrier or a disincentive to human movement across the Straits is a matter of opinion and depends on assumptions about the ability of the populations in question to make rafts or boats, or to swim across several kilometers of water, and the attractions of resources on the other side of the channel. Certainly the data suggest that it would not have been possible to make the crossing without getting wet, even at lowest sea level. However, we might argue that it would have been relatively simple to make short crossings by simple rafting or by swimming, aided by warm sea temperatures and the increased buoyancy resulting from higher salinities. Current flow in the narrowed channel was probably higher than today, but estimates of likely flow rates suggest that this is unlikely to have been a significant hazard (Bailey et al., 2007a).

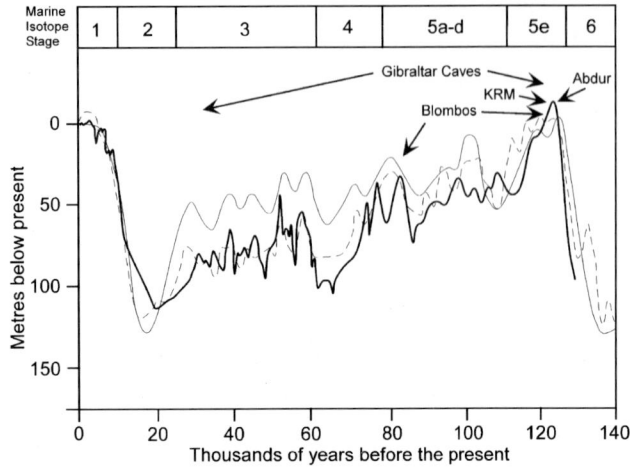

Fig. 4 Global sea-level change over the past 140 ka. The *dashed grey line* is based on deep-sea oceanic isotope records of planktonic and benthonic fauna, the *solid grey line* shows the same curve corrected for temperature effects using dated and elevated marine terraces in New Guinea, the *dark solid line* is based on isotope records of planktonic fauna from the Red Sea. Coastal archaeological sites dated to MIS 5 or earlier with marine indicators are also shown. Sea level data based on Chappell and Shackleton (1986); Shackleton (1987); Van Andel (1989); Lambeck and Chappell (2001); Siddall et al. (2003) (© G. Bailey)

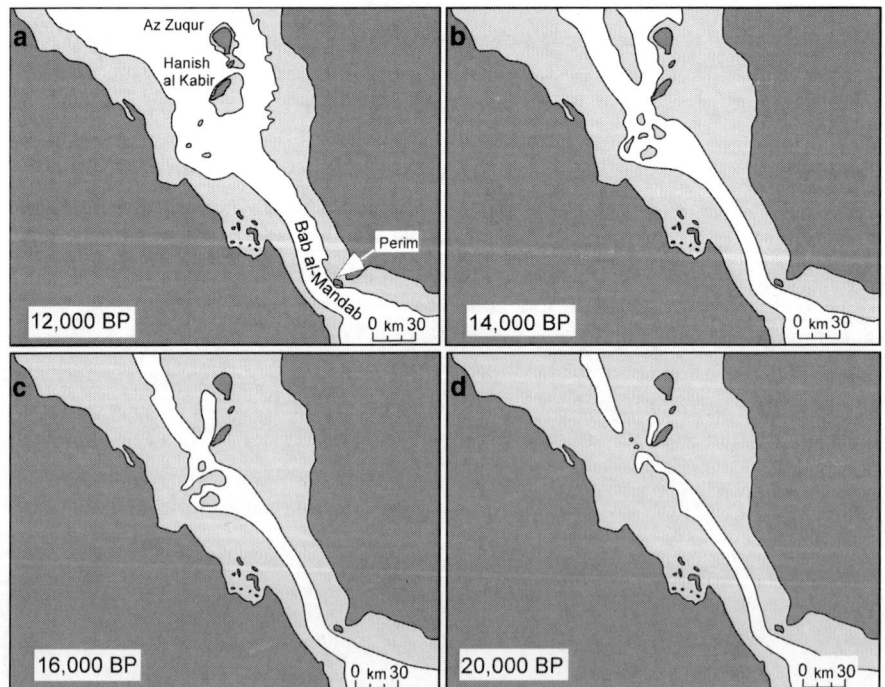

Fig. 5 Shoreline positions in the region of the Bab al Mandab Straits and the Hanish Islands at different periods of the last glacial cycle, taking account of isostatic modelling of crustal deformation (data compiled by Kurt Lambeck, © G. Bailey)

Fig. 6 Periods during the past 125,000 years when sea level conditions were most conducive to sea crossings at the southern end of the Red Sea. The upper diagram (**a**) shows the period when sea level was below −100 m according to the Red Sea isotope curve (*black*) or the global deep-sea isotope curve (*grey*), and therefore the periods when the channel was at its narrowest and windows of opportunity existed for sea crossings assuming minimal abilities to cross water. The lower diagram (**b**) shows the periods when sea level was below −50 m, using the same conventions for sea level information as in (**a**), and assumes abilities to cross somewhat wider sea channels than in (**a**). See Figure 4 for sea level curves and sources of data, Table 1 for dates and durations, and the text for further discussion of assumptions about sea-crossing abilities. © G. Bailey

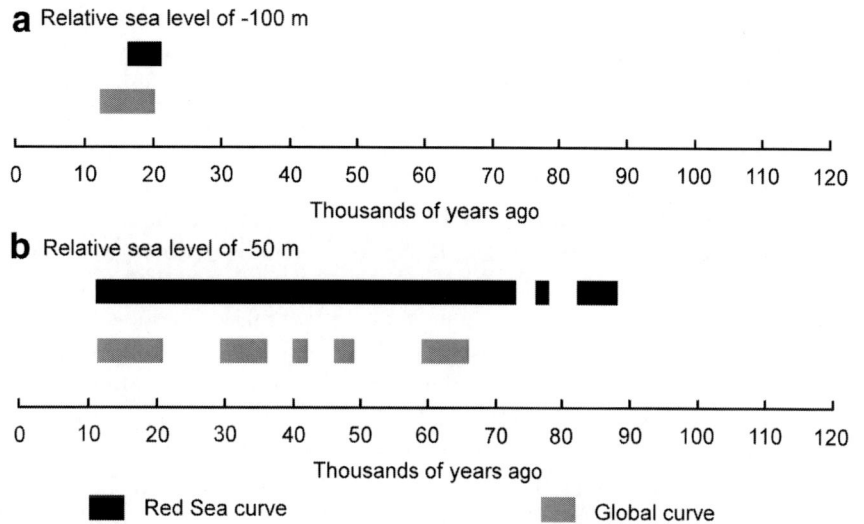

Whether the crossing could have been made at intermediate sea levels without simple rafts or boats is less certain. At a relative sea-level depth of about −50 m equivalent to the 12 ka reconstruction of Fig. 5a, island hopping via the enlarged islands of Hanish al-Kabir and Az-Zuqur would have required five sea crossings, two involving distances of at least 10 km, which would arguably have been marginal without effective rafts or boats. At a relative sea-level depth of about −70 m, equivalent to the 14 ka reconstruction of Fig. 5b, two sea crossings of ca. 10 km would have been required, which still looks marginal without some form of water transport. At a relative sea-level depth of about −90 to −100 m, equivalent to the 16 ka reconstruction of Fig. 5c, a single crossing of, at most, 5 km would have been required. Taking a conservative view of the likelihood of sea crossings by drifting, floating or active swimming, we might suggest a high probability of crossings at relative sea level depths greater than −100 m. With simple rafts or boats the window of opportunity might widen to periods when sea level was at least −50 m or deeper.

The periods when sea crossings might have taken place, using the above criteria, are shown in Fig. 6. The windows of opportunity are clearly only very approximate, both because of the assumptions involved, and because different sea level curves give different results, the local curve derived from the Red Sea giving the longest periods of potential crossing. On this basis, crossings at sea levels below −100 m could have occurred over a period of 5,000–8,000 years at the glacial maximum (Table 1). At −50 m the divergence between different sea-level curves is greater, with the possibility of crossings over a period that ranges from 19 to 70 ka. Of course, with seaworthy boats crossings might have occurred even at high sea levels. Such a possibility is naturally raised by the much longer sea journeys undertaken by the earliest immigrants to Australia and New

Table 1 Periods of potential sea crossing at the southern end of the Red Sea during the Last Glacial. Dates and durations are in thousands of years. Sea-level curves are taken from Fig. 4

Sea-level curve	Sea-level depth			
Global	−50 m	Duration	−100 m	Duration
	11–27	16	13–21	8
	29–36	7		
	40–42	2		
	46–49	3		
	59–66	7		
Total		19		
Red Sea	11–73	62	17–22	5
	76–78	2		
	82–88	6		
Total		70		13

Guinea as much as 50 ka. In that region, however, such an achievement was made possible not only by favorable winds and currents, but also by large quantities of bamboo washed down to the coast and out to sea during the monsoon season, providing a ready supply of floatable material that could easily be lashed together to provide serviceable water craft. Such circumstances probably do not apply to the Red Sea region. In any case, any form of crossing would have been easier when sea levels were lower and the channel was narrower.

The circumstances described so far apply to the last 125,000 years, the period for which we have the most detailed records of global ice history and sea-level variation. There are also good reasons to suppose that there has been little overall vertical or horizontal movement of the earth's crust in the Red Sea region resulting from rifting and tectonic movement over this time span, although there has been some local distortion because of salt tectonics (Bailey et al., 2007a). For earlier periods, however, the uncertainties multiply.

A similar cycle and amplitude of sea level variation can probably be extended back to about 900 ka with reasonable confidence, judging from the deep-sea isotope record, although the detailed pattern of sea-level variation during the glacial periods and the amount of lowering at glacial maxima may have differed in different cycles. The isotope record is difficult to interpret more precisely in terms of sea-level variation in earlier periods because it also probably includes the effects of temperature variation as well as changes in isotopic composition resulting from expansion of the continental ice sheets, and the volume of ice accumulated in earlier glaciations is known with much less certainty or not at all. However, it is worth noting that if earlier glaciations were more extensive than the last glacial, these would have produced a greater drop in sea level, and it would require only a small additional drop to produce a dry crossing across the southern end of the Red Sea, assuming crustal stability, but the Red Sea isotope evidence appears to rule out such a possibility for the past 400 ka, and there is insufficient data to pursue such speculations for earlier periods. Even so, and allowing for all the assumptions involved in extrapolating from the last glacial period, there might have been some 10 episodes during the past 900 ka when sea levels were below –100 m, representing a cumulative total of 50 to 80,000 years, and a larger number of episodes, when sea levels were below –50 m, representing a total of 190 to 700,000 years.

Before 900 ka, the deep-sea record suggests ongoing sea level fluctuations back to at least 2 Ma but a lower amplitude of sea level variation. Again the detailed implications of the isotope record are difficult to interpret, but it seems likely that they indicate a reduced drop in sea level at glacial maxima, and one that most likely did not reach the critical depth threshold of –100 m, and perhaps not even the –50 m level. Hence the likelihood of sea crossings in this earlier period without rafts or boats must be much lower, assuming that sea level change resulting from changes in global ice volumes and isostatic adjustments is the only relevant variable.

However, another variable that potentially affects the width and depth of the southern channel as we go further back into the Pleistocene is tectonic movement. The general effect of rifting is to deepen the floor of the Rift. As we have noted earlier, significant deepening of the Red Sea Rift has occurred since 5 Ma. Edgell (2006: 488) uses a figure of 15.6 mm based on recent GPS measurements for the rate of separation of the Arabian Plate from the African Plate, deducing a land bridge across the Bab al Mandab as recently as 1.4 million years ago by simple extrapolation. However, rates of movement, considered in detail in Bailey et al. (2007a), have probably not been constant over this period. Moreover, most of the movement at the southern end of the Red Sea has been taken up by deformation in the highly active Danakil depression

of the Afar rather than in the region of the Hanish and Bab al Mandab channel, which exhibits no significant current activity, and this has probably been the case for the past 2 million years (Manighetti et al., 1997; Ayele et al., 2007), leading to the conclusion that there has been little change in the geometry of the channel over this period with little impact on shoreline reconstruction. Cessation of evaporite formation after 5 million years ago is consistent with this interpretation, suggesting that the Red Sea has maintained a connection with the Indian Ocean since then. While much remains uncertain about the rate of separation between the Arabian and African Plates, the above considerations suggest that there has been relatively little change in the Bab al Mandab region during since the late Pliocene. In conclusion, current understanding of the tectonics in the southern Red Sea region does not alter the above assessment that sea crossings would have been less likely before about 900 ka.

Terrestrial Resources and Paleoclimate

If human populations had crossed the southern Red Sea at an early period, what sorts of resources would they have found there – how abundant, how easily accessible, how variable and how extensively distributed? The answer to this question can be considered under two main headings: mammalian resources and topographic constraints on their abundance and accessibility; and the timing and effects of climate change. To these we might add a third variable and that is the distribution and availability of raw materials for stone-tool manufacture. Where these are limited in occurrence, they may be a serious disincentive to successful hominin occupation of otherwise productive landscapes (Dennell, 2007). However, many different sorts of materials are quite widely available in the Red Sea region, including basaltic lavas, and a variety of other materials including ferruginous quartzite and other fine-grained siliceous and metamorphic materials. Availability of stone is unlikely to have imposed a serious limiting factor on human settlement or dispersal, although arguably basaltic lavas were a particularly favorable material, being easily accessible in appropriately sized nodules on the surface, extensively distributed, and easily worked (Fig. 7).

Land Mammals and Topographic Roughness

Information on past mammalian resources that might have offered potential resources for prehistoric hunters is confined to very limited direct evidence from paleontological or archaeological deposits and deductions from present-day species distributions (for more detail, see Turner and O'Regan, 2007). Mammals of medium to large size that exist

Fig. 7 Lava field on the Saudi Arabian coastline in the vicinity of Al Birk (photo by G. Bailey, March 2004)

today in the Arabian peninsula include oryx, gazelle, Nubian ibex, and rock hyrax, together with a suite of carnivores that includes red fox, wolf, wild cat and leopard (Harrison and Bates, 1991). These are largely limited to semi-desert and rough or mountainous country, reflecting both current climatic conditions and distributions confined by the encroachment of modern human activities.

For past conditions, there are very few archaeological or paleontological deposits to provide clues. A small collection of bones (n = 149) has been found in stratified context in the Nafud desert associated with paleo-lake conditions (Thomas et al., 1998). Finds include part of a fish maxilla, tortoise, horse (*Equus*), antelope (*Oryx*), a trace of elephant, probably the extinct form *Elephas recki*, a camelid, buffalo (*Pelorovis* cf. *Oldowayensis*), hippopotamus, and some indeterminate bovids. The combination of species together with carbon isotope measurements suggests a lake-edge setting surrounded by open savannah and semi-arid grasslands. The authors consider the assemblage to be of Lower Pleistocene date with African affinities. Zarins et al. (1979) and McClure (1978, cited in Zarins et al., 1979), refer to a Late Pleistocene fauna in central Arabia and the Rub' al Khali of *Bos primigenius*, *Bubalus* (water buffalo), hippopotamus, *Equus hemionus*, camel and gazelle. From the MIS 5e deposits of Abdur in Eritrea, indeterminate species of elephant, hippo, rhino and bovid have been recorded, but without further detail (Walter et al., 2000).

This meager evidence at least provides some clues about the sorts of species that we might expect to find in different types of topography (mountains, flat plains) and different climatic conditions. Two factors are critical to past potential: topographic conditions as they affect conditions of local fertility and access by human predators to otherwise elusive prey; and climate change.

Many of the species of animal prey that are of interest to humans are fast moving, dangerous or elusive. Without effective weapons for killing at a distance, humans are at a significant disadvantage compared to other predators. One of the

conundrums of human evolution is how early hominins from such a position of disadvantage and with weapons that were initially quite rudimentary were able to target successfully concentrated supplies of animal protein, and make that an expanding part of their diet, of their evolutionary trajectory – particularly in fuelling an extended childhood and a larger brain (cf. Aiello and Wheeler, 1995) – and of their ability to extend their habitat range. Scavenging of dead carcasses, clearly implemented at an early stage, eliminates one set of problems in gaining access to otherwise elusive prey, but poses other problems, notably the risk of falling victim to non-human hunters. One solution to this problem lies in the use of complex topography, where localized barriers, narrow or steep valleys, blind canyons, fully or partially enclosed basins of varying size, and rough terrain provide tactical opportunities for an intelligent but unspecialized predator to maneuver and trap prey, out compete other carnivores, and find protection from predators and safety for vulnerable young (King and Bailey, 1985, 2006; Bailey et al., 1993, 2000; King et al., 1994, 1997). Such topographic features are especially characteristic of tectonically active regions. In the African Rift, which is one of the largest and longest-lived tectonic structures on the planet, the particular style of tectonics results in the creation and continuous rejuvenation of complex landforms comprising near-vertical fault scarps, lake basins, numerous volcanic cones, extensive lava flows, and a swarm of minor faults and surface irregularities, all of which provide exactly these sorts of topographic opportunities for gaining tactical advantage and protection at the edges of the herbivore-rich savannah plains. As we have recently argued (Bailey et al., 2000; King and Bailey, 2006), the occurrence of some of the earliest and most concentrated fossil and archaeological evidence of human evolution in the African Rift may not be merely a coincidence of geological visibility and survival of evidence, but may instead reflect the impact of a distinctive set of topographic conditions that exercised a powerful selective impact on the human evolutionary trajectory with its emphasis on meat-eating, an extended childhood, and wide-ranging bipedalism. Moreover, the sorts of topographic features resulting from active tectonics provide not only tactical advantage in scavenging and hunting, they also create localized basins that trap sediment and water, ranging from the largest lakes of the African Rift to small bodies of water in collapsed calderas, resulting in pockets of fertility of greater or lesser extent, which can sustain plant and animal life even in otherwise relatively arid and disadvantageous climatic conditions.

If these sorts of tectonically related topographic conditions were important in Africa, the question naturally arises as to their wider distribution beyond Africa and their impact on patterns of hominin dispersal. We describe this topographic complexity as topographic 'roughness', a 'rough' surface in this context being an irregular 'corrugated' surface at any geographical scale in contrast to a smooth and flat one. Described in this way, relative roughness can be measured and mapped over large areas using satellite imagery and digital elevation data (see King and Bailey, 2006, Fig. 8a, for a simple example on a global scale using SRTM [Shuttle Radar Topography Mission] 30 data). The basis of the technique lies in the measurement of slope angles from relative height data and the mathematical transformation of this information using Fourier transforms to further smooth the data for the purpose of creating a colored or shaded map. Additional manipulations of the data can be undertaken to eliminate areas of roughness that are assumed to be inaccessible to hominin occupation because of high altitude or high latitude. Other information such as geological or climatic data can be draped over these roughness maps to help characterize the relative attractions of different areas of landscape.

A 'roughness' map is not the same as a relief map of differential elevations, and it should be emphasized that it is not roughness by itself, but the combination of rough terrain alongside extensive areas of smoother terrain capable of supporting a large herbivore biomass into which the human population can tap, which provides the key to interpretation. Fuller details of the method and its application are given elsewhere (Bailey et al., in press). We use satellite digital elevation data, which is available at different resolutions, and can be enhanced or simplified to highlight roughness at a variety of scales, ranging from mountains and large basins at one end of the spectrum to areas of rough terrain no larger than a football pitch at the other, for example a lava flow that has degraded into an irregular boulder field.

A simple, low-resolution roughness map of the Red Sea is shown in Fig. 8, with major lava fields added. Even a cursory examination shows that there is considerable variability in the distribution of roughness, with the most obvious combinations of rough terrain and smooth plains lying on either side of the Ethiopian Plateau, including the Afar Depression and the coastal region of Eritrea, the western and eastern flanks of the Arabian escarpment along its full length from Yemen in the south to the opening of the Jordan Valley in the north, and the Sinai peninsula.

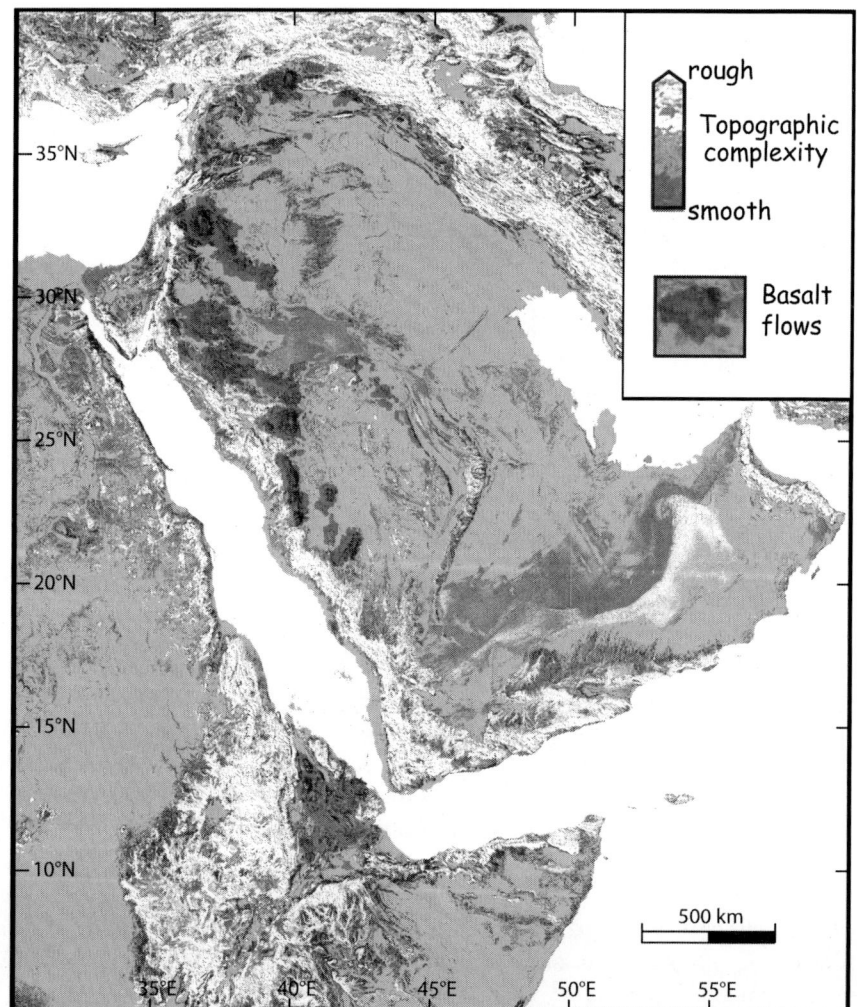

Fig. 8 Roughness map of the Red Sea region based on SRTM 30 digital elevation data. The *gray scale* ranges from white (very rough) to *dark gray* (smooth), and the black areas show the major distribution of volcanic lavas (© G. King)

These are also the areas that coincide with the main distribution of the major basaltic lava flows, and these are important not only as a source of easily accessible and workable raw material for making stone tool artifacts, but also as 'safe' areas with small scale roughness and localized patches of fertility into which populations can retreat in the face of threats from predators, competitors, or other crises. Lava fields seem at first sight to represent areas of barren and lifeless terrain, but closer inspection reveals that, in amongst the lava cones and boulder fields of degraded lava flows, they are dotted with local patches of fertility where the inherent irregularities of the surface have trapped sediment and water. One does not penetrate far into these lava fields before one encounters such minor 'oases' (Fig. 9).

In more extensive and smooth areas one might hypothesize the presence of equids and elephants, along with bovids in wetter conditions or with good surface supplies of water, buffalo and hippo near large areas of standing water such as lakes, and antelopes, gazelle and ibex in semi-arid conditions or rougher terrain.

If one were to plot the core areas likely to have provided the most favorable and enduring conditions for human settlement on the basis of this roughness map, and the most obvious pathways of dispersal, leaving aside questions about the crossing of the Red Sea, the key areas are Eritrea and Afar in the Southwest, and the flanks of the Arabian escarpment. This zone of favorable conditions continues northwards without interruption into the so-called Syrio-Jordanian Rift[3] and then eastwards along the foothills of the Taurus-Zagros arc.

There is not space here to examine more closely the distribution and interpretation of topographic roughness, or to consider its wider distribution in the Arabian peninsula. There are extensive fingers of rough terrain that extend into

Fig. 9 Local area of fertility amidst volcanic lavas in the hinterland behind Al Birk. Palm trees and lush vegetation indicate water close to the surface (photo by G. King, March 2004)

[3] Strictly speaking this is a strike–slip fault structure resulting in the creation of long and relatively narrow valleys, but with many of the same tectonic and topographic features as the African Rift, including lake basins, fault barriers and lava fields.

the central peninsula (visible in Fig. 8), and smaller-scale surface irregularities not visible at this resolution but potentially of great significance at the local scale. However, large-scale roughness tends to peter out as one moves eastwards along the southern corridor. On this basis, the primary areas of settlement and movement for earliest hominin settlement and dispersal are likely to have been on a north-south axis between the Afar and the east coast of the Red Sea, extending northwards along the Syrio-Jordanian Rift, and around the Zagros arc to the head of the Gulf region between Arabia and Iran, rather than eastwards by the more direct route along the Indian Ocean coastline. Such an interpretation strengthens the case for a crossing of the southern channel by early human populations, if not by other mammals, but it does not exclude a northerly passageway via the Sinai, and we should remember that the east coast of the Red Sea could have been accessed just as easily from the north as by sea crossings from the Southwest.

Paleoclimate

Critical to the above interpretation is a reconstruction of climatic conditions. With more abundant rainfall or surface water supplies we might expect a more diverse range of herbivores, a larger animal biomass, higher human population densities and the possibility of population expansion over a larger territory, particularly to the east of the Arabian escarpment and into central areas of the Arabian peninsula that are now semi arid or desert, and similarly on the west coast of the Red Sea. With increasing aridity, we would expect populations to thin out, to require larger home ranges and greater mobility, and in the extreme to abandon large areas of territory, and to contract and concentrate in lowland areas with ongoing supplies of water from springs or other sources, or in highland regions with relatively higher rainfall.

On a global scale, glacial periods are generally associated with conditions of increased aridity, given the amount of moisture locked up in the continental ice sheets, and interglacials with wetter conditions. The Red Sea region broadly follows this pattern with good evidence of wettest conditions during interglacial periods (or some part of them) resulting from northward movement of the Indian Ocean Monsoon (Rohling et al., 2002).

However, this is not the full story. Massive alluvial deposits attest to higher volumes of water and stream competence than today, most probably extending far back into the Pleistocene (Jado and Zötl, 1984). Sanlaville (1992) has drawn attention to evidence of higher precipitation than today in many areas, including parts of the central desert and the coastal regions of the Red Sea, in the form of alluvial sediments, travertines, tufas, calcretes, and lake deposits, and has identified four wet phases since the Last Interglacial (see also Edgell, 2006; Parker, 2009). The first, well-dated and corresponding to MIS Stage 5e, ca. 125 ka, is the wettest and is followed by a second wet phase representing a renewal of extended monsoon conditions corresponding to Stage 5a. After about 80 ka, arid conditions set in until about 35 ka, or possibly earlier, when a new and prolonged phase of increased precipitation took over and extended until about 25–20 ka. These conditions were less wet than in the previous two phases, but were widespread throughout the peninsula, and included the formation of extensive lakes in the Rub' al Khali (McClure, 1976) and in the Nafud region (Schultz and Whitney, 1986; Arz et al., 2003). Sanlaville attributes this episode to the southward movement of Mediterranean cyclones bringing winter rainfall. There then followed more arid conditions, with widespread dune building, but with brief intervening phases of increased humidity, until the onset of the Early Holocene. Between about 9 and 7 ka, the Indian Ocean Monsoon extended northwards again and the Rub' al Khali saw a renewed phase of lake formation (Parker et al., 2004, 2006; Parker 2009). Mediterranean cyclones also brought increased winter rainfall to northern Arabia and the northern Red Sea region during this period.

Clearly, then, there have been wet–dry fluctuations during both glacial and interglacial periods, though of varying amplitude and periodicity. The wettest and presumably most favorable conditions for plant and animal life on land occurred during interglacial periods of relatively high sea level, typically at sea levels slightly below the present on the evidence of the Early Holocene and MIS 5a wet phases, but an extended and relatively humid period also coincided with a large part of MIS stage 3 when sea levels ranged from –40 m to more than –60 m, immediately before the maximum regression, when crossing the southern end of the Red Sea would have been easiest, and possibly overlapping with the beginning of that low sea level stand. How far this cycle of changes can be extended back into earlier periods of the Pleistocene remains unclear, although there are certainly earlier deposits indicating equivalent conditions of climate wetter than today.

This picture of periodic and quite extensive periods of wetter climate is further enhanced if we take account of environmental conditions in the enlarged coastal region exposed by a drop in sea level. Faure et al. (2002) have argued that as sea level dropped, exposing large areas of the continental shelf, the exit of groundwater from underground aquifers would have greatly increased because of the increased hydrostatic head and the removal of the overlying mass of seawater, which would otherwise tend to inhibit stream flow. Even today some water escapes from these underground sources onto the continental shelf in the form of underwater freshwater springs, and these are well known to local fishermen. At lower sea levels, so Faure et al. (2002) argue, the landscape of the emerged coastal plain would have been transformed into coastal wetlands. Hence, plant and animal distributions

undergoing contraction in the hinterland because of increased aridity would have found new and more favorable territory into which to expand. This contrast should be qualified by the evidence for formation of massive, linear Pleistocene sand-dunes accumulated during periods of low sea level when marine sands were picked up by wind action from the former sea floor. These dunes are extensively distributed in present-day coastal areas such as the Tihama, where they create a patchwork of sandy areas with little soil formation and sparse vegetation alternating with more fertile alluvial fans or thin soils on bedrock (Munro and Wilkinson, 2007). To what extent these active dunes were spread over the territory exposed by sea-level retreat is unknown, but it seems likely that a similar patchwork existed there too with opportunities for local conditions of soil formation and concentrations of well-watered environments. If this argument is correct, then the periods when local conditions of water supply and plant and animal life were at their most favorable in coastal regions would have coincided with lowest sea levels, when dispersal across the southern end of the Red Sea and around the coastal margins of the Arabian peninsula would have been easiest.

This notion of coastal wetlands at lowered sea level is, of course, a hypothesis, but it is a testable hypothesis, and especially suitable conditions for testing it can be found in the southern basin of the Red Sea. Here the exposed shelf is relatively shallow and extensive, and would have represented an extension of new territory at lowered sea levels of up to 100 km in width on either side of the Basin (Fig. 2). In the vicinity of the Farasan Islands, even the simplified depth readings available from navigation charts indicate that the islands would have been connected to the mainland at relative sea levels of –20 to –50 m (Fig. 10). At lower sea levels, the

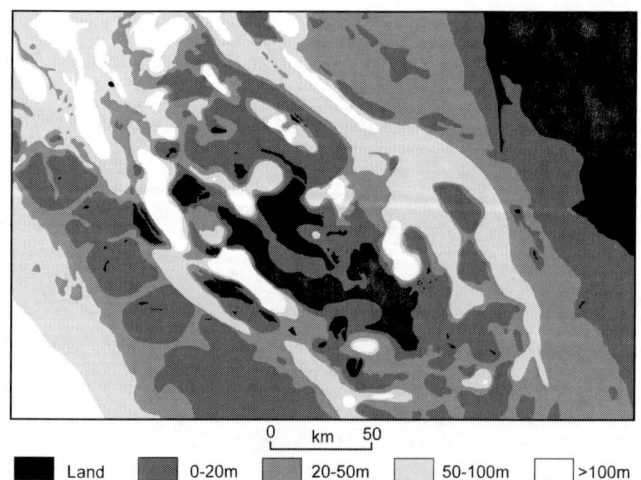

Fig. 10 General bathymetry and topography of the submerged landscape in the vicinity of the Farasan Islands. Data compiled by Garry Momber from British Admiralty chart 15. Note the complex and fragmented topography resulting from salt tectonics and the deep depressions offshore of the islands (© G. Bailey)

islands would have become part of a complex and varied coastal topography, with deep depressions capable of filling with freshwater, and other surface irregularities that would have facilitated trapping of sediments and water supplies, as well as more extensive areas of flatter coastal plain – in short a combination of rougher and smoother topography with all the consequent advantages described earlier. Moreover, this type of complex topography is also likely to create the most favorable conditions for the protection and preservation under water of terrestrial sediments and archaeological materials following inundation by sea level rise. Preliminary underwater investigations have already begun, with the use of sonar soundings and deep diving to identify traces of submerged shorelines (Bailey et al., 2007a, b), and further investigations are planned using the full range of techniques required for underwater survey, including coring, remote sensing and diving.

An insight into the impact of these combined topographic, paleoclimatic and sea-level variables is shown for the region of southwestern Saudi Arabia in the coastal region between Jizan and Al Birk (Fig. 11). This coastal region is backed by some of the highest uplands of the Arabian escarpment in the Asir highlands, providing a very wide range of altitudinal conditions over a relatively short distance. It would also have had some of the most extensive coastal lowlands exposed at lowered sea levels. There are important concentrations of sites across this region and in a variety of settings, including the sites around the lava fields of Al Birk and Ash-Shuqayk on the coast, in the coastal hinterland near Abu Arish, in the Asir Highlands near Khamis Mushayt, and on the channels draining eastwards from the Arabian

escarpment, notably at Hima and Najran and eastwards into the Rub' al Khali (Zarins et al., 1981). These sites attest both to the general attractions of the region and the diversity of ecological settings in which early sites occur. In climatic conditions wetter than today, extensive drainage channels would have provided a well-watered landscape with diverse ecologies and a full range of habitats and mammalian resources from the highest mountains down to the coastal plain, and attractive pathways for animal migration and human dispersal along river valleys on both sides of the watershed, linking the coastal region to extensive basins in the Arabian interior.

Combining this discussion of terrestrial conditions at lowered sea level with deductions from the distribution of topographic roughness and paleoclimates, Fig. 12 shows a simple hypothetical model of areas that are likely to have been most favorable to human settlement over the longest period in relation to climate and sea-level change and the other factors discussed earlier. Coastal regions on the wider coastal plains in the south clearly rate highly because of their combination of different resources, their relative proximity to the better-watered uplands of the higher escarpments, extensive areas of rough topography and basaltic lavas, and their access to more extensive areas of the continental shelf with their hypothesized wetlands during periods of low sea level. The Northwest sector rates less favorably because of the lower relief, the closer encroachment towards the Red Sea of the desert interior, and the relative lack of high escarpments that could have captured rainfall. One final variable that needs to be factored into this discussion is the availability of marine resources.

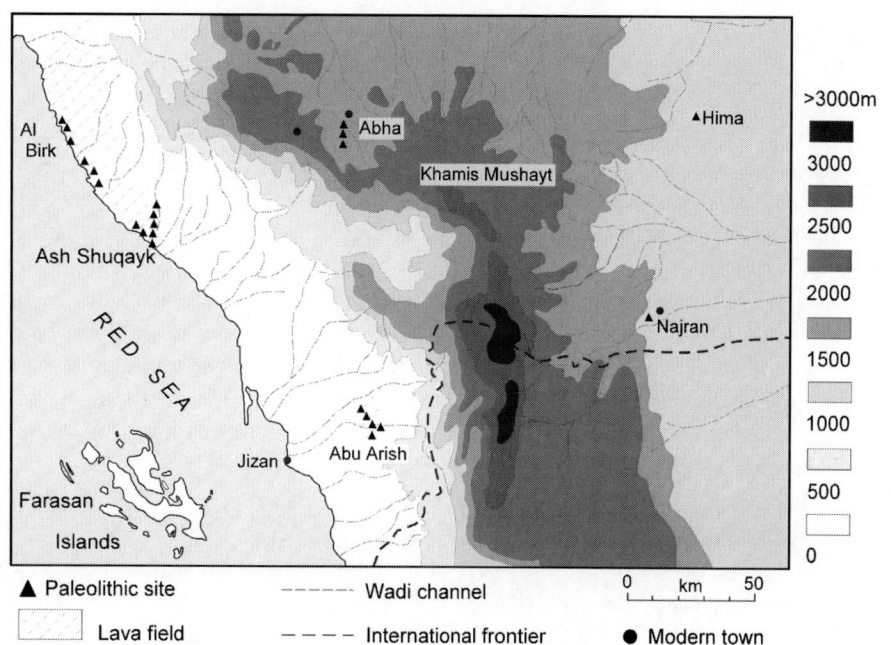

Fig. 11 The Arabian escarpment between Al Birk and Jizan showing the main concentrations of known Paleolithic sites and distribution of topographic variables (© G. Bailey)

▲ Paleolithic site   - - - - - Wadi channel

▨ Lava field   - - - - International frontier   ● Modern town

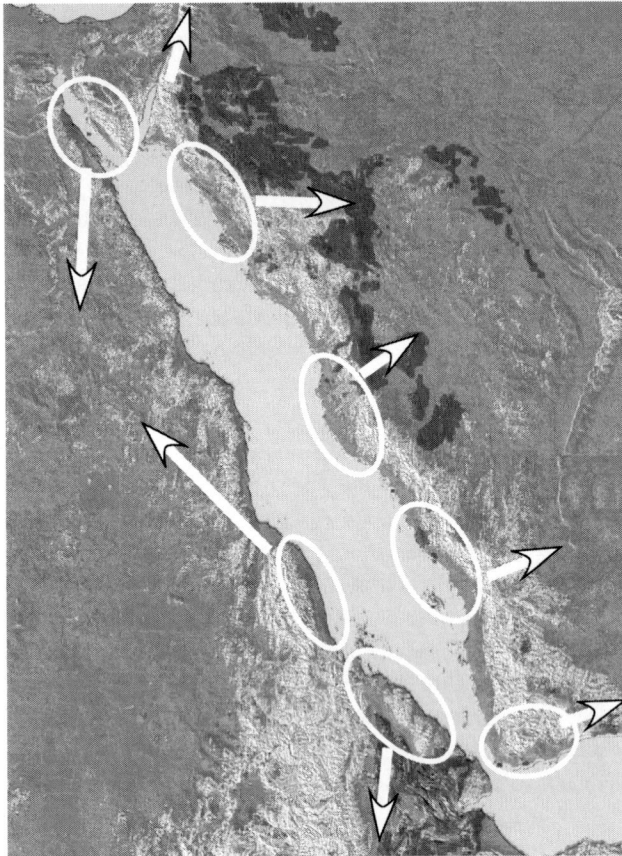

Fig. 12 Model of refugia in the Red Sea region. *Circles* indicate areas with the most varied range of resources and the most persistently favorable conditions for human settlement. *Arrows* indicate general directions of habitat extension and contraction during wetter climatic periods (© G. Bailey)

Fig. 13 Holocene shell mounds on Qumah Island in the Farasan Islands (photo by G. Bailey, March 2004)

Coastal Habitats and Marine Resources

As on land, so at the shore edge, the most favorable coastlines of the Red Sea Basin for marine resources are in the south, with its extensive areas of shallow shelf, high inputs of nutrients and more moderate salinities. The archipelagos of the Dahlak and Farasan Islands also provide the longest extent of shorelines with access to abundant intertidal and inshore resources. The numerous and substantial shell mounds of the Farasan Islands that have formed during the past six millennia in association with the present sea level give some indication of the archaeological signature that one might expect from coastal economies with a high dependence on marine resources (Fig. 13).

Even these sites, which contain many millions of shells, do not necessarily imply specialized shellfood economies or intensive exploitation of the molluscs, since large numbers of shells are needed to supply a given requirement of food, such mounds are usually accumulated quite slowly and intermittently over many centuries or millennia, and shells are notoriously liable to over-representation in archaeological deposits in comparison with other food remains. Similar substantial mounds of a similar time-range are known in their tens of thousands from many other parts of the world, and are invariably associated with broad-spectrum economies that draw on a range of marine and terrestrial resources according to local environmental circumstances (Bailey and Milner, 2002). Substantial shell mounds are notoriously rare from earlier periods, most probably in part because of lowered sea levels and the submergence of shore-edge sites, and we would not necessarily expect Pleistocene shell-midden deposits to match the size and density of those well known from the Holocene. Lower human population densities in earlier periods, lack of technological items such as containers and boats, which would have facilitated processing and consumption of larger numbers of shells in one location, and factors of differential preservation and visibility resulting from sea-level change, coastal erosion and weathering, all argue against the survival or visibility of such evidence as one goes further back into the Pleistocene. But the more recent examples do at least provide a benchmark against which to evaluate earlier evidence.

Two issues require further exploration here, the impact of lowered sea levels on resource-productivity, and the actual evidence that marine resources were exploited in earlier periods and their possible contribution to human settlement.

With reduced inflow from the Indian Ocean during periods of low sea level, salinity in the Red Sea was clearly high enough to inhibit plankton photosynthesis, at least at the maximum low sea-level stand, implying more extensive areas of marine desert than today (Hemleben et al., 1996; Fenton et al., 2000; Siddall et al., 2003). In the south, low sea levels would have reduced the extensive shallow areas of the continental shelf that form such an important nutrient supply and nursery ground for the modern fisheries, presumably with some corresponding reduction in marine productivity.

However, these effects should not be over-exaggerated. As discussed earlier, many marine organisms are already adapted to conditions of high salinity today, in salinities that may range from 40‰ to 80‰ or more (Jones et al., 1987), although the range of marine species becomes increasingly impoverished with increasing salinity, and the evidence of Red Sea endemism indicates that many persisted through the periods of lowest sea level and relative isolation from the Indian Ocean. Many marine animals can also obtain nutrients from plant matter independently of the phytoplankton food chain, particularly reef fish, molluscs, and the predators that feed on them, and these are particularly the marine resources that would have been most accessible to human gatherers on the coast edge. Nevertheless, the productivity and abundance of marine resources would almost certainly have been much reduced during low sea levels, especially in the central and northern sectors of the Red Sea, but even at a reduced level would have offered an additional advantage to human populations living in coastal areas, especially at the southern end of the Basin, in closest proximity to the Gulf of Aden.

When we turn to actual evidence of exploitation, the evidence is obscure. Many of the shorelines that might provide relevant evidence are of course under water and have yet to be explored (cf. Bailey et al., 2007a, b). Otherwise, a variety of proxies and evidence of uncertain validity have been relied on. The question of what constitutes a coastal habitat or a coastal economy in the context of human settlement is a matter of definition and is open to considerable differences of opinion. Often the terms are used as very general descriptive labels to refer to any location or archaeological site that is within reach of the present-day coastline, a reach that may range from 100 m or less to 100 km and more, a descriptive label of such generality as to be devoid of useful analytical content. A strict definition of a coastal economy is one 'devoted, at least more so than not so . . . to the exploitation of marine resources' (Beaton, 1995: 802), or, in the absence of precise quantitative measures of the marine component in subsistence, one with clear evidence of material culture items devoted to marine subsistence such as fish hooks, harpoons and boats, dense concentrations of marine food remains, or substantial settlements on the coast edge.

However, people may be attracted to coastal areas and even to the immediate shoreline for reasons that have nothing to do with marine resources, for raw materials washed up on the beach or outcropping nearby to make artifacts, or for terrestrial plant and animal food resources that often occur in coastal areas in greater diversity or abundance because of higher water tables and climatic amelioration relative to the adjacent hinterland. The presence of stone tools on or close to a present-day shoreline, unaccompanied by unequivocal evidence of food remains, by itself can tell us nothing about the nature of the subsistence economy,

especially if the artifacts are not well dated and may have been deposited during a period of lower sea level when the contemporaneous shoreline would have been displaced many kilometers seaward.

Even where organic materials are present in association with artifacts, their status as food remains may be ambiguous. This is especially a problem for sites in beach locations, where shells and other marine materials may be accumulated by natural processes, or picked up as dead specimens for use as artifacts or ornaments, requiring detailed taphonomic analyses to distinguish natural from cultural accumulations (cf. Bailey et al., 1994; Stiner, 1994: 177, 182). This is evidently a problem with the Abdur site, where the oyster shells originally claimed as evidence of food remains (Walter et al., 2000) later turned out to be a natural death assemblage (Bruggemann et al., 2004).

Sites in similar locations to the Abdur site have been recorded along the Saudi Red Sea coastline in association with raised marine terraces and lava fields, with stone tools of Middle Stone Age and Acheulean type, particularly in the lava field between Ash-Shuqayk and Al Birk (Figs. 1 and 11; Zarins et al., 1981). The Middle Stone Age material at Al Birk (also referred to as site 216–208) is located on the surface of a coral beach terrace presumed to be of Last Interglacial (MIS Stage 5) date.[4] Zarins et al. (1981) reported tools embedded in the beach deposit, but we were unable to replicate that observation in 2004 either at this site or anywhere else along this stretch of the Red Sea coastline where stone tools have been reported in association with lava fields and coral terraces, although the area has undergone considerable disturbance and damage because of bulldozing activity, road building and other development since the original surveys. It is also clear that the beach terrace at Al Birk is banked up against lava flows from the nearby volcanic cone and stratigraphically later than them (Figs. 3 and 14), not stratified beneath the lava as originally believed (see Zarins et al., 1981, plate 5A), and this is consistent with K/Ar ages of ca. 1.3 Ma for the lava cone (Bailey et al., 2007a, b).

Much has also been made of the appearance of shells and other marine indicators in South African coastal sites located on or close to the present-day shoreline with deposits dating to MIS Stage 5 (see Fig. 4), notably at the caves of Die Kelders, Klasies River Mouth and Blombos Cave, and at the open air sites of Sea Harvest and Hoedjies Punt (Avery et al., 1997; Henshilwood et al., 2001; Henshilwood and Marean, 2003; Klein et al., 2004). These date between about 130 and 75 ka and contain variable quantities of marine shells, mostly rocky shore species of limpets and mussels, often forming

---

[4]This site is referred to as Al Qamah in Bailey et al. (2007a) to distinguish it from a Holocene shell midden to the north of the town of Al Birk. It is one of a number of sites in the area and is in fact closer to Al Birk.

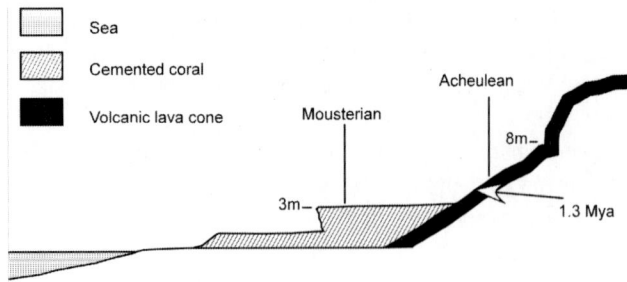

Fig. 14 A cross-section of the deposits at Al Birk (206–218). Data from Zarins et al. (1981, plate 5) and from personal observations. The section is viewed looking north, and should be compared with the photograph of the same site in Fig. 3 looking south (© G. Bailey)

layers of quite dense shell midden, and bones of seals, penguins, and fish. This concentration and visibility of marine indicators reflects the fact that the deposits in question are associated with a period of high sea level when the shoreline was nearby, while the absence of earlier material almost certainly reflects the fact that similar activities carried out during periods of low sea level along this coast would have left their mark in locations that are now far offshore and deeply underwater.

The record on this coastline has recently been extended back to 160 ka by the recently reported finds from the cave of Pinnacle Point (Marean et al., 2007). Here, however, the marine indicators comprise just 79 shells deposited at a time when the site was many kilometers inland from the contemporaneous shoreline during a period of low sea level. This is about what one would expect by way of a visible marine signature in such geographical circumstances. Whether this evidence implies much greater quantities of seafood consumed or processed near the now-submerged shoreline is a matter of debate. Similar small quantities of marine shells are present in European coastal-cave sequences with Upper Paleolithic deposits accumulated during the low sea levels of the last glacial, notably sites of northern Spain, which were up to 10 km inland at the time (Bailey and Craighead, 2003; Bailey and Milner, 2008).

There is nothing in this evidence to support claims that the African material represents the earliest appearance of marine resources in the human paleoeconomy, or that their consumption signifies the appearance of 'modern' behaviors associated with cognitive developments in later *Homo sapiens*. Such claims are at risk both of over-exaggerating the significance of marine resources, and of underestimating the problems of differential visibility of relevant evidence as one goes further back into the Pleistocene (cf. Erlandson and Fitzpatrick, 2006). All the marine resources present are consistent with collecting or scavenging on the foreshore or the intertidal zone, and with the simplest levels of technology. Similar evidence has been recovered from sites in Gibraltar and elsewhere in the Mediterranean associated with Neanderthals. At

Vanguard Cave in Gibraltar, marine mollusc shells have been recovered from Last Interglacial deposits, along with bones of sea mammals and some fish vertebrae (Finlayson et al., 2006; Stringer et al., 2008). Shells are found in other Mousterian coastal deposits in Italy and Spain, and the earliest known evidence, comprising shells and fish spines, is from the 400 ka Mediterranean open-air site of Terra Amata (De Lumley, 1966). The quantity of shells and other remains of marine foods in these Mediterranean sites appears to be less than in the African ones, but this probably reflects the fact that the marine productivity of the Mediterranean is generally-speaking relatively low, and especially in the intertidal zone, where the extent of the molluscan habitat is limited by the lack of tidal movement (Fa, 2008).

None of this is to play down the potential role of marine resources at earlier periods. However, the significance of the few finds available is clearly vulnerable to over interpretation on the flimsiest of evidence. There is no evidence whatsoever that any of these early coastal settlers in these regions during the Pleistocene were marine specialists, or so dependent on marine resources that this provided a spur to migration because of overexploitation of local supplies. Such notions are inherently implausible and there are few if any precedents for them even in the vast ethnographic and archaeological literature on coastal hunters and gatherers of recent millennia. Before the advent of specialist technology for offshore fishing and sea-mammal hunting, resources on land would have provided the mainstay of human subsistence, even if resources gathered on the shore provided added advantage to coastal populations. Equally, the belief that marine resources were ignored or avoided until the appearance of modern humans is both factually incorrect and fails to take sufficient account of the loss or invisibility of relevant archaeological evidence from earlier periods. The implication that marine resources such as molluscs required some advanced level of human cognition or technology before they were recognized or accessible as edible resources is scarcely credible, given the simplicity with which most molluscs can be collected and consumed, the evidence of their consumption by other animals including other primates, and the omnivorous instincts of humans.[5] The dominant theme of human evolution is adaptive flexibility and omnivory, and it seems more plausible to suggest that marine resources would have been exploited where and when available by human populations living in coastal areas from the earliest period, and that such resources, even if exploited at much lower levels of intensity than in later periods, would have provided some added advantage to those populations that took advantage of them, in combination with plant and animal resources on land.

---

[5] Charles Darwin, of course, famously observed that 'To knock a limpet from the rocks does not require even cunning, that lowest power of the mind' (Darwin, 1839, pp. 235–236).

Conclusion

Resource conditions in the Red Sea region and potentials for hominin settlement and movement are clearly quite variable and still subject to a number of uncertainties, especially for the earlier periods of the full time range considered here. The belief that general aridity prevailed during glacial periods and would have deterred all but ephemeral settlement is certainly an oversimplification, and indeed substantially incorrect. There have been quite lengthy episodes of wetter climate throughout the Pleistocene and probably earlier, and not only during interglacial periods. During some part of MIS 3, climate was wet enough to sustain more or less permanent lakes in the desert regions of Saudi Arabia for as much as ten millennia in the period immediately before the glacial maximum (MIS 2). On the coastward side of the Arabian escarpment it is likely that conditions were even more favorable, given the steep relief and the large number of drainage channels that could have supplied water to the coastal lowland. Nevertheless, these wet periods have alternated with lengthy arid periods as well, and much remains to be discovered about their dating and duration and the extent to which local conditions of water supply and fertility may have helped to maintain adequate conditions for human settlement during these arid periods, especially the supply of spring-water from underground aquifers on the coastal plain exposed at the maximum lowering of sea level, a period when general climatic conditions were at their most arid during the glacial–interglacial cycle. Even so, the region as a whole would probably always have lacked long-lasting supplies of surface water comparable to the great lakes of the African Rift and the Jordan Valley or major rivers such as the Nile, The Awash and the Jordan. The extent of territory accessible to human settlement is likely to have undergone regular fluctuations with changing climatic conditions, with core areas or refugia centered on the flanks of the Arabian and Ethiopian escarpments and their adjacent coastal lowlands. Otherwise, given adequate water supplies, the region would have had many attractions in terms of topography, food resources and raw materials for stone-tool manufacture.

The possibilities for human transit across the southern end of the Red Sea also remain uncertain. At no time does there appear to have been an unbroken land connection, at least over the past 400 ka, but during periods of maximum sea-level regression during glacial maxima the channel would have been sufficiently narrow to suggest a high likelihood of sea crossings by swimming or simple rafting, at least from the time of the Lower-Middle Pleistocene boundary (ca. 800–900 ka) onwards. Before that the picture is less clear. The amplitude of sea-level variation during the Lower Pleistocene or earlier seems to have been considerably less, but equally the impact of long-term tectonic processes on reconstructions of channel geometry at these earlier periods is harder to specify, though it was probably quite minor, at least back to about 2 Ma, so that the probability of sea crossings was much lower. In any case, sea crossings were probably less significant than the nature of the resources available on either side of the channel, since these could have been reached equally well by population movement from the north.

Finally, the case for marine resources as a primary factor in promoting a process of coastal colonization by early human populations whether anatomically modern or earlier remains weak. This is not to say that the hypothesis is wrong. Marine resources, especially those easily accessible close inshore, in the intertidal zone or as carcasses washed up on the beach, would undoubtedly have added to the attractions of the coastal environment, but are unlikely to have provided the basis for specialized marine subsistence economies. Moreover the hypothesis may be more appropriate to some coastal regions than others, and perhaps more so to the coastal regions of Southeast Asia with their archipelagos, rich coastal wetlands and river estuaries (cf. Bulbeck, 2007), than to the coastal environments of the Red Sea, Arabia and adjacent regions. Even so, other conditions in the coastal zone such as water supplies and terrestrial plant and animal resources are likely to have played an equally important or more important role in the attractions of the coastal zone for settlement and dispersal. Either way, the relevant evidence to test these propositions will almost certainly require sustained investigation of the underwater landscape (Bailey and Flemming, 2008), since for most of the period under discussion here sea levels have been persistently lower than the present, and the shorelines and associated paleoenvironmental and archaeological evidence of their exploitation are mostly now deeply submerged.

Acknowledgments I am indebted to Abdullah Al-Sharekh, Nic Flemming, Geoffrey King, Kurt Lambeck and Claudio Vita-Finzi for discussions of Red Sea archaeology, geology and environment, and to NERC through its EFCHED program (Environmental Factors in Human Evolution and Dispersal), the British Academy, the Leverhulme Trust, Saudi Aramco, the Saudi British Bank and Shell Companies Overseas for funding the fieldwork in the southern Red Sea that forms the foundation for this chapter.

References

Aiello LC, Wheeler P. The expensive tissue hypothesis: the brain and the digestive system in human and primate evolution. Current Anthropology. 1995;36(2):199–221.

Amirkhanov KA. The Palaeolithic of southern Arabia. Moscow (in Russian): Nauka; 1991.

Arz HW, Lamy F, Pätzold J, Müller PJ, Prins M. Mediterranean moisture source for an Early-Holocene humid period in the northern Red Sea. Science. 2003;300:118–21.

Avery G, Cruz-Uribe K, Goldberg P, Grine FE, Klein RG, Lenardi MJ, et al. Excavations at the Die Kelders Middle and Late Stone Age cave site, South Africa. Journal of Field Archaeology. 1997;24(4):263–91.

Ayele A, Jacques E, Kassim M, Kidane T, Omar A, Tait S, et al. The volcano-seismic crisis in Afar, Ethiopia, starting September 2005. Earth and Planetary Science Letters. 2007;255:177–87.

Bailey G. Earliest coastal settlement: marine palaeoeconomies, submerged landscapes and human dispersals. In: Anderson A, Barker G, editors. Global origins of seafaring. Cambridge: McDonald Institute for Archaeological Research; In press.

Bailey G, Flemming N. Archaeology of the continental shelf: marine resources, submerged landscapes and coastal environments in early prehistory. Quaternary Science Reviews; In press.

Bailey GN, Milner NJ. Coastal hunters and gatherers and social evolution: marginal or central? Before Farming: The Archaeology of Old World Hunter-Gatherers. 2002;3–4(1):1–15.

Bailey G, Milner N. Molluscan archives from European prehistory. In: Mantczak A, Cipriani R, editors. Early human impact on megamolluscs. Oxford: British Archaeological Reports International Series; 2007, In press.

Bailey GN, King GCP, Sturdy DA. Active tectonics and land-use strategies: a Palaeolithic example from northwest Greece. Antiquity. 1993;67:292–312.

Bailey G, Chappell J, Cribb R. The origin of *Anadara* shell mounds at Weipa, North Queensland, Australia. Archaeology in Oceania. 1994;29:69–80.

Bailey G, King G, Manighetti I. Tectonics, volcanism, landscape structure and human evolution in the African Rift. In: Bailey G, Charles R, Winder N, editors. Human ecodynamics: Proceedings of Association for Environmental Archaeology Conference 1998 held at the University of Newcastle upon Tyne. Oxford: Oxbow; 2000. p. 31–46.

Bailey GN, Flemming N, King GCP, Lambeck K, Momber G, Moran L, et al. Coastlines, submerged landscapes and human evolution: the Red Sea basin and the Farasan islands. Journal of Island and Coastal Archaeology. 2007b;2(2):127–60.

Bailey G, Al-Sharekh A, Flemming N, Lambeck K, Momber G, Sinclair A, et al. Coastal prehistory in the southern Red Sea basin, underwater archaeology, and the Farasan islands. Proceedings of the Seminar for Arabian Studies. 2007c;37:1–16.

Beaton JM. The transition on the coastal fringe of Greater Australia. Antiquity. 1995;69:798–806.

Beyin A. The Bab al Mandab vs the Nile-Levant: an appraisal of the two dispersal routes for early modern humans out of Africa. African Archaeological Review. 2006;23:5–30.

Beyin A, Shea J. Reconnaissance of prehistoric sites on the Red Sea coast of Eritrea, NE Africa. Journal of Field Archaeology. 2007;32:1–16.

Bonatti E. Punctiform initiation of seafloor-spreading in the Red Sea. Nature. 1985;316:33–7.

Bosence DWJ, Al-Aawah MH, Davison I, Rosen BR, Vita-Finzi C, Whitaker E. Salt domes and their control on basin margin sedimentation: a case study from the Tihama Plain, Yemen. In: Purser BH, Bosence DWJ, editors. Sedimentation and tectonics of Rift basins: Red Sea – Gulf of Aden. London: Chapman & Hall; 1998. p. 448–78.

Braithwaite CJR. Geology and palaeogeography of the Red Sea region. In: Edwards AJ, Head SM, editors. Key environments: the Red Sea. Oxford: Pergamon; 1987. p. 22–44.

Bruggemann JH, Buffler RT, Guillaume MMM, Walter RC, von Cosel R, Ghebretensae BN, et al. Stratigraphy, palaeoenvironments and model for the deposition of the Abdur Reef Limestone: context for an important archaeological site from the Last Interglacial on the Red Sea coast of Eritrea. Palaeogeography, Palaeoclimatology, Palaeoecology. 2004;20:179–206.

Bulbeck D. Where river meets sea: a parsimonious model for *Homo sapiens* colonization of the Indian Ocean Rim and Sahul. Current Anthropology. 2007;48(2):315–21.

Butzer KW. Pleistocene history of the Nile valley in Egypt and Lower Nubia. In: Williams MAJ, Faure H, editors. The Sahara and the Nile. Rotterdam: Balkema; 1980. p. 253–80.

Butzer KW, Hansen C. River and desert in Nubia. Madison, WI: The University of Wisconsin Press; 1968.

Cabrera VM, Abu-Amero KK, Larruga JM, González AM. The Arabian peninsula: gate for human migrations out of Africa or cul-de-sac? A mitochondrial DNA phylogeographic perspective. In: Petraglia MD, Rose JI, editors. The evolution of human populations in Arabia: paleoenvironments, prehistory and genetics. The Netherlands: Springer; 2009. p. 79–87.

Caton-Thompson G. Some palaeoliths from South Arabia. Proceedings of the Prehistoric Society. 1953;19:189–218.

Chappell J, Shackleton NJ. Oxygen isotopes and sea level. Nature. 1986;324:137–40.

Crassard R. The Middle Paleolithic of Arabia: the view from the Hadramawt region, Yemen. In: Petraglia MD, Rose JI, editors. The evolution of human populations in Arabia: paleoenvironments, prehistory and genetics. The Netherlands: Springer; 2009. p. 151–68.

Dabbagh A, Hotzl H, Schnier H. Farasan islands. General considerations and geological structure. In: Jado AR, Zotl JG, editors. Quaternary period in Saudi Arabia, vol. 2. New York: Springer; 1984. p. 212–20.

Darwin C. Journal of researches into the natural history and geology of the countries visited during the voyage round the world of H.M.S. 'Beagle' under command of Captain Fitz Roy. London: R.N. John Murray; 1839.

De Lumley H. Les fouilles de Terra Amata a Nice: premiers resultats. Bulletin de la Musée d' Anthropologie Préhistorique à Monaco. 1966;13:29–51.

Dennell R. "Resource-rich, stone poor": early hominin land use in large river systems of northern India and Pakistan. In: Petraglia M, Allchin B, editors. The evolution and history of human populations in South Asia. New York: Springer; 2007. p. 41–68.

Dennell R, Roebroeks W. An Asian perspective on early human dispersal from Africa. Nature. 2005;438:1099–104.

Derricourt R. Getting "out of Africa": sea crossings, land crossing and culture in the hominin migrations. Journal of World Prehistory. 2005;19(2):119–32.

Edgell HS. Arabian deserts. Nature, origin and evolution. The Netherlands: Springer; 2006.

Edwards FJ. Climate and oceanography. In: Edwards AJ, Head SM, editors. Key environments: the Red Sea. Oxford: Pergamon; 1987. p. 45–69.

Erlandson JM, Fitzpatrick SM. Oceans, islands, and coasts: current perspectives on the role of the sea in human prehistory. Journal of Island and Coastal Archaeology. 2006;1:5–32.

Fa DA. Effects of tidal amplitude on intertidal resource availability and dispersal pressure in prehistoric human coastal populations: the Mediterranean-Atlantic transition. Quaternary Science Reviews. In press.

Faure H, Roubet C. Découverte d'un biface Acheuléen dans les calcaires marins du golfe Pléistocène de l'Afar (Mer Rouge, Ethiope). Comptes Rendus Hebdomadaires des Seances de l'Académie des Sciences Paris. 1968;267:18–21.

Faure H, Walter RC, Grant DR. The coastal oasis: ice age springs on emerged continental shelves. Global and Planetary Change. 2002;33: 47–56.

Felis T, Pätzold J, Loya Y, Fine M, Nawar AH, Wefer G. A coral oxygen isotope record from the northern Red Sea documenting NAO, ENSO, and North Pacific teleconnections on Middle East climate variability since the year 1750. Paleoceanography. 2000;15: 679–94.

Fenton M, Geiselhart S, Rohling EJ, Hemleben C. Aplanktonic zones in the Red Sea. Marine Micropaleontology. 2000;40:277–94.

Fernandes CA, Rohling EJ, Siddal M. Absence of post-Miocene Red Sea land bridges: biogeographic implications. Journal of Biogeography. 2006;33:961–6.

Field JS, Lahr MM. Assessment of the southern dispersal: GIS based analyses of potential routes at Oxygen Isotope Stage 4. Journal of World Prehistory. 2005;19:1–45.

Field JS, Petraglia MD, Lahr MM. The southern dispersal hypothesis and the South Asian archaeological record: examination of dispersal routes through GIS analysis. Journal of Anthropological Archaeology. 2007;26(1):88–108.

Finlayson C, Giles Pacheco F, Rodríguez Vidal J, Fa DA, Gutierrez López M, Santiago Pérez A, et al. Late survival of Neanderthals at the southernmost extreme of Europe. Nature. 2006;443:850–3.

Flemming N, Bailey G, Courtillot V, King G, Lambeck K, Ryerson F, et al. Coastal and marine palaeo-environments and human dispersal points across the Africa-Eurasia boundary. In: Brebbia CA, Gambin T, editors. The maritime and underwater heritage. Southampton: Wessex Institute of Technology Press; 2003. p. 61–74.

Forster P, Matsumura S. Did early humans go north or south? Science. 2005;308:965–6.

Frazier JG, Bertram CB, Evans PGH. Turtles and marine mammals. In: Edwards AJ, Head SM, editors. Key environments: the Red Sea. Oxford: Pergamon; 1987. p. 288–314.

Gilmore M, Al-Ibrahim M, Murad AS. Part I: Comprehensive Archaeological Survey Program: 1 – preliminary report on the northwestern and northern regions survey 1981 (1401). Atlal, The Journal of Saudi Arabian Archaeology. 1982;6:9–23.

Girdler RW, Styles P. Two stage Red Sea floor spreading. Nature. 1974;247:7–11.

Girdler RW, Whitmarsh RB. Miocene evaporites in Red Sea cores, their relevance to the problem of the width and age of oceanic crust beneath the Red Sea. Initial Report of the Deep Sea Drilling Program. 1974;23:913–21.

Gvirtzman G. Fluctuations of sea level during the past 400,000 years: the record of Sinai, Egypt (northern Red Sea). Coral Reefs. 1994;13:203–14.

Harrison DL, Bates PJJ. The mammals of Arabia. Sevenoaks, England: Harrison Zoological Museum; 1991.

Head SM. Introduction. In: Edwards AJ, Head SM, editors. Key environments: the Red Sea. Oxford: Pergamon; 1987a. p. 1–21.

Head SM. Red Sea fisheries. In: Edwards AJ, Head SM, editors. Key environments: the Red Sea. Oxford: Pergamon; 1987b. p. 363–82.

Hemleben C, Meischner D, Zahn R, Almogi-Labin A, Erlenkeuser H, Hiller B. Three hundred eighty thousand year long stable isotope and faunal records from the Red Sea: influence of global sea level change on hydrography. Paleoceanography. 1996;11(2):147–56.

Henshilwood CS, Marean CW. The origin of modern behavior: critique of the models and their test implications. Current Anthropology. 2003;44(5):627–51.

Henshilwood CS, Sealy JC, Yates R, Cruz-Uribe K, Goldberg P, Grine FE, et al. Blombos Cave, southern Cape, South Africa: preliminary report on the 1992–1999 excavations of the Middle Stone Age levels. Journal of Archaeological Science. 2001;28(4):421–48.

Hubert-Ferrari A, King GCP, Manighetti I, Armijo R, Meyer B, Tapponnier P. Long-term elasticity in the continental lithosphere: modelling the Aden Ridge propagation and the Anatolian extrusion process. Geophysical Journal International. 2003;153:111–32.

Ingraham M, Johnson T, Rihani B, Shatla I. Preliminary report on a reconnaissance survey of the Northwestern Province (with a note on a brief survey of the Northern Province). Atlal, The Journal of Saudi Arabian Archaeology. 1981;5:59–84.

Jado AR, Zotl JG, editors. Quaternary period in Saudi Arabia, vol. 2. New York: Springer; 1994.

Jagher R. The Central Oman Palaeolithic Survey: recent research in Southern Arabia and reflection on the prehistoric evidence. In: Petraglia MD, Rose JI, editors. The evolution of human populations in Arabia: paleoenvironments, prehistory and genetics. The Netherlands: Springer; 2009. p. 139–50.

James HVA, Petraglia MD. Modern human origins and the evolution of behavior in the later Pleistocene record of South Asia. Current Anthropology. 2005;46(Suppl):S3–S27.

Johanson DC, Taieb M. Plio-Pleistocene hominid discoveries in Hadar, Ethiopia. Nature. 1976;260:293–7.

Jones DA, Ghamrawy M, Wahbeh MI. Littoral and shallow subtidal environments. In: Edwards AJ, Head SM, editors. Key environments: the Red Sea. Oxford: Pergamon; 1987. p. 169–93.

Killick A, Whalen N, James N, Morsi G, Kamal M. Saudi Arabian archaeological reconnaissance 1980. Preliminary report on the Western Province survey. Atlal, The Journal of Saudi Arabian Archaeology. 1981;5:43–58.

King G, Bailey G. The palaeoenvironment of some archaeological sites in Greece: the influence of accumulated uplift in a seismically active region. Proceedings of the Prehistoric Society. 1985;51:273–82.

King GCP, Bailey GN. Tectonics and human evolution. Antiquity. 2006;80(308):265–86.

King GCP, Bailey GN, Sturdy DA. Active tectonics, complex topography and human survival strategies. Journal of Geophysical Research. 1994;99(B10):20063–78.

King G, Sturdy D, Bailey G. The tectonic background to the Epirus landscape. In: Bailey G, editor. Klithi: Palaeolithic settlement and Quaternary landscapes in Northwest Greece: Volume 2: Klithi in its local and regional setting. Cambridge: McDonald Institute for Archaeological Research; 1997. p. 541–59.

Klein RG, Avery G, Cruz-Uribe K, Halkett D, Parkington JE, Steele T, et al. The Ysterfontein 1 Middle Stone Age site, South Africa, and early human exploitation of coastal resources. Proceedings of the National Academy of Sciences USA. 2004;101(16): 5708–15.

Lahr M, Foley R. Multiple dispersals and modern human origins. Evolutionary Anthropology. 1994;3(2):48–60.

Lambeck K. Sea-level change through the last glacial cycle: geophysical, glaciological and palaeogeographic consequences. Comptes Rendus Geoscience. 2004;336:677–89.

Lambeck K, Chappell J. Sea level change through the last glacial cycle. Science. 2001;292:679–86.

Lordkipanidze D, Bar-Yosef O, Otte M, editors. Early humans at the gates of Europe. Proceedings of the First International Symposium, Dmanisi, Tbilisi (Georgia) September 1998. Etudes et Recherches Archéologiques de l'Université de Liège (E.R.A.U.L.), Liège, 92; 2000.

Macaulay V, Hill C, Achilli A, et al. Single, rapid coastal settlement of Asia revealed by analysis of complete mitochondrial genomes. Science. 2005;308:1034–6.

Manighetti I, Tapponnier P, Courtillot V, Gruszow S, Gillot P. Propagation of rifting along the Arabia–Somalia plate boundary: the Gulfs of Aden and Tadjoura. Journal of Geophysical Research. 1997;102:2681–710.

Mannino MA, Thomas KD. Depletion of a resource? The impact of prehistoric human foraging on intertidal mollusc communities and its significance for human settlement, mobility and dispersal. World Archaeology. 2002;33:452–74.

Mastaller M. Molluscs of the Red Sea. In: Edwards AJ, Head SM, editors. Key environments: the Red Sea. Oxford: Pergamon; 1987. p. 194–214.

McClure H. Radiocarbon chronology of late Quaternary lakes in the Arabian Desert. Nature. 1976;263:755–6.

Mellars PA. Going east: new genetic and archaeological perspectives on the modern human colonization of Eurasia. Science. 2006;313: 796–800.

Munro RN, Wilkinson TJ. Environment, landscapes and archaeology of the Yemeni Tihāmah. In: Starkey J, Starkey P, Wilkinson T, editors. Natural resources and cultural connections of the Red Sea.

British Archaeological Reports International Series 1661. Oxford: Archaeopress; 2007. p. 13–33.

Omar GI, Steckler MS. Fission track evidence on the initial rifting of the Red Sea: two pulses, no propagation. Science. 1995;270:1341–4.

Oppenheimer S. Out of Eden: the peopling of the world. London: Constable; 2003.

O'Regan H. The Iberian peninsula – corridor or cul de sac? Mammalian faunal change and possible routes of dispersal in the last 2 million years. Quaternary Science Reviews. In press.

Ormond RFG, Edwards AJ. Red Sea fishes. In: Edwards AJ, Head SM, editors. Key environments: the Red Sea. Oxford: Pergamon; 1987. p. 251–87.

Parker AG. Pleistocene climate change from Arabia: developing a framework for hominin dispersal over the last 350 ka. In: Petraglia MD, Rose JI, editors. The evolution of human populations in Arabia: paleoenvironments, prehistory and genetics. The Netherlands: Springer; 2009. p. 39–49.

Parker AG, Eckersely L, Smith MM, Goudie AS, Stokes S, Ward S, et al. Holocene vegetation dynamics in the northeastern Rub' al-Khali desert, Arabian peninsula: a phytolith, pollen and carbon isotope study. Journal of Quaternary Science. 2004;19(7):665–76.

Parker AG, Goudie AS, Stokes S, White K, Hodson MJ, Manning M, et al. A record of Holocene climate change from lake geochemical analyses in southeastern Arabia. Quaternary Research. 2006;66(3):465–76.

Petraglia M. The Lower Palaeolithic of the Arabian peninsula: occupations, adaptations, and dispersals. Journal of World Prehistory. 2003;17:141–79.

Petraglia M. Mind the gap: factoring the Arabian peninsula and the Indian subcontinent into out of Africa models. In: Mellars P, Boyle K, Bar-Yosef O, Stringer C, editors. Rethinking the human revolution. Cambridge: McDonald Institute for Archaeological Research; 2007. p. 383–94.

Petraglia M, Alsharekh A. The Middle Palaeolithic of Arabia: implications for modern human origins, behaviour and dispersals. Antiquity. 2003;77(298):671–84.

Plaziat J-C, Baltzer F, Choukroi A, Conchon O, Freytet P, Orszag-Sperber F, et al. Quaternary marine and continental sedimentation in the northern Red Sea and Gulf of Suez (Egyptian coast): influences of rift tectonics, climatic changes and sea-level fluctuations. In: Purser BH, Bosence DWJ, editors. Sedimentation and tectonics of rift basins: Red Sea – Gulf of Aden. London: Chapman & Hall; 1998. p. 537–73.

Rídl J, Edens CM, Černý V. Mitochondrial DNA structure of Yemeni population: regional differences and the implications for different migratory contributions. In: Petraglia MD, Rose JI, editors. The evolution of human populations in Arabia: paleoenvironments, prehistory and genetics. The Netherlands: Springer; 2009. p. 69–78.

Rohling EJ, Cane TR, Cooke S, Sprovieri M, Bouloubassi I, Emeis KC, et al. African monsoon variability during the previous interglacial maximum. Earth and Planetary Science Letters. 2002;6304:1–15.

Ron H, Levi S. When did hominids first leave Africa? New high-resolution palaeomagnetic evidence from the Erk-El-Ahmar formation, Israel. Geology. 2001;29:887–90.

Rose JI. The question of Upper Pleistocene connections between East Africa and South Arabia. Current Anthropology. 2004;45(4):551–5.

Rose JI, Usik VI. The "Upper Paleolithic" of South Arabia. In: Petraglia MD, Rose JI, editors. The evolution of human populations in Arabia: paleoenvironments, prehistory and genetics. The Netherlands: Springer; 2009. p. 169–185.

Sanlaville P. Changements climatiques dans la péninsule Arabique durant le Pléistocène Supérieur et l'Holocène. Paléorient. 1992;18(1):5–26.

Schultz E, Whitney JW. Upper Pleistocene and Holocene lakes in the An Nafud, Saudi Arabia. Hydrobiologia. 1986;143:175–90.

Scott-Jackson J, Scott-Jackson W, Rose JI. Paleolithic stone tool assemblages from Sharjah and Ras al Khaimah in the United Arab Emirates. In: Petraglia MD, Rose JI, editors. The evolution of human populations in Arabia: paleoenvironments, prehistory and genetics. The Netherlands: Springer; 2009. p. 125–38.

Shackleton NJ. Oxygen isotopes, ice volume and sea level. Quaternary Science Reviews. 1987;6:183–90.

Siddall M, Smeed DA, Matthiesem S, Rohling EJ. Modelling the seasonal cycle of the exchange flow in Bab el Mandab (Red Sea). Deep-Sea Research. 2002;I:1551–69.

Siddall M, Rohling EJ, Almogi-Labin A, Hemleben C, Meischner D, Schmelzer I, et al. Sea-level fluctuations during the last glacial cycle. Nature. 2003;423:853–8.

Stiner MC. Honor among thieves: a zooarchaeological study of neandertal ecology. Princeton, NJ: Princeton University Press; 1994.

Taviani M. Post-Miocene reef faunas of the Red Sea: glacio-eustatic controls. In: Purser BH, Bosence DWJ, editors. Sedimentation and tectonics of rift basins: Red Sea – Gulf of Aden. London: Chapman & Hall; 1998. p. 574–82.

Tchernov E. Eurasian-African biotic exchanges through the Levantine corridor during the Neogene and Quaternary. In: von Koenigswald W, Werdelin L, editors. Mammalian migration and dispersal events in the European Quaternary. Frankfurt am Mein: Courier Forschung-Institut Senckenberg; 1992. p. 103–23.

Templeton A. Out of Africa again and again. Nature. 2002;415:45–51.

Thangaraj K, Chaubey G, et al. Reconstructing the origin of Andaman Islanders. Science. 2005;308:996.

Turner A, O'Regan H. Afro-Eurasian mammalian fauna and early hominin dispersals. In: Petraglia M, Allchin B, editors. The evolution and history of human populations in South Asia. New York: Springer; 2007. p. 23–39.

Van Peer P. The Nile corridor and the out-of-Africa model: an examination of the archaeological record. Current Anthropology. 1998;39(Suppl):115–40.

Vermeersch PM. 'Out of Africa' from an Egyptian point of view. Quaternary International. 2001;75:103–12.

Wahida G, Yasin A-T, Beech MJ, Al Meqbali A. A Middle Paleolithic assemblage from Jebel Barakah, Coastal Abu Dhabi Emirate. In: Petraglia MD, Rose JI, editors. The evolution of human populations in Arabia: paleoenvironments, prehistory and genetics. The Netherlands: Springer; 2009. p. 117–24.

Walter RC, Buffler RT, Bruggemann JJ, Guillaume MMM, Berhe SM, Negassi B, et al. Early human occupation of the Red Sea coast of Eritrea during the Last Interglacial. Nature. 2000;405:65–9.

Warren JK. Evaporites, their evolution and economics. Oxford: Blackwell; 1999.

Weikert H. Plankton and the pelagic environment. In: Edwards AJ, Head SM, editors. Key environments: the Red Sea. Oxford: Pergamon; 1987. p. 90–111.

Wendorf F. Prehistory of the Nile valley. London/New York: Academic; 1976.

Wendorf F, Marks A. Prehistory of Nubia. Dallas: Fort Burgwin Research and Southern Methodist University; 1968.

Whalen NM, Fritz GA. The Oldowan in Arabia. Adumatu. 2004;9:7–18.

Whalen NM, Sindi H, Wahidah G, Siraj-Ali JS. Excavation of Acheulean sites near Saffaqah in al-Dawadmi (1402/1982). Atlal, The Journal of Saudi Arabian Archaeology. 1983;7:9–21.

Whalen NM, Siraj-Ali JS, Davis W. Excavation of Acheulean sites near Saffaqah, Saudi Arabia, 1403 AH 1983. Atlal, The Journal of Saudi Arabian Archaeology. 1984;8:9–24.

Whalen NM, Siraj-Ali JS, Sindi H, Pease DW. A Lower Pleistocene site near Shuwayhitiyah in northern Saudi Arabia. Atlal, The Journal of Saudi Arabian Archaeology. 1986;10:94–101.

Whalen NM, Siraj-Ali J, Sindi HO, Pease DW, Badein MA. A complex of site in the Jeddah-Wadi Fatimah area. Atlal, The Journal of Saudi Arabian Archaeology. 1988;11:77–85.

Whalen NM, Wilbon PD, Pease DW. Early Pleistocene migrations into Saudi Arabia. Atlal, The Journal of Saudi Arabian Archaeology. 1989;12:59–75.

White T, Asfaw B, DeGusta D, Gilbert H, Richards DG, Suwa G, et al. Pleistocene *Homo sapiens* from Middle Awash, Ethiopia. Nature. 2003;423:742–7.

Woldegabriel G, White TD, Suwa G, Renne P, de Heinzelin J, Hart WH, et al. Ecological and temporal placement of early Pliocene hominids at Aramis, Ethiopia. Nature. 1994;371(22):330–3.

Bailey G, Lambeck K, Vita-Finzi C, Al-Sharekh A. Sea-level change and the archaeology of human dispersal in the Red Sea region. Quaternary Science Reviews. In preparation.

Zarins J, Ibrahim M, Potts D, Edens C. Saudi Arabian archaeological reconnaissance 1978. The preliminary report on the third phase of the Comprehensive Archaeological Survey Program – the Coastal Province. Atlal, The Journal of Saudi Arabian Archaeology. 1979;3:9–42.

Zarins J, Whalen N, Ibrahim M, Mursi AJ, Khan M. The Comprehensive Archaeological Survey Program. Preliminary report on the Central and Southwestern Provinces. Atlal, The Journal of Saudi Arabian Archaeology. 1980;4:9–36.

Zarins J, Al-Jawad Murad A, Al-Yish KS. The Comprehensive Archaeological Survey Program, a. The second preliminary report on the Southwestern Province. Atlal, The Journal of Saudi Arabian Archaeology. 1981;5:9–42.

Chapter 3
Pleistocene Climate Change in Arabia: Developing a Framework for Hominin Dispersal over the Last 350 ka

Adrian G. Parker

Keywords Chronology • Climate Change • Dispersal • Pleistocene

Introduction

Environmental change in Arabia has oscillated between climatic extremes throughout the Quaternary period with evidence for ancient pluvials, apparent in the lacustrine sediments, alluvial fans and gravels, paleosols, and speleothems (e.g., McClure, 1976; Schultz and Whitney, 1986; Parker et al., 2006a, 2006b; Lézine et al., 2007; Fleitmann et al., 2007). Conversely, there are numerous signals that Arabia was also subjected to extremes in aridity, most obviously manifested in the expansive sand seas comprising the Nafud, Rub' al Khali, and Wahiba deserts, as well as fracture calcites from hyperalkaline springs (Clark and Fontes, 1990) and petrogypsic soil horizons (Rose, 2006).

Evidence for small eroded lake basins comprising marl terraces and hardened evaporitic crusts, with associated freshwater shells and lithic implements scattered around the edges were reported in the Rub' al Khali during early exploration of the region (e.g., Philby, 1933; Holm, 1960; Clark, 1989). Occurrences of ancient stone tools near relict lake beds in Arabia provided the first evidence for a rich prehistoric past (Caton-Thompson, 1953; Field, 1958). To date, however, the association between humans and environment is still much in its infancy and the precise relationships between human dispersals into and across Arabia is not fully understood. Both environmental and archaeological research has made significant progress in recent years but, to date, no major synthesis has been attempted which provides the environmental backdrop for assessing hominin emergence within the Arabian peninsula. The aim of this chapter is to present an overview of the variable and shifting landscapes in Arabia during the past 350 ka (isotope stage 9 to the present) with particular emphasis on indicators of pluvial conditions (mostly lacustrine sediments and alluvial sediments along with supporting data from other proxy sources). These data provide a useful framework for understanding the role of the climate in influencing Pleistocene hominin dispersals and occupation across the Arabian Corridor – a critical geographic zone that we now know served as a conduit bridging early human populations in Europe, Africa, and Asia (Parker and Rose, 2008).

In general, the record of Arabian lakes is restricted to typical lacustrine deposits, which often consist of white calcium carbonate-rich accumulations, or finely stratified silts and clays. From their contained microfossils and geochemistry these can readily be identified as having accumulated in bodies of standing water of varied duration (although the hydrological mechanisms behind the formation of such lakes continue to be a matter for debate). These lakes are frequently the locus of considerable human Paleolithic and Neolithic activity (Zeuner, 1954; Field, 1958; McClure, 1976; Edens, 1988; Masry, 1997; Petraglia and Alsharekh, 2003; Rose, 2007).

Geography, Geology, and Climate

The Arabian subcontinent measures 2,100 km from north to south along the Red Sea coast, and nearly 2,000 km across at its maximum width from the westernmost region of Yemen to the easternmost point in Oman. The Arabian peninsula is bounded on the west by the Gulf of Aqaba and the Red Sea (Bailey et al., 2007a, 2007b), on the south by the Gulf of Aden and the Arabian Sea, and on the east by the Gulf of Oman and the Arabian Gulf. The littoral is characterized by tropical and sub-tropical ecosystems, while the basin-shaped interior is dominated by alternating steppe and desert landscapes. Three major sand seas are found in Arabia: the Rub' al Khali (600,000 km²), Nafud (72,000 km²) and Wahiba Sands (12,500 km²) (Goudie, 2003; Edgell, 2006; Parker and Rose, 2008).

A.G. Parker (✉)
Human Origins and Palaeo-Environments (HOPE) Research Group,
Department of Anthropology and Geography,
Oxford Brookes University, Oxford, OX3 0BP, UK
e-mail: agparker@brookes.ac.uk

Arabia is skirted by mountainous terrain along the western, southern and eastern edges of the peninsula (Edgell, 2006; Parker and Rose, 2008). The 'Asir Highlands run along the western flank of the Kingdom of Saudi Arabia, called the Yemen Highlands where they extend into the Republic of Yemen. This mountain chain reaches nearly 4,000 m asl in the south – the highest point on the entire peninsula; as a result, it receives up to 1,000 mm of rainfall per annum. The coastal plain of southern Arabia is bounded by the Hadramaut (in Yemen) and Nejd (in Oman) plateaus. Extending north from the Dhofar Escarpment, sedimentary beds rise sharply to an elevation of 1,000 m asl, gradually levelling off northward onto the Nejd. The entire region is comprised of uplifted Tertiary limestone that gradually slopes into the Rub' al Khali basin. The ridge of the Dhofar escarpment marks the watershed divide; southward flowing drainages are seasonally active under present conditions and drainage basins debouch into the Indian Ocean. The northward flowing drainages receive almost no stormflow, but, during pluvial cycles, the magnitude of the monsoon was sufficient to produce high-energy fluvial systems, which drained into the Arabian interior and at times into the Persian Gulf basin.

The peninsula is subject to two major weather regimes (Barth and Steinkohl, 2004). From the north come Atlantic late-winter Northwesterlies, which move eastward over the Mediterranean Sea, down the Arabian Gulf, and eventually dissipate over the Rub' al Khali desert and Musandam peninsula, bringing cool gentle winds and light precipitation (Parker et al., 2004). The second weather regime consists of summer storms brought by the Southwest Indian Ocean Monsoon system. From June to September, the highlands of Yemen and Oman receive relatively heavy rainfall as the mountainous terrain of southern Arabia traps moisture from the monsoon (Lézine et al., 1998; Glennie and Singhvi, 2002). Consequently, the 'Asir and Dhofar Mountains receive between 200 and 1,000 mm annually; while areas closer to sea level seldom collect more than 100–200 mm per year (Schyfsma, 1978).

When reconstructing past environments (Anderson et al., 2007) and the potential influence upon human dispersals and migration it must be borne in mind that the modern landscape may in many places bear little resemblance to the landscapes of the Quaternary. The impacts of tectonics, sea-level changes, equilibrium adjustment, deflation, incision, aridity and wetness on the Arabian landscape have been profound over Quaternary timescales.

Arabia is a region of extremes in climate and landscape response. The region has been exposed to immense sub-aerial processes, changing patterns in weathering, transportation (both fluvial and aeolian) and deposition. This has resulted in zones of denudation and accumulation, which will have a profound impact on the preservation of evidence for human occupation.

The Question of Chronology

In this chapter dates were compiled from a variety of paleoenvironmental proxy signals from across the Arabian peninsula (Fig. 1), which together are used to build a comprehensive database of climate change in Arabia. These data are derived from published sources as well as new evidence collected by the author in the field.

A total of 427 absolute dates are used to construct a composite sum probability density function (pdf) curve (Fig. 2) that is used to infer climatic oscillations over the past 350 ka, from isotope stage 9 to the present. It should be noted that sediments and speleothems pre-dating 350 ka have been noted, however, they are beyond the range of uranium–thorium dating (UTh) and optical stimulated luminescence (OSL), hence chronological precision is poor and thus omitted from this study. Peaks in curve represent periods where dates cluster which are inferred as phases of increased wetness, while troughs highlight few/no dates implying drier phases. It should be noted that the height of the peaks is related to the number of dates which are stacked and not the intensity of wetness. The dates compiled are recorded in the literature as representing wet periods and thus where a large number of dates overlap there is an increased probability that they represent the same period of wetness. All UTh, OSL and calibrated radiocarbon (cal. ^{14}C) dates are presented on the same timescale for comparative purposes. All radiocarbon dates were calibrated using CALIB v.5.0.1 up to 26 ka (Reimer et al., 2004) and up to 50 ka using CALPAL (Weninger and Jöris, 2008).

It should be noted that several potential chronological problems exist. All three of the dating techniques used to determine ages are prone to problems. For ^{14}C these include hard-water errors as well as the potential deposition of younger carbonates into older carbonate sediments (Sanlaville, 1992; Immenhauser et al., 2007). Samples which have not undergone full bleaching may yield OSL ages that are too old. This is potentially problematic for sites with colluvial reworking of slope sediments including rock shelters where stratified evidence for human occupation may be found. Also burrowing or bioturbation can mix sediments and introduce younger sediments into a stratigraphic sequence. UTh may suffer from the leaching of uranium, dertrital contamination or low thorium ratios.

Two examples where chronological discrepancies have been noted are highlighted below. The first problem with dating is clearly illustrated in two recent papers by Cremaschi and Negrino (2005) and Immenhauser et al. (2007). Cremaschi and Negrino (2005) dated a stratified Early Holocene rock shelter with in situ lithics from the Gebel Qara region, Dhofar, Oman. The uppermost unit was dated using UTh and yielded an age of 92 ka. Low Th ratios, however, make this an unreliable date and this is verified by the underlying ^{14}C

Fig. 1 Map of Arabian peninsula showing location of fluvial/lacustrine, speleothem and marine core samples

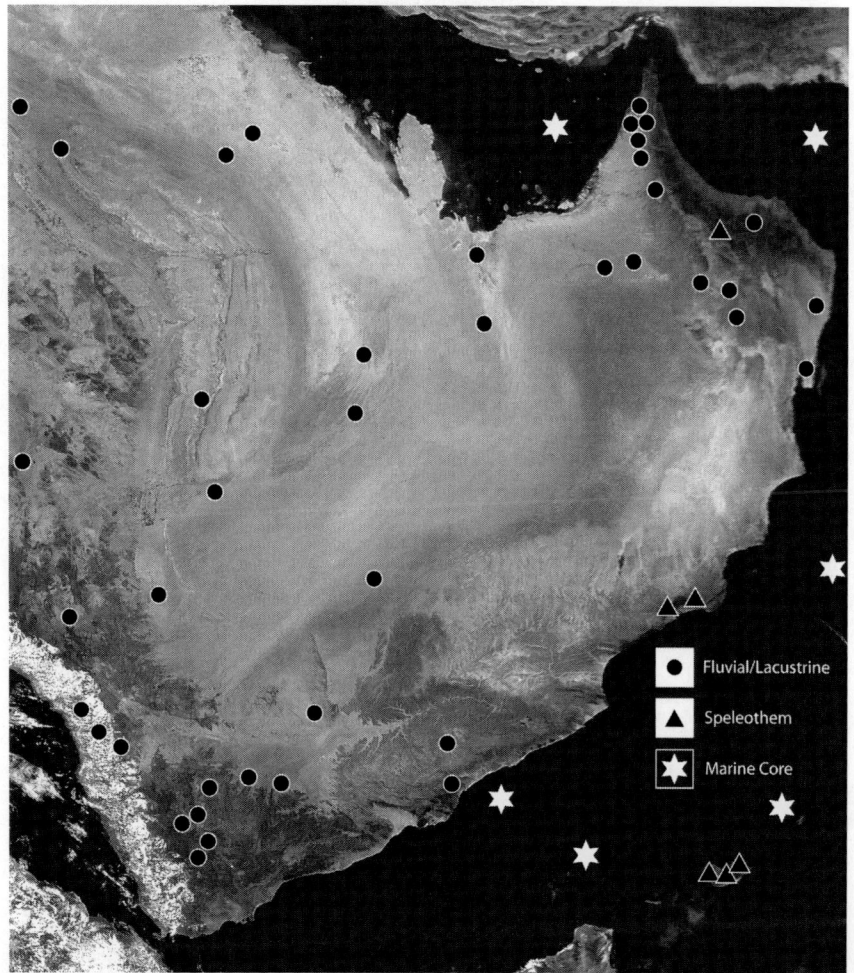

Fig. 2 Paleoclimatic record across Oxygen Isotope Stages 1–9

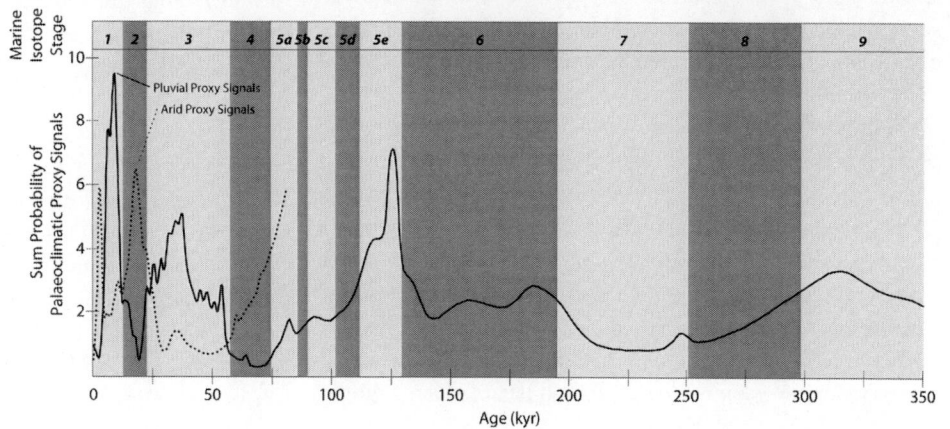

ages and archaeological materials both of which corroborated Early Holocene ages with corresponding epi-Paleolithic and early Neolithic lithics.

Immenhauser et al. (2007) analysed calcite deposits from Jebel Madar, Oman. ^{14}C dating yielded ages ranging between 27 and 23 ka. As this date range falls into a period when no speleothem growth has been recorded in Arabia, despite only

a handful of sites being dated, they assumed the ages were incorrect and implied post-depositional alteration of the calcite. The period between 10.5 and 6 ka was characterised by summer monsoon and enhanced phreatic speleothem growth. Immenhauser et al. (2007) suggested a dead carbon proportion (dcp) of ~85% was required to reach the stage 3 ages determined by the ^{14}C measurements. The calcite deposits were also dated using UTh disequilibrium which yielded ages ranging between 600 and 150 ka with a series of dates loosely clustering around 212–158 ka. Given the unknown/unidentifiable questions of detrital contamination (as raised for the ^{14}C ages) as well as the problem of U leaching these dates must be questioned as well. Given that so few speleothem sites have been dated the lack of stage 3 samples may be a premature assumption. What does remain is a vast body of data indicating that stage 3 was wet and it seems unlikely that all stage 3 ages are incorrect due to diagenetic alteration or post depositional contamination. What is raised, however, is that the stage 3 issue needs to be re-evaluated and resolved in order to fully understand the nature of the Late Pleistocene climate evolution of Arabia.

Pleistocene Climate Change in Arabia

Most of the precipitation that falls over Arabia is brought by the Southwest Indian Ocean Monsoon system (IOM), considerably more so than from Northwesterly winter storms. Consequently, the environmental fate of the region, amelioration or desiccation, rests upon the intensity of the monsoon and northwesterly systems, which has been in flux for at least the last quarter of a million years (Clemens et al., 1991; Muzuka, 2000; Fleitmann et al., 2007).

Marine cores from the Indian Ocean, Gulf of Oman, and the Arabian Sea provide a detailed history of the Southwest Indian Ocean Monsoon system throughout the Quaternary. Biogeochemical and lithogenic data from Arabian Sea cores spanning the last 350 ka also support the notion that monsoon winds were sensitive to changes in glacial boundary conditions, continental albedo, and sea-surface changes (SSTs) in the western Indian Ocean (Clemens et al., 1991).

Computer simulations have been used to estimate the average wind speed of the Southwest Monsoon during such phases of intensification. Speeds currently average around 10 m/s, while increased periods of activity saw wind speeds reaching 15 m/s. Precipitation would have been 50% greater than its present value, growing from 5 to 7.5 mm/day. Northward shifting insulation patterns drove the monsoons further into the Arabian peninsula, with evidence for seasonal storms reaching as far north as Bubiyan Island in the Arabian Gulf (Sarnthein, 1972; Kutzbach, 1981).

Analysis of various planktonic foraminiferal species (i.e., *Globigerinoides ruber*, *Globigerina bulloides*, and *Neogloboquadrina dutertrei*) frequency distribution over the last glacial cycle shows a direct correlation between paleoproductivity in the Arabian Sea, the strength of the monsoon, and the global oxygen isotope curve. Scholars note the onset of intensified monsoon episodes can lag up to 1,000 years after shifts in glacial conditions, possibly due to the threshold necessary for sufficient amounts of snow and ice to melt

and affect Indian Ocean insulation patterns (Reichart et al., 1997; Petit-Maire et al., 1999; Ivanova et al., 2003).

Stages 9–6 (350–130 ka)

Prior to stage 5 a few absolute ages exist for Middle Pleistocene contexts. In northwestern Arabia stage 9 lake sediments from the Azraq basin, Jordan, have been dated to 330 ka (Abed et al., 2008). The authors suggested that the formation of a freshwater, mega-lake Azraq was most probably due to intense Mediterranean cyclones, however, they also suggested the possible penetration of monsoonal rains as far north as 30° as an additional source of moisture. Stage 9 has been referred to as an intensely wet interglacial period.

OSL ages from lacustrine beds overlain by fluvial terrace sediments in the Wadi Dhaid, UAE, are dated to stage 9 (319 ka) and stage 7 (193 ka) respectively (Parker and Rose 2008). A stage 9 age (337 ka) was also determined from a speleothem in Magharat Qasir Hafit, UAE (Fogg et al., 2002). A stage 7 lake was dated using OSL at Al-Hisa, Jordan, where a shallow, freshwater lake formed at 182 ka (Abed et al., 2008).

These dates are in close agreement with speleothem records in Hoti Cave, northern Oman where U/Th measurements indicate periods of increased growth between 325 and 300 ka and 200 and 180 ka (Burns et al., 2001), corresponding with interglacial conditions during stages 9 and 7.

Anton (1984) speculated that the environment was hyperarid during MIS 6, given that monsoon intensity roughly tracks the global marine isotope curve. The emerging picture in Arabia indicates the situation was more varied than this initial assessment. A growing body of chronometric dates suggests there were brief pulses in precipitation. The prospect of stage 6 sub-pluvials is corroborated by optical dates on fluvial silts at Sabkha Matti (147 ka) (Goodall, 1995), two UTh measurements from freshwater mollusca within lacustrine sediments at Mudawwara, on the Jordanian/Saudi Arabian border (170 and 152 ka) (Petit-Maire et al., 1999), optically dated fluvial silts in the Wadi Dhaid, UAE (152 ka) (Parker and Rose, 2008), OSL measurements on fluvial silts recorded at the Camel Pit Site, Umm al-Qawain, UAE (174 ka), and optical measurements on evaporitic lacustrine sediments sampled from a relict interdunal sabkha in the Liwa region of the Rub' al Khali, UAE (160 ka) (Wood et al., 2003).

Stage 5 (130–74 ka)

The onset of the Last Interglacial period (stage 5e) around 130 ka was punctuated by an abrupt and drastic increase in rainfall over South Arabia that lasted until ~120 ka, followed by a second peak in precipitation corresponding with stage 5a (82–74 ka).

Researchers noted that speleothem growth was most pronounced during MIS 5e, more so than all subsequent pluvials (Burns et al., 1998, 2001). Multiple MIS 5 pluvial episodes are signalled by the aforementioned Hoti Cave speleothems, which yield U/Th dates indicating rapid growth between 135 and 120 ka (5e) and 82 and 78 ka (5a) (Burns et al., 2003). Fogg et al. (2002) recorded a UTh age of 100 ka from Qasir Hafit, UAE.

At Mudawwara, Jordan, lake sediments formed between 135 and 116 ka and 95 and 88 ka. At Mohadeb, UAE, fluvial sediments were dated to 95 ka (S. Stokes et al., nd), whilst in Jordan lake sediments at al-Hisn were dated to 82 ka (Moumani, 1996). Further evidence for increased fluvial activity comes from a series of buried alluvial fans interstratified with fluvial sands along the western edge of the Hajar Mountains in Ras al-Khaimah, UAE OSL ages obtained from these sediments indicate they were deposited at 117–108 ka (S. Stokes et al., nd).

Paleosols were noted in the ad-Dahna Desert of northern Arabia, where Late Pleistocene dunes overlie two separate pedogenic strata that could only have formed on stabilized dunes with a dense cover of vegetation which are thought to be from stage 5 (Anton, 1984). There is a network of Plio-Pleistocene bas relief gravel channels west of the Wahiba desert that is overlain by thinner fluviatile gravels tentatively associated with particularly humid episodes during MIS 5e and MIS 5a (Maizels, 1987).

Stokes and Bray (2005) obtained over 50 optical dates from megabarchan dunes in the Liwa region, which lies along the eastern margin of the Rub' al Khali. Their findings suggest a prolonged period of dune accumulation from 130 to 75 ka. This deposition was attributed to a unique combination of factors such as reduced sea levels in the Arabo-Persian Gulf that produced an abundance of sedimentary material available for transport, rise in regional groundwater levels, and vegetation cover that stabilized the dunes. The Liwa and al-Qafa ages reflect perhaps that dune deposition is in essence a reflection of interglacial conditions (fluctuating wetness and aridity) marking the cessation of aridity and increased stability due to the factors above. This may account for the near absence of LGM age dune sediments. This is in contrast with the mega-linear dunes of SW Arabia, which are Late Glacial in age (see below).

Stage 4 Aridity Onset (75–60 ka)

It has been postulated that the onset of arid conditions at the stage 5a/4 boundary may have coincided with the Toba eruption 74 ka (Ambrose, 1998; Rampino and Ambrose, 2000), although this remains a matter of debate (Gawthorne-Hardy and Harcourt-Smith, 2003). Toba ash does, however, provide a key stratigraphic marker which has been detected in Arabian Sea sediments (Schulz et al., 1998). It has not yet been identified on the Arabian peninsula, though this is likely to be related to the dearth of known dated stage 4 sites which could be sampled and tested. Toba has been suggested as a marker for a human genetic bottleneck as the onset of aridity forced humans into small population pockets (Ambrose, 1998) although this notion has been challenged recently by Petraglia et al. (2007). There are meagre terrestrial climatic data from stage 4 in Arabia. The HOPE ENV summed probability curve as well as the index of Indian Ocean Monsoon activity (Fleitmann et al., 2007) suggests this timeframe was characterized by increasingly hyperarid conditions until 50 ka. Limited evidence for aridification during MIS 4 is available from the Rub' al Khali. Dune accumulation was recorded in the Liwa region at 63 ka (Stokes and Bray, 2005) and between 60 and 50 ka (Juyal et al., 1998). In the Wahiba Sands dune deposition occurred in stage 4 until 64 ka (Preusser et al., 2002).

Stage 3 The Debated Pluvial (60–20 ka)

Archaeological and genetic studies indicate that stage 3 was an important period for the dispersal of Homo sapiens into and across Arabia (Parker and Rose, 2008; Rose and Usik, 2009). However, the paleoenvironmental records of Arabia and the surrounding regions during stage three remain contentious. First, the dating limit of ^{14}C falls within stage 3 and as described earlier there are potential problems from diagenetic changes associated with suggested stage 3 age sediments. Until recently the general consensus was that this stage was wet between 35 and 20 ka based on lacustrine sediments dated using ^{14}C dates from the sites in the Rub' al Khali (McClure, 1976; Wood and Imes, 1995) and Mundafan depression (McClure, 1976), and the Nafud desert (Garrard and Harvey 1981; Schultz and Whitney, 1986). There appear to be discrepancies between marine core evidence, lack of speleothems dating to this period on the Arabian mainland and the terrestrial lacustrine evidence. This area needs to be a major focus for future work in order to understand and address the problems identified. I think something needs to be said briefly here about what the discrepancies and problems are. For completeness this section will present the dated evidence as it stands and present the author's interpretation. This will provide a background overview which can be tested and challenged by future work.

No dune accumulation has been recorded in the Wahiba region between 64 and 23 ka and the emergence of interdunal sibakh recorded in the Liwa region of the UAE has produced 31 dates (both uncalibrated ^{14}C as well as OSL) that cluster between 46 and 22 ka supporting the notion of wetness during stage 3 (Wood and Imes, 1995; Juyal et al., 1998; Glennie and Singhvi, 2002; Lancaster et al., 2003). Wood and Imes (1995) estimated annual rainfall in the Liwa region to be at least a minimum of 250 mm during this period.

In northern Arabia a series of lake deposits in Jordan have been dated to stage 3. A lake deposit in Wadi Muqat was dated to 25 ka using OSL and was followed by an arid period between 21 and 15 ka (Abboud, 2000 cited in Abed et al., 2008). In the Jafr basin a 1,000 to 1,800 km^2 lake was reported to have formed between 27 and 25 ka. This lake then disappeared during the LGM (Huckreide and Wieseman, 1968). A high stand for Late Pleistocene Lake Hasa, Jordan, was dated between 31 and 24 ka corresponding with dated Ahmarian archaeological sites in close association with the lacustrine deposits (Schuldenrein and Clark, 1994).

If the terrestrial dating evidence from Arabia is taken at face value, investigations in the heart of the Rub' al Khali sand sea have revealed a landscape during MIS 3 that featured a series of small lakes spread across the interior (McClure, 1984). Radiocarbon measurements on mollusc shells and marls indicate the lakes reached their highest levels around 37 ka (McClure, 1976). These playas ranged from ephemeral puddles to pools up to 10 m deep, and numbered well over a thousand. They are primarily distributed along an east–west axis across the centre of the Rub' al Khali basin, covering a distance of some 1,200 km (McClure, 1984). Similar lake basins have been reported from the Ramlat as-Sabatayn desert in Yemen (Lézine et al., 1998, 2007), as well as the an-Nafud in northern Arabia (Garrard et al., 1981; Schultz and Whitney, 1986).

The Mundafan Depression, situated along the Tuwaiq Escarpment in central Arabia, provides a thick stratigraphic sequence (over 20 m) spanning stages 3–1. Fossilized faunal remains excavated within the Mundafan sediments yielded a menagerie of large vertebrates including: oryx, gazelle, auroch, wild ass, hartebeest, water buffalo, tahr, goat, wild camel, and ostrich (McClure, 1984). Most of these species belong to family Bovidae, whose survival required expansive grasslands produced by light to medium rainfall distributed evenly over the Rub' al Khali.

Ostracoda and freshwater mollusca indicative of low salinity were present at Mundafan, as well as species of foraminifera that attest to highly brackish conditions (McClure and Swain, 1974). Evidence of grasses, shrubs, and herbs are indicated by both phytoliths and dikaka-thin, tubular fragments of fossilized material scattered in the aeolian sediments around the basins. These floral fossils were formed when dissolved calcium carbonate in the water precipitated onto plants as the lake evaporated. Evidence of fish remains

is conspicuously absent from the Rub' al Khali lakes, because lakes were rarely refilled and became too alkaline too quickly to develop a population (McClure, 1984).

In addition to interior paleolakes, other signals for an MIS 3 wet-phase include depositional terraces in the Wadi Dhaid; although undated, their stratigraphic position suggests a timeframe between 35 and 22 ka (Sanlaville, 1992). Paleosols have been recorded in the ad-Dahna desert, which are stratigraphically positioned between MIS 4 and MIS 2 aeolian deposits (Anton, 1984). McClure (1984) dated a paleosol at Mundafan, KSA, to 30 ka. Two soil horizons were discovered around Ibb in the central plateau of the Yemeni highlands, characterized as molissols – soils that form on landscapes covered by savannah vegetation. A calibrated ^{14}C date of 26 ka was recorded for the lower stratum and 23 ka for the upper horizon (Brinkmann and Ghaleb, 1997).

Clark and Fontes (1990) dated calcite formations from hyperalkaline springs in northern Oman, producing radiocarbon ages between 33 and 19 ka. Carbon isotope values suggested that C3 vegetation was an important component of the local vegetation at this time. In the Emirate of Ras al-Khaimah, UAE, Sanlaville (1992b) reported two ^{14}C ages of 27 and 20 ka from terrestrial mollusca.

Marine evidence suggests that stage 3 was complex and comprised a series of fluctuations of aridity, including Heinrich events H2 to H5, as well as increased phases of monsoon intensity (Schulz et al., 1998). With respect to correlating marine and terrestrial records there are two issues which need to be resolved. The first is chronological and the second is geochemical. Calibrating ^{14}C ages from stage 3 is difficult as the calibration curve has yet to be adequately refined for this period. If the stage 3 ^{14}C ages from lacustrine sediments, paleosols and lacustrine sediments are all correct and non-diagenetically altered their calibrated ages will be much older (by as much as 10–15 thousand years). This would better fit the marine records which suggest wetter conditions between 40 and 30 rather than 30 and 20 ka. Evidence for aridity during stage 3 has largely been derived from the presence of dolomite in marine sediments as a proxy for dust derived from the Arabian peninsula (Sirocko, et al., 1991). However, it should be noted that dolomite records do not replicate patterns of aridity as denoted by the geochemical fingerprinting of dust using K, Ti, Al and Fe. Ivanova suggests that dolomite records reflect sea-level fluctuations and the peaks and troughs largely reflect the exposure of sabkha flats as the source for dolomite. This would account for the discrepancies between the marine and continental records during stage 3. Arabian Sea δ^{15}N records suggest increased monsoon strength between 60 and 30 ka. A series of abrupt arid phases punctuate this period of time corresponding with Heinrich events H4–H6. This view is corroborated by speleothem records on the island of Socotra where a stalagmite record spanning 55 to 42 ka shows rapid changes in the IOM with corresponding changes in rainfall. Burns et al. (2003) suggested that the early stage 3 records showed strong links between the Indian Ocean and North Atlantic regions.

Marine cores suggest that the onset of aridity began ca. 33 ka (Ivanochka, 2005). The lacustrine records in Arabia are best preserved in the larger lake basins at Jubba in the Nafud, Mundafan along the Tawaiq Escarpment, and al-Hasa, Jordan. These basins are not closed basin interdunal lakes but derive their water from groundwater sources in addition to rainfall. It is possible that the lakes continued to exist at these locations as the aquifer recharge may have longer lag times thus sustaining water into the arid phase. This question will require further testing.

Given the emerging body of evidence attesting to favorable conditions during parts of MIS 3, Upper Paleolithic assemblages falling within this timeframe are not surprising. A rock shelter site at Jebel Faya, Sharjah, UAE, comprising stratified materials contains occupational horizons, one of which is before ~30 ka.

Stage 2 Late Glacial Maximum (LGM) and Late Glacial (20–10 ka)

Researchers speculate that the arid-phase that set in during the Terminal Pleistocene was more arid than the peninsula had experienced since the Penultimate Glaciation, if not earlier (Anton, 1984). During this phase widespread dune mobilization and emplacement took place with material being reworked from earlier phases of dune formation with additional sediment supplied from the exposed area of continental shelf along the Indian Ocean seaboard, the bed of the Persian Gulf and parts of the Red Sea basin as well as materials derived from continental weathering. Ages obtained from dune formations in the Rub' al Khali (McClure, 1984; Dalongeville et al., 1992; Goudie et al., 2000; Parker and Goudie, 2007), an-Nafud (Anton, 1984), and the Wahiba Sands (Gardner, 1988; Glennie and Singhvi, 2002; Preusser et al., 2002) all signal a major phase of aeolian accumulation during the LGM (20–15 ka) (Fig. 2). Calcite fractures in northern Oman corroborate the evidence for increasing aridity, indicating there was considerably less moisture in the environment starting around 19 ka (Clark and Fontes, 1990). The age of the stage 2 dunes corresponds with the LGM and Late Glacial periods of intensified aridity (Sarnthein, 1978). This is also noted in marine cores (Sirocko et al., 1991; Overpeck et al., 1996; Schultz et al., 1998; Von Rad et al., 1999), which display an increased influx of dust. Major phases of dust influx derived from geochemical analyses of offshore records denote pulses of dust originating from distinct source regions (e.g., central Arabia and the Persian Gulf region). The LGM peak (Sirocko et al., 1991; Leuschner and Sirocko, 2003) from marine cores is corroborated by optical dates from Arabia and highlight the intensified northwesterly trajectories at this time (Von Rad et al., 1999).

Evidence from an increasing number of sites points towards a brief wet phase between 15 and 13 ka with the deposition of travertines at Nizwa, Oman (Clark and Fontes, 1990), sabka deposits at Liwa, UAE (Wood and Imes, 1995), alluvial fan deposits at Wadi Abu Saww, KSA (Hacker et al., 1984), and lacustrine sediments at Al Ayun, KSA (Zarins et al., 1979), and Mundafan, KSA (McClure, 1976). In addition, an undated lacustrine bed at Awafi was found under the dune sequence dated by Goudie et al. (2000) to 13–9 ka. This lacustrine bed (ca. 70 cm thick) was stratigraphically higher than the basal sands, dated to 18 ka underlying the interdune Holocene lake sediments also at Awafi. This would suggest an age between 18 and 13 ka for this phase of wetness.

It is tempting to suggest that this phase of wetness coincides with the Bölling-Allerød (BA) interstadial, which is dated between 15 and 13 ka. In the northern Hemisphere this event is related to the mass wasting of the LGM ice sheets and was a period of rapid warming. Teleconnective climatic links between the North Atlantic and the Arabia Sea records have been suggested by a number of authors (Schulz et al., 1998; Gupta et al., 2003; Leuschner and Sirocko, 2003; Fleitmann et al., 2007). The period of wetness identified in Arabia has been identified in several marine records from the Arabia sea using TOC (Schulz et al., 1998) and $\delta^{15}N$ records (Ivanova et al., 2003; Ivanochko, 2005) which correspond with the Arabian Sea Monsoon period ASM 1e–c (Schulz et al., 1998; Ivanochko, 2005) and Greenland isotope stage IS 1e–c. The BA brief interlude may have permitted brief human entry into parts of Arabia from either outside the peninsula or from refugia within Arabia (e.g., Dhofar or the now submerged Arabo-Persian Gulf basin). This point does, however, require further work in order to test this notion in full.

In central and northern Arabia the ASM 1e–c (Bölling-Allerød) was terminated by the onset of the Younger Dryas ~13.5 ka. Dune emplacement during the Younger Dryas and Early Holocene was forced by an intensified NW system and Shamal system blowing materials across the Persian Gulf region. Marine records in the northern Arabian Sea (Von Rad et al., 1999), Indian Ocean (Gupta et al., 2003) and off the coast of India (Thamban et al., 2001, 2002) also support this view, with increases in lithogenic materials derived from Arabia corresponding to the Younger Dryas and Early Holocene, peaking at 11.5 ka. At Awafi the development of dunes between 13.5 and 9.1 ka (Goudie et al., 2000) suggests that monsoon activity was low during this period and that enhanced monsoon precipitation did not migrate this far north until after 9.0 ka (Parker et al., 2004).

Stage 1 Holocene Climate Change in Arabia (10 ka–Present)

The Terminal Pleistocene hyperarid phase ended with yet another pronounced oscillation back to humid conditions at the onset of the Holocene. This pluvial phase period lasted until ~5 ka, at which time the present climatic regime was established (Overstreet et al., 1996; Cleuziou et al., 1992; Sanlaville, 1992; Brunner, 1997; Wilkinson, 1997; Stokes and Bray, 2005; Parker et al. 2004, 2006a, 2006b, 2006c).

In southern Arabia the onset of wet conditions at the stage 2/1 boundary was much earlier than in the central Arabian desert regions and the Arabo-Persian Gulf (Parker et al., 2006b). In Yemen lacustrine conditions developed by 11 ka in the Highlands and the Ramlat as-Sabatyn (Parker et al., 2006c; Davies, 2006; Lézine et al., 2007). The onset of wet conditions is related to the northwards migration of the Inter Tropical Convergence Zone (ITCZ) owing to increased heating across the northern Hemisphere (sensu deMenocal et al., 2000). Stalactite records from Southern Oman (Fleitmann et al., 2007) record the northwards movement of the ITCZ and incursion of the IOM by 10.3 ka into southern Oman and northern Oman by 9.6 ka (Neff et al., 2001). At Awafi the cessation of dune emplacement occurred at 9.0 ka (Goudie et al., 2000) and the onset of lacustrine sedimentation did not take place until 8.5 ka (Parker et al., 2004). Thus, it took ~1,800 years for the IOM to move from Southern Arabia (15°N) to Northern Arabia (25°N). This provides important information on the time lag and northwards migration and latitudinal position of the summer ITCZ and incursion of monsoon rainfall across the eastern sector of Arabia during the Early Holocene. The impact of human migration into and across Arabia during the Early Holocene would have been profoundly influenced by this variation in moisture across the peninsula.

Evidence from Awafi, UAE, indicates the dune field became stabilized and vegetated during the Early Holocene with a predominant mix of C3 grasslands and scatters of woody elements including Acacia, Prosopis and Tamarix (Parker et al., 2004). The evidence for a rich cover of grassland supports the archaeological evidence for Neolithic herding between the mountains, desert and coast during the period of maximum monsoonal rainfall (Uerpmann, 2002).

Lacustrine and speleothem records suggest the IOM weakened and retreated southwards around 5.9 ka (Neff et al., 2001; Uerpmann, 2002; Parker et al., 2004, 2006a). The retreat of the IOM led to the cessation of the Hoti cave speleothem, which records a large reduction in precipitation immediately prior to this date (Neff et al., 2001). The lakes of the central Rub' al Khali also ceased to exist beyond this point (McClure, 1976). The reduction in precipitation led to a lowering of the lake level at Awafi, unlike the central Rub'

al Khali lakes, which did not dry up completely. For the lake to have persisted it is suggested that westerly winter rainfall must have existed in the Gulf region to have maintained the lake. Change in precipitation from IOM to westerly sources is marked by a sharp change from C3 to drier adapted C4 grasslands across the dune field (Parker et al., 2004). A similar pattern of winter rainfall was postulated in the Nafud in western Arabia (Schultz and Whitney, 1986). The archaeological record indicates that the Arabian Bifacial Type/Ubaid period came to an abrupt end in eastern Arabia and the Oman peninsula at 5.8 ka and no evidence of human presence exists in the area for ~1,000 years (Uerpmann, 2002). This period has been described as the 'Dark Millennium' in the Arabian Gulf region because of the lack of known archaeological sites (Vogt, 1994; Uerpmann, 2002). In contrast to the sites on the Arabian Gulf, those on the Omani coast contnued into the 4th millennium and persisted during the dry period Uerpmann (2002). It has been suggested that climatic deterioration caused dramatic changes in semi-desert nomadism, subsistence, and settlement patterns around 5.8 ka. The number of known sites suggests that the population shrank considerably at this time and became concentrated in the few parts of Arabia which offered greater ecological diversity (Uerpmann and Uerpmann, 1996; Parker et al., 2004).

Conclusions

Human response in Arabia is inexorably linked with oscillations in the Southwest Indian Ocean Monsoon System and Northwesterly systems; the predicted timing of Pleistocene range expansions is dependent upon a firm grasp of the paleoclimatic record over the past quarter of a million years. Arabia served a unique role in the region due to these environmental extremes, in combination with its geographic position as the nexus of three continents. Arabia was a bridge connecting Africa with Eurasia (Bailey et al., 2007a, 2007b). During arid phases the bridge was discontinuous and this would have restricted or prevented any movement eastward During pluvials, Arabia facilitated genetic bottleneck releases via hunter–gatherer range expansions onto the peninsula.

These genetically-predicted expansions likely occurred during pluvial episodes in southern Arabia. A number of environmental signals have been presented that attest to periodic phases of intensified monsoon activity, leading to amelioration of the interior deserts. Significant wet-phases correlate with interglacials corresponding to isotope stages 9, 7, 5e, and 1. There is evidence for increased wetness during interstadial stages 5a and 3. During these stages retreating glacial conditions altered the Indian Ocean insolation patterns and the monsoon migrated northward into the Arabian interior.

References

Abed AM, Yasin S, Sadaqaa R, Al-Hawari Z. The paleoclimate of the eastern desert of Jordan during Marine Isotope Stage 9. Quaternary Research 2008;69:458–68

Ambrose S. Late Pleistocene human population bottlenecks, volcanic winter, and differentiation of modern humans. Journal of Human Evolution. 1998;34:623–51.

Anderson DE, Goudie AS, Parker AG. Global environments through the Quaternary: exploring environmental change. Oxford: Oxford University Press; 2007.

Anton D. Aspects of geomorphological evolution: paleosols and dunes in Saudi Arabia. In: Jado AR, Zötl JG, editors. Quaternary period in Saudi Arabia. Vol. 2. Sedimentological, hydrogeological, hydrochemical, geomorphological, and climatological investigations of Western Saudi Arabia. Wien: Springer; 1984.

Bailey GN, Al-Sharekh A, Flemming NC, Lambeck K, Momber G, Sinclair A, et al. Coastal prehistory in the southern Red Sea basin, underwater archaeology, and the Farasan Islands. Proceedings of the Seminar for Arabian Studies. 2007a;37:1–16.

Bailey GN, Flemming NC, King GCP, Lambeck K, Momber G, Moran LJ, et al. Coastlines, submerged landscapes, and human evolution: the Red Sea basin and the Farasan islands. Journal of Island and Coastal Archaeology. 2007b;2:127–60.

Barth HJ, Steinkohl F. Origin of winter precipitation in the central coastal lowland of Saudi Arabia. Journal of Arid Environments. 2004;57:101–15.

Bray H, Stokes S. Temporal patterns of arid-humid transitions in the southeastern Arabian peninsula based on optical dating. Geomorphology. 2004;59:271–80.

Brinkmann R, Ghaleb AO. Late Pleistocene mollisol and cumulic fluvents near Ibb, Yemen Arab Republic. In: Grolier MJ, Brinkmann R, Blakely JA, editors. The Wadi al-Jubah Archaeological Project: Volume 5 Environmental Research in Support of Archaeological Investigations in the Yemen Arab Republic, 1982–1987. Washington, DC: American Foundation for the Study of Man; 1997. p. 251–8.

Brunner U. Geography and human settlements in ancient southern Arabia. Arabian Archaeology and Epigraphy. 1997;8:190–202.

Burns SJ, Matter A, Frank N, Mangani A. Speleothem based paleoclimatic record from Northern Oman. Geology. 1998;26:499–502.

Burns SJ, Fleitmann D, Matter A, Neff U, Mangini A. Speleothem evidence from Oman for continental pluvial events during interglacial periods. Geology. 2001;29:623–6.

Burns SJ, Fleitmann D, Matter D, Kramers J, Al-Subbary AA. Indian Ocean climate and an absolute chronology over Dansgaard/Oeschger events 9 to 13. Science. 2003;301:1365–57.

Caton-Thompson G. Some palaeoliths from South Arabia. Proceedings of the Prehistoric Society. 1953;19:189–218.

Clark A. Lakes of the Rub' al Khali. Aramco. 1989;40:28–33.

Clark I, Fontes JC. Paleoclimatic reconstruction of northern Oman based on carbonates from hyperalkaline groundwaters. Quaternary Research. 1990;33:320–36.

Clemens S, Prell W, Murray D, Shimmield G, Weedon G. Forcing mechanisms of the Indian Ocean monsoon. Nature. 1991;353:720–5.

Cleuziou S, Inizan ML, Marcolongo B. Le peuplement pre-et protohistorique du systeme fluviatile fossile du Jawf-Hadramawt au Yemen. Paleorient. 1992;18:5–29.

Cooke GA. Reconstruction of the Holocene coastline of Mesopotamia. Geoarchaeology. 1987;2:15–28.

Cremaschi M, Negrino F. Evidence for an abrupt climatic change at 8700 ^{14}C yr B.P. in rock shelters and caves of Gebel Qara (Dhofar-Oman): Palaeoenvironmental implications. Geoarchaeology. 2005;20:559–79.

Dalongeville R, de Medwecki V, Sanlaville P. Évolution du piédmont occidental de l'Oman depuis le Pléistocène supérieur. 116e Congress National Societe Sav., Chambrey, Déserts. PICG. 1991;252:93–115.

Davies C. Quaternary paleohydrology and past climates of the Dhamar Highlands, Yemen. Quaternary Research. 2006;66:454–64.

deMenocal P, Ortiz J, Guilderson T, Sarnthein M, Baker L, Yarunsinsky M. Abrupt onset and termination of the African humid period: rapid climate responses to gradual insolation forcing. Quaternary Science Reviews. 2000;19:347–61.

Edens C. The Rub' al Khali 'Neolithic' revisited: the view from Nadqan. In: Potts D, editor. Araby the Blest: studies in Arabian archaeology. Copenhagen: Tusculanum Press; 1988. p. 15–43.

Edgell HS. Arabian deserts: nature, origins and evolution. The Netherlands: Springer; 2006.

Field H. Stone implements from the Rub'al Khali, Southern Arabia. Man. 1958;58:93–4.

Fleitmann D, Burns SJ, Mangini A, Mudelsee M, Kramers J, Villa I, et al. Holocene ITCZ and Indian monsoon dynamics recorded in stalagmites from Oman and Yemen (Socotra). Quaternary Science Reviews. 2007;26:170–88.

Fogg T, Fogg P, Waltham T. Magharat Qasir Hafit, a significant cave in the United Arab Emirates. Tribulus. 2002;12:5–14.

Gardner RAM. Aeolianites and marine deposits of the Wahiba sands: character and palaeoenvironments. Journal of Oman Studies. 1988;Special Report 3:75–94.

Garrard AN, Harvey CPD, Switsur VR. Environment and settlement during the Upper Pleistocene and Holocene at Jubbah in the Great Nafud, northern Arabia. Atlal. 1981;5:137–48.

Gawthorne-Hardy FJ, Harcourt-Smith WEH. The super-eruption of Toba, did it cause a human bottleneck? Journal of Human Evolution. 2003;45:227–30.

Glennie KW, Singhvi AK. Event stratigraphy, palaeoenvironment and chronology of SE Arabian deserts. Quaternary Science Reviews. 2002;21:853–69.

Goodall TM. The geology and geomorphology of the Sabkhat Matti region (United Arab Emirates): a modern analogue for ancient desert sediments of north-west Europe. Ph.D. dissertation, University of Aberdeen, Aberdeen; 1995.

Goudie AS. Great warm deserts of the world: landscapes and evolution. Oxford: Oxford University Press; 2002.

Goudie AS, Colls A, Stokes S, Parker AG, White K, Al-Farraj A. Latest Pleistocene dune construction at the north-eastern edge of the Rub' al Khali, United Arab Emirates. Sedimentology. 2000;47:1011–21.

Gupta A, Anderson DM, Overpeck JT. Abrupt changes in the Asian southwest monsoon during the Holocene and their links to the North Atlantic Ocean. Nature. 2003;421:354–6.

Hacker et al. Region around Jeddah: geology, geomorphology and climate. In: Jado AR, Zötl JG, editors. Quaternary period in Saudi Arabia, Vol 2. New York: Springer; 1984.

Holm DA. Desert geomorphology in the Arabian peninsula. Science. 1960;132:1369–79.

Huckreide R, Wieseman G. Der jungpleistozäne pluvial-see von El-Jafr und weitere daten zum Quartär Jordaniens. Geologica et Palaeontologica. 1968;2:73–95.

Hughen K, Southon J, Lehman S, Bertrand C, Turnbull J. Marine-derived 14C calibration and activity record for the past 50,000 years updated from the Cariaco basin. Quaternary Science Reviews. 2006;25:3216–27.

Immenhauser A, Dublyansky YV, Verwer K, Fleitman D, Pashenko SE. Textural, elemental, and isotopic characteristics of Pleistocene phreatic cave deposits (Jabal Madar, Oman). Journal of Sedimentary Research. 2007;77:68–88.

Ivanova E, Schiebel R, Singh AD, Schmiedl G, Niebler HS, Hemleben C. Primary production in the Arabian Sea during the last 135000 years. Palaeogeography, Palaeoclimatology, Palaeoecology. 2003;197:61–82.

Juyal N, Singhvi AK, Glennie KW. Chronology and paleoenvironmental significance of Quaternary desert sediment in southeastern

Arabia. In: Alsharhan AS, Glennie KW, Whittle GL, Kendall CGStC, editors. Quaternary deserts and climatic change. Rotterdam: Balkema; 1998. p. 315–25.

Kutzbach JE. Monsoon climate of the Early Holocene: climate experiment with the Earth's orbital parameters for 9000 years ago. Science. 1981;214:59–61.

Lancaster N, Singhvi AK, Teller JT, Glennie KW, Pandey VP. Eolian chronology and paleowind vectorsin the northern Rub' al Khali, United Arab Emirates. XVI INQUA Congress Programs with Abstracts, Desert Research Institute, Reno, NV, 141; 2003.

Leuschner DC, Sirocko F. Orbital insolation forcing of the Indian monsoon – a motor for global climate changes? Palaeogeography, Palaeoclimatology, Palaeoecology. 2003;197:83–95.

Lézine A, Saliège J, Robert C, Wertz F, Inizan M. Holocene lakes from Ramlat as-Sab'atayn (Yemen) illustrate the impact of monsoon activity in southern Arabia. Quaternary Research. 1998;50:290–9.

Lézine A, Tiercelin JJ, Robert C, Saliège JF, Cleuziou S, Inizan ML, et al. Centennial to millennial-scale variability of the Indian monsoon during the Early Holocene from a sediment, pollen and isotope record from the desert of Yemen. Palaeogeography, Palaeoclimatology, Palaeoecology. 2007;243:235–49.

Maizels JK. Plio-Pleistocene raised channel systems of the western Sharqiya (Wahiba), Oman. In: Frostick L, Reid I, editors. Desert sediments: ancient and modern. London: Geological Society Special Publication 35; 1987. p. 31–50.

Masry AH. Prehistory in northern Arabia. The problem of interregional interaction. London: Kegan Paul; 1997.

McClure HA. Radiocarbonchronology of Late Quaternary lakes in the Arabian desert. Nature. 1976;263:755–6.

McClure HA. Ar Rub' al Khali. In: Al-Sayari SS, Zötl JG, editors. Quaternary period in Saudi Arabia. Vol. 1: sedimentological, hydrogeological, hydrochemical, geomorphological, and climatological investigations in Central and Eastern Saudi Arabia. Wien: Springer; 1978.

McClure HA. Late Quaternary palaeoenvironments of the Rub' al Khali. Ph.D. dissertation, University of London, London; 1984.

McClure HA, Swain FM. The fresh water and brackish water fossil Quaternary Ostracoda from the Rub' al Khali, Saudi Arabia. Tunis: 6th African Micropalaeontological Colloquium; 1974.

Moumani K, Alexander J, Bateman MD. Sedimentology of the late Quaternary Wadi Hasa Marl formation of Central Jordan: a record of climate variability. Palaeogeography, Palaeoclimatology, Palaeoecology. 2003;191:221–42.

Muzuka AN. 350 kaOrganic δ^{13}C record of the monsoon variability on the Oman continental margin, Arabian Sea. Proceedings of the Indian Academy Science. 2000;109:481–9.

Neff U, Burns SJ, Mangini A, Mudelsee M, Fleitmann D, Matter A. Strong coherence between solar variability and the monsoon in Oman between 9 and 6 kyr ago. Nature. 2001;411:290–3.

Overpeck J, Anderson D, Trumbore S, Prell W. The southwest Indian monsoon over the last 18000 years. Climate Dynamics. 1996;12:213–25.

Overstreet WC, Grolier MJ. Summary of environmental background for the human occupation of the al-Jadidah basin in Wadi al-Jubah, Yemen Arab Republic. In: Grolier MJ, Brinkman R, Blakely JA, editors. Environmental research in support of archaeological investigations in the Yemen Arab Republic, 1982–1987. Washington, DC: American Foundation for the Study of Man; 1996. p. 337–429.

Parker AG, Goudie AS. Development of the Bronze Age landscape in the southeastern Arabian Gulf: new evidence from a buried shell midden in the eastern extremity of the Rub' al Khali desert, Emirate of Ras al-Khaimah, UAE. Arabian Archaeology and Epigraphy. 2007;18:232–8.

Parker AG, Rose J. Demographic confluence and radiation in southern Arabia. Proceedings of the Seminar for Arabian Studies. 2008;38: 227–44.

Parker AG, Eckersley L, Smith MM, Goudie AS, Stokes S, White K, et al. Holocene vegetation dynamics in the northeastern Rub' al-Khali desert, Arabian peninsula: a pollen, phytolith and carbon isotope study. Journal of Quaternary Science. 2004;19:665–76.

Parker AG, Wilkinson TJ, Davies C. The early-mid Holocene period in Arabia: some recent evidence from lacustrine sequences in eastern and southwestern Arabia. Proceedings of the Seminar for Arabian Studies. 2006a;36:243–55.

Parker AG, Goudie AS, Stokes S, White K, Hodson MJ, Manning M, et al. A record of Holocene climate change from lake geochemical analyses in southeastern Arabia. Quaternary Research. 2006b;66: 465–76.

Parker AG, Preston G, Walkington H, Hodson MJ. Developing a framework of Holocene climatic change and landscape archaeology for southeastern Arabia. Arabian Archaeology and Epigraphy. 2006c;17:125–30.

Petit-Maire N, Burollet PF, Ballais JL, Fontugne M, Rosso JC, Lazaar A. Paléoclimats Holocènes du Sahara septentionale, Dépôts lacustres et terrasses alluviales en bordure du Grand Erg Oriental à l'extrême – Sud de la Tunisie. Comptes Rendus. Académie des Sciences. 1999;Series 2(312):1661–6.

Petraglia MD, Alsharekh A. The Middle Palaeolithic in Arabia: Implications for modern human origins, behaviour and dispersals. Antiquity. 2003;77:671–84.

Petraglia M, Korisettar R, Boivin N, Clarkson C, Ditchfield P, Jones S, et al. Middle Paleolithic assemblages from the Indian subcontinent before and after the Toba Super-eruption. Science. 2007;317:114–6.

Philby H. Rub' al Khali: an account of exploration in the Great South desert of Arabia under the auspices and patronage of His Majesty 'Abdul 'Aziz ibn Saud, King of the Hejaz and Nejd and its dependencies. Geographical Journal. 1933;81:1–21.

Preusser F, Radies D, Matter A. A 160000 year record of dune development and atmospheric circulation in Southern Arabia. Science. 2002;296:2018–20.

Radies D, Hasiotis ST, Preusser F, Neubert E, Matter E. Palaeoclimatic significance of Early Holocene faunal assemblages in wet interdune deposits of the Wahiba Sand Sea, Sultanate of Oman. Journal of Arid Environments. 2005;62:109–25.

Rampino M, Ambrose S. Volcanic winter in the Garden of Eden: the Toba supereruption and the Late Pleistocene human population crash. Geological Society of America Special Paper. 2000;345: 71–82.

Reichart GJ, Lourens LJ, Zachariasse WJ. Temporal variability in the northern Arabian Sea Oxygen Minimum Zone (OMZ) during the last 225000 years. Paleooceanography. 1997;13:607–21.

Reimer PJ, et al. IntCal04 Terrestrial radiocarbon age calibration, 26–0 ka. Radiocarbon. 2004;46:1029–58.

Rose JI. Among Arabian Sands: defining the Palaeolithic of southern Arabia. Ph.D. dissertation, Southern Methodist University, Dallas; 2006.

Rose JI. The Arabian corridor migration model: archaeological evidence for hominin dispersals into Oman during the Middle and Upper Pleistocene. Proceedings of the Seminar for Arabian Studies. 2007;37:219–37.

Rose JI, Usik VI. The "Upper Paleolithic" of South Arabia. In: Petraglia MD, Rose JI, editors. The evolution of human populations in Arabia: paleoenvironments, prehistory and genetics. The Netherlands: Springer; 2009. p. 169–85.

Sanlaville P. Changements climatiques dans la péninsule Arabique durant le pléistocène supérieur et l'holocène. Paleorient. 1992;18:5–25.

Sarnthein M. Sediments and history of the postglacial transgression in the Persian Gulf and northwestern Gulf of Oman. Marine Geology. 1972;12:245–66.

Sarnthein M. Sand deserts during the glacial maximum and climatic optimum. Nature. 1978;272:43–6.

Schuldenrein J, Clark GA. Landscape and prehistoric chronology of west-central Jordan. Geoarchaeology. 1994;9:31–55.

Schultz E, Whitney JW. Upper Pleistocene and Holocene lakes in the An Nafud, Saudi Arabia. Hydrobiologia. 1986;143:175–90.

Schulz H, von Rad U, Erlenkeuser H. Correlation between Arabian Sea and Greenland climate oscillations for the past 110000 years. Nature. 1998;393:54–7.

Schyfsma E. Climate. In: Al-Sayari SS, Zötl JG, editors. Quaternary period in Saudi Arabia. Vol. 1: sedimentological, hydrogeological, hydrochemical, geomorphological, and climatological investigations in Central and Eastern Saudi Arabia. Wien: Springer; 1978.

Sirocko F, Sarnthein M, Lange H, Erlenkeuser H. Atmospheric summer circulation and coastal upwelling in the Arabian sea during the Holocene and last glaciation. Quaternary Research. 1991;36:72–93.

Stokes S, Bray H. Late Pleistocene eolian history of the Liwa region, Arabian peninsula. Geological Society of America Bulletin. 2005;117:1466–80.

Thamban M, Purnachandra Raoa V, Schneider RR, Grootes PM. Glacial to Holocene fluctuations in hydrography and productivity along the southwestern continental margin of India. Palaeogeography, Palaeoclimatology, Palaeoecology. 2001;165:113–27.

Uerpmann M. Structuring the Late Stone Age of southeastern Arabia. Arabian Archaeology and Epigraphy. 1992;3:65–109.

Uerpmann M. The Dark Millennium – remarks on the final Stone Age in the Emirates and Oman. In: Potts D, al-Naboodah H, Hellyer P, editors. Archaeology of the United Arab Emirates. Proceedings of the First International Conference on the Archaeology of the U.A.E. London: Trident Press; 2002. p. 74–81.

Uerpmann M, Uerpmann HP. Ubaid pottery in the eastern Gulf – new evidence from Umm al-Qaiwain (UAE). Arabian Archaeology and Epigraphy. 1996;7:125–9.

Vogt B. In search for coastal sites in prehistoric Makkan: mid-Holocene "shell eaters" in the coastal desert of Ras al Khaimah, UAE. In: Kenoyer JM, editor. From summer to Meluhha. Wisconsin Archaeological Report 3; 1994. p. 113–28.

Von Rad U, Schulz H, Reich V, den Dulk M, Berner U, Sirocko F. Multiple-monsoon-controlled breakdown of oxygen-minimum conditions during the past 30000 years documented in laminated sediments off Pakistan. Palaeogeography, Palaeoclimatology, Palaeoecology. 1999;152:129–61.

Weninger B, Jöris O. Towards an absolute chronology at the Middle to Upper Palaeolithic transition in Western Eurasia: a new Greenland Hulu Time-scale based on U/Th ages. Journal of Human Evolution. 2007; In press.

Wilkinson TJ. Holocene environments of the high plateau. Yemen. Recent geoarchaeological investigations. Geoarchaeology. 1997;12:833–64.

Wood WW, Imes JL. How wet is wet? Precipitation constraints on late Quaternary climate in the southern Arabian peninsula. Journal of Hydrology. 1995;164:263–8.

Wood WW, Rizk ZS, Alsharhan AS. Timing of recharge, and the origin, evolution, and distribution of solutes in a hyperarid aquifer system. In: Alsharhan AS, Wood WW, editors. Water resources perspectives: evaluation management and policy. Amsterdam: Elsevier; 2003. p. 295–312.

Zarins J, Ibrahim M, Potts D, Edens C. Saudi Arabian archaeological reconnaissance 1978: the preliminary report on the third phase. Atlal. 1979;3:9–42.

Zeuner FE. Neolithic sites from the Rub' al-Khali, southern Arabia. Man. 1954;54:133–6.

Chapter 4
Environment and Long-Term Population Trends in Southwest Arabia

Tony J. Wilkinson

Keywords Bronze Age • Neolithic • Paleolithic • Populations • Settlement • Yemen

Introduction

Any paper that claims to present long term population trends of Arabia or any part of it has to face the fact that demographic data is, at best, limited. Nevertheless, that southwest Arabia is a very well populated area today, and may have been so during parts of prehistory needs to be explored. This chapter therefore focuses primarily on emerging archaeological evidence that suggests that this little known, but verdant and agriculturally productive region was during much of the Holocene a significant population center. How far such a model can be projected back in time (for example back into the Paleolithic) is difficult to say, but by laying out the evidence for climatic and population cycles during the past 10,000 years or so it should be possible to suggest what might have prevailed during those earlier periods, and more importantly to seek the relevant evidence for such occupations. It is not the aim of this chapter to present a full and detailed synthesis of the archaeological sites in southwest Arabia; regional syntheses can be found in Breton (1999), Cleuziou and Tosi (1998), Durrani (Durrani (2005), Edens and Wilkinson (1998), and de Maigret (2002).

Modern southwest Arabia, mainly the Republic of Yemen and neighboring parts of the province of 'Asir in Saudi Arabia is one of the more populous parts of Arabia, and indeed was so at least as early as classical times. Nevertheless, the hazards of population estimation for this region are underscored by recent controversies concerning a Swiss study that recorded for the state of North Yemen in 1975 a total population 4,705,336 (Steffen, 1979). This figure was disputed by the government of North Yemen who initiated their own census which resulted in a population estimate of 7,146,341 only 6 years later in 1981 (Wenner, 1991: 19). Clearly with the existence of such dispari-ties today, it must be appreciated that past populations for this region cannot be estimated quantitatively. In this chapter I will therefore examine the question of relative population levels: for example how does the region compare with other parts of Arabia in the past? Was southwest Arabia an ancient population center? Might the existence of such a population center been a significant factor in the long-term movements of people through the region?

Despite their differences, the two censuses of Yemen were in agreement on the relative population distribution, which shows the greatest population densities in the moist highlands between Ta'izz and Ibb and generally high densities for much of the plateau. Interestingly, the irrigated lowlands along the ancient incense trading route (al-Jawf, Marib, Shabwa and neighboring regions) were during the twentieth century AD more sparsely populated (Y.A.R., 1977; Steffen, 1979). Moreover, in the moist area of the western escarpment, Steffen (1979: I/141) notes that population densities are significantly higher at elevations of greater than 1,500 m. It is tempting to speculate whether this is not only because the moist climatic conditions created greater potential for cultivation in these moist areas, but also because malaria was less prevalent with the result that death rates were significantly lower at high altitudes as discussed below.

Given the existence of such high population densities in the recent past it was therefore baffling that early studies of southern Arabia by Doe (1971: 134) and Lankester Harding (1964) revealed little evidence for any occupation during the Holocene before approximately 1,000 BC (de Maigret, 2002). In other words predecessors of the magnificent South Arabian civilizations of Saba, Himyar, Qataban, Ma'in, Awsan and Hadramaut were hardly evident. This archaeological void has even been explained as the result of an influx of people from the Levant and neighboring areas from which the demand for incense came. This chapter will outline how recent archaeological research has not only started to populate this void, but also how the sketchy demographic history may have related to the environmental factors, specifically to fluctuations of the Indian Ocean Monsoon.

Although the colonial period surveys provided limited information on prehistoric settlement, by 1981, the picture of

T.J. Wilkinson (✉)
Department of Archaeology, University of Durham,
South Road, Durham, DH1 3LE, UK
e-mail: t.j.wilkinson@durham.ac.uk

M.D. Petraglia and J.I. Rose (eds.), *The Evolution of Human Populations in Arabia*, Vertebrate Paleobiology and Paleoanthropology, DOI 10.1007/978-90-481-2719-1_4, © Springer Science+Business Media B.V. 2009

Yemeni prehistory had changed dramatically as a result of the discoveries by an Italian team working in the Khawlan area to the southwest of Marib. In this area of low arid mountains and valleys, surveys under the direction of de Maigret demonstrated the presence of numerous settlements dating to the so-called Yemeni Bronze Age of the third millennium BC (de Maigret, 1984). These results backed up by a number of radiocarbon dates, paleoenvironmental and economic data (de Maigret, 1990) provided an unambiguous picture of the existence of pre-Sabaean complex societies. Although these societies were less sophisticated than those of the Levant and Fertile Crescent to the north of Arabia they provided a clear statement that pre-Sabaean settlement was present in Yemen and that archaeologists had perhaps been looking in the wrong place and using inappropriate techniques. These early discoveries have subsequently been supported by surveys and excavations in various parts of the highlands and surrounding areas (Ghalab, 1990, 2005; Gibson and Wilkinson, 1995; Edens and Wilkinson, 1998; Kallweit, 1996).

A key consideration when analyzing long term archaeological settlement trends is that the record of earlier phases is often obscured (or even removed entirely) by the evidence of later cultures. Hence the famous settlements of Mesopotamian and South Arabian civilizations remain precisely because the areas in question have been long abandoned or are thinly populated today. In general, the archaeological landscape (including its component settlements) falls into two broad zones: a "landscape of destruction" (or attrition) where archaeological remains have often been destroyed or re-cycled by later populations, and a "landscape of survival" where archaeological remains are distinctive because there has been relatively little occupation and agriculture since the remains in question were abandoned (Wilkinson, 2003a). In the case of Yemen, not only are the verdant highlands relatively unexplored archaeologically, the shear scale of post-prehistoric settlement and terraced agriculture appears to have obscured and sometime erased the remains of much settlement. This is well illustrated by the moist southern mountains around Ibb and Ta'izz where some 40% of the population is today contained within around 15% of the land area (Wenner, 1991: 20). When such high populations must be supported by a relatively limited area of cultivable land, it is necessary to create land by terracing with the result that previous fields, settlements and other traces of human occupation are disturbed and incorporated into the later anthropogenic landscapes. Consequently, those areas that are most densely populated today may have been so in the past, but they may yield limited evidence for such occupations because of the destructive nature of later settlement in such restricted mountainous areas.

However, landscapes of survival and destruction form a complex spatial mosaic consisting of areas of archaeological loss or burial punctuated by occasional "windows" of survival. Moreover, the degree of feature survival will partly depend upon the scale, robustness and spiritual value of the buildings and their construction materials. In the case of prehistoric Yemen, de Maigret's discoveries were made in an extensive "landscape of survival" within an area of semi-arid terrain between the formerly irrigated lands of the Sayhad and the rain-fed highlands of the Yemen plateau. Further to the west on the high plateau, windows of preservation are smaller, but are sufficient to demonstrate that major settlements were dotted at frequent intervals across the landscape, often on isolated hill tops above the cultivated lands, or in areas that had experienced little later human activity. It is the evidence from these taphonomic windows that has complemented and extended the original discoveries of de Maigret and the Italian team.

Not only have cultural processes served to dismember parts of the archaeological record, but also deep sedimentary sequences along the major irrigated valleys of the Hadhramaut, Dhana, Beihan and others will have served to obscure pre-Sabaen activity in areas where huge depths of irrigated silts have accumulated (Orchard, 1982). This suggestion by Jocelyn Orchard is now borne out by recent discoveries of fourth and third millennium water control structures in the Wadi Sana region of the Jauf of eastern Yemen (McCorriston and Oches, 2001; McCorriston et al., 2005). This "landscape of survival", which has been virtually unoccupied over the last 5,000 years, demonstrates eloquently what can survive where ancient south Arabian irrigated agriculture was never practiced.

The Environment of Southwest Arabia During the Late Quaternary

A significant part of the Republic of Yemen as well as the districts of Jizan and 'Asir in southwest Saudi Arabia consist of mountains and plateaus with elevations between 2,000 and 3,600 m above sea level. These uplands developed on a combination of Tertiary period volcanics, and associated granites. Occasional glimpses of pre-Cambrian basement rocks occur to the northwest and southeast, whereas east of the mountainous core are a complex of semi arid mountains and plateaus developed on sedimentary and more recent volcanic rocks.

The Red Sea coastal plains of Yemen and Saudi Arabia (the Tihama, Fig. 1) which occupy the down-faulted Red Sea trough to the west of the highlands, and the equivalent plain overlooking the Indian Ocean in Yemen, are arid and hot. In contrast, the mountains behind receive the benefits of orographically amplified monsoon rains and some westerly winds, and most of the resultant rains fall in the spring and summer. Despite their aridity, the arid interior deserts of the Sayhad, and Jauf benefit from rainfall captured by the Wadis Jauf and Dhana which have their headwaters in the mountains. As a result of this process of water capture as well as increased rainfall during the Late Quaternary the Ramlat al-Sabatain became episodically the location of intermittent lakes and an earlier course of the proto Wadi Jauf and Hadhramaut

(Cleuziou et al., 1992). Although most of eastern Yemen receives less than 200 mm of rainfall, the elongate oases of the Wadi Hadramaut provided the locus for the irrigation civilization that developed in the Hadramaut during the late second millennium BC (Sedov, 1996).

Today, rainfall is up to 700 mm per annum and more in the highlands around Ibb and Ta'izz but decreases to ca. 200 mm per annum in the highlands east of Sana. Further north this figure falls to roughly 300–400 mm over much of the plateau, which at elevations of 2,000–2,700 m above sea level has a relatively cool climate for most of the year. Such a bonus of rainfall nourishes rain-fed and terraced agriculture on the plateau which merges into areas of run-off farming where rainfall is less than 300 mm (Wilkinson, 2006). Around the perimeter of this mountain core, in the Tihama, the Indian Ocean coastal plain and the Sayhad, agriculture benefits from the upland water surplus that is captured for flood runoff agriculture (Abdulfattah, 1981; Hehmeyer and Keall, 1993, Hehmeyer, 1995; Brunner, 1997; Munro and Wilkinson, 2007). Therefore the Indian Ocean monsoon should not be seen as of benefit to the highlands alone: the surrounding areas also benefit from perennial flow and flood runoff that have nourished the very specific adaptation of flood runoff agriculture. By so doing, however, many of the best areas for early farming have been obscured by subsequent accumulations of sediment up to 6 m or more in depth. Consequently, in such "self consuming" landscapes of the Tihama mountain front zone and the Sayhad, it is difficult to determine whether prehistoric occupation was ever present, or whether it has been obscured from view.

Late Pleistocene Environmental Change

In southwest Arabia the main terrestrial sources of data for Late Quaternary environmental change are:

1. Dune sands, indicative of arid conditions and plentiful sources of sand.
2. Lakes, the growth of which indicate that precipitation had exceeded evaporation over sufficient time for lakes to form.
3. Humic paleosol horizons which suggest the former existence of a moister, more verdant and less disturbed environment.

The presence of relict dunes, lakes and paleosols can then be compared with climate proxy records derived from cores drilled in the floor of the Indian Ocean (Sirocko, 1996; Zonneveld et al., 1997) or from oxygen isotope analyses of cave speleothems (Fig. 1). Of the latter the best and most appropriate for the Indian Ocean monsoon are from Qunf and Hoti Caves in Oman (Burns et al., 1998; Fleitmann et al., 2003)

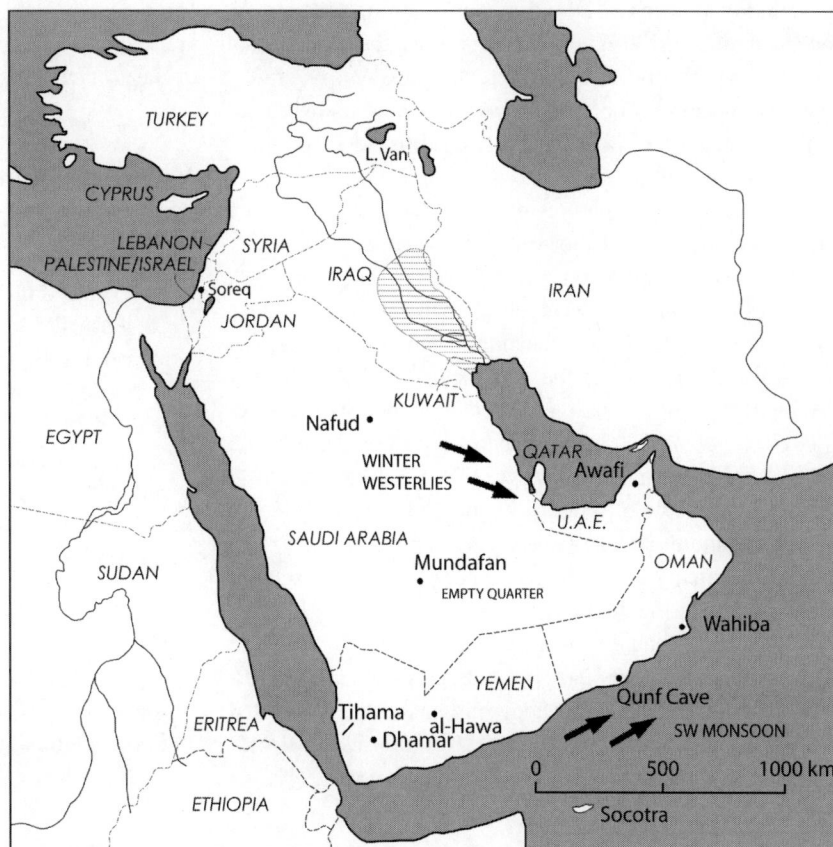

Fig. 1 Environmental sites mentioned in the text (including the key sequences of Soreq Cave and Lake Van)

as well as Dimarshim and Moomi Caves on the Island of Socotra in Yemen (Fleitmann et al., 2007; Shakun et al., 2007).

A Late Pleistocene humid interval, inferred from the relict lakes of Mundafan and the Rub' al Khali areas of southern Saudi Arabia, has been dated to the range 30–21 ka (McClure, 1976; Roberts, 1982: 242) or more recently to 34–24 ka (Anderson et al., 2007: 138). Two paleosol horizons recorded near Ibb in the Yemen highlands were radiocarbon dated to 26,150±350 and 19,290±350 BP (Brinkmann and Ghaleb, 1996: 251–259). The paleosol horizons are characteristic of mollisols developed under savannah or grassland steppe, although in this case the reversal of the radiocarbon estimates renders these dates problematic. In addition, support for this Late Pleistocene phase of increased moisture also comes from radiocarbon-dated groundwaters in the Nejd area of southern Oman and Liwa oasis in the United Arab Emirates (Quinn, 1986; Macumber et al., 1994: 94; cited in Zarins, 2001: 30; Wood and Imes, 1995). The Mundafan lake phase could be associated with the brief strengthening of the Indian Ocean Monsoon whereas the Ibb paleosol would fall towards the end of the moist interval (Leuschner and Sirocko, 2000: 251–252). However, when compared with the high-resolution, but discontinuous record from Moomi Cave, Socotra, the above observations highlight the ambiguities that result when comparing terrestrial, speleothem and oceanic records. For example, the lakes of Mundafan fall within the period for which there is no record from speleothems M1-5 and M1-at Moomi cave this is when atmospheric conditions appear to have been moist and then followed by a drier phase of low June insolation (Shakun et al., 2007: Fig. 7). In contrast the Ibb paleosol falls fully within the dry phase at Moomi Cave. Overall it seems as if dating problems as well as the complexities of local hydrological response to precipitation and evaporation regimes resulted in the above lack of correlation.

Geomorphological features of alluvial activity have been less securely dated and the resultant coarse geochronology means that inferred environmental phases overlap with both moist and dry phases as recorded in speleothems and ocean cores. For example, the continuation of the Wadi Jauf across the Ramlat al-Sabatayn to the Wadi Hadramaut, was considered by Marcolongo to have occurred at around 80 and 30 ka (Cleuziou et al., 1992: 8) whereas a phase of enhanced wadi flow in the Khawlan area southwest of Marib is thought to have occurred between roughly 40 and 18 ka (Fedele, 1990). Finally, and more speculatively, Zarins (2001) has analyzed a range of data to argue for a phase of increased alluvial activity across Arabia during the final phases of the Pliocene. In addition, in the Hadramaut (Yemen) and Dhofar (Oman) phases of Pleistocene wadi activity have been inferred from alluvial terraces at elevations of 20–30 m above present wadi levels (dated to 1.3–1.1 Ma, 900–650 ka or 400–120 ka; Zarins, 2001: 30) and from 3–10 m above wadi level (for the period 70–120 ka or simply Late Pleistocene) (Zarins, 2001: 30).

Relict channels of episodically flowing wadis suggest episodes of considerably greater water discharge, and it is possible that such phases of increased flow were associated with phases of strengthened monsoonal winds during interglacial cycles as has been recognized from deep sea cores taken from the Arabian Sea (Clemens et al., 1991). Although the dating of the above-mentioned fluvial phases is at best weak, in terms of the movement of Paleolithic peoples, such phases of strengthened wadi flow would clearly have increased the opportunity of movement through otherwise inhospitable deserts.

The Late Pleistocene moist phase was apparently terminated by a period of significantly increased aridity corresponding to the Late Glacial Maximum. For example, the upper strata of east-west mega dunes on the Tihama coastal plain were still active at 12,500 ± 1,100 and 10,100 ± 2,100 (OSL dates before 2000 AD: Munro and Wilkinson, 2007: 21). A phase of stabilization equivalent to the earlier Holocene moist interval is then represented by a humic paleosol developed over and effectively stabilizing the dune sands. To the SE of Dhamar, sands, located stratigraphically below an early to mid Holocene paleosol, although undated, suggest that the highlands were also significantly drier during the final phases of the Late Glacial period (Wilkinson, 1997). At the same time, the most elevated parts of the Yemen highlands have provided evidence for significant periglacial activity (el-Nakhal, 1993). Overall, the Glacial Maximum can be seen to have been both cooler and drier, an observation supported by data from ocean cores (Zonneveld et al., 1997) as well as the speleothem record at Moomi Cave which suggests peak dryness at ca. 23 ka (Shakun et al., 2007: 453).

Holocene Environmental Change

Although an Early Holocene moist interval has been known to have existed in Arabia for at least 35,000 years (McClure, 1976) it has now been documented over a broader geographical range using different criteria in a wide range of geographical locations. Current evidence suggests that moist conditions developed very rapidly at the beginning of the Holocene with lakes first appearing as early as 12 ka at al-Hawa in the Ramlat Sabatayn (Lézine et al., 1998, 2007). Within the highlands around Dhamar three different locations suggest the onset of moist conditions around the beginning of the Holocene:

1. Peat developed at Sedd adh-Dhra' around 10,253–10,560 cal. BP.
2. At Zeble organic sedimentation in marshes or lakes commenced ca. 9,900–10,200 cal. BP (Davies, 2006: Table 1; Parker et al., 2006).
3. At al-Adhla' a lake formed around 11,280–12,100 cal. BP (Wilkinson, 2003b: 159).

The two later dates from Sedd adh-Dhra' and Zeble approximately match the cessation of dune sedimentation in the Tihama at around 10,100 BP cited above and align closely with the isotopic record from Qunf Cave, southern Oman, where atmospheric conditions rapidly became moist from around 10,000 cal. BP (Fig. 2; Fleitmann et al., 2003). Nevertheless, the initiation of lakes, marshes and wet valley floors will depend upon a number of factors including rainfall, slope conditions and runoff as well as local hydrology, and in the twentieth century AD there continued to be a number of local small wetland areas that persisted in favorable circumstances (DHV, 1990). In addition, earlier lake marls (undated, but probably of Late

Pleistocene date) have been recognized at Zeble southeast of Dhamar and Bet Nahmi in the Qa Jahran (Davies, 2006: 460).

The Holocene phase of lake development, which apparently resulted from increased rainfall associated with a strengthened Indian Ocean Monsoon, continued until approximately 7,700 BP (at al-Hawa in the arid Ramlat al-Sabatayn), (Lézine et al., 2007: 245–246; Fig. 2). In the highlands around Dhamar, lakes were starting to dry at or slightly after 7,310–7,430 cal. BP, although paleosols, and a single radiocarbon date on freshwater molluscs at Bet Nahmi suggests the persistence of occasional lakes until 3,690–3,900 cal. BP (Davies, 2006; Parker et al., 2006: 246–247).

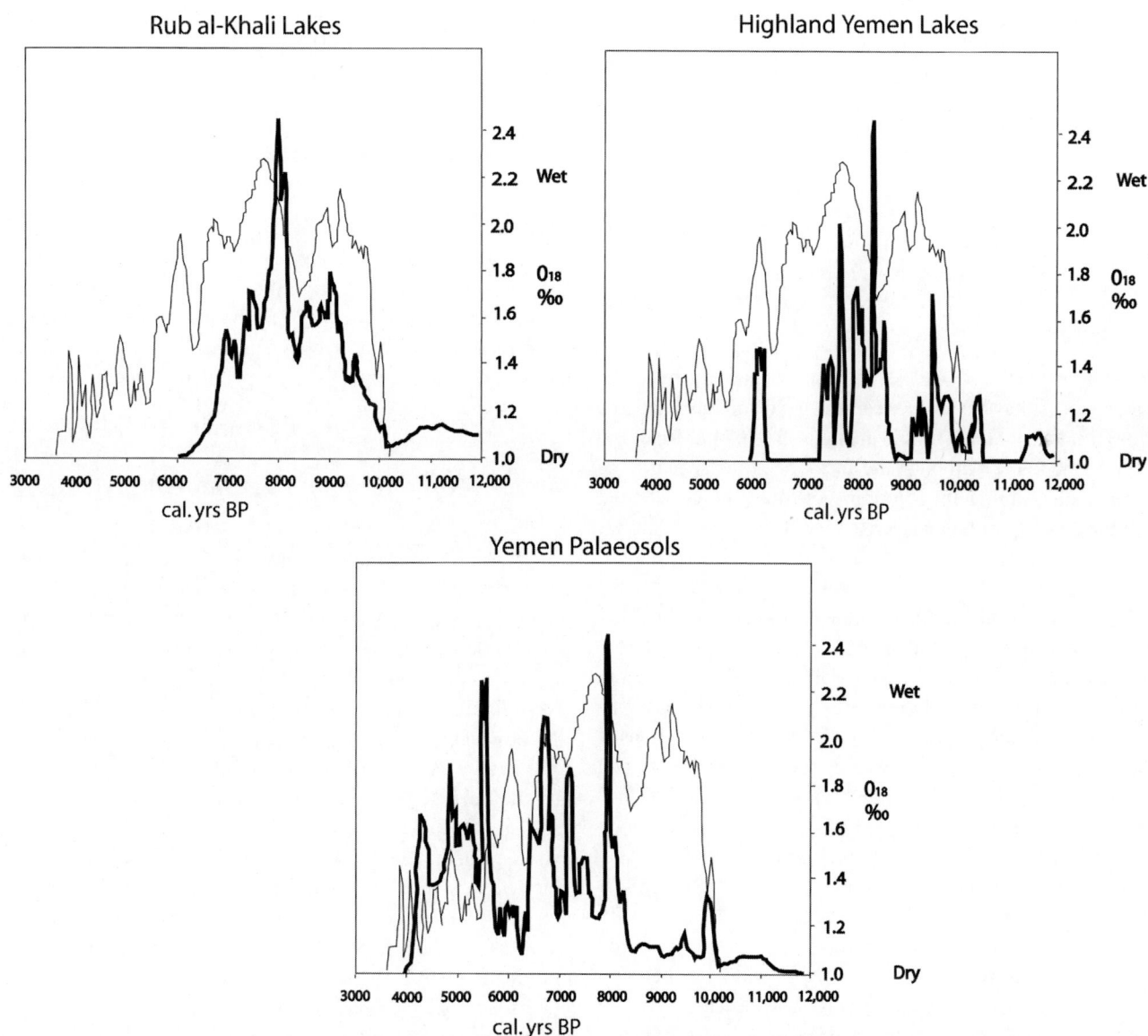

Fig. 2 Aggregated radiocarbon dates (*bold curve*) for (*top left*) lakes in the Rub' al Khali (*top right*) lakes in the Yemen highlands and (*bottom*) Yemen highland paleosols. These are compared to the climate proxy record from Qunf Cave, Oman (*light curve*; based also upon Fleitmann et al., 2003; from Parker et al., 2006)

Complementary to the record from paleolakes is that from paleosols (Fig. 2). In the highlands around Dhamar, in the Khawlan (between San'a and Dhamar), as well as in the Wadi al-Jubah area to the east, a well developed dark brown to black paleosol forms a distinctive stratigraphic marker below later Holocene anthropogenic soils. These paleosols developed mainly during the early to mid-Holocene moist period, although in some case they continued to exist into the subsequent phase of Late Holocene climatic drying. Southwest of Marib, the Thayyillah paleosol has been dated within the sixth to fifth millennia BC on the basis of a single radiocarbon date of 5,750±500 BP (uncalibrated: Fedele, 1990; Fedele and Zaccara, 2005: 219) whereas the Wadi al-Jubah paleosols have yielded 12 dates in the range 9,520±280 BP to 5,270±90 BP or perhaps as late as 4,120±75 BP (uncalibrated: Overstreet and Grolier, 1996: 363, 373 and 375 and Table 14.01 and 14.02). In the highlands around Dhamar, the Jahran paleosol falls into two overlapping classes (Wilkinson, 2005: 178), namely:

1. Relict soil horizons that lack evidence for significant human activity: these date 11,000–4,830 cal. BP (9,000–2,900 cal. BC).
2. Those with evidence for significant activity mainly in the form of occasional obsidian flakes and artifacts, animal bones, and evidence for in situ occupation. This anthropogenic unit dates from 6,890–4,080 cal. BP (that is 4,900–2,100 cal. BC).

In addition to the above, a single dated horizon from the Tihama coast has been described as a peat and has yielded a date (from a contained log) in the range 5,070–4,820 cal. BC (Keall, 2004: 43). Although the above horizons appear to be superficially similar they vary in both their humic content and the quantity of human-derived inclusions.

Although the paleosols fall roughly within the early to mid-Holocene moist interval they continued to develop after the lakes as well as the Holocene moist interval as recorded at Qunf cave (Fig. 2; e.g., Fleitmann et al., 2003). It is therefore premature to view them as being solely a product of increased moisture. Brinkmann (1996) regards these as mollisolic horizons that developed under a savannah environment whereas in the highlands, soil micromorphology suggests they developed in the presence of some tree cover, as well as in the presence of significant human activity (French, 2003: 224–234). It is possible therefore that the enhanced humic content of the paleosol, by holding more moisture, initiated a process of feedback that encouraged the retention of more humus. In other words, their initiation might result from increased atmospheric moisture and associated vegetation, but their persistence may result from a process of positive feedback.

The Jahran-Thayyillah-Jubah palesol provides a stratigraphic marker horizon that has yielded a significant amount of evidence of Neolithic activity (Overstreet and Grolier, 1996: 374–375; Edens and Wilkinson, 1998; Fedele and Zaccara, 2005). For example in the highlands traces of Neolithic occupation are occasionally found within the Jahran horizon, which is then followed by a phase of soil erosion and anthropogenic soils associated with increasingly visible and extensive Bronze Age sites (Wilkinson, 2005).

According to the spleothem record, the late Holocence drying phase in southern Arabia was gradual (Fleitmann et al., 2007: 185), whereas its initiation was very rapid, the latter being supported by some of the terrestrial records. Nevertheless, at Qunf Cave this drying trend was itself punctuated by occasional drying phases of greater severity (Fig. 2). Similar drying phases are also evident at the site of al-Hawa between 10,500 and 10,100, 9,100 and 8,400, and 8,000 and 7,700 cal. BP (Lézine et al., 2007) where they are interpreted as resulting from weak phases of the Indian Ocean Summer Monsoon, themselves correlated with ice-rafting and cooling in the North Atlantic (Lézine et al., 2007: 246–247).

Population and Settlement

Because of the dearth of archaeological information for southwest Arabia as a whole, I have chosen to present the evidence for long-term settlement from the present day back in time. This enables the increasingly attenuated record of ancient settlement to be seen from both a historical perspective and in terms of what we know of present population levels.

Modern Population and Settlement

Today southwest Arabia is one of the most densely populated parts of the Arabian peninsula, and it is only rivalled by areas of recent urban and industrial growth around for example al-Hasa/Hofuf, Riyadh, parts of the United Arab Emirates and Bahrain (Beaumont et al., 1976: Figure 5.2, pp. 184). According to the Naval Intelligence Division handbook, of the estimated six million population of Arabia, some 2.6–3.6 million (i.e., about 50%) were considered to live in Yemen and the neighboring Aden Protectorate (N.I.D. 1946: 364). Earlier estimates from the late nineteenth and early twentieth century, whether they come from official Ottoman and British colonial sources or from travellers and visiting academics, are extremely variable ranging from some 750,000 to as many as 9,000,000. Nevertheless, nine sources quoted by Grohmann (1922, 1966) provide a median population estimate for Ottoman controlled Yemen of 2,500,000, a figure that is close to that of the N.I.D. (1946).

As discussed above, although the above mentioned Swiss census estimate of 4,705,336 was disputed in terms of its absolute population estimate, its assessment of settlement pattern and structure provides a valuable perspective on modern settlement. That census demonstrated that in the former republic of North Yemen (The Yemen Arab Republic) some 41,000 settlements, or 78% of the total, contained populations of 100 persons or less (Steffen, 1979: Figure 2-62, pp. I/149). Despite the recent movement of population to modern cities such as San'a, Ibb, and Ta'izz, Yemen remains a primarily rural society dominated by villages. In fact in the highlands, the record of ancient settlement compares quite closely (at least in terms of size) with the small and medium size villages of today that contain populations of up to 500.

Overall, the highest populations (>200 persons/km^2) are found on the plateaus and mountains around Ibb and Ta 'iz. Populations in excess of 100 persons/km^2 are typical of much of the remainder of the plateau, whereas populations fall to less than 50 (and often considerably less) in most of the arid lowlands to the northeast as well as much of the Red Sea Coastal Plain (Steffen, 1979: Population Density Map). Significantly such variability is also evident at the local level and Steffen's team recorded dense populations at higher elevations along some of the major wadis leading down to the Tihama (such as the Wadi Zabid), whereas the valley floors exhibit significantly lower populations.

The Recent Historical Record

When Carsten Niebuhr (1792) reached Yemen in 1763 he described a populous and well cultivated country, and there is little to dispute this picture from the Ottoman or Rasulid records of the medieval and post medieval periods. Similarly the tenth century AD writer al-Hamdani (1938) frequently refers to an intensively cultivated landscape remarkably similar to that of today. For example, for the area of the former Himyarite capital of Zafar (south of the modern town of Yarim) al-Hamdani described some 80 "sedds" for impounding soil and water, many of which continue to be evident today (al-Hamdani, 1938; Gibson and Wilkinson, 1995: 172; Barceló et al., 2000).

The Classical Record

Unfortunately the classical authorities mainly supply us with rather vague statements about southwest Arabia, and even when these are supported by actual figures they seem to provide a degree of spurious accuracy. Hence Ptolemy's Geography (completed ca. AD 150) credits Arabia (Petraea, Deserta and Felix) with 218 "settlements" 151 of which were village size (kômai) (Hoyland, 2001: 169). Although only six "cities" are mentioned, all of these were located in south

western Arabia, which suggests that this was the most urbanized part of Arabia. Moreover, according to one version of Agatharchides: the Sabaeans "surpasses in wealth and all the various forms of extravagance not only the nearby Arabs but also the rest of mankind" (Burstein, 1989: 167). Two thousand years ago, southwest Arabia can be therefore be considered to have contained a significant and wealthy population, which we now know to have derived much of its wealth from the incense trade (Groom, 1981).

The Epigraphic Old South Arabian Record

South Arabian civilizations were literate during most of the first millennium BC and AD, and consequently they supply us with a record that complements that derived from archaeological excavations and surveys. For example, some records allude to the mobilization of large numbers of people to build public works such as the great Marib Dam. This information has been collated by Schippmann (2001) and includes:

1. Some 20,000 people who were mobilized to repair the Marib dam in 450 AD.
2. 16,000 who are stated to have been killed and 40,000 taken prisoner in battle by the Sabaeans against the residents of the kingdom of 'Ausan.
3. 45,000 casualties and 63,000 prisoners of war who are reported in a conflict against Najran (Schippmann, 2001: 10 and 120 [citing von Wissman and Höfner, 1952, and Hommel, 1926]).

Because such figures probably include both sedentary and nomadic populations, they give us little indication of either the sedentary population alone, or the proportions of sedentary to mobile populations. Nevertheless they underscore how individual kings were capable of mobilizing large numbers of people for their public works, the evidence of which is evident in the archaeological record.

South Arabian Settlement and Population: The Archaeological Record

In addition to the records from texts, population estimates can be made by comparing the size of some of the cities with their surrounding fields, the latter being evident on the ground as rectilinear grids of gullies eroded out of the channels and field boundaries of the original Sabaean irrigated fields. Such population estimates fall in the range of 30,000–50,000 (Schippmann, 2001: 12, Brunner, 1983), figures that are significantly in excess of the possible population of the 110 ha Sabaen city of Marib which probably accommodated between 10,000 and 30,000 people.

In turn Schippmann has used these already approximate estimates to arrive at figures of 310,000 to not greater than 500,000 people for the kingdom of Saba' at its height. It must be emphasized, however, that the above figures of captives and corvée labour may well be exaggerated, and the overall population estimates provided by Schippman, although useful, are little more than educated guesses. Moreover, they do not take into account the recent archaeological surveys conducted in the 1990s that demonstrate a rather densely populated plateau (Kallweit, 1996; Wilkinson et al., 1997; Wilkinson and Edens, 1999; Edens et al., 2000). Overall, Schippmann contrasts his estimated population of 500,000 for the kingdom of Saba' with a total estimated population of 1 million for the Indus civilization at its height and a modern Yemeni population of 16 million. Because the Indus civilization was probably one of the more populous parts of Eurasia during the late third and early second millennium, and North Yemen housed some 4–7 million people during the 1970s and early 1980s, these figures are not as low as Schippman implies. Thus his estimate places the kingdom of Saba' as containing perhaps half the population of the Indus civilization. This provides a hint of the importance of Saba', especially when this figure is combined with its evident and much famed wealth. Because of the danger of indulging in spurious accuracy, for the remainder of this chapter I will keep quantitative estimates to the minimum and simply attempt to show the approximate degree of settlement that prevailed.

Recent surveys demonstrate that by the late first millennium BC the elevated plateau around Dhamar was very well settled (Fig. 3). For example, for the Dhamar Survey area Lewis (2005: 138) reports some 132 sites with Himyarite ceramics, 109 with a significant occupation (Fig. 4; K. Lewis pers. comm., Dec. 5th 2007). Such sites, which range from large towns such as Masna'at Mariyah down to small farmstead-like settlements, may not all have been occupied at the same time. On the other hand, it is clear that because only a fraction of the Dhamar area has been surveyed this figure is but a fraction of the original settlement. The total number of sites, just within this one area, must have been significantly higher. Overall, it is sufficient to say that the Yemen plateau between Yarim and Sana'a and apparently to the north as well, was during the later first millennium BC and AD, very well populated with perhaps between 1 and 5 settlements per 100 km².

South of Yarim, even such estimates are impossible, because virtually none of the area has been surveyed and much of the terrain is blanketed by extensive staircases of terraced fields as well as a dense scatter of villages and houses. Although not well dated, many terraced fields in the Dhamar area have been traced back to prehistoric times (Wilkinson, 1999) a measure which probably applies to the area of Ibb and Ta'izz as well. Not only do these terraced fields imply that a large population has gone to considerable lengths to increase the area of cultivation within a mountainous area, by so doing they have significantly destroyed much of the

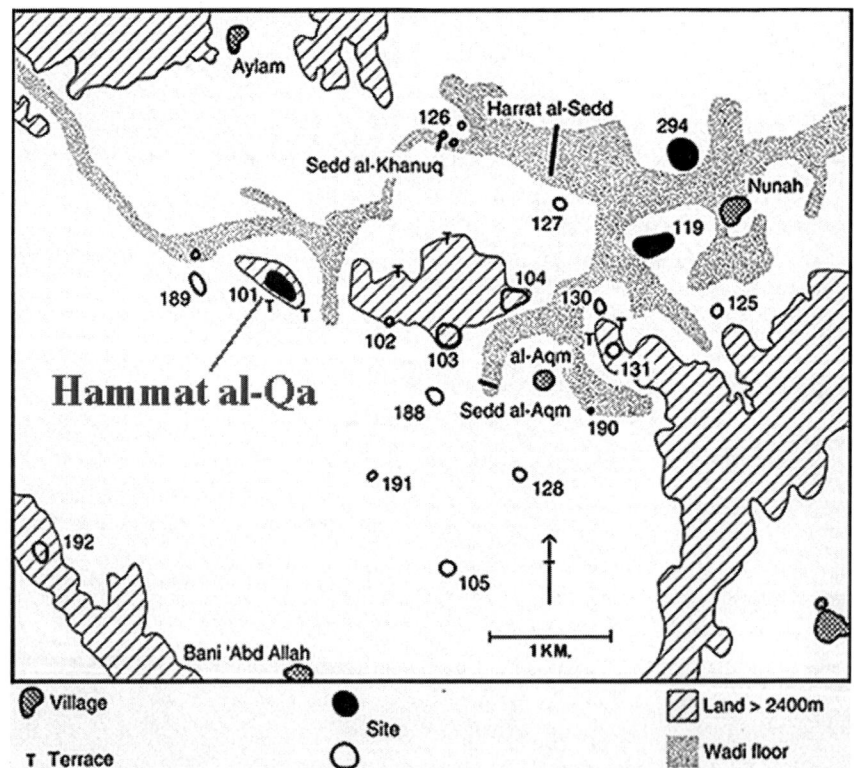

Fig. 3 Archaeological sites of all periods recorded around the site of Hammat al-Qa, near Ma'bar, Yemen, and indicative of the density of archaeological sites in the Yemen highlands. The earliest significant site is third and second millennium BC Hammat al-Qa

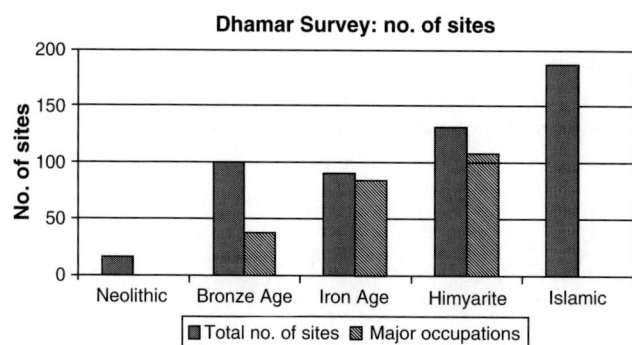

Dhamar Survey: no. of sites

Legend: ■ Total no. of sites ▨ Major occupations

Fig. 4 The number of site occupations through time in the plateau around Dhamar (total and significant occupations) (data compiled by T.J. Wilkinson and revised by Krista Lewis)

pre-existing archaeological record, therefore making population estimates from survey even more difficult than would normally be the case. At best, all we can say for the Ta'izz–Ibb area, is that such vast areas of terraced fields imply a significant population, extending back for an unknown period of time.

The Yemeni Bronze Age

Probably the most significant breakthrough in the investigation of pre-Sabaean settlement in Yemen came as a result of fieldwork conducted in the Khawlan area southwest of Marib by an Italian team, directed by de Maigret (1990). The Italian surveys, excavations and environmental investigations enabled a Yemeni "Bronze Age" to be recognized and approximately defined. Settlement took the form of sprawling sites measuring 0.1–1.0 ha in area, consisting of stone footings of sub-rectangular buildings and compounds at intervals of every 2–3 km over gravel terraces alongside tributaries of the Wafi Dhanah and other wadis. Although the Italian investigations provided clear evidence for settlement patterns of sedentary complex societies, the Khawlan sites were little more than small agro-pastoral villages or even hamlets. Nevertheless, their dates of occupation (ca. 2,500–2,000 BC) and the repertoire of associated ceramics, ground stone, and lithics provided unequivocal evidence for the existence of a ceramic-using complex society in southern Arabia, equivalent in date to the Umm an-Nar Culture of Oman. However, as de Maigret (1999) has pointed out, these small communities with their repertoire of simple material culture and rudimentary architectural forms can hardly be regarded as the fore-runners of the impressive South Arabian civilization.

At approximately the same time, the cities of the incense route were being extended back into the second millennium BC. Thus sites in the Wadis Beihan, al-Jubbah and Hadramaut as well as at Shabwa, have been shown to have early occupa-

tions in the mid-late second millennium (van Beek, 1969; Glanzman and Ghaleb, 1987: 65–66; Badre, 1991; Sedov, 1996). Moreover third or early second millennium settlement and irrigation has been documented in the Marib area, the Ramlat Sabatayn and several other areas of the Sayhad (Cleuziou et al., 1992; Brunner, 1997: 196; Vogt, 2005).

However, it is on the high plateau of Yemen that so called Bronze Age settlement of the third and second millennium BC is most evident (Gibson and Wilkinson, 1995; Kallweit, 1996; Wilkinson et al., 1997; Ghaleb, 2005). This may be because:

1. The highlands formed one of the most densely settled parts of Yemen at this time.
2. In other areas much early settlement has been obscured by later activity or by sedimentary aggradation (above; also Orchard, 1982).
3. Or that the Dhamar area is one of the few parts of Yemen to have been systematically surveyed for a long period, that is from 1994–2003 and later.

On the plateau around Dhamar Bronze Age sites dating to the third and second millennium BC occupy a range of locations: hill summits; plateau tops (often with convenient oversight of valley-floor pastures or access to accessible land: Wilkinson, 2003b: 162); lower valley side slopes as well as on valley floors that exhibit the potential for the development of terraced fields. Although some sites, such as Hammat al-Qa, were eminently defensible, others were not. Archaeological surveys have recorded a total of 101 Bronze Age sites, 37 with a significant Bronze Age component (Fig. 4; K. Lewis, pers. comm., 2007). These varied in size up to a maximum size of 15–20 ha, although the 5 ha settlement of Hammat al-Qa is more representative of the scale of settlement (Edens et al., 2000). Because the associated ceramics are somewhat generic in their form it is difficult to determine narrow chronological ranges of occupation, but sufficient radiocarbon dates have been processed to demonstrate that occupation occurred during the entire third and second millennia BC with no obvious gaps. Because we are not able to demonstrate which sites were exactly contemporaneous, the estimation of population is speculative at best. Nevertheless it is evident that settlement was remarkably common throughout the Dhamar and Ma'bar areas, a pattern that appears to continue to the north towards Sana'a and beyond (Kallweit, 1996; Ghaleb, 2005).

In the lowlands to the east of the Yemen highlands ceramic-using Bronze Age settlements become rare and even absent, whereas on the coastal plains bordering the Indian Ocean and Red Sea several large sites at Sabir, Ma'layba, Qashawba and al-Midaman (Ciuk and Keall, 1996; Vogt and Sedov, 1998; Buffa, 2002; Durrani, 2005; Khalidi, 2005) attest to at least areas of high population concentration during the second and early first millennium BC.

The Neolithic

Neolithic sites, which are noticeably fewer in number than those of the Bronze Age, have been recognized on the basis of a distinctive lithic assemblage of the "Arabian Bifacial Tradition" or an "Upland Neolithic Tradition" (Edens and Wilkinson, 1998: 62). They have a widespread, albeit sparse, distribution throughout southwest Arabia and take a variety of forms, depending upon their location:

1. Shell middens on the Red Sea coast (Tosi, 1985, 1986; Zarins and Zahrani, 1985; Zarins and Badr, 1986; Khalidi, 2005).
2. Lithic scatters (and associated stone structures in the interior) along wadis leading down to the Tihama and in the interior (Khalidi, 2005; Fedele and Zaccara, 2005: 214–216).
3. Lithic scatters in the interior deserts and Hadhramaut (Zeuner, 1954; McCorriston, 2000).
4. Several highland Neolithic sites have been found in a good stratigraphic context being contained within the brown or black humic Jahran or Tayyillah soil discussed above.
5. Stone-built huts and agricultural structures are evident in the Khawlan, eastern Yemen (McCorriston et al., 2005), Dhofar (Zarins, 2001), as well as the highlands (Edens and Wilkinson, 1998: 65).

Neolithic sites of the interior mountains and valleys are mainly small, consisting of occasional small built structures, surface lithic scatters, or simply a dispersal of lithics and bone within the paleosol. In Yemen, the best manifestation of this culture to date appears to be the Neolithic site of WTH 3 in Wadi Thayillah (Fedele, 1990; Fedele and Zaccara, 2005) which consists of about six elliptical houses of block and boulder construction (Fedele and Zaccara, 2005: 222) extending over approximately 0.5 ha. Further to the east, the Neolithic complexes in the Wadi Ghadun (Zarins, 2001: 34–51) demonstrate just how much activity can remain in the desertic areas where there has been little later occupation to disturb or remove Neolithic features.

Elsewhere, particularly in the highlands the under-representation of Neolithic sites may be because the sites are partly buried within the paleosol, or because sites have been removed or obscured by the numerous later settlement or agricultural activities. Consequently Neolithic sites are more evident within "landscapes of visibility" such as in the Jauf area of the Hadhramaut and the arid wadis southwest of Marib or in Dhofar, where the dearth of later occupation means there has been little subsequent occupation to disturb or obscure those of the Neolithic.

Overall, there is a significant difference between both the number and the scale of Neolithic and Bronze Age sites. Therefore, according to a recent re-analysis by Lewis and Khalidi there were 16 Neolithic sites versus 101 Bronze Age sites (Fig. 4; K. Lewis pers. comm., 2007). A superficial reading of the raw data on site counts from the Dhamar sur-

vey would therefore suggest that there was a real increase in population at some time around the third millennium BC, although equally this disparity might be because the number of Neolithic sites has been suppressed as a result of the site preservation factors noted above. On face value, however, it appears that all topographic zones of southwest Arabia were occupied by a low density scatter of Neolithic communities who engaged either in fishing near the coast, or animal husbandry of sheep, goat and cattle with some hunting in the interior. There is no evidence for cultivation of crops and cereals until the late fourth millennium BC (Edens, 2005: 190–191), but there is a need for considerably more research before it is clear when domestic crops were introduced.

The Paleolithic

Paleolithic settlement is even more elusive than that of the Neolithic, although several studies have now demonstrated the existence of Paleolithic sites in the Tihama, the highlands, and the interior. Owing to the sparse nature of the find spots and the poor chronological resolution of the finds, it is difficult however, to tie such occupations to glacial and interglacial cycles. Nevertheless, a recent survey in the southern highlands south of Ta'izz by Whalen and Pease (1991) added 37 sites with Acheulean traditions of pebble tool and bifacial technology. These supplement the earlier discoveries by Caton-Thompson (1953), Amirkhanov (1994), Doe (1971), Bulgarelli (1987), de Maigret (2002), and others of Paleolithic occupation. Few sites occur within a well developed environmental or stratigraphic context, although Al-Guza Cave in the Wadi Hadramaut has produced what is referred to as a pre-Acheulian industry within a stratified context (Amirkhanov, 1994: 218–220). In the highlands, an Italian team reported Lower Paleolithic lithics from the Ma 'bar plateau in the Qa Jahran Plain (Bulgarelli, 1987). Significantly this find was in the vicinity of a series of ancient lakes in which Holocene lakes apparently were preceded by those of Pleistocene date. Because of the presence of large areas of Quaternary basalt lava flows and volcanic outcrops, Yemen offers some potential for the recovery of Paleolithic sites within a dateable volcanic stratigraphy, which would enable occupations to be fixed within the long term chronology of glacial–interglacial isotopic stages. Many Paleolithic sites are little more than lithic scatters, but at Humayd al'Ayn on a plateau in the vicinity of Marib, de Maigret and Bulgarelli a Lower Paleolithic site (Bulgarelli, 1987; de Maigret, 2002: 119–120) extends over a very large area.

Gaps in the sequence of Paleolithic sites, such as a possible dearth of Upper Paleolithic sites, may simply result from a lack of focused survey, or mixing with later material (Whalen and Pease, 1991: 128; also Zarins, 2001: 33). Missing from our data base are the now submerged Paleolithic coastal plains which fringed the Red Sea and Indian Ocean coasts during the cold intervals of the Pleistocene. For example, during the Late Glacial Maximum when our scanty evidence suggests that the highlands were both significantly cooler and drier, the Red Sea coastal plains would have been considerably wider. Consequently the coastal fringe, which in the Holocene became the locus of a vigorous cultural zone of shell midden settlement, was displaced several kilometers to the west and may have been the primary locus for Late Paleolithic activity. This question is now being examined for the southwest Arabian coastal plains off the Jizan coast of Saudi Arabia (Bailey et al., 2007; Bailey, 2009).

Although it is impossible to make demographic estimates from the data on Paleolithic sites, it is now becoming evident that there was significant occupation in southwest Arabia during the Paleolithic, and therefore its location opposite the Horn of Africa makes a compelling case for a migration pathway between Africa and Eurasia (Rose, 2004).

Discussion: Human Settlement and Changing Environments

Archaeological research over the past two decades suggests that southwest Arabia, specifically the moist highlands, were well populated over the past 5,000–6,000 years and may have acted as a long-term population centre during much of the Holocene. It is difficult to say whether such a centre also existed during the Early Holocene, because we know little about the amount and scale of settlement as well as how much has survived. Although it is possible to suggest that this region might have acted some form of refuge for populations during the Paleolithic, this remains to be determined.

Table 1 sketches the regional variation in environmental conditions experienced in key geographical areas of southwest Arabia for specific phases of the last glacial–interglacial cycle. These fluctuations would have benefited different areas in different ways and at different times. Although Table 1 provides only a rough guide to some of the main trends as inferred from the proxy indicators summarized above, these variations

are sufficient to suggest that although climate did, to some degree, change synchronously from region to region, relatively minor differences between the regions might have significantly impacted local populations. Such potentially deterministic suggestions should however only be regarded as rough models: the complexities of human social and political relationships as well as the early development of technological innovations in the region (such as terraced fields and valley floor check dam agriculture), all would have resulted in a complex array of local adaptations to environmental conditions. Nevertheless, the table may provide a starting point for the analysis of long term relationships between humans and the environment.

For example, periods during which favorable niches were more restricted would have witnessed populations aggregating within a smaller number favorable habitats. This is hardly a novel concept that has already been suggested by V. Gordon Childe for the Fertile Crescent during the Neolithic Revolution under the rubric of the "oasis propinquity" concept. Thus in the Late Glacial Maximum settlement may have been focused along the coast of the now submerged coastal plains of the Red Sea coast, whereas much of the remainder of the plains may have been less favorable for settlement (or evidence has been lost from view below the alluvial fans).

One settlement model suggests that the interior of Arabia was essentially empty before 10,000 ka (Amirkhanov, 1994; McClure, 1994) so that it became populated during the earlier Holocene as a result of the in-migration of people and animals from the north. The evidence in support of such a model is slender, however, and it is best to say that we know very little of the Late Pleistocene and earliest Holocene populations of Arabia. In terms of the brief pulsations of climate evident at al-Hawa and Qunf Cave, the archaeological record is insufficiently fine grained to demonstrate how settlement responded to climate shifts. In the wetter intervals of the Early Holocene, the deserts and coastal plains appear to have supplied more favorable habitats for hunter-gatherers and early mobile pastoral communities than was the case before. As a result, populations would have been dispersed more widely over coastal plain, mountain and desert. Where the archaeological and environmental records are of higher resolution (as in the UAE) it can be argued that coastal shell middens may have been occupied during the winter, with the inhabitants moving into niches

Table 1 Sketch of regional variation in climate signatures for southwest Arabia over the last 20,000 years. Moister areas are shaded light grey

Phase	Coastal plains	Highlands	Eastern lowlands
Late Glacial Maximum	Wide and dry plain: moist coastal fringe	Cool and dry	Dry and relatively warm
Early Holocene	Narrow plain benefits from highland runoff	Moist and warmer	Moist and warmer
Later Holocene (post-3000 BC)	Narrow plain; drier	Warmer and slightly drier (but still verdant)	Warmer and drier
Iron Age (period of incense trade)	Warmer and slightly drier: Developing flood agriculture	Warmer and slightly drier (but still verdant)	Warmer and slightly drier; soil moisture enhanced by irrigation
Post-Sabaean	Warm and relatively dry: moisture from flood agriculture	Warm and relatively dry (but still verdant)	Warm and relatively dry

in the mountains during the hot weather (Uerpmann et al., 2000) or into the desert during moister periods. A strengthened monsoon, with increased summer rainfall may therefore have encouraged settlement in the deserts all year. At such times coastal locations may therefore have been abandoned as there was increased emphasis on activity in the interior (Parker et al., 2006: 250). Consequently, some sites might be abandoned during wetter intervals, whereas other sites would become occupied. Because of the spatial variability of wet and dry niches in Arabia and especially the desert margins, settlement would have varied both spatially and temporally.

As the Neolithic moist interval drew to a close in the fourth millennium BC the moist highlands of Yemen would again have constituted the most favorable area for settlement. In eastern Yemen and neighbouring Dhofar (Oman), ceramic using sedentary complex societies did not appear until the first millennium BC. Nevertheless, during the mid-Holocene phase of climate drying, settlement was present during the Neolithic and Bronze Age, inhabitants of the latter using an aceramic material culture (Zarins, 2001: 73). Some areas provide compelling evidence for a significant population decline during the arid period of the mid-Holocene (McCorriston et al., 2005: 148–150). Whether the relatively verdant highlands actually attracted populations from the surrounding arid or semi-arid lowlands, or simply were less prone to population decline is difficult to say. However, the evidence for the arrival of domesticated plants during the late fourth millennium BC (Edens, 2005) suggests that increasing agricultural activity may have resulted in higher population growth rates than would have been the case when the area was occupied by mobile pastoralists or hunter gatherers.

It is tempting to posit that population shifts in the fourth millennium BC took the form of dramatic abandonments under the stress of increasing aridification (Fig. 5), but more subtle interpretations are plausible: "Climate change altered the regional landscape of economic possibilities, obliging communities gradually to seek alternative solutions to the issue of feeding themselves" (Edens, 2002: 81).

For any one area such alternative solutions include hunter-gathering, mobile pastoralism and fully sedentary communities (as well as combinations of these). It can be argued that with increasingly dry conditions the amount of hunted animal food eaten will decline (McCorriston et al., 2005: 150 citing Keeley, 1995: 249), and it is now apparent that mobile pastoralism was a common, if not the dominant strategy, for Arabian desert communities during the later Neolithic (Uerpmann et al., 2000). Because mobile pastoralists adapt more rapidly to changing environmental conditions and entail lower population densities than sedentary agriculturalists, increased aridity will probably result in lower population densities. Ironically, increased aridity in the moister highlands, may have resulted in an increase in population, not only because communities were sedentary and adopted cereal

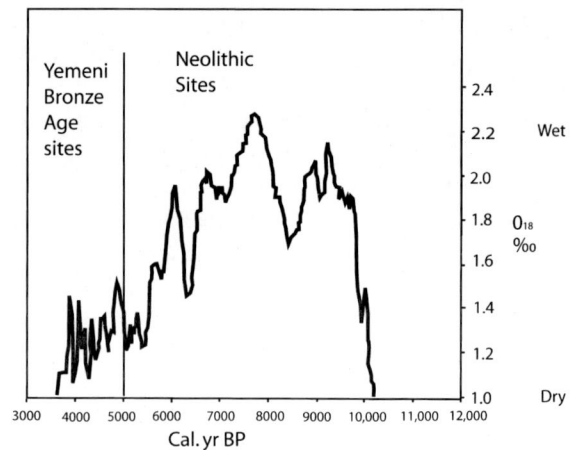

Fig. 5 Main archaeological periodization in southwest Arabia compared with the proxy climate record from Qunf Cave, Oman (based upon Fleitmann et al., 2003)

crops, but also because these environments were relatively more attractive for settlement than the interior steppe. The interior lowlands became increasingly arid and inhospitable during the fourth and third millennium BC and consequently appear to have suffered a significant population decline.

With the introduction of cereal crops, which appear to have followed by a considerable period the introduction of domestic animals, the verdant highlands experienced an increase in population. This may have been reinforced by the settling of more mobile populations from the desert who were attracted by the more favorable conditions, or in the case of transhumant communities, who simply chose to spend longer intervals resident in the highlands. Not only could the population growth that we see be accounted for by the higher agricultural productivity that was a response to the verdant conditions, the Yemen highlands may have benefited from a lower incidence of malaria which would have increased mortality rates in the lowland areas. For example early studies of the prevalence of malaria suggest that the incidence was low at elevations above 6,000 ft (ca. 2,000 m) as well as in the driest lowlands (N.I.D., 1946: 451). The combination of increased population growth, perhaps some in-migration from lowlands, and lower death rates from malaria, together would have resulted in significantly higher rates of demographic growth, which perhaps accounts for the rather dense populations evident today in the highlands.

Whereas the relatively moist conditions of the highlands enabled settlement to continue during the second and first millennium BC, the arid Sayhad communities developed or adopted flood (sayl) irrigation to support agriculture. This process appears to have commenced by the later third millennium BC in the Sayhad (Brunner, 1997: 200) but the most rapid growth appears not have taken place until around 1,000 BC when the main settlements grew and eventually outpaced their counterparts in the highlands. The appearance of flood-flow and perennial irrigation systems (sayl and ghayl) at Malayba and Sabr during the second millennium BC (Vogt et al., 2002: 25) argues for a similar trajectory of settlement and irrigation agriculture in the Indian Ocean coastal plain. Whether this was the case for the Tihama is less clear, but the presence of large settlements such as al-Hamid and Qashawba (Phillips, 1998; 2005; Durrani, 2005: 75–86) during the early first or even second millennium BC suggests a comparable trajectory of settlement growth there too.

Clearly, by the second and early first millennium BC, human communities in the Sayhad were able to over-ride the deficiencies of the local environment to establish large scale irrigation systems that made local communities less vulnerable to climatic events. On the other hand, with time, field systems became more silted up and were raised above the terrain with the result that dams also had to be raised thereby imposing new vulnerabilities into what had formerly been a productive way of supplying food for both local communities and the passing caravan trade.

In the Sayhad and neighbouring lowlands processes of positive feedback and associated growth cannot have taken place in a demographic vacuum: a sufficient "reservoir" of population would have been necessary during the initial growth stages of the incense trade in the second millennium BC to supply the labor force (voluntary or coerced) that could, in turn, fuel such growth. Such a supply probably came from three main areas:

1. From the local communities of the Sayhad, although these were not necessarily as populous as they became later.
2. Mobile pastoralists of the desert fringe and neighboring areas; these are implied to have been present by the numerous prehistoric cairn fields recorded on hills and ridges around the desert margins.
3. The moist highlands of Yemen and southern Saudi Arabia.

Overall, southwest Arabia may have formed a nucleus of early population growth, first in the mountains themselves, and second where the water shed by these mountains could be harnessed for irrigation (Fig. 6). In the desert fringe of the Sayhad, such irrigation potential must have contributed to further growth as communities became more affluent as a result of the development of the incense trade. Both "invest-

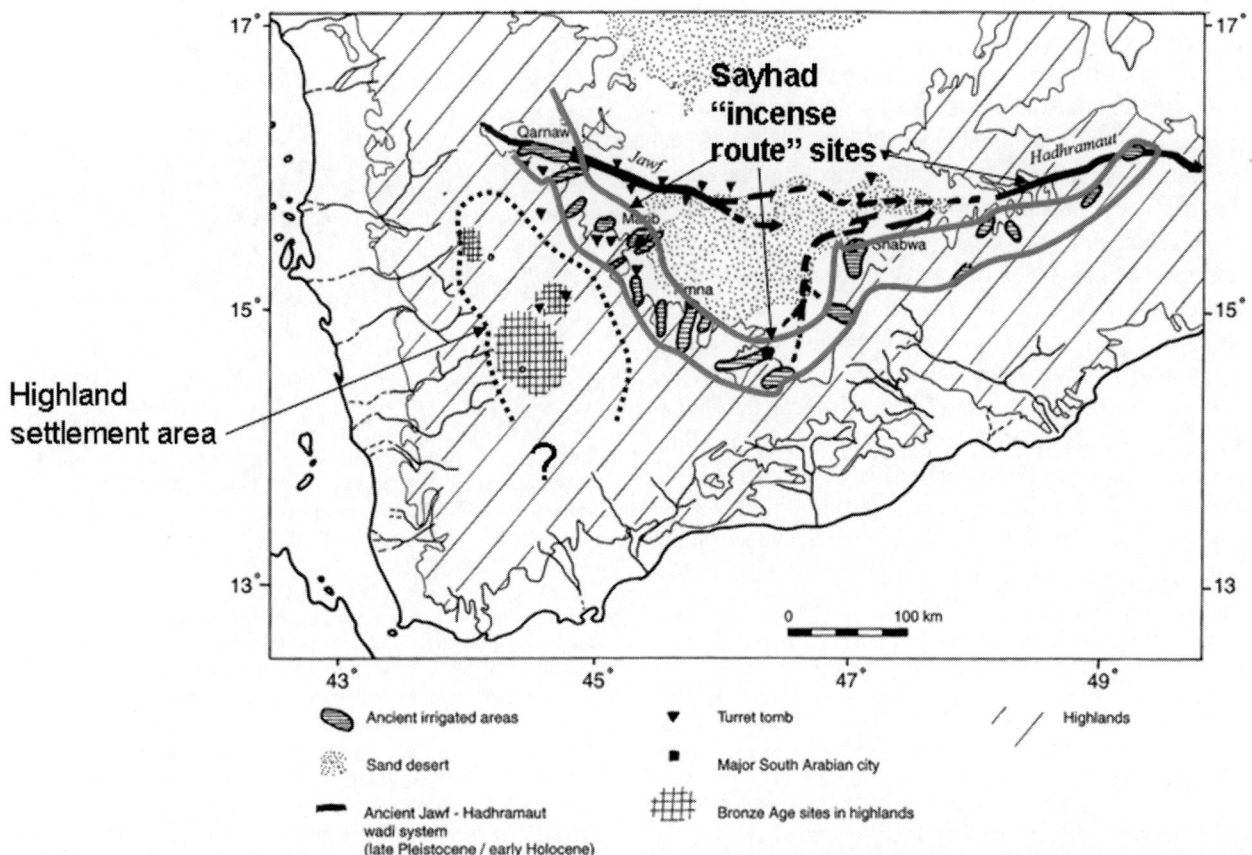

Fig. 6 The area of settlement in highland Yemen compared with the Sayhad area which formed the locus of development of the "cities of the incense route"

ment" and demand for agricultural produce must have increased significantly so that the "incense cities" appear to have benefited from a complex inter-twining of relationships in which irrigation agriculture benefited from the receipt of the profit of the incense trade and the incense trading caravans were able to rely on reliable supplies from the oases (see also de Maigret, 1999). Being marginal to this trade, the highlands may have even lost population to the rapidly growing Sayhad oasis towns, but the role of the highlands as a long-term demographic "reservoir" for southwest Arabia probably continued so that there was a flux of populations back and forth depending on the relative status of political, economic and environmental conditions in the core and fringing lowlands.

Such shifts probably also reflected changing power relationships within the dynamic and often war-torn South Arabian polities. Thus Avanzini (2003: 145) has argued that the control of the high plateau was an important factor in the development of the early kingdoms of Qataban, Saba and Himyar.

To conclude, southwest Arabia formed a significant center of population that can be traced back in time from its status as the most verdant and populous corner of Arabia during the recent past to a region of rather dense settlement back to approximately 3,000 BC. During the earlier Holocene, when conditions were moister throughout the region, communities were more dispersed hunter gatherers (and fisher gatherers on the coast) with mobile pastoralists becoming increasingly important during the Neolithic. Overall, climatic cycles had geographically varied impacts and although during the height of the Late Glacial Maximum conditions for human occupation of the highlands may have been less attractive, this would have been made up for by favorable niches along the now submerged coastlines. On the other hand during interglacials the highlands would have provided excellent resources for hunting and foraging within a rich mosaic of woodlands, marsh, lake and savannah. Nevertheless, one of the fundamental questions about southern Arabia is that precious little is known about the population of this region during the Early Holocene and Late Pleistocene, and despite the increasing evidence for environmental fluctuations, little can be said about human-environment interactions, until we have the evidence for the humans themselves.

Acknowledgments I wish to thank Carl Phillips (CNRS) and Dr. Holger Hitgen (German Archaeological Institute, in Sana') for discussing various aspects of population in southwest Arabia; Dr. Krista Lewis (University of Arkansas, Little Rock) for providing the latest counts on site numbers from the Dhamar Project; Dr. Adrian Parker (Oxford Brookes University, for supplying the originals for Fig. 2). Special thanks go to Dr. Christopher Edens (American Institute of Yemeni Studies), Professor McGuire Gibson (The Oriental Institute Chicago), and Dr Yusuf Abdullah (General Organization of Antiquities and Museums) for advice and discussion during fieldwork in Yemen.

References

Abdulfattah K. Mountain farmer and Fellah in Asir, Southwest Saudi Arabia. Erlangen: Erlanger Geographische Arbeiten Sonderban 12; 1981.

Amirkhanov H. Research on the Palaeolithic and Neolithic of Hadramaut and Mahra. Arabian Archaeology and Epigraphy. 1994;5:217–28.

Anderson DE, Goudie AS, Parker AG. Global environments through the Quaternary. Oxford: Oxford University Press; 2007.

Avanzini A. 'Centre-periphery' relations in pre-Islamic south Arabia. In: Liverani M, editor. Arid lands in Roman times. Rome: Edizioni All'Insegna Del Giglio; 2003. p. 141–7.

Badre L. Le sondage stratigraphique de Shabwa, 1976–1981. Syria. 1991;68:229–314.

Bailey G. The Red Sea, coastal landscapes, and hominin dispersals. In: Petraglia MD, Rose JI, editors. The evolution of human populations in Arabia: paleoenvironments, prehistory and genetics. The Netherlands: Springer; 2009. p. 15–37.

Bailey GN, Flemming NC, King GCP, Lambeck K, Momber G, Moran LJ, et al. Coastlines, submerged landscapes, and human evolution: the Red Sea basin and the Farasan islands. Journal of Island and Coastal Archaeology. 2007;2(2):127–60.

Barceló M, Kirchner H, Torró J. Going around Zafar (Yemen), the Banu_ Ru 'Ayn field survey: hydraulic archaeology and peasant work. Proceedings of the Seminar for Arabian Studies. 2000;30:27–39.

Beaumont P, Blake GH, Wagstaff JM. The Middle East. A geographical study. London: Wiley; 1976.

Breton J-F. Arabia Felix. Notre Dame, IN: Notre Dame University Press; 1999.

Brinkmann R, Ghaleb AU. Late Pleistocene mollisol and cumulic fluvents near Ibb, Yemen Arab Republic. In: Grolier MJ, Brinkmann R, Blakely JA, editors. The Wadi al-Jubah Archaeological Project, vol 5. Environmental research in support of archaeological excavations in the Yemen Arab Republic, 1982–1987. Washington, DC: The American Foundation for the Study of Man; 1996. p. 251–8.

Brunner U. Geography and human settlements in ancient south Arabia. Arabian Archaeology and Epigraphy. 1997;8:190–202.

Buffa V. The stratigraphic sounding at Ma 'layba, Lahj Province, Republic of Yemen. Archäologische Berichte aus dem Yemen Bd. 9. Mainz: Verlag Philipp von Zabern; 2002. p. 1–7.

Bulgarelli GM. Evidence of Palaeolithic industries in northern Yemen. In: Daum W, editor. Yemen: 3000 years of art and civilization in Arabia Felix. Frankfurt: Umschau-Verlag; 1987. p. 32–3.

Burns SJ, Fleitmann D, Matter A, Neff U, Mangini A. Speleothem evidence from Oman for continental pluvial events during interglacial periods. Geology. 1998;29(7):623–6.

Burstein SM, translator and editor. Agatharchides of Cnidus: on the Erythraean Sea. London: The Hakluyt Society, 1989.

Caton-Thompson G. Some palaeoliths from south Arabia. Proceedings of the Prehistoric Society. 1953;19:189–218.

Ciuk C, Keall E. Zabid Pottery Manual 1995. Oxford: BAR International Series 655; 1996.

Clemens S, Prell W, Murray D, Shimmield G, Weedon G. Forcing mechanisms of the Indian Ocean monsoon. Nature. 1991;353:720–5.

Cleuziou S, Tosi M. Hommes, climates et environnements de la Péninsule arabique à l'Holocène. Paléorient. 1998;23(2):121–35.

Cleuziou S, Inizan M-L, Marcolongo B. Le peuplement pré- et protohistorique du système fluviatile fossile du Jawf-Hadramawt au Yemen (d'apres l'interpretation d'images satellite, de photographes aeriennes et de prospections). Paléorient. 1992;18(2):5–29.

Davies CP. Holocene paleoclimates of southern Arabia from lacustrine deposits of the Dhamar highlands, Yemen. Quaternary Research. 2006;66:454–64.

de Maigret A. A Bronze Age for southern Arabia. East and West. 1984;34:75–106.

de Maigret A. The Bronze Age culture of Hawlan al Tiyal and al-Hada. Rome: IsMEO; 1990.

de Maigret A. The Arab nomadic people and the cultural interface between the 'Fertile Crescent' and 'Arabia Felix'. Arabian Archaeology and Epigraphy. 1999;10:220–4.

de Maigret A. Arabia Felix: an exploration of the archaeological history of Yemen. London: Stacey International; 2002.

DHV. Environmental profile Tihama. Amersfoort, the Netherlands: DHV Consultants; 1990.

Doe B. Southern Arabia. London: Thames & Hudson; 1971.

Durrani N. The Tihamah coastal plain of south-west Arabia in its regional context: c.6000 BC–AD 600. Oxford: Society for Arabian Studies Monograph No. 4. BAR International Series 1456, 2005.

Edens C. Before Sheba. In: Simpson St. John, editor. Queen of Sheba: treasures from ancient Yemen. London: British Museum Press; 2002. p. 80–85.

Edens C. Exploring early agriculture in the highlands of Yemen. In: Sholan AM, Antonini S, Arbach M, editors. Archaeological and historical studies in honour of Yusuf M. Abdullah, Alessandro de Maigret, and Christian Robin. Sana' and Naples: University of Naples: Sabaen Studies; 2005. p. 185–211.

Edens C, Wilkinson TJ. Southwest Arabia during the Holocene: recent archaeological developments. Journal of World Prehistory. 1998; 12(1): 55–119.

Edens C, Wilkinson TJ, Barratt G. Hammat al-Qa: an early town in southern Arabia. Antiquity. 2000;74:854–62.

el-Nakhal HA. The Pleistocene cold episode in the Republic of Yemen. Palaeogeography, Palaeoclimatology, Palaeoecology. 1993;100:303–7.

Fedele FG. Man, land and climate: emerging interactions from the Holocene of the Yemen highlands. In: Bottema S, Entjes-Nieborg G, Van Zeist W, editors. Man's role in the shaping of the Eastern Mediterranean landscape. Rotterdam and Brookfield: Balkema; 1990. p. 31–42.

Fedele FG, Zaccara D. Wadi al-Tayyila 3: a mid-Holocene site on the Yemen plateau and its lithic collection. In: Sholan AM, Antonini S, Arbach B, editors. Sabaean studies. Archaeological, epigraphical and historical studies in honour of Yusuf M 'Abdullah, Alessandro de Maigret, and Christian J. Robin. Naples and San 'a: University of Naples; 2005. p. 213–45.

Fleitmann D, Burns SJ, Mudelsee M, Neff U, Kramers J, Mang ini A, et al. Holocene forcing of the Indian monsoon recorded in a stalagmite from southern Oman. Science. 2003;300:1737–9.

French C. Geoarchaeology in action. London: Routledge; 2003.

Ghalab AO. Agricultural practices in ancient Radman and Wadi al-Jubah (Yemen). Ph.D. dissertation, University of Pennsylvania, Pennsylvania; 1990.

Ghaleb AO. Bronze Age sites in Bidbida, the Northeast highlands of Yemen. In: Sholan AM, Antonini S, Arbach M, editors. Archaeological and historical studies in honour of Yusuf M. Abdullah, Alessandro de Maigret, and Christian Robin. Sana' and Naples: University of Naples: Sabaen Studies; 2005. p. 279–94.

Gibson M, Wilkinson TJ. The Dhamar plain, Yemen: A preliminary study of the archaeological landscape. Proceedings of the Seminar for Arabian Studies. 1995;25:159–83.

Glanzman W, Ghaleb AO. The stratigraphic probe at Hajar ar-Rayhani. In: Glanzman WD, Ghaleb AO, editors. The Wadi al-Jubah archaeological project vol. 3: site reconnaissance in the Yemen Arab Republic, 1984: the stratigraphic probe at Hajar ar-Rayhani. Washington, DC: The American Foundation for the Study of Man; 1987. p. 5–64.

Groom N. Frankincense and Myrrh: a study of the Arabian incense trade. London: Longmans; 1981.

al-Hamdani A, Ibn Ahmad H. The antiquities of south Arabia. Translated by N.A. Faris. Princeton, NJ: Princeton University Press; 1938.

Hehmeyer I. Physical evidence of engineered water systems in MedievalZabid. Proceedings of the Seminar for Arabian Studies. 1995;25:45–54.

Hehmeyer I, Keall E. Water and land management in the Zabid hinterland. Al-'Usur al-Wusta (The Bulletin on Middle East Medievalists). 1993;5(2):25–27.

Hoyland RG. Arabia and the Arabs: from the Bronze Age to the coming of Islam. London: Routledge; 2001.

Kallweit H. Neolithische und Bronzezeitliche Besiedlungen im Wadi Dhahr, Republic Yemen. Eine Unterforschung auf der Basis vom Geländegebehungen und Sondagen, Inaugural dissertation, Albert-Ludwigs Universität, Freiburg; 1996.

Keall EJ. Possible connections in antiquity between the Red Sea coast of Yemen and the Horn of Africa. In: Lunde P, Porter A, editors. Trade and travel in the Red Sea region. Proceedings of the Red Sea Project, vol. 1 (Society for Arabian Studies Monograph 2). Oxford: BAR International Series 1269; 2004. p. 43–55.

Khalidi L. The prehistoric and early historic settlement patterns of the Tihamah coastal plain (Yemen): preliminary findings of the Tihamah Coastal Survey 2003. Proceedings of the Seminar for Arabian Studies. 2005;35:115–27.

Lankester Harding G. Archaeology in the Aden protectorates. London: HMSO; 1964.

Leuschner DC, Sirocko F. The low-latitude monsoon climate during Dansgaard-Oeschger cycles and Heinrich events. Quaternary Science Reviews. 2000;19:243–54.

Lewis K. The Himyarite site of al-Adhla and its implications for the economy and chronology of Early Historic highland Yemen. Proceedings of the Seminar for Arabian Studies. 2005;35: 129–41.

Lézine A-M, Saliège J-F, Robert C, Wertz F, Inizan ML. Holocene lakes from Ramlat as-Sab'atyn (Yemen) illustrate the impact of monsoon activity in southern Arabia. Quaternary Research. 1998;50:290–9.

Lézine A-M, Tiercelin J-J, Robert C, Saliège JF, Cleuziou S, Inizan ML, et al. Centennial to millennial-scale variability of the Indian monsoon during the Early Holocene from a sediment, pollen and isotope record from the desert of Yemen. Palaeogeography, Paleoclimatology, and Paleoecology. 2007;243:235–49.

Macumber PG, Al-Said B, Kew G, Tennakoon T. Hydrogeological implications of a cyclonic rainfall event in central Oman. In: Nash H, McCall G, editors. Groundwater quality. London: Chapman & Hall; 1994. p. 87–97.

McClure HA. Radiocarbon chronology of Late Quaternary lakes in the Arabian desert. Nature. 1976;263:755–6.

McCorriston J. Early settlement in Hadramawt: preliminary report on prehistoric occupation at Shi'b Munayder. Arabian Archaeology and Epigraphy. 2000;11:129–53.

McCorriston J, Oches E. Two Early Holocene check dams from southern Arabia. Antiquity. 2001;75:675–6.

McCorriston J, Harrower M, Oches E, Bin-'Aqil A. Foraging economies and population in the Middle Holocene highlands of southern of southern Yemen. Proceedings of the Society for Arabian Studies. 2005;35:143–54.

Munro RN, Wilkinson TJ. Environment, landscapes and archaeology of the Yemeni Tihāmah. In: Starkey J, Starkey P, Wilkinson TJ, editors. Natural resources and cultural connections in the Red Sea. Oxford: BAR International Series 1661; 2007. p. 13–33.

Naval Intelligence Division (N.I.D.). Western Arabia and the Red Sea. Geographical Handbook Series B.R. 527. London: Naval Intelligence Division; 1946.

Niebuhr C. Travels through Arabia and other countries to the east. Translated by R. Heron, 1969, Edinburgh: R Morison & Son; 1792.

Orchard J. Finding the ancient sites in southern Yemen. Journal of Near Eastern Studies. 1982;41:1–21.

Overstreet WC, Grolier MJ. Summary of environmental background for the human occupation of the al-Jadidah basin in Wadi al-Jubah, Yemen Arab Republic. In: Grolier MJ, Brinkmann R, Blakely JA, editors. Environmental research in support of archaeological investigations in the Yemen Arab Republic, 1982–1987. Washington, DC: American Foundation for the Study of Man; 1996. p. 337–429.

Parker AG, Davies C, Wilkinson TJ. The early to mid-Holocene moist period in Arabia: some recent evidence from lacustrine sequences in eastern and south-western Arabia. Proceedings of the Seminar for Arabian Studies. 2006;36:243–55.

Phillips CS. The Tihamah c.5000 to 500 BC. Proceedings of the Seminar for Arabian Studies. 1998;28:233–7.

Phillips CS. A preliminary description of the pottery from al-Hāmid and its significance in relation to other pre-Islamic sites on the Tihāmah. Proceedings of the Seminar for Arabian Studies. 2005;35:177–93.

Quinn O. Regional hydrogeological evaluation of the Najd, Sultanate of Oman. Muscat, Oman: Public Authority for Water Resources; 1986.

Roberts N. Lake levels as an indicator of near eastern palaeoclimates: a preliminary appraisal. In: Bintliff JL, van Zeist W, editors. Palaeoclimates, palaeoenvironments and human communities in the Eastern Mediterranean Region in later prehistory. Oxford: BAR International Series 133; 1982. p. 235–67.

Rose JI. The question of Upper Pleistocene connections between East Africa and South Arabia. Current Anthropology. 2004;45(4):551–5.

Schippmann K. Ancient South Arabia: from the Queen of Sheba to the advent of Islam. Princeton, NJ: Markus Wiener Publications, Princeton University Press; 2001.

Sedov AV. On the origins of the agricultural settlements in Hadramawt. In: Robin CJ, editor. Arabia antiqua. Early origins of South Arabian states. Serie Orientale Roma LXX, 1. Rome: Instituto Italiano per il Medio Estremo Oriente; 1996. p. 67–86.

Sirocko F. The evolution of the monsoon climate over the Arabian sea during the last 24000 years. Palaeoecology of Africa. 1996;24:53–69.

Steffen H. A contribution to the population geography of the Yemen Arab Republic. Wiesbaden, Switzerland: Dr. Ludwig Reichert; 1979.

Tosi M. Archaeological activities in the Yemen Arab Republic, 1985: Tihamah coastal archaeological survey. East and West. 1985;35: 363–9.

Tosi M. Archaeological activities in the Yemen Arab Republic, 1986: Neolithic and protohistoric cultures. Survey and excavation on the coastal plain (Tihamah). East and West. 1986;36:400–14.

Uerpmann M, Uerpmann H-P, Jasim SA. Stone Age nomadism in SE Arabia: Palaeo-economic considerations of Al-Buhais 18 in the Emirate of Sharjah, U.A.E. Proceedings of the Seminar for Arabian Studies. 2000;30:229–34.

Van Beek G. Hajar Bin Humeid, investigations at a pre-Islamic site in South Arabia. Baltimore, MD: The Johns Hopkins Press; 1969.

Vogt B. The great dam: Eduardo Glaser and the chronology of ancient irrigation in Ma'rib. In: Sholan AM, Antonini S, Arbach M, editors. Sabaen studies: archaeological, epigraphical and historical studies in Honour of Yusuf M. Abdullah, Alessandro de Maigret and Christian J. Robin. Naples-San'a: Universita degli Studi di Napoli; 2005. p. 505–7.

Vogt B, Sedov A. The Sabir culture and coastal Yemen during the second millennium BC: the present state of the discussion. Proceedings of the Seminar for Arabian Studies. 1998;28:261–70.

Vogt B, Buffa V, Brunner U. Ma 'layba and the Bronze Age irrigation in coastal Yemen. Archäologische Berichte aus dem Yemen Bd. 2002;9:15–26.

Wenner MW. The Yemen Arab Republic. Development and change in an ancient land. Boulder, CO: Westview Press; 1991.

Whalen N, Pease DW. Archaeological survey in SW Yemen 1990. Paleorient. 1991;17(2):127–31.

Wilkinson TJ. Holocene environments of the high plateau, Yemen. Recent geoarchaeological investigations. Geoarchaeology. 1997; 12(8):833–64.

Wilkinson TJ. Settlement, soil erosion and terraced agriculture in highland Yemen: a preliminary statement. Proceedings of the Seminar for Arabian Studies. 1999;29:183–91.

Wilkinson TJ. Archaeological landscapes of the Near East. Tucson, AZ: University of Arizona Press; 2003a.

Wilkinson TJ. The organization of settlement in highland Yemen during the Bronze and Iron Ages. Proceedings of the Seminar for Arabian Studies. 2003b;33:157–68.

Wilkinson TJ. Soil erosion and valley fills in the Yemen highlands and southern Turkey: integrating settlement, geoarchaeology and climate change. Geoarchaeology. 2005;20(2):169–92.

Wilkinson TJ. From highland to desert: the organization and landscape and irrigation in southern Arabia. In: Stanish C, editor. Agriculture, polity and society: agricultural intensification and socio-political organization. Los Angeles: UCLA, Cotsen Institute Publications; 2006. p. 38–68.

Wilkinson TJ, Edens C. Survey and excavation in the central highlands of Yemen: results of the Dhamar Survey Project 1996–1998. Arabian Archaeology and Epigraphy. 1999;10:1–33.

Wilkinson TJ, Edens C, Gibson M. The archaeology of the Yemen high plains: a preliminary chronology. Arabian Archaeology and Epigraphy. 1997;8:99–142.

Wood WW, Imes JL. How wet is wet? Precipitation constraints on late Quaternary climate in the southern Arabian peninsula. Journal of Hydrology. 1995;164:263–8.

Y.A.R. Map of population distribution, administrative division and land use in the Yemen Arab Republic, by U. Geiser and H. Steffen, Swiss Technical Co-operation Services. San'a: Berne and Central Planning Organization; 1977.

Zarins J. The land of incense: archaeological work in the Governorate of Dhofar: Sultanate of Oman 1990–1995. Muscat, Oman: Ministry of Information; 2001.

Zarins J, Badr H. Archaeological investigations in the Tihama plain II 1405/1985. Atlal. 1986;10:36–57.

Zarins J, Zahrani A. Recent archaeological investigations in the southern Tihama Plain. The sites of Athar and Sihi, 1404/1984. Atlal. 1985;9:65–107.

Zeuner FE. "Neolithic" sites from the Rub al-Khali, southern Arabia. Man. 1954;54:133–6.

Zonneveld KAF, Ganssen G, Troelstra S, Versteegh G, Visscher H. Mechanisms forcing abrupt fluctuations of the Indian ocean summer monsoon during the last glaciation. Quaternary Science Reviews. 1997;16:187–201.

Chapter 5
Mitochondrial DNA Structure of Yemeni Population: Regional Differences and the Implications for Different Migratory Contributions

Jakub Rídl, Christopher M. Edens, and Viktor Černý

Keywords MtDNA • Macrohaplogroups • Paleoclimate • Yemen

Introduction

Yemen, in the southwestern corner of Arabian peninsula, lies at the crossroads between Africa and Eurasia. Genomes of present-day Yemenis were inherited from their progenitors, and may attest to the history of the region. Molecules of DNA can, therefore, shed light on how busy this crossroads was during the past millennia. Unfortunately, Yemeni populations have been neglected in genetic literature until recently. However, from the genetic point of view, there are several important questions that cannot be addressed without detailed genetic data. Do the present-day populations of southern Arabia contain genetic traces testifying to the first migration Out-of-Africa? Can such traces survive until today? What subsequent population movements may have affected the Yemeni gene pool? What is the proportion of more ancient (Pleistocene) and more recent (Holocene) population impacts to its genetic diversity? Did the specific geographic position of Yemen influence the genetic structure of its population?

This chapter provides a review of published mitochondrial DNA data from Yemeni populations within the archaeological and paleoclimatological context. Further the implications for estimations of past migratory events as well as for future prospects are discussed.

Background

The potential of new developments in molecular biology associated with rapid characterization of DNA sequences was quickly recognized, and during the past several decades the technique has been applied to evolutionary studies. While the initial research of classic chromosomal markers relied on the comparison of allele frequencies among populations (Cavalli-Sforza et al., 1994), later analyses have dealt mainly with the sequential differences of haploid and uniparentally inherited markers. Diploid nuclear genomes consist of blocks with different histories resulting from the complex processes of recombination between maternal and paternal chromosomes (Pääbo, 2003). In contrast, mitochondrial DNA (henceforth referred to as mtDNA) is passed down to offspring almost exclusively from mothers (Sutovsky et al., 1999, 2004; Schwartz and Vissing, 2002, 2003), thus reflecting the genetic history of maternal lineages. Occasionally, this record is slightly altered by mutation. It has been shown that the mutation rate is about tenfold higher for the mtDNA molecule than for the nuclear genome (Brown et al., 1979). Moreover, mutation rate differs within mtDNA molecule and is the most pronounced in two non-coding parts called hypervariable segments (HVS-I and HVS-II) (Meyer and von Haeseler, 2003). DNA sequences with a particular set of mutations are called haplotypes. Mutations that arise within mtDNA lineages lead to branching pattern of mtDNA phylogeny. All haplotypes sharing a common ancestral type belong to a monophyletic clade called a haplogroup. Haplogroups are defined by a particular set of mutations and use to be more or less geographically specific. Thus, for example, we call L-haplogroups the sub-Saharan specific mtDNA clades because of their higher frequencies and variations in sub-Saharan Africa.

J. Rídl
Institute of Molecular Genetics of the Academy of Sciences of the Czech Republic, Prague, v.v.i., Vídeňská 1083, Prague 4, CZ-14220, Czech Republic
e-mail: ridlj@img.cas.cz

C.M. Edens
American Institute for Yemeni Studies,
P.O. Box 2658, Sana'a, Republic of Yemen
e-mail: aiysyem@y.net.ye

V. Černý (✉)
Institute of Archaeology of the Academy of Sciences of the Czech Republic, Prague, v.v.i., Letenská 4, Prague 1, CZ-11801, Czech Republic; Department of Anthropology and Human Genetics, Faculty of Science, Charles University in Prague, Prague 128 00,
e-mail: cerny@arup.cas.cz

M.D. Petraglia and J.I. Rose (eds.), *The Evolution of Human Populations in Arabia*, Vertebrate Paleobiology and Paleoanthropology, DOI 10.1007/978-90-481-2719-1_5, © Springer Science+Business Media B.V. 2009

MtDNA data place the root of all existing lineages to East Africa (Cann et al., 1987; Vigilant et al., 1991; Watson et al., 1997; Ingman et al., 2000), being the likely origin of global genetic diversity (Reed and Tishkoff, 2006). The first branching is dated to ca. 180 ka giving rise to haplogroup L0 (Torroni et al., 2006, Behar et al., 2008). At about 80 ka haplogroup L3 emerged in East Africa, and its two derivatives M and N were soon spread by human migrations to the rest of the world (Watson et al., 1997; Quintana-Murci et al., 1999; Maca-Meyer et al., 2001).

When other genetic markers are also taken into account, it seems impossible to rule out other scenarios about the origin of anatomically modern humans (e.g. Templeton, 2002; Eswaran et al., 2005; Garrigan et al., 2005; Plagnol and Wall, 2006). However, the current data suggest a major role of East Africa in the dispersal of modern humans (Relethford and Jorde, 1999; Eswaran et al., 2005; Ray et al., 2005). Although the "northern" migration route of modern humans Out-of-Africa into the Levant was initially hypothesized, recently the "southern" coastal route has been proposed as a more probable pathway for the first successful emigration of Middle Paleolithic African hunter-gatherers (Lahr and Foley, 1998; Stringer, 2000; Metspalu et al., 2004, 2006; Forster and Matsumura, 2005; Macaulay et al., 2005). They carried the haplogroups M and N (or their direct molecular ancestors) from East Africa to Eurasia. From the genetic point of view the southern route is supported mainly by the occurrence of phylogenetically old clades of haplogroup M along the coast of the Indian Ocean; in India (Metspalu et al., 2004), Andaman Islands (Thangaraj et al., 2005) and Malaysian peninsula (Macaulay et al., 2005).

However, the interpretation of the genetic data is not straightforward. Current study of haploid markers has two different approaches – the population-based and phylogenetic approaches (Forster et al., 2001; Pakendorf and Stoneking, 2005). The main difference between them arises from the unit of study. While population-based studies deal with real populations, their weak point is the absence of time perspective. On the other hand, phylogenetic studies use individual haplotypes as analytical units. So called phylogeography attempts to juxtapose the mtDNA phylogenetic scheme with the geographic distribution of clades. The main advantage is that the estimations of mutation rate allow dating the coalescence of haplogroups. However, every population does not contain only one but many haplogroups, so we cannot simply equate the branching of mtDNA lineages with a split of populations, or the coalescence time of haplogroup with a migratory event. Strictly speaking, phylogeography simply reflects the history of haplogroups. That said, it is obvious that mtDNA molecules cannot exist apart from their human vectors and migrating people carry their molecules along. We can therefore consider haplotypes as hitchhikers of human migrations.

Despite its key location for migratory events out of and back into Africa, sequences from Yemen are still quite rare. The first published mtDNA data from Yemenis were analyzed in a broader perspective focused on the branching of mtDNA tree in general (DiRienzo and Wilson, 1991). A subsequent Yemeni mtDNA dataset appeared in the extensive study published by Richards et al. (2000) that dealt with the question of Paleolithic versus Neolithic contributions to the colonization of Europe. On the other hand, Thomas et al. (2002) analyzed mtDNA sequences from Yemeni Jews in the study of matrilineal history of several Jewish communities. Another paper by Richards et al. (2003), based on the already published data, showed the different contribution of sub-Saharan mtDNA lineages to Arab and non-Arab populations of Middle East. The first detailed phylogeographic study of mtDNA haplogroups among Yemenis was aimed at estimating the amount of gene flow across Bab al Mandab strait between Ethiopia and Yemen (Kivisild et al., 2004). While Ethiopia was sampled on a finer level, the Yemeni samples were secured in Kuwait from donors who claimed their Yemeni origin (Kivisild et al., 2004). Recently, Rowold et al. (2007) sampled several populations from the Arabian peninsula (including 50 Yemeni samples without geographical specifications), Middle East and Africa in order to estimate the relative roles of Levantine versus Bab al Mandab strait corridors in different past migrations. The apparent lack of detailed genetic data from Yemen prompted us to investigate the mtDNA diversity within the country. The most recent studies of Černý et al. (2008, 2009), therefore, represents the first regional based sampling of four Yemeni populations.

Paleoclimatological and Archaeological Context

Yemen occupies the southwest corner of the Arabian peninsula, the Red Sea forming its western, and the Arabian Sea its southern, border. The Red Sea, a rifting feature that attains depths of 2500 m, is about 200–350 km wide through most of its length, but the Bab al Mandab at its southern end is only 26 km across – the island of Perim, just off the Yemeni coast, reduces the distance to only 19 km – and over 200 m deep. The strait was unlikely to have been dry even during glacial low sea-stands (Siddall et al., 2003; Fernandes et al., 2006).

Yemen's landscape is extremely diverse. Coastal plains frame the country to the west and south; Tihama (the western plain) and adjoining portions of the southern plain are 30–50 km wide, but the plain shrinks to a narrow fringe to the east. The western mountains rise abruptly from Tihama, generally reaching 2,000–3,000 m asl. This zone is typically rugged, with deeply incised drainage courses, but in-filled block-faults form large highland plains in many places. The western mountains lie at the southern end of a range that extends northward, with diminishing elevation, to Jordan. Uplands, decreasing in elevation from 1,500 asl in the west to 1,000 asl in the east but punctuated with higher peaks, face the Arabian Sea. Large sections of these southern uplands are tableland deeply incised by water courses, including the Wadi Hadramawt and its tributaries. The western mountains and southern uplands frame flatter country that trends downward to the northeast, dropping from 1,200 m asl near Ma'rib to sea level at the Persian Gulf; the Rub' al Khali sand sea (mostly in Saudi Arabia) occupies a large portion of this region, with the Ramlat al-Sab'atayn in Yemen forming an outlying sand sea entirely within Yemen.

Climate is equally diverse. Yemen today enjoys the effects of the southwest monsoon. The western mountains receive two seasons of rainfall that can average over 100 cm a year (decreasing from south to the north, and west to east), while eastern Mahrah and Dhofar (the adjacent portion of Oman) lie within the monsoon belt proper, the summer mists of which support forest growth on the slopes between the coast and the uplands. Other sections of the country are drier. The Hadramawt interior can receive 20 cm of rain in a year, but coastal plains and the interior desert are arid, with average annual rainfall on the order of 5 cm or less.

Past climates responded to changing north–south position and strength of the southwest monsoon. Abundant terrestrial evidence – radiocarbon dated lake and marsh deposits, colluvial and alluvial sequences, isotopic analysis of stalagmites, and other indicators – indicate moister conditions at 12–5.5 ka not only in the western highlands but also in Hadramawt, the Ramlat al-Sab'atayn and the Rub' al Khali, and other sections of the southern Arabian peninsula (McClure, 1976; Lézine et al., 1998; Neff et al., 2001; Wilkinson, 2005). Some terrestrial evidence from the Rub' al Khali and Dhofar also points to a moister period at ca. 36–17 ka; a paleosol in the western highlands is radiocarbon dated to >33,100 ka (Nettleton and Chadwick, 1996). But coherent older evidence is not available.

The terrestrial evidence for the younger moist episode corresponds to marine evidence for strengthening of the monsoon ca. 10–12 ka (Prell and Campo, 1986; Prell and Kutzbach, 1987; Sirocko et al., 1993; Sirocko, 1996; Zonneveld et al., 1997). The older episode finds no correspondence with the marine evidence. Instead, the marine evidence from the western Arabian Sea indicates earlier strengthened monsoons at ca. 80, 100, and 120 ka, with weak monsoons and pervasively arid conditions during ca. 60–12 ka. Severe desiccation associated with the Last Glacial Maximum during the Oxygen Isotope Stage 2 (24–12 ka) might have resulted in most of the peninsula becoming uninhabitable. It is crucial to note that population bottlenecks assumed during

the hyper arid phases may cause discontinuity in the genetic record and lead to the loss of mtDNA lineages making the ancient genetic traces of human settlements and migrations invisible (for review of oscillations in paleoclimatic conditions and their possible effects on human occupation of South Arabia see Rose, 2007).

The archaeological evidence for hominin occupation is not yet well developed for southern Arabia. Oldowan-related finds are reported from Tihama and Hadramawt, although some doubt may be entertained about the accuracy of these reports; Acheulian-related discoveries, identified mainly by the presence of handaxes, are somewhat more common (see Petraglia, 2003 for an overview). Middle Paleolithic sites – mostly surface scatters, but occasionally with some geomorphological context – are relatively widespread in southwest Arabia and elsewhere in the western and northern sections of the peninsula. The industry (or industries), assumed to start ca. 150–200 ka, have not been deeply studied, and the usual frame of reference is Levantine (or western European). Levallois techniques have a wide distribution across the peninsula, but Levallois cores are less common in individual 'assemblages' than discoidal cores or less formal flake production. Technological and typological descriptions remain general, and largely uninformative of geographical variation or of cultural links with adjacent regions, including northeast Africa (see Petraglia and Alsharekh, 2003 for an overview). For present purposes two studies have potential relevance. Aterian-like materials are reported from the Rub' al Khali (McClure, 1994), implying important technological and cultural links with North Africa. Arabian materials have been (broadly) compared to Middle Stone Age industries of east Africa, and a correlation drawn between Middle Stone Age industries and anatomically modern *H. sapiens* to propose the presence of the latter in southern Arabia (Zarins, 1998).

Identification of Late Pleistocene sites, corresponding in time (if not also in technology) to the Upper Paleolithic and Epipaleolithic of the Levant, remains contentious in Yemen and much of the Arabian peninsula: some researchers report an Upper Paleolithic blade technology (e.g. Amirkhanov, 1994) and, tentatively, an Epipaleolithic bladelet technology (e.g. Edens, 2001) in or near Yemen, while others see a largely unoccupied landscape attributable to hyperaridity, or a prolongation in time of Middle Paleolithic/Middle Stone Age technologies.

Evidence for human occupation in southern Arabian again becomes both widespread and incontrovertable only with Early and Middle Holocene (ca. 9–5 ka) sites. These sites, correlated in time with the strengthened monsoon, are widespread in southern Arabia and the rest of the peninsula. Although the currently limited archaeological visibility of human occupation after ca. 50–60 ka makes difficult assessing relative population numbers through time, a sharp increase – suggestive of immigration as well as 'natural' increase – seems probable in the Early Holocene. While the material culture of these groups is largely irrelevant to the present context, several features do entail far-ranging contacts with other regions. Chemical characterization studies suggest that much of the obsidian found in Early and Mid-Holocene archaeological sites in Yemen, and especially in Tihama, originated from as yet unidentified sources in Ethiopia and Eritrea (Francaviglia, 1995). Herding of cattle, along with some sheep/goats, figured increasingly among economic activities after 8 ka. These domesticated species were introduced from the Levant along the western Arabian uplands (or, possibly, via northeast Africa); early cattle and perhaps other domestics in southeast Arabia likely had a South Asian origin.

According to present evidence, prehistoric communities in Yemen adopted agriculture sometime between 5.5–5.0 ka, at least in the western highlands (Edens, 2005). The crop species were again mostly Levantine in origin – wheat, barley, lentils, chickpeas, etc. – that arrived via uncertain routes. East African species, notably sorghum and millets, also figured in the early crop mix perhaps by 4.5 ka and definitely by 3 ka (de Moulins et al., 2003). Agriculture appeared on the Yemeni scene at the beginning of the Bronze Age (ca. 5–3 ka), when Neolithic groups formed larger and more sedentary communities and also began producing pottery. The pottery, and especially its impressed decoration, bears generic family resemblances to roughly contemporary wares in northeast Africa and some scholars argue for Yemeni participation in a cultural network that spanned the Red Sea (e.g. Fattovich, 1997; Buffa and Vogt, 2001).

The local Bronze Age set the stage for the emergence of the South Arabian states, during the several centuries on each side of 1000 BC. In one view, based on a miscellany of linguistic, artistic and archaeological evidence, the catalyst for heightened social complexity was the arrival of groups from the north (Müller, 1988; Sedov, 1996); other interpretations look to factors stimulating secondary state formation among existing communities (e.g., Edens and Wilkinson, 1998). Whatever the causes of its origin, the South Arabian civilization produced written records, from ca. 800 BC onward, that inscribed Yemen in history. These and later sources provide rich, if not always unambiguous, testimony of Yemen's more recent interactions with populations in neighboring and more distant regions (very briefly summarized in Černý et al., 2008).

Geographic Affinities of Yemeni Populations

The first population comparison of mtDNA samples from Yemenis with neighboring populations showed their affinity to Egyptians and Ethiopians, occupying a position between the Middle Eastern and African populations (Kivisild et al.,

2004). This result seemed to correspond well with Yemen's geographical position. When data became available from sampling of different areas within Yemen (Černý et al., 2008), an interesting pattern emerged. Despite western Yemen's geographical and cultural proximity to East Africa, the western populations from Ta'izz, Tihama and Hajja cluster together with Middle Eastern and North African samples (Fig. 1). On the other hand, the population from Hadramawt to the east shows affinity to the East African populations (Černý et al., 2008). Not only does the given pattern indicate the gene flow from populations in Africa, West Eurasia and South Asia, but even more interestingly it shows that these influences are differently reflected in different regional samples.

It is worth noting that the eastern sample exhibits higher values of nucleotide diversity and pairwise differences compared to other Yemeni populations. These statistics indicate the presence of more diversified mtDNA lineages. Hadramawt has the second highest values in comparison with 37 populations from Africa, Middle East and India and falls closely to populations from East Africa in this respect (Černý et al., 2008). Thus, we can hypothesize possible scenarios where the genetic affinity of Hadramawt with East Africa reflects gene flow from Africa to South Arabia. However, it has been suggested that mere population-based comparisons are not an appropriate tool for tracing a biological origin of a population such as Yemen's, to which multiple human migrations must certainly have contributed (Richards et al., 2003; Kivisild et al., 2004). Traces of past migration may disappear due to another prevalent demographic event. The only way to analyse the different contributions involves a phylogeographic approach where ancient and derived mtDNA haplotypes are considered.

Geographical Distribution of mtDNA Lineages in Yemen

Substantial proportions of the Yemeni mtDNA gene pool can be assigned to sub-Saharan haplogroups (L-type) on one hand, and West Eurasian haplogroups (derivatives of M and N) on the other hand (Kivisild et al., 2004; Černý et al., 2008). The overall composite nature of Yemeni gene pool also supports its probable role as a recipient of gene flows from different parts of Africa and Eurasia. However, the major haplogroups exhibit different distributions among regional samples (Fig. 2) with lineages specific to sub-Saharan Africa being significantly more frequent in Hadramawt (60.0%) than in the western Yemeni populations where the frequency gradually decreases from Hajja in the north (34.3%) through Tihama (28.4%) to Ta'izz in the south (16.3%); the opposite is true for West Eurasian lineages (Černý et al., 2008).

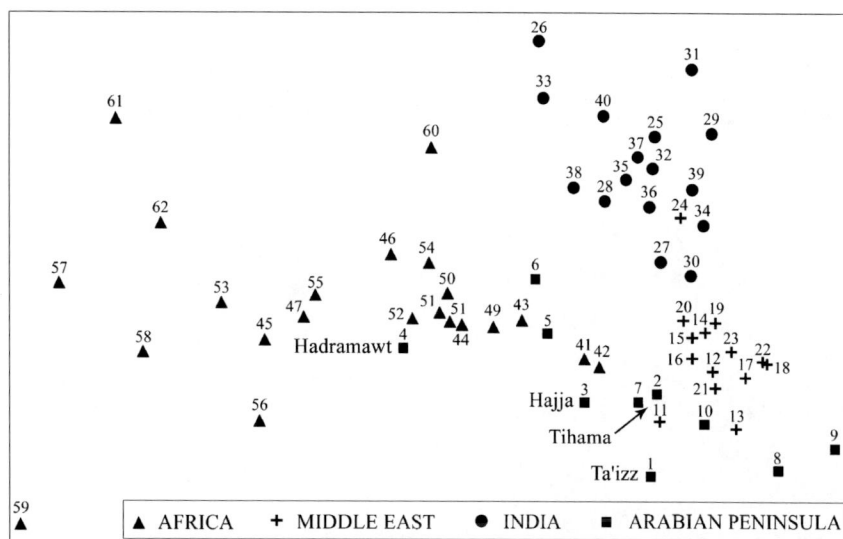

Fig. 1 Population-based comparison by means of multidimensional scaling of F_{ST} distances between 64 population samples from Arabian peninsula, Middle East, India and Africa. (1) Ta'izz, Yemen (Černý et al., 2008); (2) Tihama, Yemen (Černý et al., 2008); (3) Hajja, Yemen (Černý et al., 2008); (4) Hadramawt, Yemen (Černý et al., 2008); (5) Yemenis-Hadramawt, Yemen (Thomas et al., 2002); (6) Yemenis, Yemen (Kivisild et al., 2004); (7) Bedouins, Saudi Arabia (DiRienzo and Wilson, 1991); (8) Yemeni Jews, Yemen (Thomas et al., 2002); (9) Yemeni Jews, Yemen (Richards et al., 2000); (10) Saudi Arabs, Saudi Arabia (Abu-Amero et al., 2007); (11) Druze, Israel (Macaulay et al., 1999); (12) Iraqi, Iraq (Richards et al., 2000); (13) Iraqi, Iraq (Al-Zahery et al., 2003); (14) Kurds, Eastern Turkey (Richards et al., 2000); (15) Palestinians, Israel (Richards et al., 2000); (16) Syrians, Syria (Richards et al., 2000); (17) Northern Syrians, Syria (Vernesi et al., 2001); (18) Turks, Turkey (Richards et al., 2000); (19) Jordanians, Jordan (Richards et al., 2000); (20) Iran middle, Iran (Metspalu et al., 2004); (21) Iran northeast, Iran (Metspalu et al., 2004); (22) Iran northwest, Iran (Metspalu et al., 2004); (23) Iran southwest, Iran (Metspalu et al., 2004); (24) Karachi, Pakistan (Quintana-Murci et al., 2004); (25) Andhra Pradesh, India (Bamshad et al., 1998); (26) Assam, India (Cordaux et al., 2003); (27) Gujarat, India (Metspalu et al., 2004); (28) Himachal, India (Metspalu et al., 2004); (29) Karnataka, India (Mountain et al., 1995; Cordaux et al., 2003); (30) Kashmir, India (Kivisild et al., 1999);
(31) Kerala, India (Metspalu et al., 2004); (32) Maharashtra, India (Metspalu et al., 2004); (33) Nagaland, India (Cordaux et al., 2003); (34) Punjab, India (Kivisild et al., 1999; Cordaux et al., 2003; Metspalu et al., 2004); (35) Rajasthan, India (Metspalu et al., 2004); (36) Sri Lanka, India (Metspalu et al., 2004); (37) Tamilnadu, India (Roychoudhury et al., 2001; Cordaux et al., 2003); (38) Tripura, India (Roychoudhury et al., 2001); (39) Uttar Pradesh, India (Kivisild et al., 1999; Metspalu et al., 2004); (40) West Bengal, India (Roychoudhury et al., 2001; Metspalu et al., 2004); (41) Lower Egypt, Egypt (Krings et al., 1999); (42) Upper Egypt, Egypt (Krings et al., 1999); (43) Upper Egypt, Egypt (Stevanovitch et al., 2004); (44) Nubians, Sudan and Egypt (Krings et al., 1999); (45) Nuba, Sudan (Krings et al., 1999); (46) Nilotic, Sudan (Krings et al., 1999); (47) Dinka, Sudan (Krings et al., 1999); (48) Amhara, Ethiopia (Thomas et al., 2002); (49) Tigrais, Ethiopia and Eritrea (Kivisild et al., 2004); (50) Amhara, Ethiopia (Kivisild et al., 2004); (51) Oromo, Ethiopia (Kivisild et al., 2004); (52) Gurage, Ethiopia (Kivisild et al., 2004); (53) Turkana, Kenya (Watson et al., 1997); (54) Somali, Kenya (Watson et al., 1997); (55) Kikuyu, Kenya (Watson et al., 1997); (56) Bantu, Mozambique (Salas et al., 2002); (57) Burunge, Tanzania (Gonder et al., 2007); (58) Datog, Tanzania (Gonder et al., 2007); (59) Hadza, Tanzania (Gonder et al., 2007); (60) Sukuma, Tanzania (Gonder et al., 2007); (61) Sandawe, Tanzania (Gonder et al., 2007); (62) Turu, Tanzania (Gonder et al., 2007)

Sub-Saharan Haplogroups

L-haplogroups comprise the most ancient branching of the human mtDNA phylogeny in Africa. It is intriguing that frequency of L-haplogroups in Yemeni populations increases with the increasing distance from Bab al Mandab strait, the natural link between Africa and southern Arabia. The prevalence of L-types in eastern Yemen together with the population affinity observed between Hadramawt and East Africa raises the question of whether this pattern could reflect Paleolithic gene flow(s) from Africa. Phylogeographic analysis has shown that the most frequent L-haplotypes in Yemen belong to ancient clade L0a (Kivisild et al., 2004; Černý et al., 2008). However, it is present in the form of more recent African haplogroups such as L0a1, L0a1a and L0a2. Haplogroups L0a1 and L0a1a are suggested to

have originated in East Africa approximately 33.4 (SE 16.6) ka and 27.4 (SE 18.0) ka, respectively (dating based on HVS-I sequences) (Salas et al., 2002). These coalescence times make pre-Holocene emergence in Yemen still possible. However, several matches of these haplotypes with southeast Africa suggest more recent gene flow to Yemen. Haplogroup L0a2, which covers 17.5% of mtDNA lineages in Hadramawt (Černý et al., 2008), emerged 8.3 (SE 3.7) ka in Central or East Africa (Salas et al., 2002). Its emergence in Yemen must, therefore, post-date the Last Glacial Maximum.

Since the occurrence of substantial portion of L-haplogroups among Yemenis has been ascribed to the Arabian slave trade (Richards et al., 2003; Kivisild et al., 2004; Černý et al., 2008), the same may hold true for L0a lineages. Recent gene flow from southeastern Africa likely occurred during the

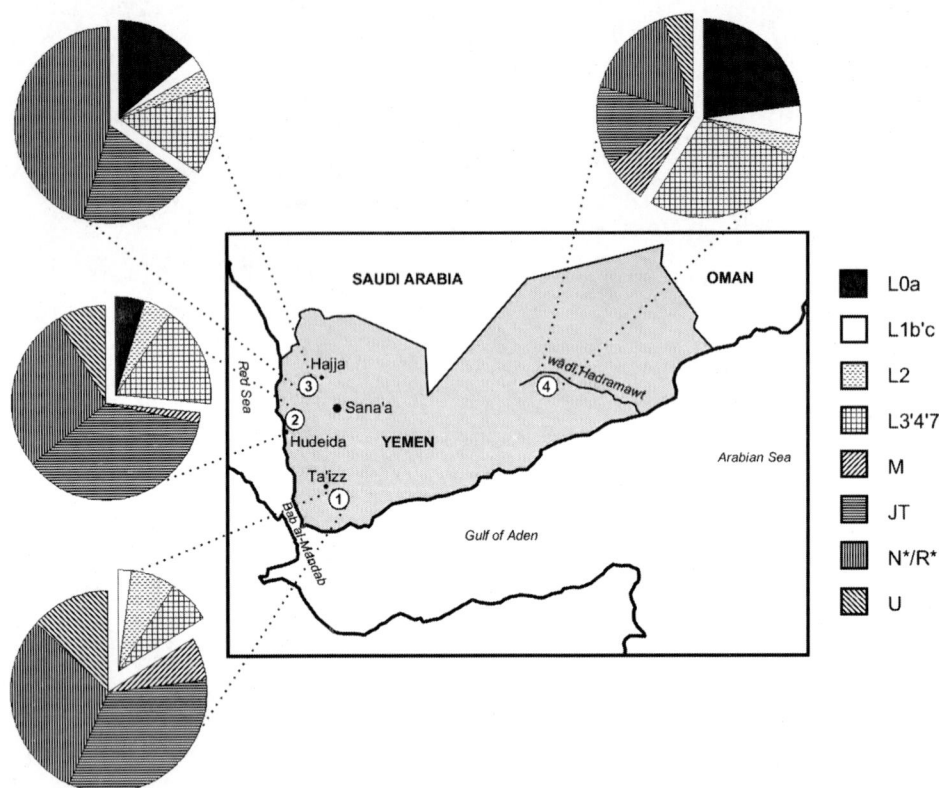

Fig. 2 Regional differences in distribution of main haplogroups among four Yemeni populations. L0a, L1b'c, L2 and L3'4'7 are sub-Saharan haplogroups. M, JT, N'R and U are Eurasian specific haplogroups. Populations are numbered as follows: (1) Ta'izz, (2) Tihama, (3) Hajja, (4) Hadramawt (data from Černý et al., 2008)

last 2500 years, predominantly mediated by women according to comparison of different signals from maternal (mtDNA) and paternal (Y chromosome) gene pools (Richards et al., 2003). The very similar impact of East African slave trade has been observed also in Makrani population from Pakistan, where the frequency of African L-haplogroups are substantially elevated in contrast to Y chromosome variants (Quintana-Murci et al., 2004). Thus, these findings demonstrate the geographic span of the Arabian slave trade from East and Southeast Africa along the south coast of Arabian peninsula and as far as the Indian subcontinent. Interestingly, the Makrani sample also exhibits an elevated value of pairwise differences compared to other populations from Southwest and Central Asia (Quintana-Murci et al., 2004). If at least some L-haplotypes in the Makrani and Yemeni gene pools (especially in Hadramawt) reflect recent gene flow rather then ancient emergence, the higher values of pairwise differences may therefore result from assimilation of female slaves carrying phylogenetically different sub-Saharan lineages. Sex-biased assimilation in favor of women is proving to have significant effect on gene pool of target populations.

Some other sub-Saharan mtDNA lineages have been sampled by one study, but missed by others. For example Černý et al. (2008) show that some haplogroups such as L1b, L1c1, L3b, L3d2 and L3e2b, present in low frequencies in Yemen, are rare in East Africa but frequent in western Africa from where they have likely been spread via North Africa to Yemen. Perhaps the most interesting example is the case of haplogroup L6. It has been described by Kivisild et al. (2004) as the most frequent haplogroup in the sample of Yemeni donors from Kuwait. This lineage derived from the sub-Saharan mtDNA phylogeny more than 20 ka measured by whole genomes comparisons (Behar et al., 2008). Moreover, this clade only rarely occurs in Ethiopia without exact matches with Yemen and it has not yet been observed elsewhere in the world. Kivisild et al. (2004) mentioned a scenario under which the L6 haplogroup might represent a trace of initial migration of anatomically modern humans out of Africa. But that scenario would require strong isolation of given population resulting in absence of L6 haplogroup in other populations. Another explanation of the absence of L6 elsewhere in the world may simply be insufficient sampling to date (Kivisild et al., 2004). Other studies (Richards et al., 2000, 2003; Rowold et al., 2007) as well as the more intense regional sampling (Černý et al., 2008) failed to recover this haplogroup among Yemenis, so the L6 haplogroup remains "[...] an enigmatic link between the southwestern Arabian gene pool with that of East Africa" (Kivisild et al., 2004: 767).

Macrohaplogroups M and N

Macrohaplogroups M and N are the early offshoots from African L3 and encompass virtually all mtDNA diversity in Eurasia, Australia, Oceania and Americas (Watson et al., 1997; Quintana-Murci et al., 1999; Maca-Meyer et al., 2001). As to haplogroup M there is uncertainty whether it has emerged in East Africa or en route out of Africa. The greatest diversity within the M haplogroup, and the oldest coalescence time estimations to date, are reported from India (Krings et al., 1999; Quintana-Murci et al., 1999; Metspalu et al., 2004). The M haplogroup is also frequent in Ethiopia (Quintana-Murci et al., 1999), but only as M1 clade with more recent coalescence time (Kivisild et al., 2004; Olivieri et al., 2006).

Southern Arabia might have acted as a connecting link in the first dispersal out of, or subsequently also in early migrations back to, Africa. In this light, the sampling of Yemeni populations for mtDNA analyses might be of great value for the question of the origin and dispersal of modern humans. Present data, however, show that haplogroup M is present in very low frequencies in Yemen (Rowold et al., 2007; Černý et al., 2008), and represented mainly by recently derived haplotypes with exact or close matches in Indian populations, reflecting a recent gene flow from India (Kivisild et al., 2004; Černý et al., 2008). The only exception is haplogroup M1, observed in two samples (4.7%) from southwestern Yemen (Černý et al., 2008) and also at low frequency in the Yemeni sample of Rowold et al. (2007). However, it should be noted that the haplogroup M1 is quite polymorphic in Yemen (Rowold et al., 2007; Černý et al., 2008). Taken together with an expansion in East Africa estimated to the Middle or early Upper Paleolithic, Rowold et al. (2007) hypothesize the pre-Holocene dispersal via Bab al Mandab, with both directions being equally possible. However, based on recently published analysis of 51 complete M1 sequences, southwestern Asian origin and back migration to Africa through Levant some 40–45 ka seems more likely (Olivieri et al., 2006; see also González et al., 2007).

While the haplogroups derived from M are frequent among populations from South and East Asia, Australia and America, haplogroup N and its early derivative R clade are predominant in western Eurasia (Richards et al., 2000). Together with their derivatives (haplogroups that belong to J, T, K and U) they have been observed in substantial proportion in the overall Yemeni gene pool (Kivisild et al., 2004) with higher frequencies especially in populations of western Yemen (Černý et al., 2008).

Haplogroup J is mainly represented in Yemen by derivatives J1, J1b and J1c when data from Kivisild et al. (2004), Černý et al. (2008) and Rowold et al. (2007) are taken together. The distribution of J1b in the Middle East and its coalescence time, estimated approximately to 15.5 ± 5.0 ka, suggest a Middle Eastern expansion during the time from the Last Glacial Maximum to the early Neolithic, although the exact origin is unresolved (Rowold et al., 2007). This time span possibly comprises the emergence of J1b in Yemen.

The occurrence of J* in five individuals from western Yemen, four of which bear the ancestral motif and show an exact match with one haplotype present both in Ethiopia and Egypt (Černý et al., 2008), is noteworthy. This pattern was not encountered by other studies (e.g. Kivisild et al., 2004; Rowold et al., 2007). Therefore, Rowold et al. (2007) suggest Middle Paleolithic to Neolithic dispersal of J* to Egypt via Levant, while the occurrence of this clade in Ethiopia is left unresolved (Rowold et al., 2007). On the other hand, the J* haplotypes more recently recovered in Yemen make the dispersal to Africa via the Bab al Mandab strait equally possible.

Similar information is provided by haplogroup R0a. This clade – previously known as (preHV)1 (see Torroni et al., 2006 for reclassification) – was initially detected in higher frequencies in Yemeni Jews (Thomas et al., 2002; Richards et al., 2003), but not among Arab populations from Yemen (Richards et al., 2003; Kivisild et al., 2004). Given its coalescence to other Middle Eastern haplogroups, the origin of R0a in the Middle East has been suggested (Richards et al., 2000, 2003). The presence of R0a in Ethiopia and somewhat different motifs of several Ethiopian R0a haplotypes may indicate relatively old migration back to East Africa (Kivisild et al., 2004), although its precise pathway is unclear (Rowold et al., 2007). Recently a phylogeny of 13 complete mtDNA R0a sequences from Saudi Arabia was reconstructed and, taken together with 255 published HVS-I sequences, the coalescence time of about 19 ± 7.0 ka was estimated (Abu-Amero et al., 2007). This date is in agreement with time estimations proposed by Rowold et al. (2007) with one exception of United Arab Emirates sample, which exhibited older time but with somewhat broader interval (37.8 ± 12.1 ka) (Rowold et al., 2007). Since Abu-Amero et al. (2007) showed that most HVS-I sequences of haplogroup R0a from Arabia were likely recent derivatives, they proposed a minor role of the Arabian peninsula in the pre-Holocene diversification of R0a lineages (Abu-Amero et al., 2007). On the other hand, the most recent regional sampling has led to the recovery of the significant proportion of R0a lineages in Yemen, mainly from western parts of the country but also on Soqotra island (Černý et al., 2008, 2009). Thus, the highest frequency of R0a ancestral types reported to date in western Yemeni populations suggests pre-Holocene occurrence of R0a haplogroup in Yemen, and even raises the possibility of the initial expansion in southwestern Arabia and subsequent dispersal around Middle East and westward to Ethiopia via the Bab al Mandab strait. It is interesting that the coalescence time estimates correspond to the culmination of the Last Glacial Maximum, when the extreme aridity in southern Arabia is assumed. However, when the confidence interval is taken into account (approximately 26–12 ka; Abu-Amero et al.,

2007), the slightly later expansion during the climate amelioration is also possible.

Conclusions

Yemen shows complex population history, as one would expect according to its geographical position on an intercontinental crossroads. Regional pattern of mtDNA diversity suggests the gene flow from African, West Eurasian and South Asian populations, but even more interestingly it shows that these influences are differently reflected in different regional samples. On the other hand, the in situ evolution of some lineages and their expansion from Yemen also seems possible, particularly according to the signal from haplogroup R0a.

This review presents the benefits of more intense geographic sampling strategy. The regional mtDNA data clearly show mtDNA differences among Yemeni populations as documented by population comparison and the frequencies of various haplogroups (Černý et al., 2008). Interestingly, western Yemeni populations show elevated frequencies of Eurasian specific lineages and exhibit affinity to Middle Eastern populations, while the eastern sample represented by Wadi Hadramawt predominantly contains sub-Saharan haplogroups and falls closer to East Africa in population and phylogeographic comparison. Thus it disrupts the expectations based on cultural and geographical proximity of western Yemen to East Africa. More detailed sampling may also reveal some previously missed lineages, thus providing greater resolution to the estimations of preferred usage of some migratory pathways.

The majority of mtDNA diversity in Yemen likely postdates the LGM and some specific recent gene flow may result from Arabian slave trade (L-haplogroups) and from contacts with India (M lineages except M1). The available mtDNA data today show no traces of the initial migration(s) out of Africa. These traces might have been erased by population bottlenecks during Late Pleistocene climate deterioration or might have become lost because of more recent and probably stronger gene flows. However, the emergence of some haplogroups in Yemen may coincide with or originate prior to the LGM. Human populations, therefore, could survive during terminal Pleistocene hyperaridity. While the current data are rather tentative, it is possible that additional sampling and whole mtDNA genome sequencing could reveal yet unrecorded traces of ancient migrations.

Acknowledgments We thank Daniel Sosna, for many comments on the manuscript, and Martin Hájek, for help with multidimensional statistical analysis and MDS plotting. The project was supported by the Andrew W. Mellon Foundation through the Council of American Overseas Research Centers and the American Institute for Yemeni Studies, and by the Ministry of Education of the Czech Republic (project KONTAKT ME 917).

References

Abu-Amero KK, Gonzalez AM, Larruga JM, Bosley TM, Cabrera VM. Eurasian and African mitochondrial DNA influences in the Saudi Arabian population. BMC Evolutionary Biology. 2007;7:32.

Al-Zahery N, Semino O, Benuzzi G, Magri C, Passarino G, Torroni A, et al. Y-chromosome and mtDNA polymorphisms in Iraq, a crossroad of the early human dispersal and of post-neolithic migrations. Molecular Phylogenetics and Evolution. 2003;28:458–72.

Amirkhanov K. Research on the Palaeolithic and Neolithic of Hadhramaut and Mahra. Arabian Archaeology and Epigraphy. 1994;5:217–28.

Bamshad MJ, Watkins WS, Dixon ME, Jorde LB, Rao BB, Naidu JM, et al. Female gene flow stratifies Hindu castes. Nature. 1998;395:651–2.

Brown WM, George M, Wilson AC. Rapid evolution of animal mitochondrial DNA. Proceedings of the National Academy of Sciences USA. 1979;76:1967–71.

Buffa V, Vogt B. Sabir – cultural identity between Saba and Africa. In: Eichmann R, Parringer H, editors. Migration und Kulturtransfer. Bonn; 2001. p. 437–50.

Cann RL, Stoneking M, Wilson AC. Mitochondrial DNA and human evolution. Nature. 1987;325:31–6.

Cavalli-Sforza LL, Menzzoni P, Piazza A. The history and geography of human genes. Princeton, NJ: Princeton University Press; 1994.

Černý V, Mulligan CJ, Rídl J, Žaloudková M, Edens C, Hájek M, et al. Regional differences in the distribution of the sub-Saharan, West Eurasian and South Asian mtDNA lineages in Yemen. American Journal of Physical Anthropology. 2008;136:128–37.

Cordaux R, Saha N, Bentley GR, Aunger R, Sirajuddin SM, Stoneking M. Mitochondrial DNA analysis reveals diverse histories of tribal populations from India. European Journal of Human Genetics. 2003;11:253–64.

de Moulins D, Phillips C, Durrani N. The archaeobotanical record of Yemen and the question of Afro-Asian contacts. In: Neumann K, Butler A, Kahlheber S, editors. Food, fuel and fields, progress in African archaeobotany. Köln: Heinrich-Barth-Institut; 2003.

DiRienzo A, Wilson AC. Branching pattern in the evolutionary tree for human mitochondrial DNA. Proceedings of the National Academy of Sciences USA. 1991;88:1597–601.

Edens C. A bladelet industry for southwestern Arabia. Arabian Archaeology and Epigraphy. 2001;12:137–42.

Edens C. Exploring early agriculture in the highlands of Yemen. In: Sholan A, Antonini S, Arbach M, editors. Sabaean studies, archaeological, epigraphical and historical studies in honour of Yusuf M. 'Abdallah, Alessandro de Maigret and Christain J. Robin on the occasion of their 60th birthdays. Naples: Torcoliere; 2005. p. 185–211.

Edens C, Wilkinson T. Southwest Arabia during the Holocene: recent archaeological developments. Journal of World Pehistory. 1998;12:55–110.

Eswaran V, Harpending H, Rogers AR. Genomics refutes an exclusively African origin of humans. Journal of Human Evolution. 2005;49:1–18.

Fattovich R. The contacts between southern Arabia and the Horn of Africa in late prehistoric and early historical times: a view from Africa. In: Avanzini A, editor. Profumi d'Arabia. Rome: "L'Ermia" di Bretschnieder; 1997. p. 273–86.

Fernandes CA, Rohling EJ, Siddall M. Absence of post-Miocene Red Sea land bridges: biogeographic implications. Journal of Biogeography. 2006;33:961–6.

Forster P, Matsumura S. Did early humans go north or south? Science. 2005;308:965–6.

Forster P, Torroni A, Renfrew C, Rohl A. Phylogenetic star contraction applied to Asian and Papuan mtDNA evolution. Molecular Biology and Evolution. 2001;18:1864–81.

Francaviglia V. Il existait déjà au Néolithique un commerce d'obsidienne à travers la mer Rouge. Revue d'Archéometrie. 1995;19:1–7.

Garrigan D, Mobasher Z, Kingan SB, Wilder JA, Hammer MF. Deep haplotype divergence and long-range linkage disequilibrium at Xp21.1 provide evidence that humans descend from a structured ancestral population. Genetics. 2005;170:1849–56.

Gonder MK, Mortensen HM, Reed FA, de Sousa A, Tishkoff SA. Whole-mtDNA genome sequence analysis of ancient African lineages. Molecular Biology and Evolution. 2007;24:757–68.

González AM, Larruga JM, Abu-Amero KK, CShi Y, Pestano J, Cabrera VM. Mitochondrial lineage M1 traces an early backflow to Africa. BMC Genomics. 2007;8:223.

Ingman M, Kaessmann H, Pääbo S, Gyllensten U. Mitochondrial genome variation and the origin of modern humans. Nature. 2000;408:708–13.

Kivisild T, Bamshad MJ, Kaldma K, Metspalu M, Metspalu E, Reidla M, et al. Deep common ancestry of Indian and western-Eurasian mitochondrial DNA lineages. Current Biology. 1999;9:1331–4.

Kivisild T, Reidla M, Metspalu E, Rosa A, Brehm A, Pennarun E, et al. Ethiopian mitochondrial DNA heritage: tracking gene flow across and around the Gate of Tears. American Journal of Human Genetics. 2004;75:752–70.

Krings M, Salem AH, Bauer K, Geisert H, Malek AK, Chaix L, et al. mtDNA analysis of Nile river valley populations: a genetic corridor or a barrier to migration. American Journal of Human Genetics. 1999;64:1166–76.

Lahr MM, Foley RA. Towards a theory of modern human origins: geography, demography, and diversity in recent human evolution. Yearbook of Physical Anthropology. 1998;41:137–76.

Lézine A-M, Saliège J-F, Robert C, Wert ZF, Inizan M-L. Holocene lakes from Ramlat as-Sab'atayn (Yemen) illustrate the impact of monsoon activity in southern Arabia. Quaternary Research. 1998;50:290–9.

Maca-Meyer N, González AM, Larruga JM, Flores C, Cabrera VM. Major genomic mitochondrial lineages delineate early human expansions. BMC Genetics. 2001;2:13.

Macaulay V, Richards M, Hickey E, Vega E, Cruciani F, Guida V, et al. The emerging tree of west Eurasian mtDNAs: a synthesis of control-region sequences and RFLPs. American Journal of Human Genetics. 1999;64:232–49.

Macaulay V, Hill C, Achilli A, Rengo C, Clarke D, Meehan W, et al. Single, rapid coastal settlement of Asia revealed by analysis of complete mitochondrial genomes. Science. 2005;308:1034–6.

McClure HA. Radiocarbon chronology of late Quaternary lakes in the Arabian desert. Nature. 1976;263:755–6.

McClure HA. A new Arabian stone tool assemblage and notes on the Aterian industry of North Africa. Arabian Archaeology and Epigraphy. 1994;5:1–16.

Metspalu M, Kivisild T, Metspalu E, Parik J, Hudjashov G, Kaldma K, et al. Most of the extant mtDNA boundaries in south and southwest Asia were likely shaped during the initial settlement of Eurasia by anatomically modern humans. BMC Genetics. 2004;5:26.

Metspalu M, Kivisild T, Bandelt H-J, Richards M, Villems R. The pioneer settlement of modern humans in Asia. In: Bandelt H-J, Macaulay V, Richards M, editors. Human mitochondrial DNA and the evolution of Homo sapiens. Berlin: Springer; 2006. p. 181–99.

Meyer S, von Haeseler A. Identifying site-specific substitution rates. Molecular Biology and Evolution. 2003;20:182–9.

Mountain JL, Hebert JM, Bhattacharyya S, Underhill PA, Ottolenghi C, Gadgil M, et al. Demographic history of India and mtDNA-sequence diversity. American Journal of Human Genetics. 1995;56:979–92.

Müller W. Outline of the history of ancient southern Arabia. In: Daum W, editor. Yemen, 3000 years of art and civilization in Arabia Felix. Frankfurt: Umschau-Verlag; 1988. p. 49–54.

Neff U, Burns SJ, Mangini A, Mudelsee M, Fleitmann D, Matter A. Strong coherence between solar variability and the monsoon in Oman between 9 and 6 ka ago. Nature. 2001;411:290–3.

Nettleton W, Chadwick O. Late Quaternary, redeposited loess-soil developmental sequences, South Yemen. Geoderma. 1996;70:21–36.

Olivieri A, Achilli A, Pala M, Battaglia V, Fornarino S, Al-Zahery N, et al. The mtDNA legacy of the Levantine early Upper Palaeolithic in Africa. Science. 2006;314:1767–70.

Pääbo S. The mosaic that is our genome. Nature. 2003;421:409–12.

Pakendorf B, Stoneking M. Mitochondrial DNA and human evolution. Annual Review of Genomics and Human Genetics. 2005;6:165–83.

Petraglia MD. The Lower Palaeolithic of the Arabian peninsula: occupations, adaptations and dispersals. Journal of World Prehistory. 2003;17:141–79.

Petraglia MD, Alsharekh A. The Middle Palaeolithic of Arabia: implications for modern human origins, behaviour and dispersals. Antiquity. 2003;77:671–84.

Plagnol V, Wall JD (2006) Possible ancestral structure in human populations. PLoS Genetics, preprint.

Prell WL, Campo EV. Coherent response of Arabian sea upwelling and pollen transport to late Quaternary monsoonal winds. Nature. 1986;323:526–8.

Prell W, Kutzbach J. Monsoon variability over the past 150000 years. Journal of Geophysical Research. 1987;92:8411–25.

Quintana-Murci L, Semino O, Bandelt H-J, Passarino G, McElreavey K, Santachiara-Benerecetti AS. Genetic evidence of an early exit of Homo sapiens from Africa through eastern Africa. Nature Genetics. 1999;23:437–41.

Quintana-Murci L, Chaix R, Wells RS, Behar DM, Sayar H, Scozzari R, et al. Where west meets east: the complex mtDNA landscape of the southwest and Central Asian corridor. American Journal of Human Genetics. 2004;74:827–45.

Ray N, Currat M, Berthier P, Excoffier L. Recovering the geographic origin of early modern humans by realistic and spatially explicit simulations. Genome Research. 2005;15:1161–7.

Reed FA, Tishkoff SA. African human diversity, origins and migrations. Current Opinion in Genetics and Development. 2006;16: 597–605.

Relethford JH, Jorde LB. Genetic evidence for larger African population size during recent human evolution. American Journal of Physical Anthropology. 1999;108:251–60.

Richards M, Macaulay V, Hickey E, Vega E, Sykes B, Guida V, et al. Tracing European founder lineages in the near eastern mtDNA pool. American Journal of Human Genetics. 2000;67:1251–76.

Richards M, Rengo C, Cruciani F, Gratrix F, Wilson JF, Scozzari R, et al. Extensive female-mediated gene flow from sub-Saharan Africa into near eastern Arab populations. American Journal of Human Genetics. 2003;72:1058–64.

Rose J. The Arabian corridor Migration Model: archaeological evidence for hominin dispersals into Oman during the Middle and Upper Pleistocene. Proceedings of the Seminar for Arabian Studies. 2007;37:219–37.

Rowold D, Luis J, Terreros M, Herrera R. Mitochondrial DNA geneflow indicates preferred usage of the Levant Corridor over the Horn of Africa passageway. Journal of Human Genetics. 2007;52:436–47.

Roychoudhury S, Roy S, Basu A, Banerjee R, Vishwanathan H, Rani MVU, et al. Genomic structures and population histories of linguistically distinct tribal groups of India. Human Genetics. 2001;109: 339–50.

Salas A, Richards M, De la Fe T, Lareu M-V, Sobrino B, Sánchez-Diz P, et al. The making of the African mtDNA landscape. American Journal of Human Genetics. 2002;71:1082–111.

Schwartz M, Vissing J. Paternal inheritance of mitochondrial DNA. The New England Journal of Medicine. 2002;347:576–80.

Schwartz M, Vissing J. New patterns of inheritance in mitochondrial disease. Biochemical and Biophysical Research Communications. 2003;310:247–51.

Sedov V. On the origin of the agricultural settlements in Hadramawt. In: Robin C, editor. Arabia antiqua, early origins of South Arabian states. Rome: IsMEO; 1996. p. 67–86.

Siddall M, Rohling EJ, Almogi-Labin A, Hemleben C, Meischner D, Schmelzer I, et al. Sea-level fluctuations during the last glacial cycle. Nature. 2003;423:853–8.

Sirocko F. The evolution of the monsoon climate over the Arabian sea during the last 24000 years. Palaeoecology of Africa and the Surrounding Islands. 1996;24:53–69.

Sirocko F, Sarnthein M, Erlenkeuser H, Lange H, Arnold M, Duplessy J. Century-scale events in monsoonal climate over the past 24000 years. Nature. 1993;364:322–4.

Stevanovitch A, Gilles A, Bouzaid E, Kefi R, Paris F, Gayraud RP, et al. Mitochondrial DNA sequence diversity in a sedentary population from Egypt. Annals of Human Genetics. 2004;68:23–39.

Stringer C. Coasting out of Africa. Nature. 2000;405:24–7.

Sutovsky P, Moreno RD, Ramalho-Santos J, Dominko T, Simerly C, Schatten G. Ubiquitin tag for sperm mitochondria. Nature. 1999;402:371–2.

Sutovsky P, Van Leyen K, McCauley T, Day BN, Sutovsky M. Degradation of paternal mitochondria after fertilization: implications for heteroplasmy, assisted reproductive technologies and mtDNA inheritance. Reproductive BioMedicine Online. 2004;8:24–33.

Templeton AR. Out of Africa again and again. Nature. 2002;416:45–51.

Thangaraj K, Chaubey G, Kivisild T, Reddy AG, Singh VK, Rasalkar AA, et al. Reconstructing the origin of Andaman islanders. Science. 2005;308:996.

Thomas MG, Weale ME, Jones AL, Richards M, Smith A, Redhead N, et al. Founding mothers of Jewish communities: geographically separated Jewish groups were independently founded by very few female ancestors. American Journal of Human Genetics. 2002;70:1411–20.

Torroni A, Achilli A, Macaulay V, Richards M, Bandelt H-J. Harvesting the fruit of the human mtDNA tree. Trends in Genetics. 2006;22:339–45.

Vernesi C, Di Benedetto G, Caramelli D, Secchieri E, Simoni L, Katti E, et al. Genetic characterization of the body attributed to the evangelist Luke. Proceedings of the National Academy of Sciences USA. 2001;98:13460–3.

Vigilant L, Stoneking M, Harpending H, Hawkes K, Wilson AC. African populations and the evolution of human mitochondrial DNA. Science. 1991;253:1503–7.

Watson E, Forster P, Richards M, Bandelt HJ. Mitochondrial footprints of human expansions in Africa. American Journal of Human Genetics. 1997;61:691–704.

Wilkinson T. Soil erosion and valley fills in the Yemen highlands and southern Turkey: integrating settlement, geoarchaeology and climate change. Geoarchaeology. 2005;20:169–92.

Zarins J. View from the south: the greater Arabian peninsula. In: Henry D, editor. The prehistoric archaeology of Jordan. Oxford: Archaeopress; 1998. p. 179–94.

Zonneveld K, Ganssen G, Troelstra S, Versteegh G, Visscher H. Mechanisms forcing abrupt fluctuation of the Indian ocean summer monsoon during the last deglaciations. Quaternary Science Reviews. 1997;6:187–201.

Chapter 6
The Arabian peninsula: Gate for Human Migrations Out of Africa or Cul-de-Sac? A Mitochondrial DNA Phylogeographic Perspective

Vicente M. Cabrera, Khaled K. Abu-Amero, José M. Larruga, and Ana M. González

Keywords Dispersals • Macrohaplogroup • MtDNA

Introduction

The reconstruction of the origin and spread of modern humans has been a multidisciplinary enterprise. Archaeological records and genetic inferences (Stringer and Andrews, 1988), have given strong support to the model of a single recent origin of modern humans in Africa around 200 ka (McDougall et al., 2005). Subsequent dispersals out of Africa replaced, in relatively short time, the archaic humans living in Eurasia (Pääbo et al., 2004). However, the dates of this exit and the routes taken to spread out of Africa are currently debatable topics. On the basis of modern human fossils in the Levant, dated around 120 ka (Valladas et al., 1988), a northern route by land across the Sinai peninsula was proposed. The lack of fossil continuity in the area prompted researchers to consider it as an unproductive exit. A later successful exit around 45 ka using the same corridor has received stronger archaeological support (Lahr and Foley, 1994). A second, maritime, southern route across the Bab al Mandab strait and afterwards coasting Arabia, India, Southeast Asia to reach the Sahul has also been proposed as a complementary or alternative exit gate (Stringer, 2000). Recent archaeological findings in coastal Eritrea dated about 125 ka (Walter et al., 2000) have been taken as support of an earlier exit age for the southern route (Stringer, 2000).

Phylogenetic analysis using autosomal gene frequency data were consistent with the out of Africa theory and with both, the southern and northern, dispersals out of Africa (Nei and Roychoudhury, 1993). Later studies using uniparental markers also agreed with dual dispersals. The phylogeography of Y chromosome binary haplotypes suggested that derived M216 and M174 haplotypes represent a southern route of dispersal from East Africa to India and beyond, whereas the M89 derived haplotypes represent a Eurasian colonization from the Levantine corridor (Underhill et al., 2001). In a similar vein, the first phylogeographic analysis using complete mitochondrial DNA genomic sequences confirmed that only two founder female mitochondrial lineages, named M and N, left Africa about 70–50 ka. Based on the geographic distribution of these lineages with M predominant in southern and eastern regions of Eurasia and N mainly in western and central Eurasia, it was proposed that M lineages expanded by the coastal southern route and N by the continental northern route (Maca-Meyer et al., 2001). However, the late detection of ancestral N lineages in south and Southeast Asia (Palanichamy et al., 2004; Macaulay et al., 2005) and in Australia (Ingman and Gyllensten, 2003) weakened the mitochondrial hypothesis (Tanaka et al., 2004). In addition, as the founder ages of M and N are very similar, it was hypothesized that both lineages were carried out in a unique migration (Forster et al., 2001), and, even more, that the southern coastal trail was the only route, being the western Eurasian colonization the result of an early offshoot of the southern radiation in India (Oppenheimer, 2003; Macaulay et al., 2005).

Under these suppositions, the Arabian peninsula has gained crucial importance to test the existence of an early southern route out of Africa across the Bab al Mandab strait. Regrettably, there is a lack of adequate hominin fossil record for this region and the archaeological material, although relatively abundant, has few reliable age estimates (Petraglia and Alsharekh, 2003). Until this situation changes, genetic inferences, gathered from phylogenetic and phylogeographic studies on the current populations of the Arabian peninsula seems to be an alternative option. In the following chapter we will review the most recent genetic information obtained from the peninsula using mitochondrial DNA as a temporal and spatial tracer.

K.K. Abu-Amero (✉)
College of Medicine, King Saud University,
P.O. Box 245, Riyadh 11411, Riyadh, Saudi Arabia
e-mail: abuamero@gmail.com

V.M. Cabrera, J.M. Larruga, and A.M. González
Genética, Biología, Universidad de La Laguna La Laguna,
38271, Tenerife, Spain
e-mail: vcabrera@ull.es; jlarruga@ull.es; amglez@ull.es

M.D. Petraglia and J.I. Rose (eds.), *The Evolution of Human Populations in Arabia*, Vertebrate Paleobiology and Paleoanthropology,
DOI 10.1007/978-90-481-2719-1_6, © Springer Science+Business Media B.V. 2009

Mitochondrial DNA Characteristics

Mitochondrial DNA (mtDNA) is still the most used genetic marker in molecular evolution and in population studies. Before the in vitro DNA polymerase amplification was discovered, mtDNA was one of a few molecules amenable for the analysis of variation detectable by restriction fragment length polymorphisms (RFLPs). This was due to its small size (16.6 kb), circular structure, cytoplasmic localization, and high copy number per cell (hundreds to thousands) compared with the two copies for nuclear autosomes. These characteristics allowed its relatively easy isolation and the direct visualization of their electrophoretic RFLP profiles. Even today, in spite of the PCR improvements, its high copy number makes it the most useful DNA molecule for the analysis of fossil samples.

Another valuable property of mtDNA is its high mutation rate, several orders of magnitude higher than that of nuclear genes. This means that initially identical molecules accumulate different mutations in the time frames of the Paleolithic, Neolithic and even historic time. Furthermore, as it has only maternal inheritance, and as all molecules in an individual are alike, it has a non-recombining haploid genetics. For these reasons, differences between mtDNA sequences are only due to mutation. As time passes, mutations accumulate sequentially along less and less related molecules that constitute independent lineages known as haplotypes or lineages.

Relationships among lineages can be estimated by phylogenetic networks where mutations are classified in hierarchical levels. Old basal mutations are shared for clusters of lineages, defined as haplogroups, clusters or clades, whereas the most recent ones, at the tips, characterize individual haplotypes.

As mutations have a time probability to appear, it is possible, under the assumption of molecular neutrality, to date clusters transforming the average number of mutations accumulated in its haplotypes to time by multiplying this average with the mutation rate. Combining the number of different haplotypes and their relative frequencies in a cluster it is possible to obtain a measure of its diversity in a population, in a region or in a continent and comparing these diversities it is possible to infer the most probable geographic origin of that cluster. Moreover, when haplotypes of a subcluster are only detected in a region it is also possible to calculate the time when it expanded in that region if, in addition, the ancestral haplotype is only found in another region, the latter will be considered the source population of the former secondary expansion. Using these calculations it has been possible for instance, to roughly determine the time back to the most recent common ancestor of all mitochondria in extant human populations. To know that all the maternal lineages existing in Eurasia had an African origin, because all the Eurasian clusters coalesce into two macrohaplogroups (M and N) that

are sister clusters of all the L3 African clusters that share with them an ancestral root, with branches of similar age into Africa. These calculations have also been used to infer the maternal genetic structure of Arabia, the most probable origin of their mtDNA lineages and the age of their expansions in this region.

To infer the mtDNA structure of the Arabian peninsula and to assess its role in the southern route, we have analyzed 1,129 Arabian partial sequences assorted into haplogroups by their HVSI/II sequence motifs and diagnostic RFLPs (Abu-Amero et al., 2008 and references within). To resolve cases of difficult haplogroup diagnosis 15 individuals had to be complete or nearly complete genome mtDNA sequenced. The majority of the Arabian lineages have been assorted into well known African and Eurasian haplogroups albeit with different frequencies and heterogeneous geographic distributions (Abu-Amero et al., 2007, 2008).

Macrohaplogroup L in Arabia

The presence of mtDNA lineages belonging to the sub-Saharan Africa macrohaplogroup L in Eurasia and America is mainly explained as the result of the historic and infamous slave trade. In the Arabian peninsula, the incidence of L lineages differs according to country. The highest frequency is found in Yemen (38%), then in Oman and Qatar (16%) and drops to 10% in Saudi Arabia and UAE (Abu-Amero et al., 2008). The most probable source of these sub-Saharan Africa lineages is the geographically closest East African border. However, in that large region it is possible to distinguish at least a northern area conformed by Egypt, Nubia, Sudan, Ethiopia and Somalia in which, at mtDNA level, the L3 haplogroups have significantly greater frequencies than in the southern area represented by Kenya, Tanzania and Mozambique where, in compensation, the most ancestral haplogroup L0 has comparatively higher frequencies.

For presumably recent contacts, a common way to measure the relative gene flow between areas is to count the number of exact haplotype matches shared. For instance, 98% of the shared lineages between Yemen and Africa can be explained by direct eastern Africa influences as only 2% of them are exclusive matches with western Africa. In a similar vein, 88% of the shared sub-Saharan Africa lineages present in Saudi Arabia are also with East Africa, 5% are exclusive matches with western Africa and 7% exclusive with the Near East, implying a more varied source of sub-Saharan African influences in Saudi Arabia compared to Yemen (Abu-Amero et al., 2008 and references within). In addition, the entire eastern African component in Yemen could be explained by south-eastern input as 46% of the matches are exclusive of this area and the remaining 64% shared by both areas, without exclusive matches with the northeast. However, for Saudi Arabia, the exclusive north-eastern (25%) and south-eastern (29%) components are rather similar with the remaining 46% of the East African matches present in both areas. As the Arab slave trade had greater impact on southern African areas, these data could be explained supposing greater traffic with Yemen than with Saudi Arabia. However, earlier contacts, between Arabian and north-eastern Africa historic kingdoms could have had relatively stronger genetic impact in Saudi Arabia compared to Yemen. Another hint of the relative independence of the Yemeni and Saudi sub-Saharan Africa genetic pools is the main and nearly exclusive presence in each country of single haplotypes of different and rare north-eastern African clades. Phylogenetically, L6 is a sister clade of the ancient and widespread African clade L2 (Kivisild et al., 2004). Outside Yemen, it has only been detected twice in Ethiopia (Kivisild et al., 2004) and once in Saudi Arabia (Abu-Amero et al., 2008), but it has a frequency of 12% in Yemen although only one haplotype was the main responsible (86%) of this frequency. On the other hand, L5 is also a phylogenetically ancestral clade that has a sparse north-eastern African distribution. It has not been detected in Yemen, however six Saudi Arab sequences (1%) belonged to the L5a1 subclade (Abu-Amero et al., 2008), and five (83%) were represented by a single haplotype (Abu-Amero et al., 2008). Most probably both lineages arrived at the Arabian peninsula from north-eastern Africa by two independent events, and expanded in rather endogamic and isolated populations. This supposition is congruent with the significant genetic structure found in the Arabian peninsula (Abu-Amero et al., 2008). Based on the lack of matches between Arabia and Africa, it has been suggested that haplogroup L6 might have originated from the same out-of-Africa migration that carried haplogroup M and N to Eurasia. This seems not to be the case for the L5a1 subclade because the main L5a1 Saudi haplotype has exact matches in Egypt, Ethiopia and Kenya. In any case, due to ancient or more recent African gene flows, the fact that any of both lineages has not spread into surrounding areas implies that, since their arrivals, the Arabian peninsula has acted more as a cul-de-sac than as a demic source of later migrations.

particularly abundant in Ethiopia (20%). From there, frequencies significantly diminish forming decreasing gradients westwards and southwards. It has been proposed that the presence of M1 in Africa and surrounding Mediterranean areas can be explained as result of two expansion centers situated in East and Northwest Africa which are marked by the radiation of subhaplogroups M1a and M1b respectively (Olivieri et al., 2006; González et al., 2007). Although the coalescence age of M1 is Paleolithic it seems that the most important expansions occurred in Neolithic times when the Sahara was a more hospitable region. Some authors consider that the presence of M1 in Africa supports the idea that macrohaplogroup M originated in eastern Africa and was carried towards Asia with the out of Africa expansion (Quintana-Murci et al., 1999), others think that the distribution of M1 in Africa traces an early human backflow to this Continent from Asia (Maca-Meyer et al., 2001; Olivieri et al., 2006; González et al., 2007).

In Arabia, M lineages account for 7% of the total and half of them belong to the M1 African clade. M1 frequencies are significantly greater in western Arabian regions than in the East (Abu-Amero et al., 2008). As the majority of the M1 haplotypes in Arabia belong to the East African M1a subclade, it seems that, likewise L lineages, the M1 presence in the Arabian peninsula signals a predominant East African influence since the Neolithic onwards.

The majority of the resting M lineages found in Arabia has matches or are related to Indian clades. In addition, some M sequences point to rare links with more remote geographic regions as Central Asia, West New Guinea and even Australia (Abu-Amero et al., 2008). Although more ancient connections cannot be discarded, it seems that this rare M component in the Arabian populations could be the result of trade and military links among those regions in Arabia during and after the British role. As all the M lineages found in Arabia belong to haplogroups that have deeper roots and diversities in other geographic regions, its presence in the Arabian peninsula is better explained as external genetic inputs. Therefore, there are no traces of autochthonous M lineages in Arabia that could support the exit of modern humans from Africa across the Bab al Mandab strait.

Macrohaplogroup M in Arabia

Macrohaplogroup M is particularly abundant and diverse in South and Southeast Asia, reaching frequencies above 60% in some regions (Metspalu et al., 2004). However, it is practically absent in western Asia (Quintana-Murci et al., 2004). In Africa, only one autochthonous basal branch of M, named M1, has been detected (Quintana-Murci et al., 1999). In this continent it has a predominant northern distribution. M1 is

Macrohaplogroup N in Arabia

A sole branch of macrohaplogroup N, named R, encompasses the overwhelming majority of the N clades. It will be treated in the next section. The resting sister branches of R, that sprout directly off the N trunk, have an irregular geographic distribution. Western (N1, W, X) and northern (A, N9, Y) Asian clades have moderate frequencies in their respective geographic ranges (Quintana-Murci et al., 2004; Tanaka et al., 2004; Abu-Amero et al., 2008). On the contrary, basic N clades in India are sparse (Palanichamy et al., 2004) and even rarer in Southeast Asia (Friedlaender et al., 2005; Macaulay et al., 2005; Hill et al., 2006). However, they are predominant and highly diverse in Australia (van Holst Pellekaan et al., 2006) the utmost limit of the out-of-Africa exit.

Only branches N1a, N1b, N1c, I, W, X2 of the western Eurasian N clades have been detected in Arabia (Kivisild et al., 2004; Abu-Amero et al., 2007; Rowold et al., 2007) albeit in low individual frequencies. The majority of these Arabian lineages reflect genetic inputs into the peninsula from adjacent areas. For instance, haplogroup X has two well defined branches of north African (X1) and Eurasian (X2) adscription (Reidla et al., 2003), but all the X haplotypes found in Saudi Arabia (Abu-Amero et al., 2008) and Yemen (Kivisild et al., 2004) belong to the Eurasian branch, which discards an East African introduction. The geographical distribution of the Arabian I and W lineages points to an eastern provenance across Iran (Abu-Amero et al., 2008). Haplogroups N1b and N1c are moderately represented in Arabia although their highest diversities are in Iran and Turkey respectively pointing to eastern and northern contributions to the Arabian genetic pool. However, the N1a haplogroup deserves a more detailed analysis. First, frequencies in Arabia (7% in Yemen, 4% in Saudi Arabia) are higher than in surrounding areas. Second, diversities in the peninsula are the highest in the geographic range of N1a (Table 1). Third, Arabian haplotypes are present in the most ancient nodes of the N1a network (Fig. 1), and in all its main expansions. Hence, it may be concluded that the Arabian peninsula was within the nuclear area that originated the first and subsequent N1a dispersions. Adding a Tanzanian N1a (Gonder et al., 2007) and an Italian (Gasparre et al., 2007) to the N1a tree of complete sequences recently published (Derenko et al., 2007), it can be deduced that N1a haplotypes carrying the 16147G transversion are ancestral compared to those with the 16147A mutation. This fact gives a root, marked with a star, to the N1a network constructed with worldwide HVSI sequences (Fig. 1). It seems that the N1a ancestor

migrated from west-central Asia, the most probable cradle of the N1 expansion, southwards to Arabia where a secondary radiation occurred affecting East Africa, and southwest and South Asia (Fig. 1). One of these lineages suffered a transition in position 16147G giving place to the 16147A clade (N1a1) that also expanded in the western range of the preceding 16147G wave. In time, at the northern edge of the 16147A clade dispersion, perhaps in southern Russia, a new mutation, 16,320, in the HVSI region, defined a new clade named N1a1a that originated the biggest expansion in all directions, reaching, southwards, the Mediterranean area, and, again, Arabia, Iran and India and northwards Siberia and Europe. In all these areas new and more geographically localized subclusters emerged, such as that characterized by the 16,189 transition in Central Siberia (Fig. 1). Today, N1a is a minor cluster in its whole range but it seems that it was more abundant in Central Europe in Neolithic times (Haak et al., 2005) and in the Altaian region around 3,000 years ago (Ricaut et al., 2004). Depending on the mutation rate chosen and on the coding or regulatory region used, coalescence times for these dispersions varied broadly, oscillating between 40 and 20 ka for the whole N1a cluster and around 25–11 ka for the N1a1a subcluster. Possibly, these expansions took place during interstadial favorable episodes. In any case, this detailed analysis of the N1a haplogroup has demonstrated the existence of late Paleolithic human expansions in Arabia. However, as the entire sister branches of N1a had a northern origin, these demographic expansions are more the result of secondary back-migration than of primary radiations in Arabia after the out-of-Africa exit.

Macrohaplogroup R in Arabia

Macrohaplogroup R derives from the N trunk by two additional mutations (gain of 12,705 and loss of 16,223 transitions). Similar to the other macrohaplogroups, it also shows a notable geographic structure with different branches characteristic of different areas. In Western Asia seven main clades (R0a, HV, H, V, U, J and T) nearly capture all its diversity. In India the R radiation was particularly impressive, and many lineages are still pending full characterization (Metspalu et al., 2004; Palanichamy et al., 2004). Haplogroups B and F are the most conspicuous R representatives in Southern and Eastern Asia and, different P branches, in New Guinea and Australia. Although represented by different clades, a notable characteristic of R is that it is widespread and abundant everywhere.

In the Arabian peninsula, except for a few R1, R2 and U2 haplotypes of clear Indian origin, the bulk of its R

Table 1 Number of individuals (N_i), number of different haplotypes (N_h) and nucleotide diversity by 1,000 with error ($\pi \pm s$), in several geographic areas

Area	N_i	N_h	$\pi \pm s$
Northeast Africa	16	9	5.602 ± 3.891
Southwest Asia	16	9	8.756 ± 5.540
NC-Asia	8	6	3.137 ± 2.733
Arabia	21	12	13.353 ± 7.772
Europe	42	23	12.164 ± 6.993

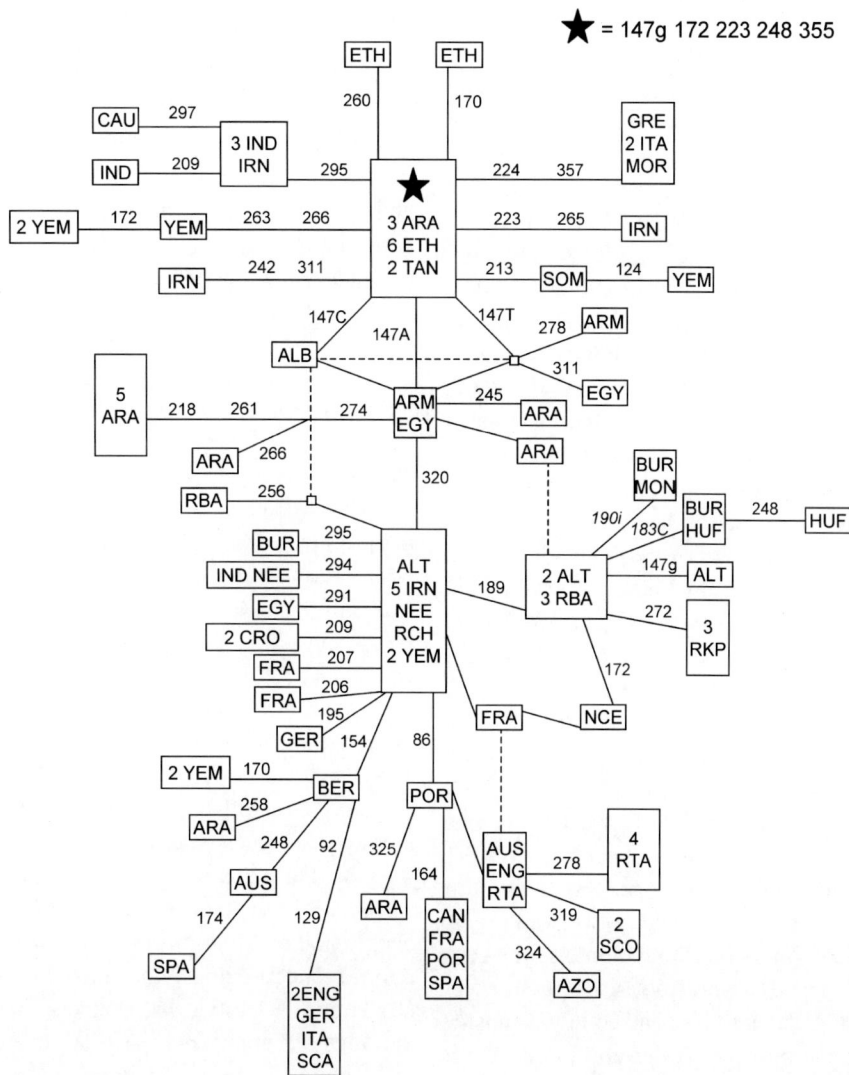

Fig. 1 Reduced median network (Bandelt et al., 1999) relating N1a HVSI sequences. The ancestral motif (star) differs from rCRS at the indicated positions. Numbers along links refer to nucleotide position minus 16,000. Broken lines are less probable links and/or recurrent mutations. Size of boxes is proportional to the number of individuals included. Codes are ALB, Albanian; ALT, Altaian; ARA, Arab; ARM, Armenian; AUS, Austrian; AZO, Azorean; BER, Berber; BUR, Buryat; CAN, Canarian;

CAU, Caucasian; CRO, Croatian; EGY, Egyptian; ENG, English; ETH, Ethiopian; FRA, French; GER, German; GRE, Greek; HUF, Hungarian fossil; IND, Indian; IRN, Iranian; ITA, Italian; MON, Mongolian; MOR, Moroccan; NCE, North-central European; NEE, North-east European; POR, Portuguese; RCH, Chuvasch Russian; RBA, Bashkirs Russian; RKP, Komi-Permyaks Russian; RTA, Tatar Russian; SCA, Scandinavian; SCO, Scottish; SOM, Somali; SPA, Spaniard; TAN, Tanzanian

lineages (70%) belong to Western Asian clades (Abu-Amero et al., 2008). From their relative geographic distributions in Arabia, the number of exact matches with surrounding areas, and relative diversities in the different regions, it is possible to assign a geographic provenance to the majority of the R haplotypes found in Arabia. It has been demonstrated that U6 had an old implantation in North Africa (Maca-Meyer et al., 2003) and that is the most probable origin of the few U6 haplotypes detected in the Arabian peninsula. The majority of the U and K representatives in Arabia (U3, U4, U7, K) show greater frequencies in the eastern and southern Arabian regions supporting an eastern

entrance through Iran. Nevertheless, the major portion of R Arabian lineages (60%) had a most probable northern source. In this respect the distribution of haplogroup H is a good example. It is the most frequent clade in Europe (45%) and Near East (25%) however in the Arabian peninsula its mean frequency, around 9%, is moderate. In fact, H frequencies significantly diminished with latitude from Turkey to Yemen (Abu-Amero et al., 2007). Haplogroup T shows a similar trend presenting its lower frequencies in the southern Yemen and Oman countries. Other minor lineages in Arabia, as those belonging to the European U2e and U5 clades and the infrequent U9, could also reach the Arabian peninsula from northern areas. Due to the lack of clear founder subclades in Arabia for these lineages, and to the difficulty of differentiating successive gene flows or expansions, because the most recent migration could carry both early and derivate lineages, it is impossible to accurately gauge their entrance times in the Peninsula. However, for the two most frequent R clades in Arabia R0a (Abu-Amero et al., 2007) and J1b (Abu-Amero et al., 2008), phylogenetic and phylogeographic analysis have allowed to date different expansion events. The time to the most recent common ancestor (TMRCA) for both R0a and J1b clades was calculated around 20 ka (Abu-Amero et al., 2007, 2008). However, whereas for the latter the ancestral motif was present in the Near East as much as in Arabia, suggesting that the peninsula played an active role in the Paleolithic spread of J1b, the R0a first radiation had a main Near East origin because its ancestral motif was barely present in Arabia. However, recent data from Yemen (Černý et al., 2008) raises the possibility that R0a also had a Paleolithic spread in southern Arabia. The successive most important radiation of both clades, signed by the R0a1a and J1b1a1 subclades, had, again, similar Neolithic ages around 10 ka (Abu-Amero et al., 2007, 2008). In both cases there was a shortage or absence of their ancestral motifs in Arabia discarding this area as a radiation center. However, whereas the R0a1a wave reached Arabia from the Near East, J1b1a1 occupied northern areas, including Europe, being absent in the Arabian peninsula. It seems that at least two well represented subclades had Arabia as their radiation origin (Abu-Amero et al., 2008). The J1b one, rooted by the 16,136 transition has a TMRCA around 11 ka and could be considered the southern branch of the J1b1a1 Neolithic northern expansion. Nevertheless, the R0a Arabian branch, defined by the 16,304 transition, is only about 4,000 years old which situates its expansion in the Bronze Age (Abu-Amero et al., 2008). From the above data it may be concluded that the Arabian peninsula was mainly a receiver of mitochondrial immigrations. Even in favorable climatic conditions population densities should be low enough to convert this region in a demographic expansive centre. Finally, the lack of ancestral R lineages in Arabia left this region without any genetic support to the proposed southern route across the Bab al Mandab strait of modern humans. Although this lack of genetic evidence can be attributed to the total extinction of the ancestral mtDNA lineages that once, hypothetically, crossed southern Arabia, and the single coastal migration model has general support (Stringer, 2000; Metspalu et al., 2004; Forster and Matsumura, 2005; Macaulay et al., 2005; Thangaraj et al., 2005), any alternative model that might explain the first successful Eurasian dispersion of modern humans without involving Arabia should be taken into consideration.

Mitochondrial Footsteps of the Old World Human Colonization

When Maca-Meyer et al. (2001) formulated the hypothesis of two dispersals from Africa based on the phylogeny and phylogeography of complete mtDNA sequences, the presence in South and Southeast Asia and in Australia of lineages belonging to the macrohaplogroup N, that were not R derivates, had not yet been detected. Based on the distribution of the two macrolineages with N prevalent in Western Asia and M predominant in South and East Asia, it was proposed that M and N were, respectively, the mitochondrial signals of the already proposed southern and northern routes (Nei and Roychoudhury, 1993). The simultaneous presence in India, Malaysia, and Australia of N, M, and R lineages has prompted other researchers (Forster et al., 2001; Kivisild et al., 2003; Hudjashov et al., 2007) to propose that there was only a single coastal southern route out of Africa. On this supposition an ancestral L3 split into haplogroups M, N and R out of Africa and, after that, was lost by genetic drift. Then the three lineages traveled together eastwards coasting South Arabia, India and Southeast Asia reaching Australia. Moreover, the colonization of West Eurasia has been explained as an offshoot from the Southern route discrediting the two dispersals hypothesis. However, as it was previously suggested (Tanaka et al., 2004), this new scenario does not satisfactorily explain the mitochondrial haplogroup phylogeographic distributions. With some modifications, the two routes model previously proposed better explains it. First of all, M and N are two independent lineages because, as all the other L3 branches in Africa, they directly spread from the common L3 trunk. Second, we consider the coalescence age of L3 as the lowest bound of the out-of-Africa exit (Maca-Meyer et al., 2001). This frame could anticipate the Eurasian colonization to as early as 100–80 ka which coincides with an interglacial stage, an optimum period to leave Africa across to the then humid and hospitable Sinai peninsula. Most probably, during this favorable period several small groups of modern humans ventured out-of-Africa through this peninsula following afterwards northern and southern corridors signaled by their preys and avoiding regions where competition with other hominins, as the Neanderthals, could be strong. It is worthwhile mentioning that this date is coincidental with the first paleontological evidence of modern human presence in the Near East (Valladas et al., 1988; Mercier et al., 1993). Mitochondrial lineages carried by these colonizers were not yet ripe M and N lineages but their L3 ancestors. Under this supposition the M and N ancestors could have left Africa independently. Around 60 ka glacial conditions returned, strongly affecting the descendants of those wandering groups that suffered

Fig. 2 Geographic dispersal routes of major human expansions in the Old World

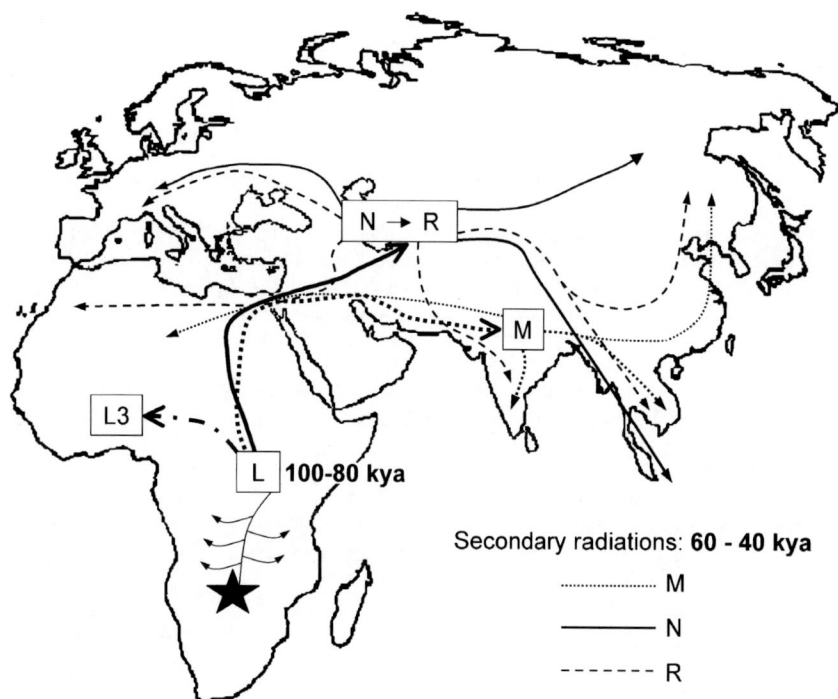

important bottle-necks with the subsequent loss of lineages, in such a way that only the direct ancestors of all the present day M and N branches lasted. As this selective process occurred well inside Asia, not in Africa, it is not necessary to invoke a very fast diaspora to explain the simultaneous existence of ancestral M and N lineages in areas geographically as distant as India, Southeast Asia or Australia. Glacial conditions forced human bands in the North going southwards and those in the South, to avoid deserts, searching for more hospitable regions. The phylogeographic distribution of M and N haplogroups points to northern and southern populations as the bearers of N and M lineages respectively. It is evident that R is an ancestral branch of N that signals a primary radiation of N in Asia. After its apparition, some R branches spread southwards to India where they met M, and others, along with N lineages, dispersed to Southeast Asia avoiding India in their southern migrations. Clearly, this model also better explains the early presence of modern humans in Australia. Due to phylogenetic considerations we think that at least two migratory waves reached Australia. The first one carrying mainly N ancestral lineages and the second, that also affected Papua New Guinea, bringing R and M lineages (Fig. 2). Around 45 ka, coinciding with an interstadial substage of the Würm Glacial, favorable climatic conditions allowed secondary dispersions including, westwards, Europe, the Near East and northern Africa, southwards India, and northwards Siberia (Fig. 2). With the

exception of an M1 African branch, these waves brought to West Asia and North Africa only N and R lineages. It is not easy to explain the absence of ancestral M lineages in Western Eurasia if they were the result of an Indian offshoot, as in India around of 60% of its lineages belong to different M haplogroups. Furthermore, the fact that the Eurasian N and R lineages are not derived from Indian clades is also against its Indian origin. As there is no evidence of African haplogroups in Eurasia that could be dated to that epoch, we think that the proposed out-of-Africa exit around 45 ka across the Sinai peninsula had little impact or did not exist at all. This scenario leaves the Bab al Mandab corridor unnecessary as the genetic studies on Arabia suggest. It has been argued that to reach Australia their colonizers had to have seafaring experience, as it was necessary to cross the Bab al Mandab strait. But it could most probably be acquired later in Asian tropical regions rich in wood and wide rivers than in the, under glacial period, desert regions of the Horn of Africa and Arabia.

References

Abu-Amero KK, González AM, Larruga JM, Bosley TM, Cabrera VM. Eurasian and African mitochondrial DNA influences in the Saudi Arabian population. BMC Evolutionary Biology. 2007;7:32.

Abu-Amero KK, Larruga JM, Cabrera VM, González AM. Mitochondrial DNA structure in the Arabian peninsula. BMC Evolutionary Biology. 2008;8:45.

Bandelt H-J, Forster P, Röhl A. Median-joining networks for inferring intraspecific phylogenies. Molecular Biology and Evolution. 1999;16:37–48.

Černý V, Mulligan CJ, Rídl J, Žaloudková M, Edens CM, Hájek M, et al. Regional differences in the distribution of the sub-Saharan, West Eurasian, and South Asian mtDNA lineages in Yemen. American Journal of Physical Anthropology. 2008;136:128–37.

Derenko M, Malyarchuk B, Grzybowski T, Denisova G, Dambueva I, Perkova M, et al. Phylogeographic analysis of mitochondrial DNA in northern Asian populations. American Journal of Human Genetics. 2007;81:1025–41.

Forster P, Torroni A, Renfrew C, Röhl A. Phylogenetic star contraction applied to Asian and Papuan mtDNA evolution. Molecular Biology and Evolution. 2001;18:1864–81.

Forster P, Matsumura S. Human evolution: did early humans go north or south? Science. 2005;308:965–6.

Friedlaender J, Schurr T, Gentz F, Koki G, Friedlaender F, Horvat G, et al. Expanding southwest Pacific mitochondrial haplogroups P and Q. Molecular Biology and Evolution. 2005;22:1506–17.

Gasparre G, Porcelli AM, Bonora E, Pennisi LF, Toller M, Iommarini L, et al. Disruptive mitochondrial DNA mutations in complex I subunits are markers of oncocytic phenotype in thyroid tumors. Proceedings of the National Academy of Sciences USA. 2007; 104:9001–6.

Gonder MK, Mortensen HM, Reed FA, de Sousa A, Tishkoff SA. Whole-mtDNA genome sequence analysis of ancient African lineages. Molecular Biology and Evolution. 2007;24:757–68.

González AM, Larruga JM, Abu-Amero KK, Shi Y, Pestano J, Cabrera VM. Mitochondrial lineage M1 traces an early human backflow to Africa. BMC Genomics. 2007;8:223.

Haak W, Forster P, Bramanti B, Matsumura S, Brandt G, Tänzer M, et al. Ancient DNA from the first European farmers in 7500-year-old Neolithic sites. Science. 2005;310:1016–8.

Hill C, Soares P, Mormina M, Macaulay M, Meehan W, Blackburn J, et al. Phylogeography and ethnogenesis of Aboriginal Southeast Asians. Molecular Biology and Evolution. 2006;23:2480–91.

Hudjashov G, Kivisild T, Underhill PA, Endicott P, Sanchez JJ, Lin AA, et al. Revealing the prehistoric settlement of Australia by Y-chromosome and mtDNA analysis. Proceedings of the National Academic of Sciences USA. 2007;104:8726–30.

Ingman M, Gyllensten U. Mitochondrial genome variation and evolutionary history of Australian and New Guinean aborigines. Genome Research. 2003;13:1600–6.

Kivisild T, Rootsi S, Metspalu M, Mastana S, Kaldma K, Parik J, et al. The genetic heritage of the earliest settlers persists both in Indian tribal and caste populations. American Journal of Human Genetics. 2003;72:313–32.

Kivisild T, Reidla M, Metspalu E, Rosa A, Brehm A, Pennarum E, et al. Ethiopian mitochondrial heritage: tracking gene flow across and around the gate of tears. American Journal of Human Genetics. 2004;75:752–70.

Lahr MM, Foley RA. Multiple dispersals and modern human origins. Evolutionary Anthropology. 1994;3:48–60.

Maca-Meyer N, González AM, Larruga JM, Flores C, Cabrera VM. Major genomic mitochondrial lineages delineate early human expansions. BMC Genetics. 2001;2:13.

Maca-Meyer N, González AM, Pestano J, Flores C, Larruga JM, Cabrera VM. Mitochondrial DNA transit between West Asia and North Africa inferred from U6 phylogeography. BMC Genetics. 2003;4:15.

Macaulay V, Hill C, Achilli A, Rengo C, Clarke D, Meehan W, et al. Single, rapid coastal settlement of Asia revealed by analysis of complete mitocondrial genomes. Science. 2005;308:1034–6.

McDougall I, Brown FH, Fleagle JG. Stratigraphic placement and age of modern humans from Kibish, Ethiopia. Nature. 2005; 433:733–6.

Mercier N, Valladas H, Bar-Yosef O, Vandermeersch B, Stringer C, Joran L. Thermoluminescence date for the Mousterian burial site of Es Skhul, Mt Carmel. Journal of Archaeological Science. 1993;20:169–74.

Metspalu M, Kivisild T, Metspalu E, Parik J, Hudjashov G, Kaldma K, et al. Most of the extant mtDNA boundaries in South and Southwest Asia were likely shaped during the initial settlement of Eurasia by anatomically modern humans. BMC Genetics. 2004;5:26.

Nei M, Roychoudhury AK. Evolutionary relationships of human populations on a global scale. Molecular Biology and Evolution. 1993;10:927–43.

Olivieri A, Achilli A, Pala M, Battaglia V, Fornarino S, Al-Zahery N, et al. The mtDNA legacy of the Levantine early Upper Palaeolithic in Africa. Science. 2006;314:1767–70.

Oppenheimer S. Out of Eden: the peopling of the world. London: Constable; 2003.

Pääbo S, Poinar H, Serre D, Jaenicke-Despres V, Hebler J, Rohland N, et al. Genetic analyses from ancient DNA. Annual Review of Genetics. 2004;38:645–79.

Palanichamy MG, Sun C, Agrawal S, Bandelt H-J, Kong QP, Khan F, et al. Phylogeny of mitochondrial DNA macrohaplogroup N in India, based on complete sequencing: implications for the peopling of South Asia. American Journal of Human Genetics. 2004;75:966–78.

Petraglia MD, Alsharekh A. The Middle Palaeolithic of Arabia: implications for modern human origins, behaviour and dispersals. Antiquity. 2003;77:671–84.

Quintana-Murci L, Semino O, Bandelt H-J, Passarino G, McElreavey K, Santachiara-Benereceti AS. Genetic evidence of an early exit of Homo sapiens sapiens from Africa through eastern Africa. Nature Genetics. 1999;23:437–41.

Quintana-Murci L, Chaix R, Well RS, Behar DM, Sayar H, Scozzari R, et al. Where west meets east: the complex mtDNA landscape of the southwest and Central Asian corridor. American Journal of Human Genetics. 2004;74:827–45.

Reidla M, Kivisild T, Metspalu E, Kaldma K, Tambers K, Tolk HV, et al. Origin and difussion of mtDNA haplogroup X. American Journal of Human Genetics. 2003;73:1178–90.

Ricaut FX, Keyser-Tracqui C, Bourgeois J, Crubezy E, Ludes B. Genetic analysis of a Scytho-Siberian skeleton and its implications for ancient Central Asian migrations. Human Biology. 2004;76:109–25.

Rowold DJ, Luis JR, Terreros MC, Herrera RJ. Mitochondrial DNA geneflow indicates preferred usage of the Levant corridor over the Horn of Africa passageway. Journal of Human Genetics. 2007;52:436–77.

Stringer CB. Coasting out of Africa. Nature. 2000;405:24–7.

Stringer CB, Andrews P. Genetic and fossil evidence for the origin of modern humans. Science. 1988;239:1263–8.

Tanaka M, Cabrera VM, González AM, Larruga JM, Takeyasu T, Fuku N, et al. Mitochondrial genome variation in eastern Asia and the peopling of Japan. Genome Research. 2004; 14:1832–50.

Thangaraj K, Chaubey G, Kivisild T, Reddy AG, Singh VK, Rasalkar AA, et al. Reconstructing the origin of Andaman Islanders. Science. 2005;308:996.

Underhill PA, Passarino G, Lin AA, Shen P, Mirazón Lahar M, Foley RA, et al. The phylogeography of Y chromosome binary haplotypes and the origins of modern human populations. Annals of Human Genetics. 2001;65:43–62.

Valladas H, Reyss JL, Joron JL, Valladas H, Bar-Yosef O, Vandermeersch B. Thermoluminescence dating of Mousterian 'Proto-Cro-Magnon' remains from Israel and the origin of modern man. Nature. 1988;331:614–6.

van Holst Pellekaan SM, Ingman M, Roberts-Thomson J, Harding RM. Mitochondrial genomics identifies major haplogroups in Aboriginal Australians. American Journal of Physical Anthropology. 2006;131:282–94.

Walter RC, Buffler RT, Bruggeman JH, Guillame MM, Berhe SM, Negassi B, et al. Early human occupation of the Red Sea coast of Eritrea during the Last Interglacial. Nature. 2000;405:65–9.

Chapter 7
Bayesian Coalescent Inference from Mitochondrial DNA Variation of the Colonization Time of Arabia by the Hamadryas Baboon (*Papio hamadryas hamadryas*)

Carlos A. Fernandes

Keywords Colonization • Hamadryas Baboon • Mammal Dispersals • Afro-Arabian Zoogeography

Background

Even after its separation from Africa, around the Miocene–Pliocene transition, the ineludible importance of Arabia in the history of biotic movements between Africa and Eurasia has been verified by accumulating data from biogeographic (Delany, 1989), paleontological (Thomas et al., 1998), genetic (Kivisild et al., 2004; Abu-Amero et al., 2008), and archaeological studies (Petraglia and Alsharekh, 2003; Beyin, 2006; Rose, 2007; Petraglia et al., 2009). What we don't know for most of the (inferred) species dispersals between the two continents since the Miocene, are the details about routes, timings, and the role of the Arabian peninsula in these events. The difficult challenge now is to uncover those details for each species dispersal between Africa and Eurasia. For instance, for any given species, was Arabia a swift shortcut, a prolonged stopover, a dead end, the remaining refuge of a receding expansion into Eurasia, or a mere bystander of a migration exclusively via the northern Levantine corridor?

The two main post-Miocene routes proposed (Tchernov, 1992; Cavalli-Sforza et al., 1993; Lahr and Foley, 1994) for species movements between Africa and Eurasia are: (1) through the Sinai peninsula, or (2) across the Bab al Mandab Strait in the southern Red Sea. The second is commonly equated with a land bridge that would have emerged during Pleistocene sea-level lowstands (Thunell et al., 1988; Delany, 1989; Robinson and Matthee, 1999; Walter et al., 2000; Mithen and Reed, 2002; Shefer et al., 2004; Werner and Mokady, 2004; Wildman et al., 2004; Winney et al., 2004; Froukh and Kochzius, 2007). Alternatively, as in the case of the Out-of-Africa migration

of modern humans for which models involving a single, or at least major, dispersal event along the "southern route" are increasingly supported by genetic and archaeological evidence (Quintana-Murci et al., 1999; Maca-Meyer et al., 2001; Underhill et al., 2001; Forster, 2004; Rose, 2004; Macaulay et al., 2005), the precise nature of the dispersal mechanism across the southern Red Sea is usually left unspecified (e.g., by means of a land bridge or using watercraft). The fact is that present paleoceanographic and paleoecological data are seemingly incompatible with the existence of Red Sea land bridges since the Miocene (Rohling et al., 1998; Siddall et al., 2003, 2004; Fernandes et al., 2006). During the last 470,000 years, sea level has remained more than 15 m above the level of the (Hanish) sill and the Bab al Mandab Strait has been at least 5 km wide, even at the most severe sea-level lowstands (Fernandes et al., 2006). Bailey et al. (2007) question how accurately the depth and width of the channel can be inferred at the Last Glacial Maximum (LGM, ~20 ka; Lambeck et al., 2002) and arguably suggest that, even without dry pathways, the southern Red Sea at the LGM, and perhaps for most of the Late Pleistocene, would not pose a significant barrier to crossings by humans or other mammals (see also Bailey, 2009). Nonetheless, any dispersal across the southern Red Sea would still have required swimming, rafting or the use of watercraft (Derricourt, 2005; Bailey et al., 2007; Rose, 2007). Moreover, it is worth mentioning that at the proposed time (~70 ka) for the Out-of-Africa human migration at the root of all living non-Africans (Macaulay et al., 2005), the width and water depth of the southern Red Sea would not have been as reduced as they were in the LGM (Fernandes et al., 2006).

Here, I review data on the phylogeography of the hamadryas baboon (*Papio hamadryas hamadryas*) and the origin of its Arabian populations (Wildman et al., 2004; Winney et al., 2004), to assess the likelihood of each of the proposed routes in the colonization of Arabia by this species, and for Afro-Arabian faunal dispersals in general. For this purpose, I apply sophisticated and recently developed Bayesian coalescent approaches for the estimation of divergence times from genetic divergence to the data sets in Wildman et al. (2004) and Winney et al. (2004), and relate the results with paleoenvironmental and paleontological evidence.

C.A. Fernandes (✉)
Centro de Biologia Ambiental, Universidade de Lisboa,
C2 2.5.41, 1749-016, Campo Grande, Lisboa, Portugal
e-mail: CaFernandes@fc.ul.pt

M.D. Petraglia and J.I. Rose (eds.), *The Evolution of Human Populations in Arabia*, Vertebrate Paleobiology and Paleoanthropology, DOI 10.1007/978-90-481-2719-1_7, © Springer Science+Business Media B.V. 2009

Human Genetic Data and Out-of-Africa Routes and Times

Phylogenetic reconstructions of human mitochondrial DNA (mtDNA) haplotypes have revealed that the extant mtDNA variation outside Africa can be assigned to two basal haplogroups, M and N, which stem of an African-specific haplogroup L3. It has been estimated that L3 appeared ~85 ka, and both M and N ~65 ka (Macaulay et al., 2005; Torroni et al., 2006).

Due to the presence of an M-type clade (M1) in Africa, with high frequency in eastern Africa, it was initially suggested that haplogroup M could be a decisive genetic indicator for the origin, timing, and route of the Out-of-Africa migration. An eastern African origin for M was therefore proposed at the time (Quintana-Murci et al., 1999). However, recent growing evidence strongly supports the case for a south Asian origin of M, with the sole M-type clade in Africa (M1) being the result of a back migration 45–40 ka via the Sinai land bridge (Richards et al., 2003; Olivieri et al., 2006; González et al., 2007). The fact that haplogroup N, the Eurasian sister clade of M and with a very similar age to M, has no sign of an African origin, further supports a non-African origin for M (Olivieri et al., 2006). Rowold et al. (2007) argue that the intercontinental movement of M1 between the Middle and Upper Paleolithic has been preferentially through the southern Red Sea passageway, but were unable to clarify the direction of the gene flow. It is indeed possible that M1 haplotypes have passed from eastern Africa to Arabia (Abu-Amero et al., 2008), but this is not incompatible with an Asian origin for M1 and their earlier entrance in Africa through the Sinai (González et al., 2007). Moreover, even if the flow of M1 haplotypes from eastern Africa to Arabia is confirmed to have occurred around the Middle to Upper Paleolithic transition, this clearly postdates and does not herald the Out-of-Africa migration.

The lack of structure for many of the basal daughter haplogroups of M and N among Europe, south Asia, east Asia, and Oceania, the product of extensive sharing of M and N founders, is best explained by a fast colonization pace of the world (Macaulay et al., 2005; Sun et al., 2006; Hudjashov et al., 2007; Underhill and Kivisild, 2007). This rather swift dispersal is almost mandatory after ~65 ka when haplogroups M and N start to take shape, apparently around the Indus Valley (Quintana-Murci et al., 2004) to explain the observed lack of structure.

The current view (Macaulay et al., 2005) proposes that, after the birth of L3, ~85 ka, the Out-of-Africa migrants would have initiated their journey 75,000–65,000 years ago by crossing the southern Red Sea and then traverse southwestern Asia to reach the Indus Valley ~65 ka. It is interesting to note, however, that the scenario of a southern Red Sea itinerary is more an assumption, given the seemingly required

rapid rate of dispersal for the Out-of-Africa expansion, than unequivocal evidence derived from genetic comparisons between eastern Africa and Arabia (Kivisild et al., 2004; Luis et al., 2004; Rowold et al., 2007; Abu-Amero et al., 2008; Cerný et al., 2008).

The Horn of Africa and Arabia have a history of cultural, economic, and social connections over millennia that promoted gene flow between the two regions. This is confirmed by the close genetic affinities between the two regions detected by several surveys (Richards et al., 2003; Kivisild et al., 2004). Interestingly, almost all of this genetic similarity seems to be the product of Upper Paleolithic and Neolithic gene flow, a significant part the product of slave trade from Africa, while unambiguous evidence for older migrations, particularly from Africa to Arabia and with ages compatible with the estimated timing for the Out-of-Africa dispersal, remains absent. In fact, some of the data not only shows that the genetic history of the populations in the Arabian peninsula is rather complex, unsurprisingly given the position of Arabia at a crossroads of three continents, but also that, at least for the moment, it is open to alternative interpretations and hypotheses (Kivisild et al., 2004; Abu-Amero et al., 2008; Cerný et al., 2008).

Both mitochondrial and Y chromosome DNA surveys (Luis et al., 2004; Rowold et al., 2007) point to a fundamentally exclusive use of the Levantine corridor in human dispersals between the Upper Paleolithic and the Neolithic. It seems therefore that, even if the crossing of the southern Red Sea could have been the first step of the Out-of-Africa expansion that ended in global occupation, for some reason it was rarely used up to the last few millennia.

The truth is that traces of older dispersals out of Africa are likely to have been erased by the several subsequent migrations in both directions, all the more so because those early demic movements into Eurasia surely encompassed limited numbers of individuals and/or episodes, hence bound to leave weak genetic signatures (Rowold et al., 2007).

Pleistocene Climates and Mammal Dispersals from Africa to Southwest Asia

Since the Middle Miocene, the Earth has been experiencing a general cooling trend, accompanied by aridification, that intensified at the end of the Pliocene and particularly at 1 Ma, with the start of the rapid Middle to Late Pleistocene glacial–interglacial cycles (Zachos et al., 2001; DeMenocal, 2004).

Biotic movements and interchange are bound to be limited during protracted arid periods. Yet, with the onset of the Pleistocene and despite the increasingly important role as a barrier played by the Saharan belt, faunal migrations between sub-Saharan east Africa and the Levant continued to take place, as illustrated by the 'Ubeidiya Formation dated to ≈1.4 Ma (Tchernov, 1992), the An Nafud fauna dated to ≈1.2 Ma (Thomas et al., 1998), the sites of Evron Quarry and Latamne dated to ~900 ka (Tchernov et al., 1994; Ron et al., 2003), the Gesher Benot Ya'akov site dated to ~780 ka (Goren-Inbar et al., 2000), the presence of *Bos* in the Asbole assemblage dated to 800–600 ka (Alemseged and Geraads, 2000; Geraads et al., 2004), fossiliferous beds of Oumm Qatafa dated to ~213 ka (Porat et al., 2002), and the assemblage of Qafzeh dated to 130–100 ka (Grün et al., 2005), albeit with a low frequency. Notwithstanding that around 1 Ma the Sahara begins to occupy, during peaks of cooling and aridity, an area comparable to its modern extent (Larrasoaña et al., 2003), it should be emphasized that this apparent low rate can be an underestimation, since there is still a large stratigraphic gap regarding the faunal succession during most of the Middle Pleistocene in the southern Levant (Tchernov, 1992).

The faunal assemblage of Qafzeh Cave falls within a long period of recurrent wet episodes, alternated with long droughts, in the southern Negev desert between 140 and 110 ka (Vaks et al., 2007), and also coincides with increased monsoonal precipitation in the eastern Sahara (Crombie et al., 1997; Moeyersons et al., 2002; Frumkin and Stein, 2004; Osmond and Dabous, 2004; Smith et al., 2004). From the same southern Negev Desert speleothems, a shorter and less intense interval of pluvial phases was identified at ~90 ka (Vaks et al., 2007), again synchronous with increased monsoonal intensity at the Sahara Desert (Fontes and Gasse, 1991; Yan and Petit-Maire, 1994; Szabo et al., 1995; Dabous et al., 2002; Frumkin and Stein, 2004) and could have been yet another climatic window for species movements via the Sinai land bridge.

The eastern Sahara registers many wet episodes between 100 and 80 ka (Dabous et al., 2002), and markedly shorter and more sporadic ones between 40 and 24 ka(Yan and Petit-Maire, 1994; Moeyersons et al., 2002). Reported evidences of extra Late Pleistocene mesic periods in the Sahel and western Sahara (Rognon, 1987; Fontes and Gasse, 1991; Yan and Petit-Maire, 1994) are not detected by surveys in Egypt and northern Sudan, possibly because eastern Sahara is the driest region in northern Africa (Szabo et al., 1995).

Although Arabia did not escape the global trend for increased aridity during glacial periods, with conditions in general drier and cooler than today, there is evidence for at least two pluvial interludes in the region within the last glacial period. The first, 82–78 ka (Burns et al., 2001; Fleitmann et al., 2003), i.e., during the Dansgaard-Oeschger interstadial 21 (D/O IS 21) (Schulz et al., 1998), was a return to extended monsoon conditions similar to the ones last observed in the first half of the previous interglacial (Burns et al., 2001; Fleitmann et al., 2003). The second was a prolonged phase

from 35 to 25 ka (Woods and Imes, 1995; Glennie and Singhvi, 2002; Bray and Stokes, 2004), though less wet than the former, of increased precipitation throughout the peninsula, leading to the formation of extensive lakes in the Rub' al Khali (McClure, 1976) and An Nafud (Schulz and Whitney, 1986).

In the last 340,000 years, i.e., since Oxygen Isotope Stage (OIS) 9, besides the first half of the Last Interglacial (130–120 ka), wet episodes during OIS 5c (~100 Ka) and OIS 5a (~80 ka) and a period between 35 and 24 Ka there are two additional time intervals for which reported pluvial phases in eastern Sahara and Arabia seem to overlap extensively. They both fall within interglacial periods, respectively at 320–300 ka (Szabo et al., 1995; Fleitmann et al., 2003) and at 200–180 ka (Crombie et al., 1997; Fleitmann et al., 2003; Osmond and Dabous, 2004).

In summary, despite a few differences between the regional records, probably due to local environmental characteristics, dating uncertainty, and/or sampling disparity, the humid/arid variations during the last glacial–interglacial cycle are roughly similar between the eastern Sahara, the Sinai peninsula, and Arabia.

An important point, noted by Vaks et al. (2007), is that this climatic agreement declines significantly as we move further north into the Levant. Indeed, whereas central and northern parts of the Negev, like northern and central Israel, mainly receive their rainfall from the eastern Mediterranean Sea, moist episodes in the southern Negev are generally the result of tropical monsoons migrating northwards (Kahana et al., 2002; Amit et al., 2006). This effectively decouples the humid/arid periods between the northern boundary of the Saharo-Arabian Desert and the Levant north of the southern Negev (Vaks et al., 2006). It is therefore plausible to envisage species migrations between Africa and Arabia, via the Sinai land bridge, that would not expand into the Levant if taking place during periods in which monsoonal episodes concurred in the eastern Sahara and Arabia while the northern Negev and essentially the whole of the Mediterranean Levant were arid (Vaks et al., 2006).

Zoogeography of Arabian Mammals and Afro-Arabian Dispersal Routes

A main argument against the hypothesis that the majority of the Afrotropical mammalian species currently, or until recent times, inhabiting Arabia arrived through the Sinai land bridge during wet episodes of the Middle Pleistocene, Late Pleistocene, and Holocene, is the larger number of Afrotropical species in the southwest than in the southeast of the peninsula, which would suggest the former region as the predominant gate into Arabia for dispersals from Africa (Delany, 1989). Based on the biogeographic affinities of the mammalian species in Arabia that are restricted to the south of the peninsula, Delany (1989) indicated that in southwest Saudi Arabia and Yemen the Afrotropical element is markedly stronger than in Oman.

However, if we constrain this comparison to non-flying mammals, by removing the bat species, such difference is hardly significant: five species in southwest Saudi Arabia and Yemen (hamadryas baboon, genet, white-tailed mongoose, African grass rat, and rock rat) and four species in Oman (Somali shrew, genet, white-tailed mongoose, and African grass rat). Several bat species were able to traverse water barriers much larger than the Red Sea (Juste et al., 2004; Chen et al., 2006; Salgueiro et al., 2007). Accordingly, Delany (1989) acknowledges that only for bat species southern Arabia appears to act as a corridor linking Africa and Iran (Delany, 1989). The apparent asymmetry between western and eastern south Arabia, which in any case becomes almost insignificant when we remove the bats from the comparison, might be consequence of the lowland area between the Dhofar and Hajar Mountains being a biogeographic filter for some mammal species, as seemingly it was for most animal groups other than mammals (Delany, 1989), and not related to the course through which the Afrotropical species reached Arabia.

Origin and Age of the Hamadryas Baboon in Arabia

Baboons (*Papio hamadryas*) comprise a series of parapatric subspecies widely distributed in sub-Saharan Africa, outside of the lowland equatorial forest belt. The five commonly recognized subspecies are chacma, Guinea, yellow, olive (or anubis), and hamadryas, and they share a common mitochondrial ancestor at ≈1.8 Ma (Newman et al., 2004).

The hamadryas baboon is the only subspecies whose range extends beyond the African continent, being found both in the Horn of Africa (Somalia, Ethiopia, Djibouti, and Eritrea) and in Arabia, specifically in western Yemen and southwestern Saudi Arabia. The questions of how and when hamadryas baboons invaded Arabia are pertinent because their absence from the known Pleistocene fossil record of the Levant (Tchernov, 1992) is at variance with an expansion from Africa to Arabia via Sinai. Thus, besides being an excellent case study in Afro-Arabian biogeography, it can be an illuminating model to the way a certain distant cousin, *Homo sapiens sapiens*, left Africa in a journey ending in worldwide distribution.

In two genetic studies on how the hamadryas baboons could have reached Arabia, Wildman et al. (2004) and Winney et al. (2004) discussed three alternative scenarios. The Arabian populations could be remnants of a past continuous range, or long distance dispersal, around the Red Sea, albeit the afore-

mentioned absence of *Papio* fossils from the Levant seems incongruent with this hypothesis. Alternatively, Egyptians of the Dynastic and Ptolemaic periods could have introduced them accidentally, since it is known that baboons were imported to Egypt at the time. Finally, they could have entered in Arabia by a postulated land bridge across the southern Red Sea, which would have been intermittently emerged during Pleistocene sea-level lowstands.

From a phylogenetic analysis of mtDNA haplotypes of Arabian and African hamadryas, of the other baboon subspecies, and of other papionins (gelada baboon, mandrill, red-capped mangabey, and Barbary macaque) used as outgroups, Wildman et al. (2004) estimated the timing of the divergence between Arabian and African hamadryas populations. Since no statistically significant differences in the likelihood scores of the trees were found for the tree with or without an enforced molecular clock, the genetic distances, corrected by the model for DNA sequence evolution best-fitting the data as determined by logarithmic likelihood ratio tests, were used to estimate divergence times between clades. To convert genetic distances, or any other measure of population/ species genetic differentiation, into divergence times we need an estimation of the mutation rate of the DNA region under study. Based on fossil and genetic data suggesting a split between *Papio* and *Theropithecus* (gelada baboon) at ≈4 Ma (Delson et al., 2000; Harris, 2000), the approximate mutation rate of the analyzed mtDNA region (comprising a fragment at the 3′ end of the *ND4* gene, the tRNA genes for histidine (His), serine (Ser), and leucine (Leu), and a fragment at the 5′ end of the *ND5* gene) was estimated at 1.5% per Ma. This is in general agreement with reported rates for the same mtDNA region in other primates (Brown et al., 1982; Hayasaka et al., 1996). In their estimation of the divergence time of the Arabian hamadryas clades, Winney et al. (2004) employed a similar approach but used a different mtDNA fragment, a segment of the D-loop hypervariable region I (HVR1) comparable to the one sequenced by Hapke et al. (2001) in hamadryas baboons from Eritrea. The mutation rate used by Winney et al. (2004), 15–20% per Ma, is a D-loop-specific transition mutation rate calibrated by assuming a human-chimpanzee split at 5–7 Ma (Jensen-Seaman and Kidd, 2001).

Wildman et al. (2004) found that the haplotypes from Arabian hamadryas baboons do not form a single monophyletic clade, which would point to a single period for the colonization of Arabia. Instead, they constitute two well-separated clades, Clade IIA and a subclade of Clade IIB (see their Fig. 2), a clade also including haplotypes from African hamadryas, suggesting at least two successful colonization episodes of Arabia. From now onwards, to facilitate the discussion, the latter subclade is here designated by IIB3. Winney et al. (2004) detected a similar pattern (see their Fig. 3), with the Arabian haplotypes falling within Clade 1, exclusive to Arabia, and within Clade 2, which also comprised African hamadryas haplotypes.

Assuming that the Arabian populations share a direct common ancestry with the ones actually sampled in Africa, colonization times can be very roughly approximated by the divergence times of the two Arabian clades. Wildman et al. (2004) estimated these as $400,000 \pm 78,000$ years for Clade IIA and $109,000 \pm 49,000$ years for Clade IIB3. Based on maximum likelihood genetic distances between Arabian and African haplotypes, Winney et al. (2004) estimated an age of $380,000 \pm 64,000$ years for Clade 1 and an age of at least $103,000 \pm 17,000$ years for subclade 2K (see below).

These results are markedly incongruent with any suggestion that ancient Egyptians, or other people in historic times, introduced hamadryas baboons in Arabia. In contrast, they do not bear on the plausibility of the other two alternative scenarios: invasions across the southern Red Sea versus a past continuous range or long distance dispersal around the Red Sea. This difficulty closely resembles the situation concerning the debate on the most likely route for the Out-of-Africa expansion of modern humans. In face of the absence of fossil remains from critical areas/periods and of conclusive archaeological spatial patterns or temporal successions favoring any particular hypothesis, genetic data could be the decisive argument but the fact is that the currently available evidence cannot rule out a dispersal around the Red Sea (Stringer, 2000, 2002).

Despite acknowledging that divergence time estimates are necessarily approximate, affected by statistical errors that are difficult to eliminate, the fact is that the means obtained by Wildman et al. (2004) and Winney et al. (2004) for what can be hypothesized as the colonization times of Arabia by hamadryas baboons do not coincide with episodes in which the Red Sea could have been shallow and narrow (Fernandes et al., 2006), at least enough to entertain suggestions like the one for a facilitated passage at the OIS 2 (Bailey et al., 2007).

Traditional phylogenetic methods may lack power when applied to divergence time estimation of recently evolved taxa and such cases may be better suited to a coalescent framework, in which the evolutionary vagaries of individual genetic lineages can be traced (Russell et al., 2008). Here I apply sophisticated and recently developed Bayesian coalescent methods to the data sets in Wildman et al. (2004) and Winney et al. (2004) and derive a wealth of additional estimates for the age of the Arabian hamadryas clades. They might corroborate or question the previous estimates and, more importantly, their combination might point to consistent colonization dates that should be seen as more accurate and reliable, given that they are inferred by several independent statistical approaches.

Depending on which time intervals these congruent estimates fall, it might be possible to hypothesize the most likely scenario for the colonization(s) of Arabia by the hamadryas baboon. For instance, if they correspond to phases of a noticeably shallower and narrower southern Red Sea, then the direct route between the Horn of Africa and Arabia is a feasible candidate. Conversely, if they agree with time intervals in which the Red Sea was not strikingly shallower and narrower than today and in which the arid barriers between eastern Africa and Arabia could have been highly reduced, then episodic range continuity or dispersals around the Red Sea might be the best explanation.

Reanalysis of the Colonization Time of Arabia Using Bayesian MCMC Methods

I revisited the mtDNA sequence data sets analyzed by Wildman et al. (2004) and Winney et al. (2004), which are available from GenBank. Accession numbers are AY212061-AY212105, AY488130-AY488132, and NC002764 for the data set studied by Wildman et al. (2004) (see their Table 1 for information on taxa, geographic origin, and haplotypes of the mtDNA sequences), and AY247443-AY247551 and AF275384-AF275475 for the data set studied by Winney et al. (2004) (see their Fig. 1 for information on taxa, geographic origin, and haplotypes of the mtDNA sequences).

Wildman et al. (2004) and Winney et al. (2004) confirmed that the DNA sequence data sets were uncontaminated by nuclear copies of mtDNA, free of significant homoplasy, and suitable for statistical analyses that assume DNA sequences evolving without a significant impact of natural selection. To allow comparisons between the divergence times in Wildman et al. (2004) and the ones obtained here, and given that $\approx 1.5\%$ per Ma is also the expected weighed average mutation rate for a mtDNA fragment with 457 bp of the $ND4$ gene, 239 bp of the $ND5$ gene, and 200 bp of tRNA genes (Pesole et al., 1999), I applied the mutation rate used by Wildman et al. (2004), and also by Newman et al. (2004) in a molecular systematics assessment of $Papio$ using the same mtDNA fragment, when estimating divergence times with their data set. In contrast, to the HVR1 data set investigated by Winney et al. (2004) I applied Tamura and Nei's modal rate of D-loop evolution 7.5% per Ma (Tamura and Nei, 1993), which assumes that humans and chimpanzees diverged at 4–6 Ma. This mutation rate value has been used in HVR1 studies of apes (Thalmann et al., 2005; Eriksson et al., 2006) and is consistent with pedigree-derived HVR1 mutation rate estimates, when these are corrected for gender and for the probability of intraindividual fixation (Santos et al., 2005). Still, by multiplying them by 0.429 (=7.5/17.5), the results here can be easily compared with those of Winney et al. (2004). I considered a generation time for baboons of 12 years (Rogers and Kidd, 1996) in the calculations of divergence times.

Table 1 Colonization time estimates for each of the Arabian clades. Divergence time estimates and confidence intervals are in thousands of years. Effective Sample Sizes (ESS) for the parameter *t* (IM) and TMRCA (BEAST) are given for each simulation or combination of simulations

Clade	Method[a]	Divergence time estimate[b]	Confidence interval[c]	ESS
IIA	IM	322,715	[135,381–?]	1,535
IIA	IM	336,177	[82,480–?]	1,524
IIA	IM	336,367	[118,695–?]	1,351
IIA	IM	334,850	[124,005–?]	2,992
IIB3	IM	109,025	[33,940–?]	20,640
IIB3	IM	97,459	[26,925–?]	17,662
IIB3	IM	97,400	[26,900–?]	16,192
IIB3	IM	117,937	[26,166–?]	28,551
1	IM	224,381	[142,095–?]	1,123
1	IM	215,810	[157,524–?]	6,688
1	IM	213,143	[157,522–?]	6,319
1	IM	208,952	[157,520–?]	3,560
2 K	IM	[d]	–	–
IIA	BEAST	542,000	[32,180–1,352,000]	4,120
IIB3	BEAST	481,000	[84,850–1,045,000]	2,154
1	BEAST	279,000	[138,000–449,000]	6,532
2 K	BEAST	78,340	[10,980–167,000]	19,700

[a]For IM each line corresponds to a single simulation, while for BEAST each line corresponds to a combination of four simulations.
[b]Estimates are population divergence time estimates for IM and TMRCA estimates for BEAST.
[c]Confidence intervals are 90% HPD intervals for IM and 95% HPD intervals for BEAST.
[d]Estimates were consistently ≈300,000, which is seemingly incompatible with the TMRCA estimates for Clade 2 K ≈ 80,000, and therefore considered unreliable.

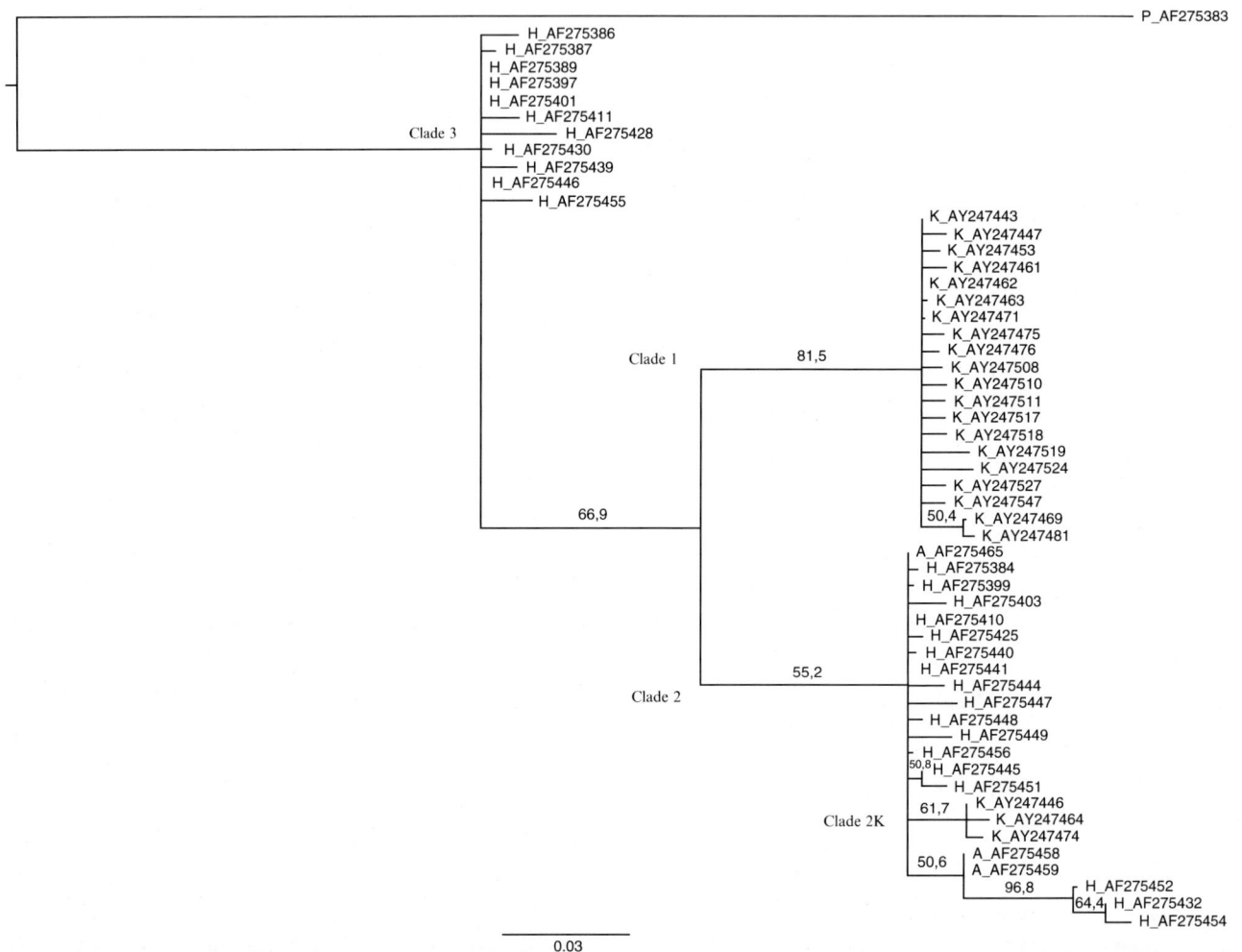

Fig. 1 Phylogeny of the D-loop haplotypes analyzed by Winney et al. (2004), rooted with a Guinea baboon haplotype described by Hapke et al. (2001). Shown is the 50% consensus tree and the numbers above branches are bootstrap values (1,000 replicates). Haplotypes are identified by taxon and geographic origin (H – African hamadryas, K – Arabian hamadryas, A – anubis baboons, P – Guinea baboon), and GenBank accession numbers. The clades identified by Winney et al. (2004) (Clades 1, 2, and 3) are indicated. Clade 2 K is here introduced because it is useful for the discussion

Figure 1 shows a maximum likelihood (ML) phylogenetic tree of the mtDNA haplotypes analyzed by Winney et al. (2004). The ML analyses were performed using TREEFINDER (Jobb et al., 2004; version June 2008). The best-fit evolutionary model was selected by the Akaike Information Criterion with correction for small sample size (AICc) (Posada and Buckley, 2004), and used for phylogeny reconstruction. The selected model was of the type Hasegawa-Kishino-Yano (HKY) (Hasegawa et al., 1985). Non-parametric bootstrapping (1,000 replicates) was employed to evaluate nodal support among clades. The 50% consensus tree was visualized and edited with Fig Tree (Rambaut, 2008; version 1.1.2). A similar ML analysis was performed on the haplotypes reported by Wildman et al. (2004), except that the data was partitioned by dividing the haplotype alignment into site classes, by genes (*ND4*, *ND5*, and the tRNA genes) and by codon position in the two protein-coding genes. The best-fit evolutionary models for each partition were selected as above and, for all partitions, the selected models were of the type HKY. Thus, I also used a HKY model in all the analyses described below to estimate divergence times. The recovered topology (not shown) was essentially the same as the one obtained by Wildman et al. (2004), except for minor variations that are an expected result of some methodological differences. Likewise, the tree in Fig. 1 was only generated to assist the discussion below, not to assess the original phylogenetic reconstruction.

IM is a computer program that applies the Isolation with Migration model (Nielsen and Wakeley, 2001) to genetic data drawn from a pair of closely related populations or species (Hey and Nielsen, 2004). One of the appealing attributes of IM is that it allows implementation of a version of the basic model in which the descendant populations, of an ancestral population that has split into two, are free to undergo changes in size. The model has then seven demographic parameters that can capture many of the phenomena that may occur when one population splits into two: the splitting event may be ancient or recent, the ancestral and the two descendant populations may differ in size, there may have been migration between the two descendant populations since the split that originated them, and this gene flow may have occurred more in one direction than the other. Although the software infers all seven parameters by default, here I am mostly interested in the population divergence time parameter (*t*). An additional interesting attribute of IM is that it produces marginal, not joint, posteriors, so uncertain inference for one parameter is not reflected in uncertain inference for another parameter.

Each program run begins by simulating a genealogy that is repeatedly updated to a new value following specific criteria. The simulation is of the type called Markov chain Monte Carlo (MCMC) and the acceptance-rejection criterion is the Metropolis–Hastings algorithm (Nielsen and Wakeley, 2001;

Hey and Nielsen, 2004). The program generates random samples from the posterior probability distribution. At intervals, a record is made of the current status of the simulation, and over the course of a sufficiently long run the distribution of recorded values is expected to approximate the posterior probability density of those values. To evaluate if the program has run for long enough, and the sample has converged to the true posterior probability distribution, a measure of the independence of the recorded values over the course of the run is estimated. This measure, Effective Sample Size (ESS), should have values greater than 200 for any estimated parameter. Also, it is recommended that at least three runs, using different seed numbers for the random number generator, be carried out. Beforehand, a number of preliminary runs are required to find a range of prior distributions that include all or most of the range over which the posterior density is non-trivial. Results can be trusted if the three or more runs show high ESS values and generate similar stationary distributions.

To infer the colonization times of Arabia, I estimated the divergence times between Clades IIA and IIC and between Clade IIB3 and the African haplotypes H5 and H10 in Clade IIB (see Fig. 2 in Wildman et al. (2004)). Similarly, divergence times were estimated between Clades 1 and 2 (with Clade 2 K removed) and between Clade 2 K and the other (African) haplotypes in Clade 2 (see Fig. 1). For each of the four estimations, corresponding to two inferred colonizations of Arabia in each of the analyzed data sets, and after optimizing MCMC settings through pilot simulations, four replicate simulations were run using the program IM (version March 2007). Simulations used six Metropolis-coupled chains, with 15 chain swap attempts per step, and were run for five million steps, the first 100,000 steps discarded as "burn-in". Genealogies were updated ten times per step. Priors for the fraction of the ancestral population that did not colonize Arabia, the parameter *s* ($0 < s < 1$), were bounded between 0.5 (assumption of a symmetric split) and 0.99 (assumption that Arabia was founded by a minority of the ancestral population).

The peaks of the resulting posterior distributions were taken as maximum-likelihood estimates of the parameter *t*. Credibility intervals were recorded as the 90% highest posterior density (HPD) interval, which represents the shortest span that includes 90% of the probability density of *t* (Table 1). Note that the upper bounds for the credibility intervals are not given (this is indicated by a question mark) since they critically depend on the assumed prior for the maximum value of *t*, and the curve slowly decreases to zero after the mode of *t* (Johnson et al., 2007; Guillaumet et al., 2008).

BEAST (Bayesian Evolutionary Analysis by Sampling Trees) (Drummond and Rambaut, 2007), a program for the Bayesian analysis of molecular sequences related by an evolutionary tree, has also Metropolis–Hastings MCMC as its core algorithm for coalescent-based estimation of phylogenetic and population genetics parameters. In particular, BEAST allows using relaxed molecular clock models, which do not assume a constant mutation rate across lineages, to estimate divergence times (Drummond et al., 2006).

I used BEAST (version 1.4.7) to infer the haplotype divergence time of each of the four Arabian clades (IIA and IIB3; 1 and 2 K). Estimates were generated under a HKY model with a parameter for the proportion of invariable sites among haplotypes (I) and a parameter (α) for the shape parameter of the gamma distribution modeling the rate heterogeneity among sites. For species-level phylogenies, coalescent priors are generally inappropriate and, consequently, the Yule tree prior was chosen. The starting tree was a UPGMA tree reconstructed by the program. Other priors were left with their default prior distributions. The MCMC operators (the proposal distribution) were auto-optimized by the program during initial runs, and non-trivial runs followed its recommendations. Depending on the data set under analysis, the mean substitution rate was set to 0.015 mutations per site per million years or to 0.075 mutations per site per million years. Two different models of rate variation among branches were investigated, the strict clock and the uncorrelated lognormal-distributed relaxed molecular clock. We found that the data should not be treated as compliant with a global clock rate, because the estimate of the *ucld.stdev* parameter, as well as of the coefficient of variation parameter, in simulations using the lognormal clock ranged between 0.5 and 1.5.

For each replicate, the MCMC was run for 30 million steps and sampled every 500 steps, following a discarded burn-in of 150,000 steps. Convergence to the stationary distribution of individual and combined replicates was confirmed using the application Tracer 1.4 (Drummond et al., 2006). For the two presumed invasions of Arabia by hamadryas baboons, estimates of the age of each of the respective clades, and the associated 95% HPD intervals, were obtained from the posterior distribution of parameters approximated by the combination of four independent MCMC analyses (Table 1). Since BEAST does not correct for shared ancestral polymorphism, it estimates the time to the most recent common ancestor (TMRCA) of the lineages/clades and not of the populations they were sampled from. Still, as long we are aware of this difference, comparisons of the parameters TMRCA and *t* can be informative (Edwards and Beerli, 2000; Arbogast et al., 2002; Griswold and Baker, 2002; Leschen et al., 2008). For instance, TMRCA values for both Clades IIA and IIB3 trace their genealogies back to the time of origin of the hamadryas baboon (Wildman et al., 2004) and suggest a large ancestral population size in eastern Africa. Also, the IM estimates for the foundation of the Arabian Clade 2 K were consistently $\approx$300,000 years ago, which is seemingly irreconcilable with TMRA estimates $\approx$80,000 years ago, and therefore the IM results for this clade were seen as unreliable. Apparently, the genetic divergence at the HVR1 between Clade 2 K and the African haplotypes in

Clade 2 is too small to allow identification of the true posterior probability distribution of the parameter *t*.

Colonization time estimates derived from each of the two data sets were clearly different. This is unsurprising, given that Wildman et al. (2004) and Winney et al. (2004) used different samples, most likely representing different populations either in Africa or in Arabia, and mtDNA fragments with markedly distinct mutation rates. In fact, since the estimated ages of the Arabian founders did not agree between the two data sets, it is not necessary that the two observed Arabian clades in each of the data sets sign the same two inferred invasions of Arabia. Yet, considering that both data sets coincide in detecting exactly the same number of migrations into Arabia in the patterns of genetic variation among the extant populations of hamadryas baboons, it is worthwhile to ponder that they might indeed represent the same dispersal events, with the discrepancies in divergence time estimates between data sets mostly the result of differences in the mutation rate. In this context, it is worth mentioning that Arbogast and Slowinski (1998) showed that the substitution rate for protein-coding genes in apes might be as high as 2.6% per Ma. Assuming that the mutation rate used for the HVR1 data set is accurate, and correcting the divergence time estimates derived from the other data set by multiplying them by 0.577 (=1.5/2.6), IM results from both data sets become strikingly similar. The same holds for the TMRCA estimates of Clades IIA and 1, but not of Clades IIB3 and 2 K although this might be possibly due to sampling issues.

The uncertainty in the estimates of divergence time, reflected by the large confidence limits in the results of both IM and BEAST simulations, highlights an important caveat of single-locus studies and the need for an evaluation based on multilocus data sets.

The divergence time estimates obtained with IM, from either data set, underscore the impression that, apparently, Arabia was colonized by hamadryas baboons during periods in which the current record for the depth and width of the southern Red Sea, inferred from several lines of paleoenvironmental evidence (Fernandes et al., 2006), is hard to reconcile with any thoroughfare connecting Africa to Arabia.

Still across the Red Sea, another hypothesis could be that the Arabian founders resulted from successful landings of sweepstake rafting events, which may be feasible given the timescales involved and the relatively small width of the Red Sea. This mode of dispersal is recognized to have played a role in a number of long-distance transmarine migrations by medium to large-sized mammals (Rogers et al., 2000; van den Bergh et al., 2001; Yoder et al., 2003; de Queiroz, 2005). A difficulty with this type of scenarios is that, unless all the alternatives are highly improbable, it is not straightforward to validate (or falsify) them. Yet, the fact that the genetic diversity seems to be as large in Arabian hamadryas as in African hamadryas (Wildman et al., 2004; Handley et al., 2006) argues against strong bottlenecks in the colonization of Arabia and, for that reason, against extremely small founder numbers. The star-like genealogy of Clade 1 in Winney et al. (2004) indicates that the Arabian population may have been through a bottleneck long after its foundation, but the African population likely suffered equivalent losses of genetic variation during similar, and possibly synchronous with pleniglacial conditions, demographic crashes, and both populations thus show comparable levels of genetic variation.

Given that, in fact, the genetic results do not allow discarding the hypothesis of a colonization of Arabia via the Sinai land bridge, the only argument against it rests on the apparent absence of *Papio* fossils in the Pleistocene record of the Levant. It is important to keep in mind, however, that fossil records need to be interpreted with caution. For instance, the bontebok (*Damaliscus pygargus*) has its last know appearance at Olduvai Bed IV (≈700,000 years ago), but it is still extant in Africa (O'Regan et al., 2005). In particular, the cercopithecid fossil identified from 'Ubeidiya has been assigned to *Theropithecus* (Belmaker, 2002), but it is by no means certain that this is the best attribution (Hughes et al., 2008). Moreover, there is still a large stratigraphic gap regarding the faunal composition in the southern Levant during most of the Middle Pleistocene (Tchernov, 1992; Lahr and Foley, 1998). Short-term range shifts and dispersals, during interstadials or even within short climatic optima that peak interglacial periods, might also be ephemeral enough to leave no fossil traces (Lahr and Foley, 1998). Finally, recent paleoclimatic data (Vaks et al., 2006, 2007) suggests distinct and independent climatic histories south and north of the Negev desert and it is, therefore, not necessary that all species dispersals between Africa and Arabia have included an expansion into the Mediterranean Levant in their itinerary.

In fact, the IM results are compatible with scenarios of range expansion northwards to the Sinai during interglacial OIS 9e (≈330,000 years ago) or interglacial OIS 7c (≈220,000 years ago), and during the second half of interglacial OIS 5e (120,000–110,000 years ago) or at the end of OIS 5a (≈80,000 years ago), probably along the Red Sea shoreline, then into the Arabian peninsula, most likely through an infiltration route alongside the Red Sea coastal mountains that ends in southwestern Arabia (Larsen, 1984). For each of the two postulated colonizations, the accepted dates depend on which data set is considered and on which mutation rate is assumed for the mtDNA fragment used by Wildman et al. (2004). In any case, the alternative dates broadly coincide with suggested humid periods in both the eastern Sahara and Arabia (Szabo et al., 1995; Crombie et al., 1997; Fleitmann et al., 2003; Osmond and Dabous, 2004). Subsequent cooler and drier phases would have caused range contractions to a distribution similar to what we observe today, in Africa with a northern limit imposed by the Nubian Desert. They would also have left surviving pockets in southwestern Arabia,

probably in the refuge offered by the sub-montane strip west of the Sarawat and Kaur Mountains, a region where ecological conditions are likely to have been consistently more favorable during arid periods in the peninsula.

It is interesting that, as noted by Wildman et al. (2004) for Yemen, the most divergent (from African counterparts) Arabian haplotypes (i.e., Clade IIA) were found to the south of latitude 15° N, whereas the least divergent (i.e., Clade IIB3) were recovered to the north of 15° N. This observation is seemingly more in accordance with the expected spatial distribution of haplotypes in Arabia if colonizers from Africa actually arrived from the north in two separate phases, the descendants of the first event occurring more southerly in Arabia, than with an invasion of Arabia by a southern Red Sea route. Curiously, Winney et al. (2004) detected an opposite north–south distribution pattern in Saudi Arabia, with Clade 2 K apparently restricted to their southernmost sampling location (Abha), which led them to argue against a northern colonization via Sinai. Sample sizes per location (four locations in total; see their Fig. 1) were notably unbalanced, being for example larger by an order of magnitude at the southernmost location (N=72) than at the northernmost location (N=6), but the authors suggested that sample sizes provided sufficient power to detect Clade 2 K north of Abha. Still, even if analyses of additional samples corroborate the pattern, one possible explanation is that Clade 2 K has been lost north of southwestern Arabia by genetic drift accompanying range contraction during periods of arid climate. Alternatively, accepting that the second colonization (Clade 2 K) possibly occurred across the southern Red Sea, if a land bridge had been the cause it is reasonable to assume faunal movements across it in both directions. However, neither Winney et al. (2004) nor Wildman et al. (2004) found any African representative of the older Arabian clades (respectively, Clade 1 and Clade IIA), an unexpected result for a land bridge hypothesis. Lastly, it is worth noting that the absence of a clear latitudinal structure or gradient of Clade 1 does not favor any specific hypothesis for the first recorded invasion of Arabia by hamadryas baboons.

Here, I intersected phylogeographic (Wildman et al., 2004; Winney et al., 2004), paleoenvironmental (Siddall et al., 2003; Fernandes et al., 2006; Vaks et al., 2006, 2007), and paleontological (Tchernov, 1992) evidence with recent statistical approaches to the issue of estimating divergence times from genetic differentiation to show, using the hamadryas baboons as a case study, that it is still too early to assume a southern Red Sea route as a central dogma in Afro-Arabian biogeography of the Plio-Pleistocene. It is clear that we need more data from all the aforementioned disciplines, in particular more genetic surveys of Afro-Arabian species to allow comparisons of phylogeographic patterns and colonization time estimates.

It is also expected that progress in our understanding about the routes and times of the faunal movements between

Africa and Arabia will shed light on the details of the Out-of-Africa migration of our species, even more now when evidence favors Arabia as its first station. Indeed, as underlined by Tchernov (1992), every recorded Pleistocene expansion of Afrotropical species into southwest Asia is associated with a hominin dispersal.

Acknowledgments   I would like to thank Michael Petraglia for the invitation and encouragement to contribute to this book, and for waiting patiently for the manuscript to actually arrive. I am also grateful to Sarah Elton and one anonymous referee for their insightful comments that helped considerably in improving the manuscript, to Dinis Pestana for statistical advice, and to Mónica Rodrigues for help with the references. I acknowledge support from the Fundação para a Ciência e a Tecnologia, grant BPD/20,317/2004.

References

Abu-Amero KK, Larruga JM, Cabrera VM, González AM. Mitochondrial DNA structure in the Arabian peninsula. BMC Evolutionary Biology. 2008;8:45.

Alemseged Z, Geraads D. A new Middle Pleistocene fauna from the Busidima–Telalak region of the Afar, Ethiopia. Comptes Rendus de l'Académie des Sciences – Earth and Planetary Sciences. 2000;331:549–56.

Amit R, Enzel Y, Sharon D. Permanent Quaternary hyperaridity in the Negev, Israel, resulting from regional tectonics blocking Mediterranean frontal systems. Geology. 2006;34:509–12.

Arbogast BS, Slowinski JB. Pleistocene speciation and the mitochondrial DNA clock. Science. 1998;282:1955.

Arbogast BS, Edwards SV, Wakeley J, Beerli P, Slowinski JB. Estimating divergence times from molecular data on population genetic and phylogenetic time scales. Annual Review of Ecology and Systematics. 2002;33:707–40.

Bailey GN, Flemming NC, King GCP, Lambeck K, Garry M, Lawrence JM, et al. Coastlines, submerged landscapes, and human evolution: the Red Sea basin and the Farasan islands. The Journal of Island and Coastal Archaeology. 2007;2:127–60.

Belmaker M. First evidence of *Theropithecus* sp. in the southern Levant. Israel Journal of Zoology. 2002;48:165.

Beyin A. The Bab al Mandab vs the Nile-Levant: an appraisal of the two dispersal routes for early modern humans out of Africa. African Archaeological Review. 2006;23:5–30.

Bray HE, Stokes S. Temporal patterns of arid-humid transitions in the south-eastern Arabian peninsula based on optical dating. Geomorphology. 2004;59:271–80.

Brown WM, Prager EM, Wang A, Wilson AC. Mitochondrial DNA sequences of primates: tempo and mode of evolution. Journal of Molecular Evolution. 1982;18:225–39.

Burns SJ, Fleitmann D, Matter A, Neff U, Mangini A. Speleothem evidence from Oman for continental pluvial events during interglacial periods. Geology. 2001;29:623–6.

Cavalli-Sforza LL, Menozzi P, Piazza A. Demic expansions and human evolution. Science. 1993;259:639–46.

Cerný V, Mulligan CJ, Rídl J, Zaloudková M, Edens CM, Hájek M, et al. Regional differences in the distribution of the sub-Saharan, west Eurasian, and south Asian mtDNA lineages in Yemen. American Journal of Physical Anthropology. 2008;136:128–37.

Chen S, Rossiter SJ, Faulkes CG, Jones G. Population genetic structure and demographic history of the endemic Formosan lesser horseshoe bat (*Rhinolophus monoceros*). Molecular Ecology. 2006;15:1643–56.

Crombie MK, Arvidson RE, Sturchio NC, Alfy ZE, Zeid KA. Age and isotopic constraints on Pleistocene pluvial episodes in the Western desert, Egypt. Palaeogeography, Palaeoclimatology, Palaeoecology. 1997;130:337–55.

Dabous AA, Osmond JK, Dawood YH. Uranium/Thorium isotope evidence for ground-water history in the Eastern desert of Egypt. Journal of Arid Environments. 2002;50:343–57.

de Queiroz A. The resurrection of oceanic dispersal in historical biogeography. Trends in Ecology and Evolution. 2005;20:68–73.

Delany MJ. The zoogeography of the mammal fauna of southern Arabia. Mammal Review. 1989;19:133–52.

Delson E, Terranova CJ, Jungers WL, Sargis EJ, Jablonski NG, Dechow PC. Body mass in Cercopithecidae (Primates, Mammalia): estimation and scaling in extinct and extant taxa. American Museum of Natural History Anthropological Papers. 2000;83:1–159.

deMenocal PB. African climate change and faunal evolution during the Pliocene–Pleistocene. Earth and Planetary Science Letters. 2004;220:3–24.

Derricourt R. Getting "Out of Africa": sea crossings, land crossings and culture in the hominin migrations. Journal of World Prehistory. 2005;19:119–32.

Drummond AJ, Rambaut A. BEAST: Bayesian evolutionary analysis by sampling trees. BMC Evolutionary Biology. 2007;7:214.

Drummond AJ, Ho SY, Phillips MJ, Rambaut A. Relaxed phylogenetics and dating with confidence. PLoS Biology. 2006;4:1–12.

Edwards SV, Beerli P. Gene divergence, population divergence, and the variance in coalescence time in phylogeographic studies. Evolution. 2000;54:1839–54.

Eriksson J, Siedel H, Lukas D, Kayser M, Erler A, Hashimoto C, et al. Y-chromosome analysis confirms highly sex-biased dispersal and suggests a low male effective population size in bonobos (*Pan paniscus*). Molecular Ecology. 2006;15:939–49.

Fernandes CA, Rohling EJ, Siddall M. Absence of post-Miocene Red Sea land bridges: biogeographic implications. Journal of Biogeography. 2006;33:961–6.

Fleitmann D, Burns SJ, Mudelsee M, Neff U, Mangini A, Matter A. Changing moistures sources over the last 330000 years in northern Oman from fluid-inclusion evidence in speleothems. Quaternary Research. 2003;60:223–32.

Fontes JC, Gasse F. Chronology of the major palaeohydrological events in NW Africa during the late Quaternary: PALHYDAF results. Hydrobiologia. 1991;214:367–72.

Forster P. Ice ages and the mitochondrial DNA chronology of human dispersals: a review. Philosophical Transactions of the Royal Society of London. Series B, Biological Sciences. 2004;359:255–64.

Froukh T, Kochzius M. Genetic population structure of the endemic four-line wrasse (*Larabicus quadrilineatus*) suggests limited larval dispersal distances in the Red Sea. Molecular Ecology. 2007;16:1359–67.

Frumkin A, Stein M. The Sahara-East Mediterranean dust and climate connection revealed by strontium and uranium isotopes in a Jerusalem speleothem. Earth and Planetary Science Letters. 2004;217:451–64.

Geraads D, Alemseged Z, Reed D, Wynn J, Roman DC. The Pleistocene fauna (other than Primates) from Asbole, lower Awash valley, Ethiopia, and its environmental and biochronological implications. Géobios. 2004;37:697–718.

Glennie KW, Singhvi AK. Event stratigraphy, paleoenvironment and chronology of SE Arabian deserts. Quaternary Science Reviews. 2002;21:853–69.

González AM, Larruga JM, Abu-Amero KK, Shi Y, Pestano J, Cabrera VM. Mitochondrial lineage M1 traces an early human backflow to Africa. BMC Genomics. 2007;9:2164–8.

Goren-Inbar N, Feibel CS, Verosub KL, Melamed Y, Kislev ME, Tchernov E, et al. Pleistocene milestones on the Out-of-Africa corridor at Gesher Benot Ya'aqov Israel. Science. 2000;289:944–7.

Griswold CK, Baker AJ. Time to most recent common ancestor and divergence times of populations of common chaffinches (*Fringilla coelebs*) in Europe and North Africa: insights into Pleistocene refugia and current levels of migration. Evolution. 2002;56:143–53.

Grün R, Stringer C, McDermott F, Nathan R, Porat N, Robertson S, et al. U-series and ESR analyses of bones and teeth relating to the human burials from Skhul. Journal of Human Evolution. 2005;49:316–34.

Guillaumet A, Crochet PA, Pons JM. Climate-driven diversification in two widespread *Galerida* larks. BMC Evolutionary Biology. 2008;8:32.

Handley LJL, Hammond RL, Emaresi G, Reber A, Perrin N. Low Y chromosome variation in Saudi-Arabian hamadryas baboons (*Papio hamadryas hamadryas*). Heredity. 2006;96:298–303.

Harris EE. Molecular systematics of the Old World monkey tribe Papionini: analysis of the total available genetic sequences. Journal of Human Evolution. 2000;38:235–56.

Hasegawa M, Kishino H, Yano T. Dating of the human-ape splitting by a molecular clock of mitochondrial DNA. Journal of Molecular Evolution. 1985;22:160–74.

Hayasaka K, Fujii K, Horai S. Molecular phylogeny of macaques: implications of nucleotide sequences from an 896-base pair region of mitochondrial DNA. Molecular Biology and Evolution. 1996;13:1044–53.

Hey J, Nielsen R. Multilocus methods for estimating population sizes, migration rates and divergence time, with applications to the divergence of *Drosophila pseudoobscura* and *D. persimilis*. Genetics. 2004;167:747–60.

Hudjashov G, Kivisild T, Underhill PA, Endicott P, Sanchez JJ, Lin AA, et al. Revealing the prehistoric settlement of Australia by Y chromosome and mtDNA analysis. Proceedings of the National Academy of Sciences USA. 2007;104:8726–30.

Hughes JK, Elton S, O'Regan HJ. *Theropithecus* and "Out of Africa" dispersals in the Plio-Pleistocene. Journal of Human Evolution. 2008;54:43–77.

Jensen-Seaman MI, Kidd KK. Mitochondrial DNA variation and biogeography of eastern gorillas. Molecular Ecology. 2001;10:2241–7.

Jobb G, Von Haeseler A, Strimmer K. TREEFINDER: a powerful graphical analysis environment for molecular phylogenetics. BMC Evolutionary Biology. 2004;4:18.

Johnson JA, Dunn PO, Bouzat JL. Effects of recent population bottlenecks on reconstructing the demographic history of prairie-chickens. Molecular Ecology. 2007;16:2203–22.

Juste J, Ibáñez C, Muñoz J, Trujillo D, Benda P, Karatas A, et al. Mitochondrial phylogeography of the long-eared bats (Plecotus) in the Mediterranean Palaearctic and Atlantic islands. Molecular Phylogenetics and Evolution. 2004;31:1114–26.

Kahana R, Ziv B, Enzel Y, Dayan U. Synoptic climatology of major floods in the Negev desert, Israel. International Journal of Climatology. 2002;22:867–82.

Kivisild T, Reidla M, Metspalu E, Rosa A, Brehm A, Pennarun E, et al. Ethiopian mitochondrial DNA heritage: tracking gene flow across and around the gate of tears. American Journal of Human Genetics. 2004;75:752–70.

Lahr M, Foley R. Towards a theory of modern human origins: geography, demography, and diversity in recent human evolution. American Journal of Physical Anthropology. 1998;Suppl 27:137–176.

Lambeck K, Yokoyama Y, Purcell T. Into and out of the Last Glacial Maximum: sea-level change during Oxygen Isotope Stages 3 and 2. Quaternary Science Reviews. 2002;21:343–60.

Larrasoaña JC, Roberts AP, Rohling E, Winklhofer MR, Wehausen R. Three million years of monsoon variability over the northern Sahara. Climate Dynamics. 2003;21:689–98.

Larsen TB. The zoogeographical composition and distribution of the Arabian butterflies (Lepidoptera; Rhopalocera). Journal of Biogeography. 1984;11:119–58.

Leschen RAB, Buckley TR, Harman HM, Shulmeister J. Determining the origin and age of the Westland Beech (*Nothofagus*) gap, New Zealand, using fungus beetle genetics. Molecular Ecology. 2008;17:1256–76.

Luis JR, Rowold DJ, Regueiro M, Caeiro B, Cinnioglu C, Roseman C, et al. The Levant versus the Horn of Africa: evidence for bidirectional corridors of human migrations. American Journal of Human Genetics. 2004;74:532–44.

Maca-Meyer N, González AM, Larruga JM, Flores C, Cabrera VM. Major genomic mitochondrial lineages delineate early human expansions. BMC Genetics. 2001;2:13.

Macaulay V, Hill C, Achilli A, Rengo C, Clarke D, Meehan W, et al. Single, rapid coastal settlement of Asia revealed by analysis of complete mitochondrial genomes. Science. 2005;308:1034–6.

McClure HA. Radiocarbon chronology of late Quaternary lakes in the Arabian desert. Nature. 1976;263:755–6.

Mithen S, Reed M. Stepping out: a computer simulation of hominid dispersal from Africa. Journal of Human Evolution. 2002;43:433–62.

Moeyersons J, Vermeersch PM, Peer PV. Dry cave deposits and their palaeoenvironmental significance during the last 115 ka, Sodmein cave, Red Sea mountains, Egypt. Quaternary Science Reviews. 2002;21:837–51.

Newman TK, Jolly CJ, Rogers J. Mitochondrial phylogeny and systematics of baboons (*Papio*). American Journal of Physical Anthropology. 2004;124:1–17.

Nielsen R, Wakeley J. Distinguishing migration from isolation: a Markov chain Monte Carlo approach. Genetics. 2001;158:885–96.

O'Regan HJ, Bishop LC, Lamb A, Elton S, Turner A. Large mammal turnover in Africa and the Levant between 1.0 and 0.5Ma. In: Head MJ, Gibbard PL, editors. Early-Middle Pleistocene transitions: the land-ocean evidence, vol 247. London: Geological Society Special Publications; 2005. p. 231–249.

Olivieri A, Achilli A, Pala M, Battaglia V, Fornarino S, Al-Zahery N, et al. The mtDNA legacy of the Levantine early Upper Palaeolithic in Africa. Science. 2006;314:1767–70.

Osmond JK, Dabous AA. Timing and intensity of groundwater movement during Egyptian Sahara pluvial periods by U-series analysis of secondary U in ores and carbonates. Quaternary Research. 2004;61:85–94.

Pesole G, Gissi C, De Chirico A, Saccone C. Nucleotide substitution rate of mammalian mitochondrial genomes. Journal of Molecular Evolution. 1999;48:427–34.

Petraglia MD, Alsharekh A. The Middle Palaeolithic of Arabia: implications for modern human origins, behaviour and dispersals. Antiquity. 2003;77:671–84.

Porat N, Chazan M, Schwarcz H, Horwitz LK. Timing of the Lower to Middle Paleolithic boundary: new dates from the Levant. Journal of Human Evolution. 2002;43:107–22.

Posada D, Buckley TR. Model selection and model averaging in phylogenetics: advantages of Akaike information criterion and Bayesian approaches over likelihood ratio tests. Systematic Biology. 2004;53:793–808.

Quintana-Murci L, Semino O, Bandelt HJ, Passarino G, McElreavey K, Santachiara-Benerecetti AS. Genetic evidence of an early exit of *Homo sapiens sapiens* from Africa through eastern Africa. Nature Genetics. 1999;23:437–41.

Quintana-Murci L, Chaix R, Wells RS, Behar DM, Sayar H, Scozzari R, et al. Where west meets east: the complex mtDNA landscape of the Southwest and Central Asian corridor. American Journal of Human Genetics. 2004;74:827–45.

Richards M, Rengo C, Cruciani F, Gratrix F, Wilson JF, Scozzari R, et al. Extensive female-mediated gene flow from sub-Saharan Africa into near eastern Arab populations. American Journal of Human Genetics. 2003;72:1058–64.

Robinson TJ, Matthee CA. Molecular genetic relationships of the extinct ostrich, *Struthio camelus syriacus*: consequences for ostrich introductions into Saudi Arabia. Animal Conservation. 1999;2:165–71.

Rogers J, Kidd KK. Nucleotide polymorphism, effective population size, and dispersal distances in the yellow baboons (*Papio hamadryas cynocephalus*) of Mikumi National Park, Tanzania. American Journal of Primatology. 1996;38:157–68.

Rogers RR, Hartman JH, Krause DW. Stratigraphic analysis of upper cretaceous rocks in the Mahajanga basin, northwestern Madagascar: implications for ancient and modern faunas. Journal of Geology. 2000;108:275–301.

Rognon P. Late Quaternary climatic reconstruction for the Maghreb (North Africa). Palaeogeography, Palaeoclimatology, Palaeoecology. 1987;58:11–34.

Rohling EJ, Fenton M, Jorissen FJ, Bertrand P, Ganssen G, Caulet JP. Magnitude of sea-level lowstands of the past 500000 years. Nature. 1998;394:162–5.

Ron H, Porat N, Ronen A, Tchernov E, Horwitz LK. Magnetostratigraphy of the Evron member-implications for the age of the Middle Acheulian site of Evron Quarry. Journal of Human Evolution. 2003;44:633–9.

Rose J. The question of Upper Pleistocene connections between East Africa and South Arabia. Current Anthropology. 2004;45:551–5.

Rose J. The Arabian corridor migration model: archaeological evidence for hominin dispersals into Oman during the Middle and Upper Pleistocene. Proceedings of the Seminar for Arabian Studies. 2007;37:1–19.

Rowold DJ, Luis JR, Terreros MC, Herrera RJ. Mitochondrial DNA gene-flow indicates preferred usage of the Levant corridor over the Horn of Africa passageway. Journal of Human Genetics. 2007;52:436–47.

Russell AL, Goodman SM, Cox MP. Coalescent analyses support multiple mainland-to-island dispersals in the evolution of Malagasy *Triaenops* bats (Chiroptera: Hipposideridae). Journal of Biogeography. 2008;35:995–1003.

Salgueiro P, Ruedi M, Coelho MM, Palmeirim JM. Genetic divergence and phylogeography in the genus *Nyctalus* (Mammalia, Chiroptera): implications for population history of the insular bat *Nyctalus azoreum*. Genetica. 2007;130:169–81.

Santos C, Montiel R, Sierra B, Bettencourt C, Fernandez E, Alvarez L, et al. Understanding differences between phylogenetic and pedigree-derived mtDNA mutation rate: a model using families from the Azores islands (Portugal). Molecular Biology and Evolution. 2005;22:1490–505.

Schulz E, Whitney JW. Upper Pleistocene and Holocene lakes in the An Nafud, Saudi Arabia. Hydrobiologia. 1986;143:175–90.

Schulz H, Von Rad U, Erlenkeuser H. Correlation between Arabian Sea and Greenland climatic oscillations of the past 110000 years. Nature. 1998;393:54–7.

Shefer S, Abelson A, Mokady O, Geffen E. Red to Mediterranean sea bioinvasion: natural drift through the Suez Canal, or anthropogenic transport? Molecular Ecology. 2004;13:2333–43.

Siddall M, Rohling EJ, Almogi-Labin A, Hemleben C, Meischner D, Schmelzer I, et al. Sea-level fluctuations during the last glacial cycle. Nature. 2003;423:853–8.

Siddall M, Smeed DA, Hemleben C, Rohling EJ, Schmelzer I, Peltier WR. Understanding the Red Sea response to sea level. Earth and Planetary Science Letters. 2004;225:421–34.

Smith JR, Giegengack R, Schwarcz HP, McDonald MMA, Kleindienst MR, Hawkins AL, et al. A reconstruction of Quaternary pluvial environments and human occupations using stratigraphy and geochronology of fossil-spring tufas, Kharga Oasis, Egypt. Geoarchaeology. 2004;19:407–39.

Stringer C. Coasting out of Africa. Nature. 2000;405:24–7.

Stringer C. Modern human origins: progress and prospects. Philosophical Transactions of the Royal Society of London. Series B, Biological Sciences. 2002;357:563–79.

Sun C, Kong QP, Palanichamy MG, Agrawal S, Bandelt HJ, Yao YG, et al. The dazzling array of basal branches in the mtDNA macrohaplogroup M from India as inferred from complete genomes. Molecular Biology and Evolution. 2006;23:683–90.

Szabo BJ, Haynes CV, Maxwell TA. Ages of Quaternary pluvial episodes determined by uranium-series and radiocarbon dating of lacustrine deposits of Eastern Sahara. Palaeogeography, Palaeoclimatology, Palaeoecology. 1995;113:227–42.

Tamura K, Nei M. Estimation of the number of nucleotide substitutions in the control region of mitochondrial DNA in humans and chimpanzees. Molecular Biology and Evolution. 1993;10(3):512–526.

Tchernov E. The Afro-Arabian component in the Levantine mammalian fauna – a short biogeographical review. Israel Journal of Zoology. 1992;38:155–92.

Tchernov E, Horowitz L, Ronen A, Lister A. The faunal remains from Evron Quarry in relation to other Lower Paleolithic hominid sites in the Southern Levant. Quaternary Research. 1994;42:328–39.

Thalmann O, Serre D, Hofreiter M, Lukas D, Eriksson J, Vigilant L. Nuclear insertions help and hinder inference of the evolutionary history of gorilla mtDNA. Molecular Ecology. 2005;14:179–88.

Thomas H, Geraads D, Vaslet D, Memesh A, Billiou D, Bocherens H, et al. First Pleistocene faunas from the Arabian peninsula: an Nafud desert, Saudi Arabia. Comptes Rendus de l'Académie des Sciences – Earth Planetary Sci. 1998;326:145–52.

Thunell RC, Locke SM, Williams DF. Glacio-eustatic sea-level control on Red Sea salinity. Nature. 1988;334:601–4.

Torroni A, Achilli A, Macaulay V, Richards M, Bandelt HJ. Harvesting the fruit of the human mtDNA tree. Trends in Genetics. 2006;22:339–45.

Underhill PA, Kivisild T. Use of Y chromosome and mitochondrial DNA population structure in tracing human migrations. Annual Review of Genetics. 2007;41:539–64.

Underhill PA, Passarino G, Lin AA, Shen P, Mirazón LM, Foley RA, et al. The phylogeography of Y chromosome binary haplotypes and the origins of modern human populations. Annals of Human Genetics. 2001;65:43–62.

Vaks A, Bar-Matthews M, Ayalon A, Matthews A, Frumkin A, Dayan U, et al. Paleoclimate and location of the border between Mediterranean climate region and the Saharo-Arabian Desert as revealed by speleothems from the northern Negev Desert, Israel. Earth and Planetary Science Letters. 2006;249:384–99.

Vaks A, Bar-Matthews M, Ayalon A, Matthews A, Halicz L, Frumkin A. Desert speleothems reveal climatic window for African exodus of early modern humans. Geology. 2007;35:831–4.

Van den Bergh GD, de Vos J, Sondaar PY. The late Quaternary palaeogeography of mammal evolution in the Indonesian Archipelago. Palaeogeography, Palaeoclimatology, Palaeoecology. 2001;171: 385–408.

Walter RC, Buffler RT, Bruggemann JH, Guillaume MMM, Berhe SM, Negassi B, et al. Early human occupation of the Red Sea coast of Eritrea during the Last Interglacial. Nature. 2000;405:65–9.

Werner NY, Mokady O. Swimming out of Africa: mitochondrial DNA evidence for late Pliocene dispersal of a cichlid from Central Africa to the Levant. Biological Journal of the Linnean Society. 2004;82:103–9.

Wildman DE, Bergman TJ, al-Aghbari A, Sterner KN, Newman TK, Phillips-Conroy JE, et al. Mitochondrial evidence for the origin of hamadryas baboons. Molecular Phylogenetics and Evolution. 2004;32:287–96.

Winney BJ, Hammond RL, Macasero W, Flores B, Boug A, Biquand V, et al. Crossing the Red Sea: phylogeography of the hamadryas baboon, *Papio hamadryas hamadryas*. Molecular Ecology. 2004;13:2819–27.

Won Y-J, Hey J. Divergence population genetics of chimpanzees. Molecular Biology and Evolution. 2005;22:297–307.

Woods WW, Imes JL. How wet is wet? Precipitation constraints on late Quaternary climate in the southern Arabian peninsula. Journal of Hydrology. 1995;164:263–8.

Yan Z, Petit-Maire N. The last 140 ka in the Afro-Asian arid/semi-arid transitional zone. Palaeogeography, Palaeoclimatology, Palaeoecology. 1994;110:217–33.

Yoder AD, Burns MM, Zehr S, Delefosse T, Veron G, Goodman SM, et al. Single origin of Malagasy Carnivora from an African ancestor. Nature. 2003;421:734–7.

Zachos J, Pagani M, Sloan L, Thomas E, Billups K. Trends, rhythms, and aberrations in global climate 65 Ma to present. Science. 2001;292:686–93.

Chapter 8
Acheulean Landscapes and Large Cutting Tools Assemblages in the Arabian peninsula

Michael D. Petraglia, Nick Drake, and Abdullah Alsharekh

Keywords Acheulean • Dawādmi • Giant Cores • Large Cutting Tools • Quarry • Saffāqah • Wadi Fatimah

Introduction

The expansion of Acheulean populations into the Arabian peninsula is a topic of some importance in human evolutionary studies as it provides information about dispersal routes and the adaptive capabilities of early humans. The presence of Acheulean sites in Arabia provides definitive evidence for the dispersal of populations from their African source. And, indeed, the recovery of characteristic tool types such as handaxes, cleavers and picks, provides solid evidence for Acheulean expansion in new territories. Moreover, the identification of spatially dispersed and sometimes dense concentrations of Acheulean sites provides information concerning hominin landscape behaviors and activities. The aim of this chapter is to review two key Acheulean site complexes in Saudi Arabia, those identified along the Wadi Fatimah near the Red Sea, and those found along hillslopes near the modern town of Dawādmi in the center of the peninsula.

The results of the archaeological investigations conducted along the Wadi Fatimah and at Dawādmi are of importance to modern investigations as these site complexes currently represent the most convincing evidence for an Acheulean presence in Arabia. In these areas, surveys identified a large number of Acheulean sites across land surfaces and certain sites produced dense artifact accumulations with a range of tool types. At Dawādmi, Whalen and colleagues (1981, 1988) conducted the first systematic surface collections and test excavations of Acheulean sites in Arabia. These investigations were a significant achievement as lithic assemblages were sytematically collected and described and inferences about paleoecological settings were made. Though this work was sometimes mentioned in the broader literature on the Acheulean (e.g., Bar-Yosef, 1998), few Paleolithic archaeologists working outside of Arabia have paid serious attention to this research despite the potential importance of these Acheulean sites. The lack of international interest is probably the result of a combination of factors, including the absence of multidisciplinary studies to determine the age and formation of the sites and publication in regional journals, which were difficult to access for many researchers. The investigators also worked in relative isolation, without placing these rather spectacular occurrences in behavioral context, as Isaac (1984) and others were doing in Africa. Indeed, virtually no interdisciplinary work was conducted on these sites, thus the findings were not properly placed in temporal and paleoenvironmental context, a problem still plaguing knowledge about the Arabian Acheulean.

Though detailed technological, geomorphological and paleoenviornmental studies have yet to be performed on the Wadi Fatimah and Dawādmi sites, we believe that these localities are of great value and that they should be brought to the attention of the international community. While we have not been engaged in rigorous study of these sites, some new information about these sites may be offered based on our critical re-examination of the published literature, analysis of site settings through remote sensing, visit to the Dawādmi localities for observations of site contexts and lithic assemblages, and preliminary study of artifacts in the National Museum (Riyadh). Based on this collective information, some inferences are offered concerning paleoecological settings, the nature of stone tool technology and hominin dispersal processes.

M.D. Petraglia (✉)
Research Laboratory for Archaeology and the History of Art
School of Archaeology, University of Oxford,
South Parks Road, Oxford, OX1 3QY, UK
e-mail: michael.petraglia@rlaha.ox.ac.uk

N. Drake
Department of Geography, King's College London,
Strand, London, WC2R 2LS, UK
e-mail: n.drake@kcl.ac.uk

A. Alsharekh
Department of Archaeology and Museology, King Saud University,
P.O. Box 2456, Riyadh, 11451, Kingdom of Saudi Arabia
e-mail: alsharekh@hotmail.com

The Acheulean of the Wadi Fatimah

A total of 32 Acheulean sites were identified along the Wadi Fatimah (Whalen et al., 1981, 1988). The network of tributaries and jebels found throughout this area was considered to be attractive for occupation by early humans. Examination and placement of the site area on satellite imagery indicates that the localities occur along one of the largest drainages in the western province of Saudi Arabia (Fig. 1). The main channel of the wadi originates in the Asir mountains and proceeds in a southwesterly direction to the coastal plain of the Tihama bordering the Red Sea. The Fatimah series of mountain chains border the northern and southern banks of the wadi, consisting of basalt jebels with some andesite and rhyolite hills, as well as occasional andesite dikes. Valleys between jebels varied in width, but all were bisected by small tributaries that flowed into the Wadi Fatimah. The north bank of the Wadi Fatimah was considered more attractive to hominin occupation in comparison to the south bank due, in part, to the aerial exposure of stone tool sources. The north bank had exposures of basalt, andesite, diabase, and rhyolite, all used to manufacture tools. The 32 Acheulean sites all occur within the immediate vicinity of the small wadis and jebel exposures, and 31 of these localities were within 3 km of the highest rim situated just north of the sites.

Though the surveyors indicated that the sites occurred along a wadi network, their rendering of site contexts was schematic (Fig. 2a). Our placement of the sites on a digital elevation model (DEM) indicates that the stream network is more complex than originally illustrated (Fig. 2b). Figure 2b indicates that the stream network is more extensive and the channel network is different than previously described. More apparent on this image are the headwaters, emanating from the high ground. It should be borne in mind, of course, that these tributaries are not necessarily Pleistocene in age; however, the fact that many of the sites are not transported shows that they have not suffered much transformation and thus the landscape at the time of site formation must have been generally similar to that evident today.

With respect to the history of site discovery, initial survey of the area identified Site 210-162, which consisted of 14 artifacts mostly made of andesite (i.e., described as a biface, a polyhedron, a unifacial chopper, a knife, scrapers, flakes) (Whalen et al., 1981). A follow-up survey was conducted by Whalen and colleagues, resulting in the identification of 31 additional Acheulean sites (210-340 to 343; 210-348 to 371; 210-374 to 376) (Whalen et al., 1988). A total of 2,227 artifacts was collected during the survey, the assemblages containing typical Acheulean tool types, described as handaxes, cleavers, picks, bifaces, discoids, polyhedrons, and spheroids. As noted by the investigators, the majority of the cores and bifaces had deep and expanding flake scars, indicating the use of hard hammer percussion.

Typologically, the assemblages were considered to be "Middle Acheulean" on the basis of the presence of some

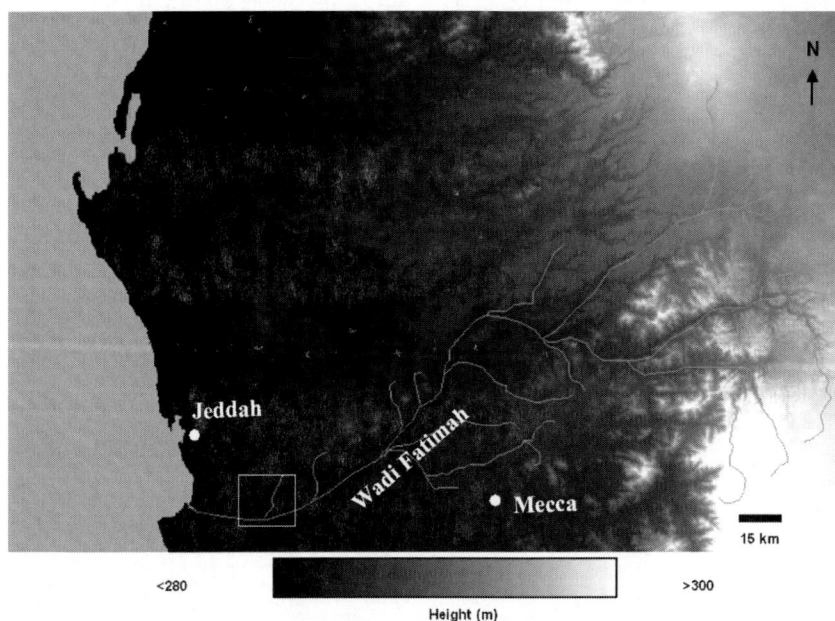

Fig. 1 Shuttle Radar Topography Mission 90 m resolution (STRM-90) digital elevation model (DEM) of the Wadi Fatimah. Black depicts low elevations and white denotes high areas. Most of the large river systems evident in the DEM have been digitized and overlain in white. The Wadi Fatimah emerges from the Asir Mountains and flows into the Red Sea. The location of the Wadi Fatimah sites is just upstream of the coast. Such wadi systems would have provided the opportunity for hominins to travel into the mountainous zone, seeking raw materials along exposed surfaces and to traverse across a number of ecological settings

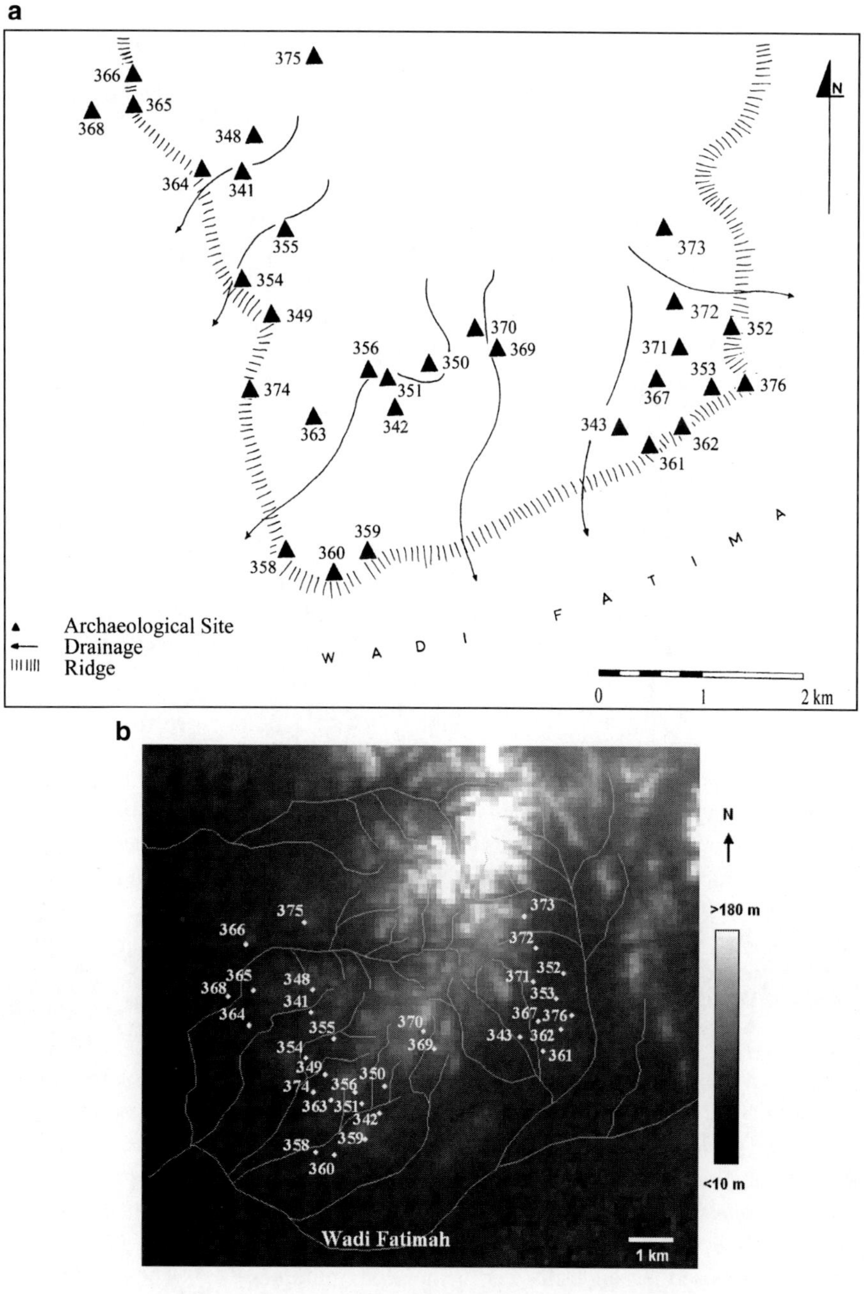

Fig. 2 Comparison of Acheulean site contexts along the Wadi Fatimah based on original survey and a DEM. (**a**) Original depiction of Acheulean site distribution relative to the Wadi Fatimah (after Whalen et al., 1988: Plate 72). (**b**) STRM-90 DEM of the Acheulean site distribution at Wadi Fatimah. The river channels are marked in *grey* and the archaeological sites are marked as *numbers*

flakes considered to be made by the soft hammer technique, the low percentage of certain "older" tool types (polyhedrons, spheroids, trihedral picks) and the recovery of several Levallois flakes and blades. In this respect, the Wadi Fatimah assemblages were thought to share close parallels with those from the Saffāqah locality near Dawādmi (discussed below) (Whalen et al., 1983, 1984).

The most diagnostic tool type found in the Wadi Fatimah was the handaxe, where 28 specimens were collected from 14 sites. Handaxes varied in shape, and were described as lanceolate, amygdaloid, ovate, and subcordiform types. The handaxes often showed deep flake scars, sinuous edges, and irregular cross-sections, and few were highly symmetrical with straight edges or thin cross-sections, though in some examples, some overall shape and symmetry was achieved. The handaxes tended to have deep-flake scars, sinuous edges and irregular unbalanced sections. Often found co-occurring with handaxes, a total of 28 cleavers was found on 12 sites and 27 picks were found on 17 sites. The 83 handaxes, cleavers and picks were accompanied by another 124 bifaces.

Interestingly, a large number of choppers (n = 203) and scrapers (n = 268) were also recovered, indicating the variation in on-site activities. The identification of 116 cores indicates the primary nature of stone tool manufacture on the sites.

Three sites (210-356, 357, 358) had high percentages of choppers and scrapers, with lower percentages of flakes and cores. Two other sites (210-359, 367) had the highest percentage of scrapers and lowest percentage of flakes, with a high rank in bifaces and discoids. An intermediate group (210–370, 371) was high in cores, bifaces, and scrapers.

Site 210-340 contained a small number of artifacts (n = 29) over a large area (300 × 400 m) whereas others (210-349, 350, 355, 365, 374) were denser stone tool manufacturing areas. Based on differences in tool types and the relatively low artifact densities, the investigators concluded that the sites represented variable activities over a relatively short period of time.

In two test pits placed by the investigators at Site 210-351, calcareous nodules were retrieved. Though little information is available from the publication, the nodules produced an age "in the range of 200,000 years" by Uranium–Thorium (Whalen et al., 1988:78). It is probable that these ages were obtained by McMaster University, similar to those produced for Dawādmi (see below). Since the ages were obtained from calcareous nodules that probably formed after site occupation, these chronometric results should be taken as a minimum age.

The Acheulean of Dawādmi

Acheulean sites were identified near the present-day town of Dawādmi in the central province of Saudi Arabia (Zarins et al., 1980). Dawādmi is located on the eastern edge of the Arabian Shield, in a zone known as the Nejd peneplain. The bedrock consists of andesites, granites, schists, and slates that are dissected and faulted. Secondary extrusive dikes largely composed of andesite were emplaced into these rocks. These dikes are usually more resistant to erosion than the surrounding bedrock and thus often form long, isolated ridges that protrude from the surrounding plains. Somewhat perplexing in terms of the location of the Dawādmi sites is the fact that the localities occur in the interior of the Arabian peninsula far from any potential route that may have been along the coasts (Fig. 3). Based on the geographic position of these sites, close to tributaries of two very large paleo-river systems, it is highly probable that Acheulean hominins traversed the then active wadi systems to penetrate deeply into the province.

With respect to the unfolding of site discovery in the region, the first archaeological survey resulted in the identification of some major sites, labelled 206-76 and 206-68 (Zarins

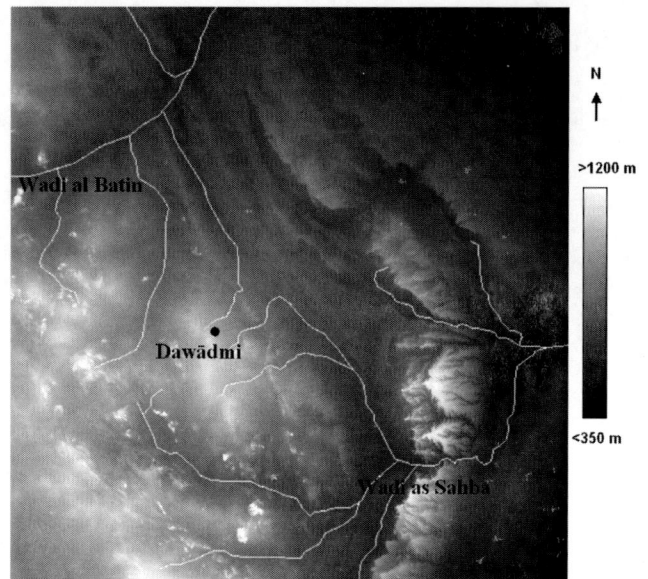

Fig. 3 STRM-90 DEM of the Dawādmi region showing the relationship between topography and the large scale wadi network. A key point is that the Acheulean sites occur in the interior of the peninsula, along a drainage system. Acheulean hominins apparently were able to disperse deep into interior zones by travelling along wadi systems

et al., 1980). The sites were situated on the north side of an andesite dike, Site 206-76 on the western end and Site 206-68 on the eastern end of the dike. The andesite dike is 15–20 m in width, and intruded along a fissure created by faulting of the granite, creating a linear feature measuring 2.5 km in length with a crest 62 m in height. Near the western terminus, close to Site 206-76, the dike disappears underground to re-emerge 100 m farther west, forming a rhyolite outcrop 32 m in height. The transition zone between the black andesite and the pink rhyolite occurred as the dike declined toward the desert floor prior to its submergence into the desert sands. In its eastern terminus, 300 m past 206-68, the dike descended beneath the desert floor, reappearing as an outcrop 800 m farther east.

The lithic scatters at Sites 206-76 and 206-68 were large, each measuring approximately 150 × 200 m. Stone tools were distributed downslope on colluvial surfaces, descending in a northerly direction from the dike propelled by sheetwash and gravity. During the survey of 206-76, a 30 × 30 m grid was established by the investigators, and a controlled surface collection was made, resulting in the recovery of 3,256 artifacts.

Follow-on survey was carried out within a 5 km range of the Dawādmi sites, resulting in the identification of 24 additional Acheulean localities (Whalen et al., 1984) (Fig. 4a). Between 206-68 and 206-76, Whalen and his colleagues described the vestiges of "waterfalls" situated about 400 m apart. The water flow evidence was based on discoloration and polishing of the gneissic outcrops. The evidence for

Fig. 4 Comparison of the Dawādmi site area based on published map and satellite image. (**a**) Original depiction of Acheulean site distribution (Whalen et al., 1984, Plate 17). (**b**) Quickbird 2.4 m spatial resolution satellite image of the Dawādmi site. The image shows the distribution of sites relative to bedrock outcrops, the andesite dike, and the wadi system. Note close relationship of archaeological sites relative to the andesite dike

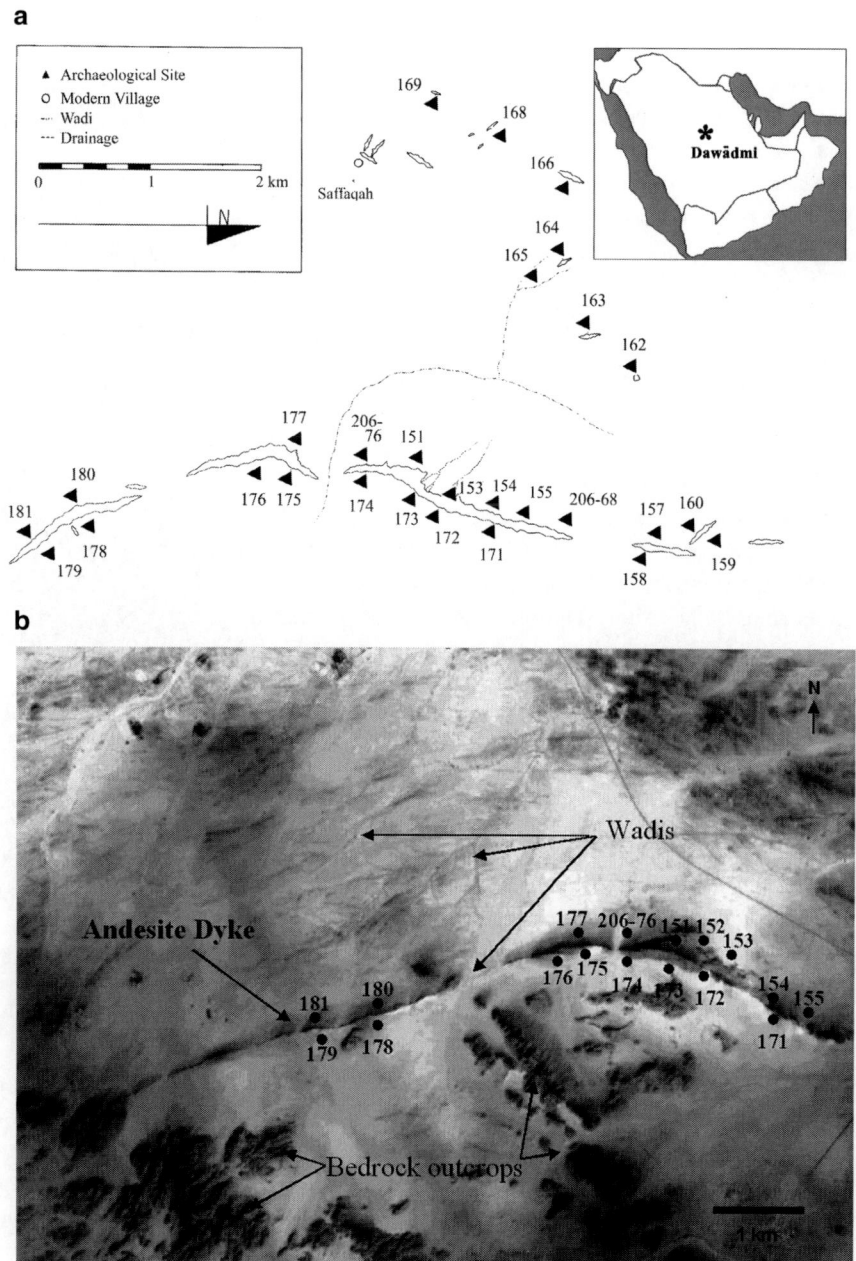

water flow and the basin-like topography of the valley was interpreted as support for a low-lying lake.

Inspection of this area in 2002 by the senior author indicated that the paleoenvironmental inferences were problematic as the low-lying areas did not show any obvious signs of lacustrine deposits. The "waterfalls" described by the investigators are interpreted here as evidence for spring activity. The springs would have formed when the groundwater table was recharged during an extended humid period, inducing lateral near surface groundwater flow through fractures in the bedrock. Some of this flow was likely blocked by the dikes and thus the water in their vicinity was forced to the surface forming lines of springs along the base of the dikes. SRTM DEM images reveal no closed basins in which such lakes could form. Remote sensing imagery shows numerous stream channels in association with the localities (Fig. 4b) throughout the region suggesting that springs fed a large river system rather than a lake.

Two sites (206-151, 206-153) were situated at the features we interpret as springs immediately below the andesite dike.

Site 206-151 was at the corner of the dike and directly behind the stained and weathered bedrock about 25 m above the desert plain. The site occurred on sloping and stepped terraces, and artifacts were recovered in an area measuring 40 × 75 m. Many of the artifacts were reported as cores and flakes with some finished tools. A second site, 206-153, occurred in an area with stained, smoothed, and weathered gneiss, bordered by narrow, stepped terraces. The site was 20–25 m above the desert floor, and measured 30 × 200 m in spatial extent. The artifacts were thought to represent a "workshop." Four additional Acheulean sites were situated on the southern slope of the dike (206-171 to 174) and at the eastern margin (206-157 to 160). Site 206-159 was among the largest of the sites, and measured 120 × 350 m horizontally. On the north side of the valley, at the base of small hills or low ridges, were seven sites (206-162 to 166, 206-168, 206-169). Site 206-162 was distinguished from the majority of the sites as it included small, thin symmetrical handaxes with shallow flake scars unlike most other sites. The investigators argued that the more refined technology on handaxes and cleavers at this site demonstrated the application of soft hammer percussion, thereby indicating a potentially younger age. West of 206-76 were seven Acheulean sites (206-175 to 181). Site 206-177 was situated at the foot of the dike, and was large, measuring 700 m in spatial extent. The stone tools were distributed onto the colluvium on the north side of the dike, extending circumlinearly around its eastern perimeter.

The presence of multiple "Middle Acheulean" sites at Dawādmi was considered to be the result of favorable resource conditions, including the presence of a lake, a complement of plants and animals, and the inexhaustible supply of fine grained andesite for stone tool manufacture. Although not systematically studied, the survey indicated that the sites varied in their stone tool composition and activity. For instance, Site 206-159 had a large number of polyhedrons and spheroids, whereas Site 206-68 had the abundant evidence for stone tool manufacture. The overall similarity of Acheulean tools between the sites indicated little overall technological change, although Site 206-162, was an exception as it possessed small, thin, and highly symmetrical handaxes. Regardless of specific interpretation, the Dawādmi survey is of clear significance, revealing laterally extensive Acheulean activities and landscape behaviors on par with those identified in other parts of the Old World.

The Saffāqah Excavations (206-76, 206-68)

Based on the possibility that subsurface remains existed at Dawādmi, testing was initiated, forming the first excavation of an Acheulean site on the Arabian peninsula (Whalen et al., 1983, 1984) (Fig. 5). Excavations conducted at Sites 206-76

Fig. 5 Excavation trench placed by Whalen at Saffāqah, 206-76. Note the density of clasts in the walls of the excavation trench. Numerous artifacts were recovered from the subsurface contexts. Note the close proximity of the hillslope and the scree surface in the background

and 206-68 indicated that the maximum depth of artifact-bearing deposits was 90 cm. The soil was described by the excavators as a yellow-red laterite, derived from chemical weathering of the granite. However, our review of the excavated trenches show no evidence of laterization; indeed, the deposit seems to be a colluvial deposit near the base of the slope of the dike. The seemingly random distribution of artifacts in the deposit suggests that they were not deposited by primary hominin activities but were transported from previous positions on the dike. Subsurface artifact density was low in comparison to the surface collections, leading the investigators to infer the operation of deflationary processes. Calcareous concentrations adhering to the underside of artifacts were sampled for uranium–thorium dating. Six dates from two laboratories (USGS, McMaster University) were obtained, producing a range from 61 to 204 ka (Whalen et al., 1984: 22). The calcareous deposits were thought to form during wet intervals when the granitic gneiss disintegrated and released dissolved carbonates in the soil. Therefore, the substances dated on the artifacts are postdepositional adherents, and the dates they yielded represent a minimal artifact age.

Although the investigators typed the stone tools as "Middle Acheulean" and assigned a date of 200 ka to the assemblages (Whalen et al., 1988: 78–79), it is probable that the typological and chronometric age estimates are not compatible.

During the systematic survey and after two seasons of excavations, a total of 11,360 artifacts was recovered from 206-76, and 2,444 artifacts was collected from 206-68 (Zarins et al., 1980; Whalen et al., 1983, 1984). The 206-76 artifact tally represents one of the largest controlled Acheulean collections in the Middle East. The preferred raw material was andesite in accordance with the presence of the dike, followed by granite, quartz, and rhyolite. Artifacts were classified as handaxes, cleavers, knives, picks, trihedrals, bifaces, polyhedrons, spheroid, discoids, choppers, core axes, scraper forms, and small tools (borers, burins, notches, chisels, knives), cores, flakes, chunks, blades, and hammerstones (Fig. 6). The sites were considered "Middle Acheulean" based on the recovery of tridherals and polyhedrons that occur in earlier sites, and the presence of some Levallois flakes, which typically occur in later sites. The dominant technique was hard hammer percussion, the bifaces showing large and deep flake scars, sinuous edges, thick cross-sections, and cortical butts. Most bifaces were made from cores, but occasionally they were made on large, side struck flakes. Handaxe shapes were described as cordiform, subcordiform, lanceolate, and ovate. The bifaces showed little attempt at highly balanced symmetry, hence thin bifaces with many flake scars was absent. Cores were rarely prepared, and mostly large nuclei of irregular shape with a few flake removals.

Scrapers did not adhere to a fixed pattern, and were mostly side- and end-scrapers, with some transverse and a few disc and convergent types. Borers, burins, notches, chisels, and small flake knives were also noted.

At 206-76, the investigators found instances of spatial co-occurrences of tools, suggesting the performance of specific functional activities. Stone tool types were assigned to functional activities (e.g., picks assigned to plant gathering/processing; polyhedrons, choppers and cores assigned to bone splitting and smashing). The investigators performed a cluster analysis on the stone tool categories to isolate groups of tools with a high degree of covariance. Their cluster analysis derived seven groups of statistically correlated tool types, which were inferred to relate to specific activities (i.e., butchering and meat slicing; bone splitting and smashing; hide scraping and processing; plant gathering/processing; stone tool manufacturing; wood working; bone working). The investigators analyzed their data in 10 cm arbitrary units, viewing changes in tool types to reflect functional variations through time (Table III in Whalen et al., 1984). While this analysis formed a centerpiece of their functional arguments for site activity (as was also the case at Wadi Fatimah, Whalen et al., 1988), these inferences remain tenuous given the lack of clear spatial patterning, the absence of organic remains, and the lack of visible tool use-wear. Moreover, the functional assignments given to certain tool types is weak, and remains speculative.

Although the Saffāqah investigators do note that "stone tool manufacture" was an important activity, this activity is seriously underplayed in the publications. Visit to this locality

Table 1 Comparison of Acheulean artifacts from Wadi Fatimah and Dawādmi (from Whalen et al., 1988: Table 1). Note that the Wadi Fatimah are surface collections and those from Dawādmi are excavated contexts

Artifact type	Wadi Fatimah No. (%)	Dawādmi No. (%)
Handaxe	28 (1.26)	52 (0.99)
Cleaver	28 (1.26)	69 (1.32)
Pick	27 (1.21)	71 (1.35)
Polyhedron	5 (0.22)	9 (0.17)
Spheroid	5 (0.22)	1 (0.02)
Discoid	36 (1.62)	4 (0.08)
Biface	124 (5.57)	44 (0.84)
Chopper	203 (9.12)	186 (3.56)
Scraper	268 (12.03)	541 (10.34)
Knife	37 (1.66)	106 (2.03)
Borer	15 (0.67)	67 (1.28)
Notch	52 (2.33)	217 (4.15)
Chisel	2 (0.09)	42 (0.80)
Burin	4 (0.18)	41 (0.79)
Core	116 (5.21)	337 (6.44)
Chunk	57 (2.56)	381 (7.28)
Flake	1,194 (53.62)	3,009 (57.53)
Blade	22 (0.99)	47 (0.90)
Hammerstone	4 (0.18)	7 (0.13)
Total	2,227 (100)	5,231 (100)

Fig. 7 Panaromic view of the surrounding landscape at the Saffāqah hillcrest

in 2002 by the senior author showed materially dense and aerially broad distributions of large cores, large flakes, and bifaces in variable state of manufacture. Indeed, the quarrying aspect of the site is perhaps the most notable behavioral feature. Unfortunately, however, no detailed study has been initiated on the Dawādmi sites to understand key elements and variations in manufacturing methods and reduction sequences, and how these may relate to hominin activities.

The tool types from Dawādmi were compared to those collected from the Wadi Fatimah, leading Whalen et al. (1988) to conclude there were close resemblances in typology and technology between the areas (Table 1), which appears to be a reasonable inference at this stage of investigation.

Discussion

Settings and Environments

The Wadi Fatimah and Dawādmi sites are located in topographic contexts which offered Acheulean hominins certain advantages. Elevated spots in both areas provided Acheulean hominins with wide fields of view, allowing for visual inspection of the surrounding landscape. The Wadi Fatimah sites were situated on elevated hillslopes on the north bank of the main river channel, thus providing for views of the opposite hillslopes as well as areas upstream and downstream along the main river channel. The viewsheds from the Dawādmi

sites were panoramic at the top of one hillslope crest (Fig. 7). From such a vantage point, hominins were able to inspect the surrounding landscape up to several kilometers in distance, and in many different directions. At the base of the hillslope, where many sites were located, the outlying plains and river courses were probably visible. At such vantage points, hominins may have been able to identify plant resource patches, locate standing water and track animal movements.

The Wadi Fatimah and Dawādmi sites were situated in zones which offered a fresh water supply in the form of springs and stream systems. Analysis of satellite images indicates that the archaeological localities were located close to stream networks and springs, in close proximity to running and standing water sources. The Wadi Fatimah river channel itself would have been an ideal environment as it is one of the largest valleys draining into the Red Sea. The river measures approximately 190 km from the headwaters to its mouth, and has a wide floodplain where running water and standing water pools must have been available. The wadi would have attracted diverse animal and plant communities along its banks as it traversed upland and lowland settings. Thus, hominins probably wandered across several ecological zones as they traveled up its channel. The Dawādmi "waterfalls" are likely spring heads emanating from the contact between basement rocks and intrusive dikes. Satellite images confirm the presence of drainage headwaters along the hillslopes as well as larger channels on the lower lying plains.

Lithic resources were abundant at both the Wadi Fatimah and Dawādmi, providing a ready supply of raw material for tool-dependent hominins. Andesite and rhyolite were available in both areas, and these clearly attracted Acheulean hominins based upon the identification of a large number of localities in close proximity to the sources. The andesite dikes were particularly prominent landscape features, as raw materials occurred over spatially extensive areas. These raw materials provided the raw material necessary for the manufacture of large cutting tools from cores or as struck flakes from boulder-sized clasts.

A combination of physical and biological factors (i.e., viewsheds, fresh water supply in the form of streams and springs, high plant and animal biomass, spatially extensive stone sources) made Wadi Fatimah and Dawādmi attractive places for Acheulean hominins. The identification of early stage stone tools obviously implies the manufacture of tools in these locations, but the identification of a wealth of artifacts in various spots in the landscape also implies the transport of raw material sources. It is probable that a variety of activities occurred in these various spots as part of landscape habits.

Stone Tool Quarrying Behavior and Giant Cores

Many surveys conducted in Saudi Arabia identified the presence of Acheulean "factory sites". Whalen's excavations at Dawādmi were the first to demonstrate the nature of some of these stone tool manufacturing loci. Though manufacturing activities were identified, little information was retrieved about the stone tool reduction sequences, including processes related to raw material procurement and selection, reduction strategies, tool shaping, and transport. While these localities have not yet been subjected to detailed technological studies, some observations may be offered based upon field observations in the Dawādmi sites and a review of the collections in the National Museum (Riyadh).

During a surface walkover of the Saffāqah localities by the senior author, it was immediately apparent that stone tool manufacturing activities occurred in particular spots along the hillslopes. Confirming Whalen's observations, stone tools were manufactured from andesite, which occurred as dikes and as weathered clasts along the scree slopes (Fig. 8). In association with these natural clasts was a high density of waste which occurred on surfaces along the slopes.

Examination of open excavation trenches and review of Whalen's publications indicated that artifacts were contained in thick deposits. As shown in Table 1, a large number of cores and associated debitage were recovered from the excavations. Though Whalen identified numerous cores, no further information on these important items was available. During surface walkover in 2002, a number of cores of various sizes were readily identified. A significant observation was the occurrence of large cores with struck faces (Fig. 9). These large cores fall within the range of giant core technologies identified in other parts of the world (e.g., Madsen and Goren-Inbar, 2004). The Saffāqah excavations also produced a large number of tool forms, including handaxes, cleavers and picks, together with items such as choppers, scrapers, and notches, indicating the desire to manufacture and use a variety of tool types on-site. The Saffāqah artifacts were made from cores and from flakes struck from clasts that fell within cobble and boulder size categories.

The quarrying evidence at Dawādmi is significant, as this adds to our growing body of information about stone tool procurement and reduction behaviors in the Acheulean. The stone tool procurement and reduction behaviors identified at Dawādmi are reminiscent to those identified at the Isampur Quarry in the Hunsgi-Baichbal valley of India (Petraglia et al., 2005). At Isampur, a siliceous limestone was procured from slabs of various sizes, which were reduced on the spot. The reduction debris at Isampur are similar to those from Saffāqah in the sense that both quarries contain an abundant amount of primary stage reduction debris, including cores and other waste products, as well as a variety of shaped tool types, including large cutting tools, and retouched and utilized

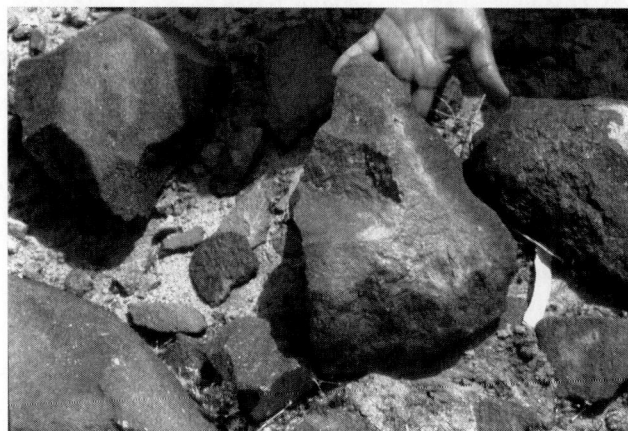

Fig. 8 Large early stage biface identified along the hillslope at Saffāqah, 206-68. Note the large and bold flake scars on the biface

Fig. 9 Giant core identified along the hillslope at Saffāqah, 206-68. Note the multiple large and deep flake removals on opposite faces of the core

flakes. Differences between the areas include the identification of non-quarry sites on the valley floor in the Hunsgi-Baichbal valley, whereas at Dawādmi the localities are located along the hillslopes and near rock outcrops. Dawādmi differs more substantially in comparison to other Acheulean landscapes where quarries have been identified. In the Upper Karoo region of South Africa, for example, few shaped tools were left near the quarries; hence, large cutting tools were transported and discarded in localities some distance away from dikes and spring heads (Sampson, 2006). Though raw material from dikes were exploited in both South Africa and Arabia, the spatial distribution of dike formation differed, probably influencing Acheulean landscape behaviors. In South Africa, the dolerite dikes with flakeable hornfels occur in numerous spatially restricted zones where 300 quarries were identified, whereas at Dawādmi, the dikes occur as spatially extensive outcrops over a long distance.

The reduction of giant cores at Dawādmi provides evidence that Arabian hominins, like those in other regions, sought large clasts in order to manufacture large cutting tools from flake blanks. In this stage of research, it is not yet clear how the stone tool reduction methods in Arabia compare to those in other regions, though it appears that large cutting tools, such as handaxes and cleavers, were important end-products. While the landscape behaviors of the Acheulean hominins in Arabia parallel the situations found in other places, the physical configuration of the raw material outcrops varied in each region, indicating that hominins adjusted their tool-making requirements to fit the local circumstances.

Inter-regional Comparison of Acheulean Large Cutting Tools

The Dawādmi and Wadi Fatimah assemblages were classified as Acheulean assemblages based on the recovery of characteristic tool types and technologies. The stone tool industries were considered to be similar to other regions, particularly those in the Levant and in Africa. Although Whalen made a plausible inference that the localities were Acheulean based on technological observations, no comparative research has yet taken place. Though detailed technological analysis has not been performed, some linear metric measurements of the Arabian LCTs have been obtained by us, useful for basic comparisons. Recently, we have performed two comparative studies, one examining LCT variation within the Acheulean (Shipton and Petraglia, 2009) and the other on an evaluation of LCT variation across the Movius Line (Petraglia and Shipton, 2008). From a comparative perspective, the LCT data from Arabia fell comfortably within the Acheulean range of variation, though there were some regional variations in tool sizes, as demonstrated here.

Table 2 Mean and median elongation of bifaces, with the total scores for each region in bold (from Shipton and Petraglia, 2009: Table 2)

	N	Mean elongation	Median elongation
Olduvai Gorge Bed II	21	2.0087	1.9680
Kariandusi	58	1.7693	1.7475
Olorgesailie DE89A	63	1.7521	1.7241
Olorgesailie H9AM	13	1.9376	2.0672
Olorgesailie I3	62	1.5425	1.5309
Olorgesailie FB	16	1.5964	1.6036
Olorgesailie DE89C	23	1.7625	1.7442
East Africa	**232**	**1.7247**	**1.7209**
Dawādmi 206-76	49	1.7711	1.7000
Wadi Fatimah	35	1.4795	1.5238
Arabia	**84**	**1.6496**	**1.6111**
Hunsgi V	151	1.6324	1.6250
Hunsgi II	34	1.6835	1.6250
Gulbal II	17	1.5863	1.6458
Mudnur VIII	9	2.1567	2.2727
Yediyapur I	21	1.5462	1.4722
Yediyapur IV	20	1.6608	1.6667
Yediyapur VI	66	1.5369	1.5000
Fatehpur V	31	1.4946	1.4773
Anagwadi	25	1.7008	1.6522
Godavari	10	1.3191	1.2933
Teggihalli II	31	1.5282	1.5233
India	**302**	**1.609**	**1.6000**
Acheulean Total	**650**	**1.6618**	**1.6479**

With respect to mean elongation (length to breadth ratio), the African and Arabian bifaces were similar, and these tended to be more elongated that the Indian ones (Table 2). A statistical difference in biface elongation was demonstrated between the African and Indian assemblages, though no statistical difference could be demonstrated between the Arabian and Indian bifaces. The biface elongation data in Arabia showed that those from Dawādmi tended to have larger length to breadth ratios when compared to those from Wadi Fatimah. The Wadi Fatimah assemblages, in fact tended to have low elongation values in comparison to many other Acheulean assemblages. Examination of linear biface length measurement showed that the Dawādmi bifaces were, on average, longer than those from Wadi Fatimah (Table 3).

With respect to biface refinement (thickness to breadth ratio), statistical tests indicated that bifaces from the three Acheulean regions have similar refinement scores. Examination of the biface refinement data within Arabia shows that those from Dawādmi tended to be thicker relative to breadth when compared to Wadi Fatimah. With respect to handaxe refinement, the Dawādmi has higher values in comparison to Wadi Fatimah, though the Arabian handaxes fall near the average mean of all Acheulean assemblages (Table 4). Examination of the small number of cleavers available from Arabia shows that the implements fall near the mean range of all Acheulean assemblages (Table 5). The handaxes and cleavers from Arabia thus do not group with the localities with the most refined or least refined implements.

Table 3 A comparison of the mean length of bifaces (after Shipton and Petraglia, 2009: Table 3)

	N	Mean length (mm)
Olduvai Gorge Bed II	21	195.39
Kariandusi	58	157.94
DE89A Olorgesailie	63	180.76
H9AM Olorgesailie	13	199.77
I3 Olorgesailie	62	97.95
FB Olorgesailie	16	98.81
DE89C Olorgesailie	69	158.70
Dawādmi 206-76	49	162.87
Wadi Fatimah	35	141.86
Hunsgi V	151	143.51
Hunsgi II	34	162.90
Gulbal II	17	147.14
Mudnur VIII	9	227.78
Yediyapur I	21	123.13
Yediyapur IV	20	132.94
Yediyapur VI	66	127.89
Fatehpur V	31	126.82
Teggihalli II	31	121.54
Anagwadi	25	137.24
Godavari	10	114.00
Acheulean Total	**531**	**145.54**

Comparative analysis indicated that the Wadi Fatimah and Dawādmi assemblages comfortably fall within the Acheulean range of variation, though differences in tool morphology were observable between localities in and outside of Arabia. It is probable that some of this variation is due to raw material types and initial clast size. Though Whalen considered the Arabian assemblages to be "Middle Acheulean," our data cannot support such an assertion based on the small number of measurements we have at our disposal, and problems with evaluating Acheulean assemblages without chronometric dates.

Dispersal Processes

Surveys conducted in various parts of the Arabian peninsula have identified a number of Acheulean localities (Petraglia, 2003). Though at first glance, the number of sites is impressive, it is also apparent that there are many gaps in our knowledge about Acheulean site distribution. The lack of identified sites in certain areas may be due to survey coverage, but in some cases, the lack of sites may also signal a limited timeframe in which Acheulean hominins were present in the region. Given times of severe aridity in the Early and Middle Pleistocene, it is probable that hominin occupations were not prolonged over long periods. Thus, the presence of Acheulean

Table 4 Handaxe refinement data. Data given in mm (from Petraglia and Shipton, 2008: Table 2)

Locality	N	Mean	SD	Min	Median	Max
Africa						
Kariandusi	35	0.4842	0.11413	0.29	0.4888	0.96
Olduvai Bed II	17	0.6821	0.13960	0.45	0.6674	0.89
Olorgesailie DE89A	60	0.4505	0.10854	0.20	0.4465	0.70
Olorgesailie FB	15	0.5988	0.14022	0.42	0.5588	0.89
Olorgesailie H9AM	10	0.3732	0.09270	0.27	0.3869	0.57
Olorgesailie I3	57	0.5597	0.11733	0.38	0.5532	0.93
Europe						
High Lodge	63	0.4426	0.13308	0.23	0.3943	0.90
Middle East						
Azraq Lion Spring	42	0.5170	0.12586	0.12	0.5064	0.82
Dawādmi 207–76	27	0.5700	0.14493	0.33	0.5556	0.91
Wadi Fatimah	15	0.5226	0.09328	0.39	0.5238	0.71
India						
Anagwadi	15	0.5969	0.07745	0.48	0.5778	0.81
Fatehpur V	11	0.5193	0.12683	0.30	0.5556	0.71
Gulbal II	12	0.5128	0.06397	0.40	0.5000	0.63
Hunsgi II	18	0.5510	0.10119	0.31	0.5774	0.67
Hunsgi V	45	0.5637	0.11432	0.33	0.5714	0.86
Mudnur VIII	9	0.5825	0.12580	0.33	0.5455	0.78
Teggihalli II	9	0.4683	0.13168	0.27	0.4276	0.69
Yediyapur I	10	0.4565	0.09061	0.38	0.4365	0.67
Yedyiapur IV	11	0.5370	0.13176	0.40	0.5000	0.86
Yediyapur VI	21	0.5145	0.11147	0.33	0.5000	0.71
Acheulean Total	**503**	**0.5180**	**0.13180**	**0.12**	**0.5000**	**0.96**

Table 5 Cleaver refinement data. Data given in mm (from Petraglia and Shipton, 2008: Table 5)

Locality	N	Mean	SD	Min	Median	Max
Africa						
Kariandusi	19	0.4334	0.07043	0.34	0.4183	0.58
Olduvai Bed II	2	0.5811	0.03085	0.56	N/A	0.60
Olorgesailie DE89A	3	0.3758	0.04639	0.32	0.3968	0.41
Olorgesailie H9AM	3	0.3335	0.07304	0.26	0.3333	0.41
Olorgesailie I3	5	0.5027	0.04124	0.45	0.5089	0.54
Europe						
High Lodge	5	0.4933	0.14957	0.36	0.4896	0.72
Middle East						
Azraq Lion Spring	11	0.5029	0.10037	0.33	0.4916	0.73
Dawādmi 206-76	2	0.5067	0.24513	0.33	N/A	0.68
Wadi Fatimah	6	0.5067	0.11429	0.35	0.5000	0.69
India						
Anagwadi	10	0.5451	0.1109	0.40	0.5357	0.73
Fatehpur V	11	0.3675	0.6915	0.25	0.3636	0.44
Godavari	9	0.5135	0.15852	0.26	0.4932	0.73
Gulbal II	2	0.4222	0.03143	0.40	N/A	0.44
Hunsgi II	13	0.4848	0.10793	0.25	0.5000	0.69
Hunsgi V	52	0.5106	0.08681	0.36	0.5000	0.71
Teggihalli II	19	0.4705	0.10891	0.27	0.4670	0.67
Yediyapur I	6	0.3981	0.14263	0.22	0.3542	0.63
Yediyapur IV	6	0.5489	0.10243	0.44	0.5357	0.67
Yediyapur VI	17	0.4651	0.12067	0.22	0.5000	0.67
Acheulean Total	**202**	**0.4792**	**0.10975**	**0.22**	**0.4759**	**0.73**

sites in the peninsula are probably tied to relatively short term events.

Given the location of identified Acheulean sites, it may be surmised that hominins initially spread along the coasts of Arabia. Soon after, hominins appear to have utilized river valleys to spread into the interior, such as along the Wadi Fatimah, where they could travel into and across various ecological zones over long distances, in this case potentially traversing ca. 190 km (Fig. 10). The presence of Acheulean sites along rivers far into the interior, such as found at Dawādmi, indicates that hominins migrated long distances along the river systems. The movement of hominins in the case of Dawādmi may have begun near the Red Sea, at the outflow of the Wadi al Hamd. Once traversing along the Wadi al Hamd, Acheulean hominins likely encountered the headwaters of the Wadi al Batin, and continued their travels further inland. Though we do not know the specific routes of these dispersals and the exact location of these paleodrainages, it is clear that the hominins were able to make use of these major drainage networks. Thus, dispersing hominins could potentially travel from the Red Sea zone to the alluvial zones in the Arabian Gulf by utilizing corridors such as the Wadi al Batin, the Wadi as Sahba and the Wadi Dawasir. After these expansion events, and during arid stages, Acheulean population sizes must have either shrunk considerably, or groups may have even gone extinct. The ability of hominins to survive under the most adverse, arid periods

must have been impossible or nearly impossible, unlike in other parts of the world, where refugia in river valleys and basins were almost always present (e.g., Petraglia, 2005; Korisettar, 2007). Hominin re-expansions into Arabia must have occurred after the return to more favorable, moist periods. Thus, it is likely that Acheulean site distributions, such as identified at Wadi Fatimah and Dawādmi, must represent short term population expansions during particular stages. The implications of our research indicates that many paleodrainages, noted by geographers (e.g., Edgell, 2006), must have been used. These drainages should be the location of future surveys.

It is not yet possible to map the specific routes that Acheulean hominins took while entering Arabia, though populations may have uitilized the Bab al Mandab and the Sinai (Petraglia, 2003). Though populations may have had to deal with open water bodies as they crossed the Bab al Mandab (see Bailey, 2009), this is not impossibile, as indicated by the presence of Middle Pleistocene hominins in island southeast Asia (e.g., Morwood et al., 1998). Once present in Arabia, it is probable that hominins spread into Iran, where we know little about Acheulean presence, and into the Indian subcontinent, where Acheulean sites are plentiful (Petraglia, 2005).

Unfortunately, the chronology of Acheulean occupation in Arabia has yet to be initiated. Based upon our knowledge about Acheulean occuaptions in the Levant, it is possible that

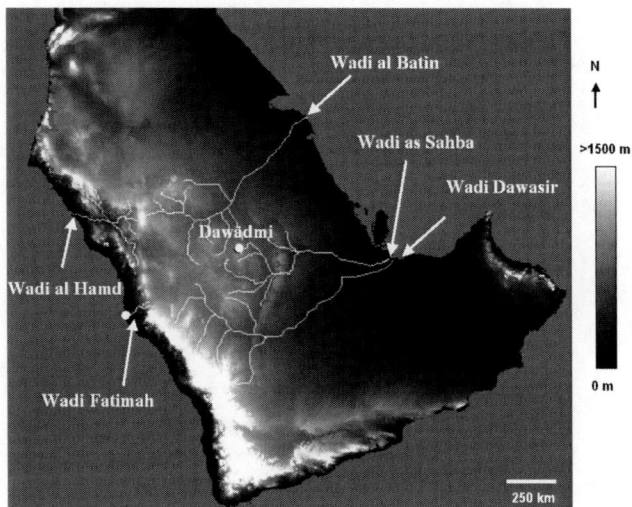

Fig. 10 STRM 1 km resolution DEM of the Arabian peninsula showing the location of the Wadi Fatimah and Dawādmi sites. Most of the large river systems evident in the DEM have been digitized and overlain in white. The figure demonstrates that a large stream network covers much of the peninsula. The headwater of the streams occurs in elevated or mountainous zones and traverses many topographic and ecological settings. Main streams such as the Wadi al Batin, the Wadi as Sahba and the Wadi Dawasir drain large areas. These main channels were presumably active at times during the Pleistocene and they were likely routes for hominin dispersals, as demonstrated by the Dawādmi sites which are located in the interior of the peninsula. During the Pleistocene sea level lowstands these channels would have flowed into a much larger river system (now the Arabian Gulf) thus allowing hominins to disperse into low-lying alluvial plains to the north and towards the Zagros mountains of Iran and the Makran coast of Pakistan. The location of numerous large river mouths along the Red Sea coast provided hominins the opportunity to easily penetrate into the uplands and the mountainous zones and across a variety of ecological settings. There are a number of other rivers and streams along the Red Sea coast, in addition to the Wadi al Hamd and the Wadi Fatimah, which suggest that this may have provided a more common dispersal route than currently realized

hominins spread into Arabia as early as 1.4 Ma. Otherwise, it is probable that the majority of Acheulean sites in Arabia are part of a post-800 ka spread of hominins which is more prevalent in the Levant and other regions. In this regard, the high incidence of cleaver forms and giant cores in Dawādmi is technologically similar to those found at the 800 ka-old assemblages of Gesher Benot Ya'aqov (Goren-Inbar et al., 2000).

Conclusion

This overview indicates that Arabia has much to offer concerning Acheulean adaptations, landscape use and dispersal processes. The Wadi Fatimah and Dawādmi localities have been chosen for discussion as they are Arabia's most convincing and spectacular examples of Acheulean sites and site

complexes. These Acheulean sites provide significant information about tool making and tool-using behaviors at particular locations as well as across landscapes. Though concentration has been placed on two of Arabia's best known Acheulean sites complexes, it is without doubt that many more localities exist in the region, which will yield valuable insights about behavioral and evolutionary processes. Interdisciplinary research, new field studies and detailed technological studies have enormous potential for better understanding hominin behaviors and how the Arabian contexts compare to regions where Acheulean hominins were present.

Acknowledgments Funding for Petraglia's initial research in the Kingdom of Saudi Arabia was supplied by a Fulbright Senior Specialists Program Fellowship. We are grateful to the Ministry of Antiquities and Museums and the US Embassy in Riyadh for facilitating research and the National Museum for allowing access to the artifact collections.

References

Bailey G. The Red Sea, coastal landscapes, and hominin dispersals. In: Petraglia MD, Rose JI, editors. The evolution of human populations in Arabia: paleoenvironments, prehistory and genetics. The Netherlands: Springer; 2009. p. 15–37.

Bar-Yosef O. Early colonizations and cultural continuities in the Lower Palaeolithic of western Asia. In: Petraglia MD, Korisettar R, editors. Early human behaviour in global context: the rise and diversity of the Lower Palaeolithic record. London: Routledge; 1998. p. 221–79.

Edgell HS. Arabian deserts: nature, origin and evolution. Dordrecht: Springer; 2006.

Goren-Inbar N, Feibel CS, Verosub KL, Melamed Y, Kislev ME, Tchernov E, et al. Pleistocene milestones on the Out-of-Africa corridor at Gesher Benot Ya'aqov, Israel. Science. 2000;289:944–7.

Isaac GL. The archaeology of human origins: studies of the Lower Pleistocene in East Africa 1971–1981. Advances in World Archaeology. 1984;3:1–87.

Korisettar R. Toward developing a basin model for Paleolithic settlement of the Indian subcontinent: geodynamics, monsoon dynamics, habitat diversity and dispersal routes. In: Petraglia MD, Allchin B, editors. The evolution and history of human populations in South Asia. The Netherlands: Springer Academic; 2007. p. 69–96.

Madsen B, Goren-Inbar N. Acheulian giant core technology and beyond: an archaeological and experimental case study. Eurasian Prehistory. 2004;2(1):3–52.

Morwood MJ, O'Sullivan PB, Aziz F, Raza A. Fission-track ages of stone tools and fossils on the east Indonesian island of Flores. Nature. 1998;392:173–6.

Petraglia MD. The Lower Paleolithic of the Arabian peninsula: occupations, adaptations, and dispersals. Journal of World Prehistory. 2003;17:141–79.

Petraglia MD. Hominin responses to Pleistocene environmental change in Arabia and South Asia. In: Head MJ, Gibbard PL, editors., Early-Middle Pleistocene transitions: the land-ocean evidence. London: Geological Society; 2005. Special Publications, 247. p. 305–319.

Petraglia MD, Shipton C. Large cutting tool variation west and east of the Movius line. Journal of Human Evolution. 2008;55:962–6.

Petraglia MD, Shipton C, Paddayya K. Life and mind in the Acheulean: a case study from India. In: Gamble C, Porr M, editors. The hominid individual in context: archaeological investigations of Lower and

Middle Palaeolithic landscapes, locales and artefacts. London: Routledge; 2005. p. 197–219.

Sampson CG. Acheulian quarries at hornfels outcrops in the Upper Karoo region of South Africa. In: Goren-Inbar N, Sharon G, editors. Axe age: Acheulian toolmaking from quarry to discard. London: Equinox Publishing; 2006. p. 75–107.

Shipton C, Petraglia MD. Inter-continental variation in Acheulean bifaces. In: Norton CJ, Braun D, editors. Asian paleoanthropology: from Africa to China and beyond. The Netherlands: Springer; 2009. in press.

Whalen N, Killick A, James N, Morsi G, Kamal M. Saudi Arabian archaeological reconnaissance 1980: B. Preliminary report on the Western Province survey. Atlal. 1981;5:43–58.

Whalen NM, Sindi H, Wahida G, Siraj-ali JS. Excavation of Acheulean sites near Saffaqah in ad-Dawādmi 1402–1982. Atlal. 1983;7: 9–21.

Whalen N, Siraj-Ali JS, Davis W. 1 – Excavation of Acheulean sites near Saffaqah, Saudi Arabia, 1403 AH 1983. Atlal. 1984;8: 9–24.

Whalen NM, Siraj-Ali J, Sindi HO, Pease DW, Badein MA. A complex of sites in the Jeddah–Wadi Fatimah area. Atlal. 1988;11:77–85.

Zarins J, Whalen N, Ibrāham M, Jawad Mursi AA, Khan M. Comprehensive archeological survey program, preliminary report on the Central and Southwestern Provinces survey, 1979. Atlal. 1980;4:9–36.

Chapter 9
A Middle Paleolithic Assemblage from Jebel Barakah, Coastal Abu Dhabi Emirate

Ghanim Wahida, Walid Yasin Al-Tikriti, Mark J. Beech, and Ali Al Meqbali

Keywords Abu Dhabi Emirate • Jebel Barakah • Middle Paleolithic • Technology

Introduction

Until recently, our knowledge of the Paleolithic period in Arabia has been limited. Occasional Paleolithic tools were collected and reported early in the last century, such as the discovery of a Lower Paleolithic handaxe from central Arabia (Cornwall, 1946). Geological teams exploring Arabia for its mineral wealth reported on the identification of Acheulean implements (Field, 1971; Overstreet, 1973). In the late 1970s, knowledge concerning the Paleolithic of Arabia began to change as archaeologists began a systematic, five year comprehensive program to survey various provinces of Saudi Arabia. A large number of archaeological sites, of varying periods, were discovered across the country. As a result of survey efforts, nearly 200 Acheulean and Middle Paleolithic sites were discovered in the central, western and south-western provinces. Of special importance were three old sites, namely, Shuwayhitiyah in the north, site 226-63 near Najran in the south and Tathlith in the southwest of Saudi Arabia. These sites were thought to belong to an early part of the Pleistocene on typological grounds (Whalen and Pease, 1992). In addition, important research into the Middle Paleolithic along the Red Sea coast has progressed and a

possible Lower Paleolithic site has been reported in central Saudi Arabia (Petraglia and Alsharekh, 2003; Alsharekh, 2007). The only in situ, dated site excavated in Arabia is that of Saffāqah, near Dawādmi in central Saudi Arabia (Whalen et al., 1983; Petraglia et al., 2009). Uranium–thorium dating has placed Acheulean artifacts to a minimum of 200 ka (Whalen et al., 1982). In southern Yemen, the discovery of five pre-Acheulean sites has been claimed within the Hadhramaut Mountains (Whalen et al., 1982).

Archaeological work on the Paleolithic of the Persian Gulf began in the early 1990s. A number of international expeditions discovered Pleistocene sites in Abu Dhabi Emirate (McBrearty, 1993) and in Sharjah in the United Arab Emirates (Scott-Jackson and Scott-Jackson, 2006; Uerpmann et al., 2006, 2009; Scott-Jackson et al., 2009), as well as in neighboring Oman (Rose, 2004, 2005, 2007; Rose and Usik, 2009).

Genetic studies have recently been introduced in Arabia and evolutionary geneticists have begun to appreciate the major role that Arabia must have played in the origin of modern humans. New genetic evidence has highlighted the significance of the Arabian peninsula as a corridor for early human migration to and from Africa (Abu-Amero et al., 2007, 2008; Petraglia, 2007; Cabrera et al., 2009; Rídl et al., 2009).

Although prehistoric research in the Arabian peninsula is still in its infancy, the present book is a sign of the importance of prehistory on the peninsula. It will hopefully encourage more archaeological work in this vast and vital area bridging Africa and southwest Asia. New Paleolithic evidence discovered at Barakah, on the Gulf, promises to provide a wealth of data to explore questions surrounding Paleolithic occupation of the eastern end of the peninsula.

Geomorphology of Jebel Barakah

Jebel Barakah is located on the west coast of Abu Dhabi Emirate, overlooking the sea between Jebel Dhannah and the Qatar peninsula (Fig. 1). The coastline of Abu Dhabi is

G. Wahida (✉)
McDonald Institute for Archaeological Research,
University of Cambridge, Downing Street, Cambridge CB2 3ER, UK
e-mail: ghanimwahida@hotmail.com

W.Y. Al-Tikriti and A.A. Meqbali
Historic Environment Department, Abu Dhabi Authority for Culture and Heritage (ADACH), Al Ain National Museum, P.O. Box 15715, Al Ain, United Arab Emirates
e-mail: wyasin11@yahoo.com; aly_elmegbali@yahoo.com

M.J. Beech
Historic Environment Department, Abu Dhabi Authority for Culture and Heritage (ADACH), P.O. Box 2380, Abu Dhabi, United Arab Emirates
e-mail: mark.beech@cultural.org.ae

Fig. 1 Location of Jebel Barakah in the
Western Region of Abu Dhabi Emirate
(after Whybrow and Hill, 1999)

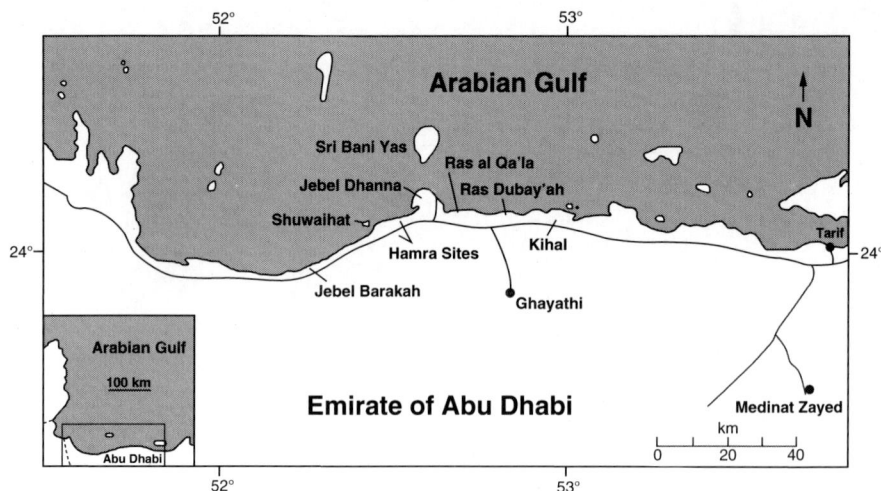

Fig. 2 Jebel Barakah looking north

generally low and dominated by sabkha (salt land) with occasional sand hills and low grass vegetation. Jebel Barakah, at 62.6 m above sea level, is the highest point along this stretch of coastline. It is an isolated outcrop composed of red sandstone (originally wind-blown sand) and thin bands of conglomerate (originally water-transported, wadi pebbles). The outcrop, oval in shape, occupies a low plateau of 2.5 km from north to south and 2 km from east to west (Fig. 2). The international road to Saudi Arabia and Qatar divides the plateau into two sections. The larger northern section is the most important as it has yielded all the Upper Miocene fossils discovered at Barakah, as well as Paleolithic artifacts. The southern section is disturbed, and partly occupied by new installations. Newly opened tracks have been built in the northern section to serve the modern observatory built by the army at the northern side of the Jebel. Construction of this structure resulted in the exposure of the upper sections of the Baynunah geological formation.

The Jebel, a small outcrop with a narrow flat summit and sloping surfaces, occupies about 1 km² of the north-western

side of the plateau. Like most of the outcrops in the western region of Abu Dhabi Emirate, the Jebel is capped by narrow, flat summits and covered with a layer of deflated cherts. The eastern most end of the Plateau is high ground, separated from another similar high ground to the west by low ground which seems to have been formed by water and natural erosion. A lower ground surface with pronounced outcrops separates these two areas from the Jebel. The low and wide gullies, formed by rain, slope down towards the sea (Fig. 2).

The exterior edges of the plateau are indicated by a series of pronounced cliffs formed by gushes of rain water. The international highway (Abu Dhabi-Qatar) cuts the southern part of the plateau, a large distance from the Jebel itself. The Jebel is the last elevated area as you head westwards towards Sabkhat Matti.

Prior to the recent archaeological discoveries, Jebel Barakah was probably best known for its Late Miocene fossil remains (Whybrow and Hill, 1999). Part of the sea cliff contains the type section for the Baynunah Formation, which covers the Shuwayhat Formation (Whybrow, 1989; Whybrow et al., 1999). At most outcrops of the Baynunah Formation, which covers the Shuwayhat Formation, the sequence is capped by a thick layer of resistant tabular chert-flint (cryptocrystalline siliceous rocks produced by diagenetic solution).

The lithic material from Jebel Barakah was first reported by McBrearty (1993, 1999). She noted that a large number of artifacts occurred on the level bluffs on the southeast side of the Jebel (McBrearty, 1999: 378). The artifacts lie directly on Baynunah Formation rocks; upslope they are overlain by a thin superficial layer of soft unconsolidated sediment derived from the exposures of the Baynunah Formation above. McBrearty also reported that the Barakah artifacts demonstrate a highly consistent and formalized flaking method, being composed almost entirely of radial cores and the flakes derived from them. All 16 cores collected by McBrearty are radial or high-backed radial form. There was no trace of any blade element. The aim of this chapter is to introduce new findings from Jebel Barakah, providing evidence to support that the site represents a Middle Paleolithic locus.

Jebel Barakah: Archaeological Localities

The lithic material provided here and currently under study by the first author was collected by staff members of Abu Dhabi Authority for Culture and Heritage (ADACH). The lithic material from Localities One to Three have already been discussed (Wahida et al., 2008). Two added Localities (Four and Five) with more material were discovered in January 2008, together with further material being collected from Locality Two.

Locality One (BRK0001), is situated on the north-west and western side of the Jebel, between the sea cliffs and the first line of ridges up the slope, a distance of about 300 m. Artifacts were scattered on a thin layer of soft soil derived from the exposure of the Baynunah Formation outcrops. Much of the lithic artifacts along the cliffs must have been eroded away to the Gulf along its substantial cliffs. Upon further study this season, it was noted that this Locality covers a wider area than initially anticipated. It covers the western section and extends beyond the Jebel from the northern side.

Locality Two (BRK0002), is located to the south and southwest side of the Jebel, and descends southwards away from the Jebel. A few artifacts were collected from the western section of this locality in 2007. Additional lithics have been collected from the eastern section of Locality Two this season though the total count of artifacts is small. Locality Three (BRK0003) lies to the eastern slope of the Jebel providing a small number of artifacts. McBrearty's lithic material may have come from this locality. Locality Four (BRK0004) lies to the east of the Jebel and unlike Localities One to Three, it is separated from it by low-lying, flat ground. It is an irregular, long and narrow outcrop, extending north-east south-west and rising about 4 m above sea level (ASL). The irregular surface of the outcrop, which extends about 200 m, consists of soft soil mixed with quantities of chert-flints and small gravels.

Locality Five (BRK5) is a long plateau measuring about 250 m long with a triangular shape, and it is located about 400 m to the north of Locality Four. The site which is only about 120 m away from the beach represents a peninsula of wide and flat surface, surrounded by two wide gullies from the east and west. Its elevation is about 4 m ASL and has a low ground extension at the north eastern side with an elevation of 3 m ASL. Both sides of the peninsula have been extensively damaged by rain erosion. Stone artifacts have been collected from both areas but were more prominent on the main peninsula. It should be noted here that subsequent to the initial reconnaissance that a small number of artifacts were discovered to the east of Locality Five and south-east of Locality Two.

The Lithic Assemblage

The five localities at Barakah appear to represent a single techno-typological industry. It should be stressed here that study of the Barakah assemblage is still in its preliminary stages and further analysis has been planned, including detailed artifact analysis and comparable study with other sites in the region.

The lithic collection strategy was determined by erosion and deflation that the five localities had suffered. Laying down a grid for a systematic collection of artifacts would have been of little use. Instead, a system of latitudinal and longitudinal coordinates for each artifact was obtained by Global Position System (GPS). In cases where a number of implements were located within a 5 m radius, one reading was obtained for the group as they lay within the possible area of error of the system. For fear of possible looting of artifacts, it would be unwise to publish those readings before the new Antiquity Law of the Abu Dhabi Emirate is in force (Beech, 2006). It is for this reason that the precise position of artifacts are only recorded with general dots on the map (Fig. 3).

The Barakah artifacts were made of good quality flints, but had black to blue-black patina. The artifacts were unlike those found by McBrearty and more numerous. Beyond the

Fig. 3 Jebel Barakah, Localities One–Five, with distribution of artifacts

radial cores, McBrearty collected 218 objects, of which eight are modified flakes as tools. McBrearty suggested several dates for the Barakah assemblage, including probably the Acheulean, the Middle Stone Age assemblages and Mid- to Late-Holocene (McBrearty, 1999). The first of these three dates were based on the presence of radial and high backed radial cores from which the flakes originated. The youngest age was probably based on the presence of two broken implements: one a bifacial tip, and the other a flake with unifacial trimming. McBrearty is credited for her identification of the Middle Paleolithic artifacts although her limited collection of tool types gave her limited space for other conclusions to be drawn. McBrearty provided an excellent outline of the paleoenvironmental history of the Western Region of Abu Dhabi Emirate, including Barakah, to which little can be added. Although an attempt was made to locate McBrearty's material, at the time of publication the material cannot unfortunately be located.

Technology and Typology

The main technique of core reduction at Barakah was the prepared core method by radial flaking known as the Levallois technique. This technique requires the working face of the core to be specially prepared beforehand, allowing a predetermined flake of probable shapes to be detached. The underside of the core was partially flaked off around the age and this was the case with all radial cores. The lithic assemblage was dominated by radial, high backed radial and discoid cores that reflect a tendency towards Levallois centripetal core strategy (Fig. 4a–f). The other technique was the bipolar whereby two flakes were struck off from two opposed ends of an elongated Levallois core. A third technique was probably that of Nubian Method Type One, where one Levallois flake core, oval in shape, had the last flake struck off from the thinner distal end. Two earlier removals from the thicker proximal end were probably part of the preparation

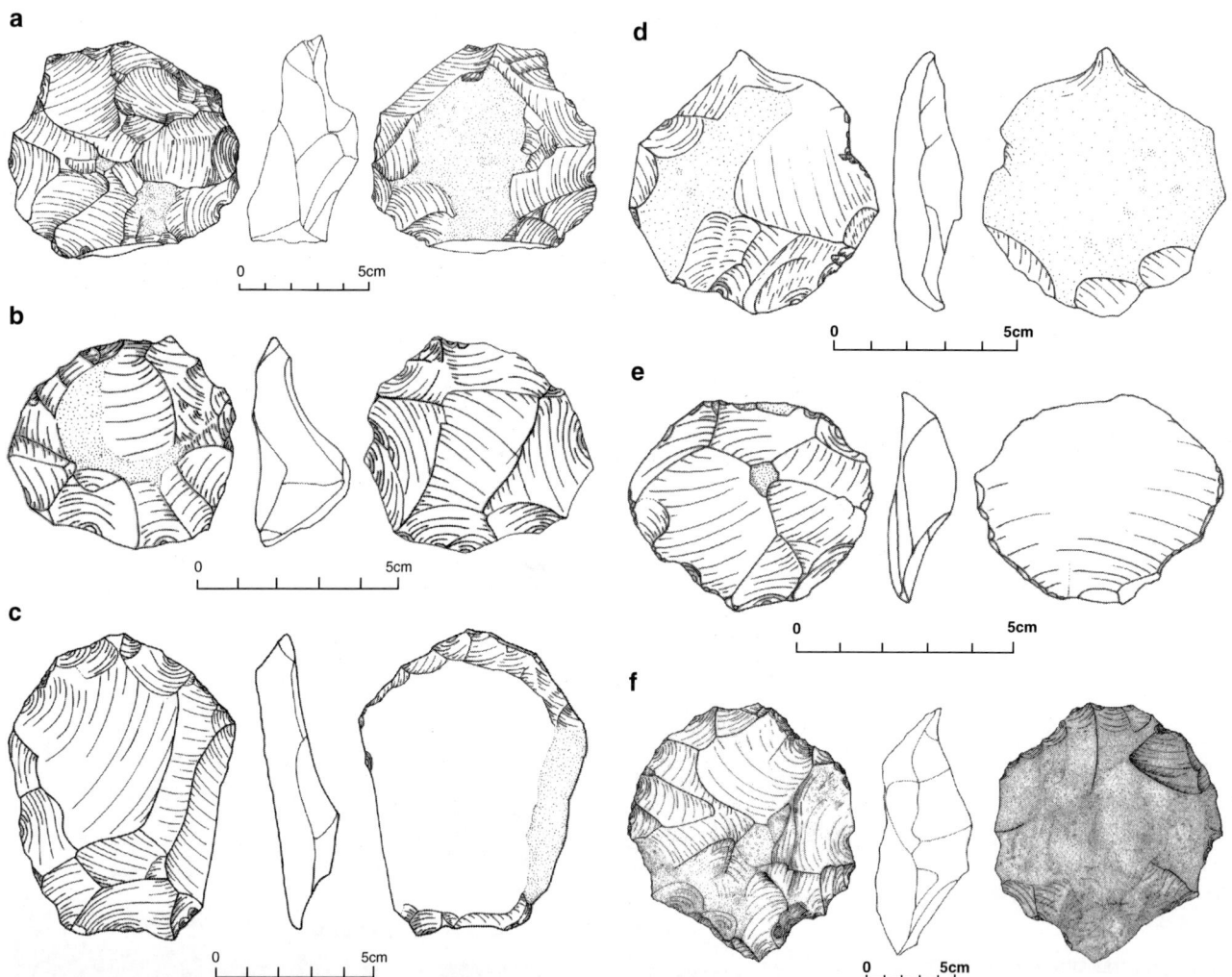

Fig. 4 (**a**) Bifacial centripital core; (**b**) high-backed radial core; (**c**) elongated bifacial core; (**d**) pointed bifacial discoidal core; (**e**) unifacial centripital radial core; (**f**) bifacial centripital radial core

technique (Fig. 5). Applying the Levallois technique of obtaining as many flakes as possible, the original large nodule of raw material was reduced in size, such that no more desired flakes were possible. As McBrearty (1999) noted, we agree that the assemblage displayed a very consistent and formalized flaking method, being composed almost entirely of radial cores and the flakes derived from them.

Among the 158 specimens collected from Localities Two, Four and Five, 49 radial, high backed radial or discoid cores were found. These cores were distributed as follows: Locality Two, 17, of which ten were cores. Among the 97 specimens collected from Locality Four, 28 were cores. Locality Five produced 44 specimens, of which 11 were cores. One bipolar Levallois core was found in Locality Four, and one Levallois flake core was found in Locality Five, bringing the total number of cores to 51.

The smallest radial core comes from Locality Four, and measures 4.1 × 4.0 × 1.4 cm, whereas the largest radial core, comes from Locality Three, and measures 13.2 × 12.3 × 5.2 cm. One handaxe was found in Locality Five. The base was broken towards the proximal end and would have been of cordiform type if complete. Combined shallow flaking and sinuous retouch have been applied to both sides, with the original cortex remaining on both sides, in the area closer to the proximal end. The retouch was confined mainly to a single side of the handaxe. A hard hammer was probably applied in the primary flaking and a soft hammer was likely used to produce the final flaking and retouching (Fig. 6).

Apart from some diagnostic types, the 19 registered tools included two side-scrapers (Fig. 7), 11 notches (Fig. 8), one denticulate (Fig. 9), two borers (Fig. 10) and several points (Fig. 11). One side-scraper, a bifacially retouched fragment on a thin tabulated flint, was found at Locality Five. The ventral retouch is shorter than that on the dorsal surface. Apart from the retouched area, the remainder of the fragment had cortex.

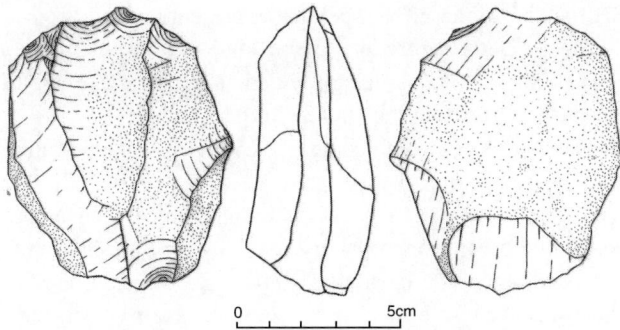

Fig. 7 Bifacial sidescraper

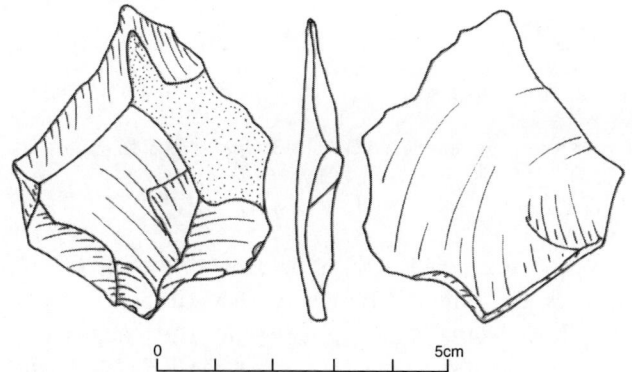

Fig. 8 Dorsally directed notch

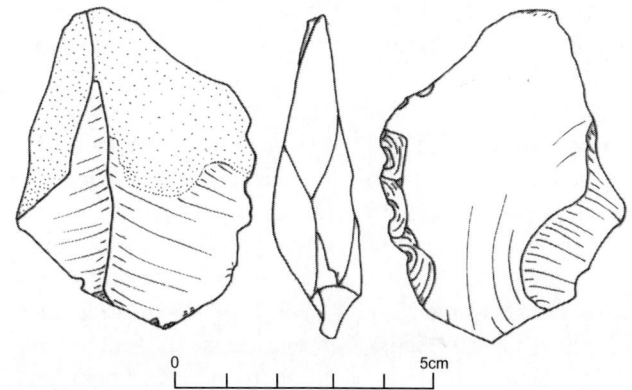

Fig. 5 Levallois flake core (probably of Nubian method type one)

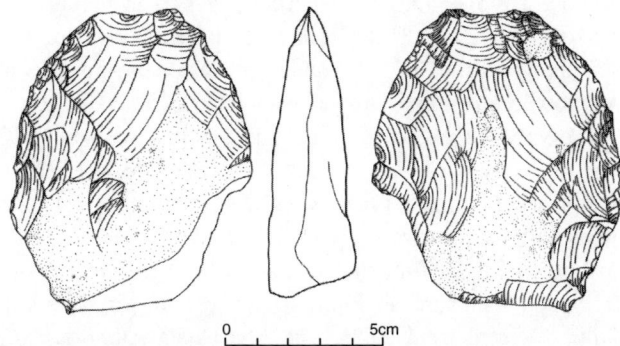

Fig. 6 Bifacial handaxe

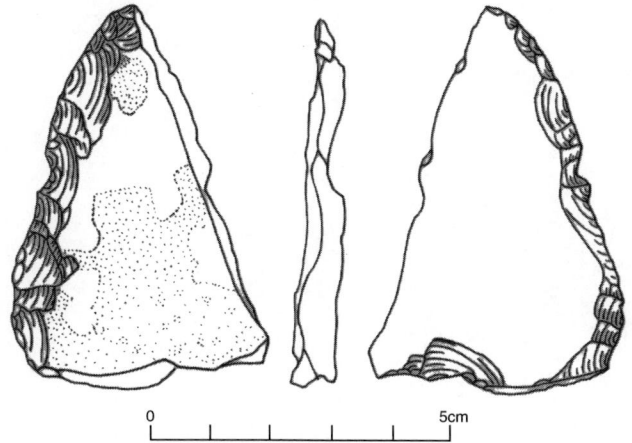

Fig. 9 Denticulate

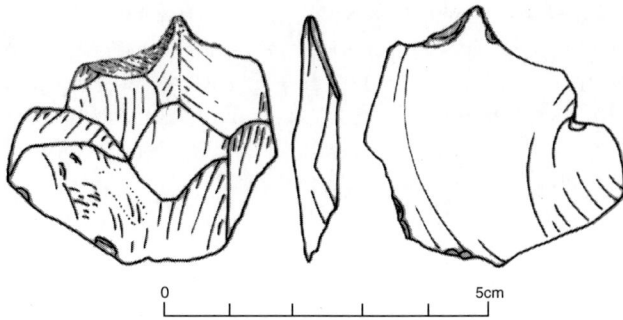

Fig. 10 Borer

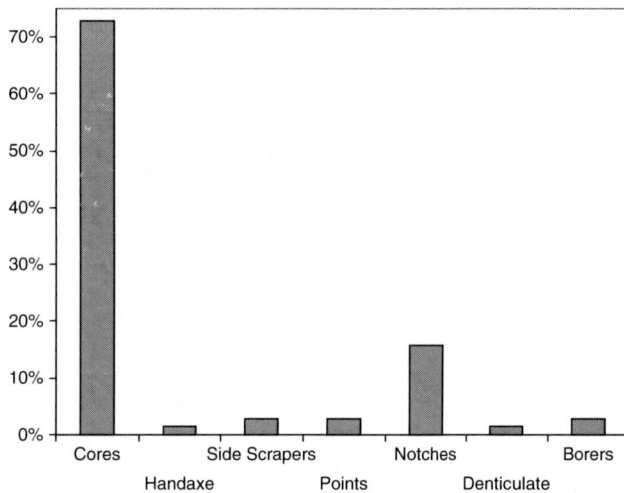

Fig. 11 The percentage of cores and tool types at Jebel Barakah

Conclusions

It may be confidently stated that the Barakah assemblage belongs to the Middle Paleolithic period. This conclusion is supported by the presence of the Levallois centripetal radial strategy, and its resultant radial and discoidal cores, the presence of two Levallois flake cores, one of probably Nubian Method Type One and one bipolar as well as one typical handaxe of cordiform type. The assemblage also included one bifacial side-scraper fragment similar to those found in Nubian Mousterian Type B (Marks, 1968). The absence of blade elements and blade manufacturing techniques may suggest that the Barakah flake assemblage belongs to the early Middle Paleolithic. In this respect it is worth mentioning here that early Mousterian assemblages in the Levant (Shea, 2007), the Middle Stone Age of Africa (van Peer and Vermeersh, Van Peer and Vermeersch, 2007; McBrearty, 2007) and the Middle Paleolithic of Arabia (Petraglia and Alsharekh, 2003; Petraglia, 2007) all had among their components blades and retouched tools such as endscrapers, points and burins. Obviously, the Barakah assemblage lacks any such 'Upper Paleolithic' elements. In some parts of the world Middle Paleolithic industries developed out of the Late Acheulean. At Barakah, there is no sign of any Acheulean elements.

The Barakah assemblage complements the recent discovery of Middle Paleolithic material elsewhere in the United Arab Emirates, Oman and Yemen. The stratified materials of Paleolithic industry, discovered at Jebel Faya in Sharjah Emirate has been dated by OSL dating to around 85 ka. Bedrock is still about 2 m below the present level of excavations (see Uerpmann et al., 2006; Marks, 2009; Uerpmann et al., 2009). This date has provided an approximate age of Paleolithic origins in the United Arab Emirates.

The Middle Paleolithic sites of southern Arabia complement the migration theory from Africa into Asia. If these sites were vestiges of the early migrants, this evidence supports the short crossing route theory along the Bab al Mandab waterway into Asia (Petraglia, 2007). The Barakah assemblage would present the most eastern extension of migrants into Arabia, probably during one of the pluvial phases associated with MIS 5. Paleoenvironmental conditions from southern Arabia indicate at least three pluvial conditions were associated with MIS 5e, 5a and 3 (Rose, 2004). Earlier climatic conditions in the Arabian peninsula during MIS 6 were too arid to support hunter-gatherer populations.

The very high ratio of cores indicates that Barakah was most probably used as a raw material workshop for a short interval of time, as artifact types were consistent and not mixed with other later tool types. It should be noted here that the Gulf during the time in question was a large river-system valley and the Barakah hominins were living in a world totally different from today. The Abu Dhabi Authority for Culture and Heritage (ADACH) is currently striving to protect

The notch concavities were made mainly by a single blow, and lack any form of deliberate retouch. The notch may be dorsally or ventrally directed or straight. These implements were an important component within the Barakah assemblage. Microwear and refitting studies (Keeley, 1977, 1980; Cahen, et al., 1979) showed that similar tools had one or more functions; including woodworking, splitting bone for the extraction of marrow and fashioning bone tools, hide cutting and piercing, butchering of animals and the preparation of plant food.

The number of primary flakes from the three localities (Two, Four and Five) is 110 in total, including specimens (complete and broken) that lack deliberate retouch. Three flakes have sharp edges or wide distal ends suitable for cutting or scraping. Three others have probably use-retouch on their sides. Nine flakes have their long axis shorter than their breadths. This small number of flakes is not unusual since their manufacture technique depends on the shape of the core and the force of the blow on the platform. Two of them have dorsal cortex.

important archaeological and paleontological sites throughout the Emirate of Abu Dhabi. The discovery of the first Middle Paleolithic site in the Abu Dhabi Emirate should place the site, with its already known fossil rich Late Miocene deposits, at the highest level of protection.

Acknowledgments Thanks go to His Excellency Mohammed Khalaf Al Mazrouei, Director General of the Abu Dhabi Authority for Culture and Heritage, to Dr. Sami El-Masri, Director of the Strategic Planning Office and to Mr. Mohammed Amer Al-Neyadi, Director of Historic Environment at ADACH, for supporting our work. We are grateful to Professor Paul Mellars (Cambridge University, UK) for sharing his much valued opinions on the Barakah material. Finally, warm thanks are due to Hans-Peter Uerpmann, Margarethe Uerpmann and to Anthony Marks for generously discussing a number of their ideas and thoughts during our visit to their archaeological headquarters and to the Jebel Faya rockshelter site in Sharjah emirate.

References

Abu-Amero KK, González AM, Larruga JM, Bosley TM, Cabrera VM. Eurasian and African mitochondrial DNA influences in the Saudi Arabian population. BMC Evolutionary Biology. 2007;7:32.

Abu-Amero KK, Larruga JM, Cabrera VM, González AM. Mitochondrial DNA structure in the Arabian peninsula. BMC Evolutionary Biology. 2008;8:45.

Alsharekh A. An early Lower Paleolithic site from central Saudi Arabia. 2007. Poster presentation at the Seminar for Arabian Studies 2007.

Beech M. Rethinking the Neolithic of south-eastern Arabia: new results from excavations on Marawah Island and Umm az-Zamul, Abu Dhabi Emirate, UAE. In: Orchard J, Orchard G, editors. Archaeology of the Arabian peninsula through the ages. Muscat: Ministry of Heritage and Culture; 2006. p. 111–41.

Cabrera VM, Abu-Amero KK, Larruga JM, González AM. The Arabian peninsula: gate for human migrations Out of Africa or cul-de-sac? A mitochondrial DNA phylogeographic perspective. In: Petraglia MD, Rose JI, editors. The evolution of human populations in Arabia: paleoenvironments, prehistory and genetics. The Netherlands: Springer; 2009. p. 79–87.

Cahen D, Keeley LH, Van Noten FL. Stone tools, toolkits, and human behavior in prehistory. Current Anthropology. 1979;20(4):661–83.

Cornwall PB. A Lower Palaeolithic hand-axe from Central Arabia. Man. 1946;46(Nov–Dec.):144.

Field H. Contribution to the anthropology of Saudi Arabia. Miami: Field Research Projects; 1971.

Keeley LH. The functions of Palaeolithic flint tools. Scientific American. 1977;237(5):108–26.

Keeley LH. Experimental determination of stone tool uses: a microwear analysis. Chicago, IL: University of Chicago Press; 1980.

Marks AE. The Mousterian industries of Nubia. In: Wendorf F, editor. The prehistory of Nubia: Volume 1. Dallas: Southern Methodist University Press; 1968. p. 194–314.

Marks AE. The Paleolithic of Arabia in an inter-regional context. In: Petraglia MD, Rose JI, editors. The evolution of human populations in Arabia: paleoenvironments, prehistory and genetics. The Netherlands: Springer; 2009. p. 295–308.

McBrearty S. Lithic artefacts from Abu Dhabi's western region. Tribulus: Bulletin of the Emirates National History Group. 1993;3(1):13–4.

McBrearty S. Earliest tools from the Emirate of Abu Dhabi, United Arab Emirates. In: Whybrow P, Hill A, editors. Fossil vertebrates of Arabia with emphasis on the Late Miocene fauna, geology, and paleoenvironments of the Emirate of Abu Dhabi, United Arab Emirates. New Haven, CT/London: Yale University Press; 1999. p. 373–88.

McBrearty S. Down with the revolution. In: Mellars P, Boyle K, Bar-Yosef O, Stringer C, editors. The human revolution revisited. Cambridge: McDonald Institute Archaeological Publications; 2007. p. 133–51.

Overstreet WC. Contributions to the prehistory of Saudi Arabia. Miami: Field Research Projects; 1973.

Petraglia MD. Mind the gap: factoring the Arabian peninsula and the Indian subcontinent into Out of Africa models. In: Mellars P, Boyle K, Bar-Yosef O, Stringer C, editors. The human revolution revisited. Cambridge: McDonald Institute Archaeological Publications; 2007. p. 383–94.

Petraglia MD, Alsharekh A. The Middle Palaeolithic of Arabia: implications for modern human origins, behaviour and dispersals. Antiquity. 2003;77(298):671–84.

Petraglia MD, Drake N, Alsharekh A. Acheulean landscapes and large cutting tool assemblages in the Arabian peninsula. In: Petraglia MD, Rose JI, editors. The evolution of human populations in Arabia: paleoenvironments, prehistory and genetics. The Netherlands: Springer; 2009. p. 103–16.

Rídl J, Edens CM, Černý V. Mitochondrial DNA structure of Yemeni population: regional differences and the implications for different migratory contributions. In: Petraglia MD, Rose JI, editors. The evolution of human populations in Arabia: paleoenvironments, prehistory and genetics. The Netherlands: Springer; 2009. p. 69–78.

Rose JI. The question of Upper Pleistocene connections between East Africa and South Africa. Current Anthropology. 2004;45:551–5.

Rose JI. Archaeological investigations in the Sultanate of Oman, 2004. Archaeological Researches in Ukraine 2003–2004; 2005. p. 339–342. (co-authored; in Russian).

Rose JI. The role of the Saharo-Arabian arid belt in the modern human expansion. In: Bicho N, Thacker P, editors. From the Mediterranean basin to the Portuguese Atlantic shore. Papers in honor of Anthony E. Marks. Actas do IV Congresso de Arqueologia peninsular (Faro, 14–19 September 2004). Far: Universidade do Algarve; 2007.

Rose JI, Usik VI. The "Upper Paleolithic" of South Arabia. In: Petraglia MD, Rose JI, editors. The evolution of human populations in Arabia: paleoenvironments, prehistory and genetics. The Netherlands: Springer; 2009. p. 169–85.

Scott-Jackson J, Scott-Jackson WB, Jasim S. Middle Palaeolithic – or what? New sites in Sharjah, United Arab Emirates (UAE). Proceedings of the Seminar for Arabian Studies. 2007;37:277–9.

Scott-Jackson J, Scott-Jackson W, Rose JI. Paleolithic stone tool assemblages from Sharjah and Ras al Khaimah in the United Arab Emirates. In: Petraglia MD, Rose JI, editors. The evolution of human populations in Arabia: paleoenvironments, prehistory and genetics. The Netherlands: Springer; 2009. p. 125–38.

Shea JJ. The boulevard of broken dreams: evolutionary discontinuity in the Late Pleistocene Levant. In: Mellars P, Boyle K, Bar-Yosef O, Stringer C, editors. The human revolution revisited. Cambridge: McDonald Institute Archaeological Publications; 2007. p. 219–32.

Uerpmann HP, Jasim SA, Handel M, Schmitt J. Excavations at different sites along the eastern slopes of the Jebel Faya. http://www.urgeschichte.uni-tuebingen.de/index.php?id=292 (2006).

Uerpmann H-P, Potts DT, Uerpmann M. Holocene (re-) occupation of Eastern Arabia. In: Petraglia MD, Rose JI, editors. The evolution of human populations in Arabia: paleoenvironments, prehistory and genetics. The Netherlands: Springer; 2009. p. 205–14.

Van Peer P, Vermeersch PM. The place of northeast Africa in the early history of modern humans: new data and interpretations on the Middle Stone Age. In: Mellars P, Boyle K, Bar-Yosef O, Stringer C, editors. The human revolution revisited. Cambridge: McDonald Institute Archaeological Publications; 2007. p. 187–98.

Wahida G, Yasin Al-Tikriti W, Beech M. Barakah: a Middle Palaeolithic site in Abu Dhabi Emirate. Proceedings of the Seminar for Arabian Studies. 2008;38:55–64.

Whalen NM, Pease DW. Early mankind in Arabia. Aramco World. July–August 1992;43(4):16–23. http://www.saudiaramcoworld. com/issue/199204/early.mankind.in.arabia.htm (1992).

Whalen NM, Sindi H, Wahida G, Sirag-Ali J. Excavation of Acheulean sites near Saffaqah in Ad-Dawadmi. Atlal. 1982;7:9–21.

Whalen N, Siraj-Ali J, Davis W. Excavation of Acheulean sites near Saffaqah, Saudi Arabia. Atlal. 1983;8:9–24.

Whybrow PJ. New stratotype; the Baynunah formation (Late Miocene), United Arab Emirates: lithology and palaeontology. Newsletters on Stratigraphy. 1989;21:1–9.

Whybrow PJ, Hill A, editors. Fossil vertebrates of Arabia with emphasis on the Late Miocene faunas, geology, and paleoenvironments of the Emirate of Abu Dhabi, United Arab Emirates. New Haven, CT/ London: Yale University Press; 1999.

Chapter 10
Paleolithic Stone Tool Assemblages from Sharjah and Ras al Khaimah in the United Arab Emirates

Julie Scott-Jackson, William Scott-Jackson, and Jeffrey I. Rose

Keywords Levallois • Ras al Khaimah • Sharjah • Survey • Technology • UAE

Background

Over the past 20 years a virtual moribundity has descended on Paleolithic research in the region of the Persian Gulf. This predicament arose as a direct consequence of the reassessment of Holger Kapel's lithic 'Group' classifications in his 'Atlas of the Stone Age Cultures of Qatar' (Kapel, 1967) by the French team working in Qatar during 1976–1978 (see Inizan, 1980). Group A, which Kapel had tentatively assigned to the Paleolithic, was categorized by the French as Neolithic effectively curtailing Paleolithic research in the Persian Gulf region as the re-evaluation of Group A was seen by many to demonstrate a general absence of the Paleolithic in the entire region and furthermore, suggesting that any lithics found in the Gulf area would almost certainly not be Paleolithic. A view which was strengthened, certainly in the United Arab Emirates, following field surveys in Sharjah Emirate by various French Archaeological Missions between 1984 and 1988 (see Boucharlat et al., 1984; Cauvin and Calley, 1984; Calley and Santoni, 1986; Millet, 1997) and further investigations between 1990 and 1992 (Briand et al., 1992). The result of these investigations was the discovery of numerous prehistoric lithic assemblages. Briand and colleagues state in their 1992 report, "We have already carried out a certain number of studies which show that the lithic industry in the area of Mleiha, as in all the Emirate of Sharjah, dates back to the

sixth and fourth millennia, though in most cases it does not present a well-defined typology… Without going into detail, we may say that all the petrographic examples found among the tools whether from the interior or the from the coast of Sharjah, may be found near the sites…but, the fabrication of the tools which we know at Sharjah could have been carried out using raw materials from local outcrops". They also add: "The fact that we find in the Emirate of Sharjah all the petrographic components encountered in the stone tools does not mean that all the lithic industry recorded locally comes automatically from this emirate. It only means that men of the fifth and fourth millennia could find nearby all the materials necessary to [for] the debitage and to [for] their knapping. Inversely, even if imports from afar took place, they could not explain all the local lithic industry".

So often, thorough investigations generated problematic data, and for the French researchers these were no exception as they concluded that the Sharjah Emirate lithic assemblages dated back to the fourth, fifth and sixth millennia (thereby making them post-Paleolithic) although generally, in their words, the lithics "did not present a well defined typology". Furthermore, (as noted above) "they could not explain all the local lithic industry".

Negativity on this scale has a tendency to lose its impact over time as new techniques are developed and new ideas arise. For Julie Scott-Jackson and William Scott-Jackson it was the hypothesis of a 'southern' migration route of hunter-gathers out of Africa via the Arabian peninsula to the Far East during the Middle Paleolithic (Petraglia and Alsharekh, 2003; Forster and Matsumura, 2005) that inspired initial investigations in the limestone areas of Sharjah Emirate (the main focus of attention) and in a small area of Ras al Khaimah (UAE) during 2006 and 2007 (Fig. 1). Karstic environments, such as limestone, have the potential to yield in-situ Paleolithic sites which may be found on the highest parts of ridges or hill-tops, retained in depressions or fissures.

The retention of these ancient sites on such high-levels is due to simultaneous geomorphological processes (both chemical and mechanical) operating on the limestone over geological time. The development of depressions and/or fissures is effected by the dissolution of the underlying limestone.

J. Scott-Jackson (✉) and W. Scott-Jackson
PADMAC Unit, Institute of Archaeology, University of Oxford, 36 Beaumont Street, Oxford, OX1 2PG, UK
e-mail: julie.scott-jackson@arch.ox.ac.uk;
william.scott-jackson@arch.ox.ac.uk

J.I. Rose
Department of Anthropology and Geography, Oxford Brookes University, Oxford, OX3 0BP, UK
e-mail: jeffrey.i.rose@gmail.com

M.D. Petraglia and J.I. Rose (eds.), *The Evolution of Human Populations in Arabia*, Vertebrate Paleobiology and Paleoanthropology, DOI 10.1007/978-90-481-2719-1_10, © Springer Science+Business Media B.V. 2009

Fig. 1 Topographic map showing the relationship between the location of newly discovered Upper Pleistocene manufacturing sites in Sharjah Emirate, UAE and the proposed southern route out of Africa (after Forster and Matsumura, 2005)

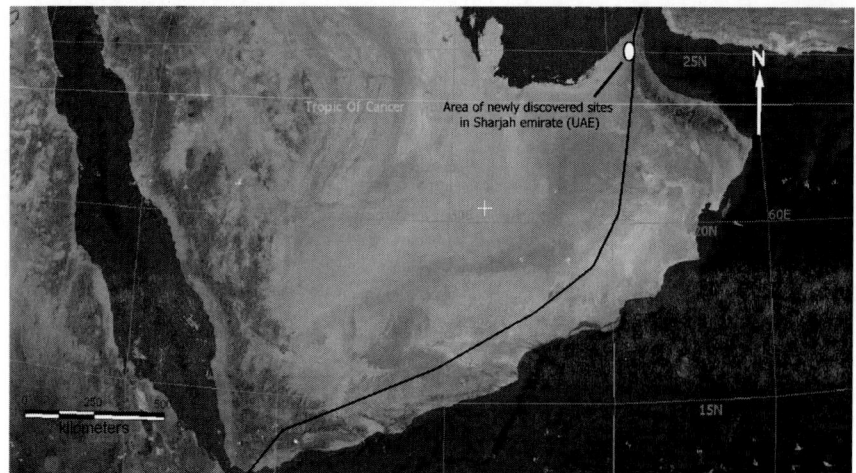

Fig. 2 Comparable topographic locations of the newly discovered sites in Sharjah and Ras al Khaimah Emirates (UAE)

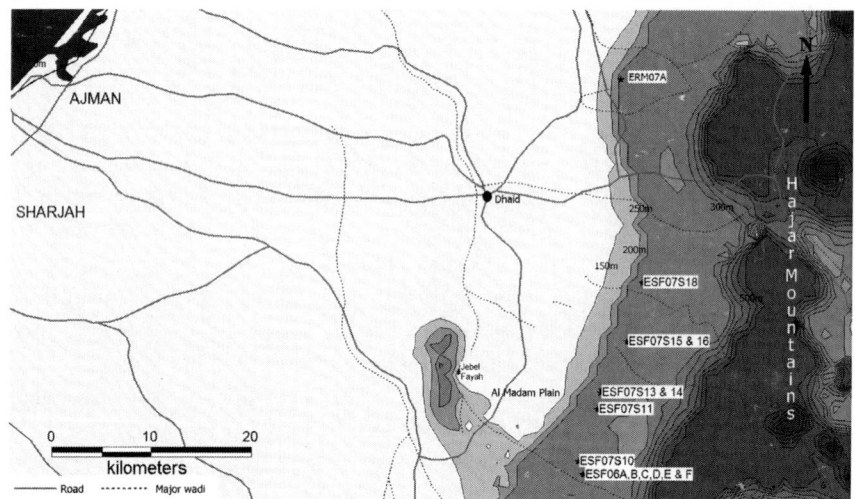

Surface weathering, often by the action of water, and aeolian processes, which include erosion, transport and deposition of materials by wind, contribute to the Paleolithic assemblage becoming incorporated into the infilling deposits. Later perhaps, the assemblage could be exposed once again by deflation. On some sites with multiple occupations these processes can lead to the creation of palimpsests.

The integrity of Paleolithic surface-scatters on high-level sites is therefore, somewhat different to many found at low-levels. For although these high-level scatters may represent different occupations over time (palimpsests), they are not in a derived context. They are, in a geomorphological sense, essentially in-situ, that is to say, they are clearly places where Paleolithic people made these stone-tools. This in-situ category is not the same however, as the archaeological definition of 'primary context'; a situation where it can be proved conclusively that the artifact/finds were in the same position when found, as they were when they were originally deposited by Paleolithic people on Paleolithic

landsurfaces. Worldwide, very few Paleolithic sites indeed are in this category.

The use of the geomorphological in-situ definition here therefore, acknowledges that taphonomic changes could have occurred to the surface of the site by a variety of processes. Deflation for example, and perhaps also the movements of artifacts horizontally (due to wind and flows); vertical movement when sedimentation in arid environments may not occur and bioturbation and anthropomorphic activities. Also the loss of part or all of a high-level in-situ Paleolithic site is the result of slope destabilization by erosional processes whereby materials are removed from the sides of the hills and ridges. Over time, the highest areas of these hills and ridges change shape and are reduced in size (for a detailed discussion see Scott-Jackson, 2000; Scott-Jackson et al., 2007, 2008).

The result of the 2006 and 2007 fieldwork in the UAE was the new discovery of many well delineated Paleolithic surface-sites at high-levels on the limestone ridges that have outcropping seams of red chert. These limestone ridges flank the western anticlines of the Hajar Mountains in Sharjah and Ras al Khaimah (ridges such as these extend through the UAE and south into Oman). For Paleolithic people these ridge-top sites were ideal locations for the manufacture of stone-tools from the readily available seams of good knappable red chert and to observe the movements of animals and perhaps other hunters in the wadis below (similar topographic locations of Middle Paleolithic sites have been recorded for other areas of the Arabian peninsula (e.g., Smith, 1977; Amirkhanov, 1994; Rose, 2004; Alsharekh, 2006)). The distribution pattern of these Paleolithic sites (Fig. 2) supports the 'southern' route hypothesis but it is also conceivable that the sites represents evidence of other patterns of migration including that of Paleolithic people coming back into the Arabian peninsula from all points east.

The 2006–2007 Field Investigations: Sites and Artifacts

The number and size of the Paleolithic assemblages at the various sites in UAE during the 2006 and 2007 field investigations exceeded all our expectations. To maintain site integrity many artifaces were examined at the various sites and the general size of assemblages observed and noted (very few artifacts, therefore, were removed from the sites for essential off-site examination). It is therefore judicious to stress that the tools and debitage considered here are but a very small percentage indeed of the total of Paleolithic artifacts which make up the various assemblages. To minimize confusion in the discussions which now follow, sites in the Fili area of Sharjah are annotated thus (F). Also, (06) and (07) annotate the year in which the site was discovered. The total number of sites discovered by the end of 2007 were:

1. Four in-situ high-level sites: ES(F)06A; ES(F)06D; ES(F)07F; ES07S14.
2. One eroded high-level site: ES07S15.
3. Nine discrete lithic scatters (found at lower levels): ES(F)06B; ES(F)06C; ES(F)07E; ES07S10; ES07S11; ES07S13; ES07S16; ES07S18; ERM07A.

All the lithic assemblages are made of red chert, generally they have a dark reddish brown patina (Munsell color 2.5 yr 3/4) or strong brown patina (Munsell color 7.5 yr 5/6). When newly fractured, the chert is a light reddish-yellow color (Munsell color 7.5 yr 7/6). Although the patinated and unpatinated chert both have a Munsell color *Hue* 7.5 and *Chroma* 6, it is important to note that the Munsell color *Values* of 5 and 7 are very different indeed.

The sites and associated lithic assemblages are now discussed in order of discovery and topographical/ geographical relationships.

Field Investigations in 2006

Site ES(F)06A

This prolific in-situ stone-tool manufacturing site (in the Fili area of Sharjah) is on the highest part of the limestone ridge at ~279 m above mean sea-level (AMSL). The ridge has been subject to a certain amount of slope erosion, attested to by extensive lithic scatters on the steep sides (Fig. 3). From this site there are long views (to a distance of ~10 km) across the Al Madam plain to the west, the foothills of the Hajar mountains in the east and into the wadis below. All the lithics are made from the outcropping seams of red chert.

Site ES(F)06B

At ~263 m AMSL, this is a discrete lithic scatter (made from red chert) on a lower terrace to the east of ES(F)06A.

Site ES(F)06C

At ~264 m AMSL, this is a discrete lithic scatter (made from red chert) on a lower terrace to the east of ES(F)06A.

Field Investigations in 2007

Site ES(F)06D

Discovered in 2006 and investigated in 2007 this site is at ~276 m AMSL on a prominent hilltop that has steep sides. It is situated on the same ridge as ES(F)06A but ~100 m to the

Fig. 3 Site ESF06A and surrounding area

Fig. 4 Sites in the locale of ESF06A and
ESF06D

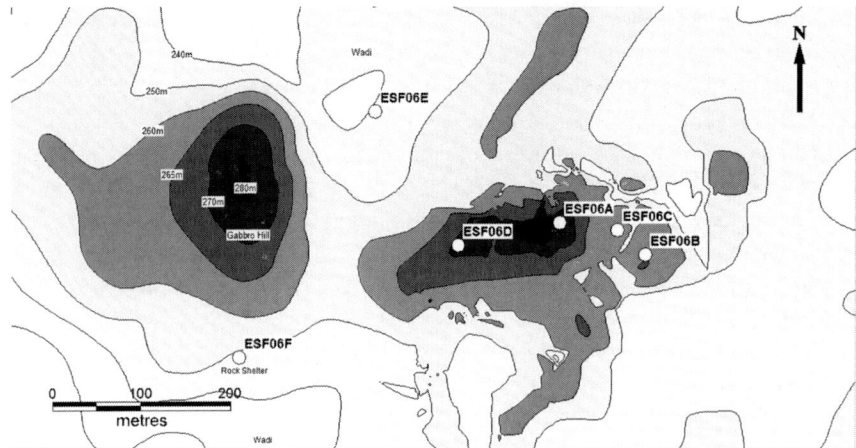

west and separated by a low eroded section of the ridge
(Fig. 4). The site per se covers an area of ~8 m × 5 m which
is composed of large red chert boulders that outcrop from a
sandy silt deposit. Numerous lithics made from red chert litter
the surface of this high-level site and on the slopes directly
below. As the site is on the same ridge as ES(F)06A there are
similar long views to the west across the Al Madam plain (to
a distance of ~10 km), the foothills of the Hajar mountains in
the east and into the wadis below.

Site ES(F)07E

This distinctive glossy dark red-brown lithic scatter was
observed on a low wadi terrace at ~249 m AMSL. In this area
the wadi is ~1 km wide with a narrow incised channel winding
through it. The lithics were in a discrete concentration on the
southeast side of the channel. This wadi terrace site is ~267 m
to the northwest of the high-level site ES(F)06A.

Site ES(F)07F and 'Gabbro Hill'

Situated on a rocky outcrop at ~252 m AMSL above a small
cave (Fig. 5), the site is so positioned to overlook a large
wadi immediately to the south and smaller wadis to the east
and west. The assemblage from this site is characterized by
blades and blade cores made of red chert. Chert outcrops in
the immediate vicinity show possible signs of large flakes
being removed for the manufacture of stone-tools. The site is
~300 m to the southeast of ES(F)06A and on the lower slopes
of the large rounded foothill, which runs north–south (orthogonal to the ES(F)06A ridge).

To the north, 'Gabbro' hill rises to ~279 m AMSL and is
covered with a black gabbro boulder train. Amongst the boulders, occasional chert lithics were observed, but no sign of chert
raw material, this scatter is ~200 m to the north of ES(F)07F.

Fig. 5 ESF07F site area and small cave

Site ES07S10

Lithic scatters made of chert were found in a derived context
at various locations on rounded hills (height range 230–250 m
AMSL) north of Fili, ~2 km northwest of ES(F)06A.

Site ES07S11

Lithic scatters made of chert were found at ~252 m AMSL
on a low rounded hill, ~8 km north of ES(F)06A. These lithics
are on heavily eroded slopes and not in-situ.

Site ES07S13

An extensive lithic scatter made of chert was found at a height of
~240 m AMSL on a low rounded hill, ~2 km northeast of ES0711.
These lithics are on heavily eroded slopes and not in-situ.

Site ES07S14

This prolific site, which appears to be an in-situ manufacturing site, is centred on a small plateau on the top of a limestone ridge at ~238 m AMSL. Just below the site there is an outcrop of good quality chert. The lithics observed here were mainly large flakes made from chert; there was also a small gabbro boulder (~25 cm in diameter) that exhibited wear consistent with use as a hammer stone. There is no source of gabbro on the hill or in the immediate vicinity. Slope erosion has resulted in extensive lithic scatters on the slopes and on the track at the base of the hill. This site has similarities to ES(F)06A and ES(F)06D, both of which are ~9.2 km to the SSW. The site is ~370 m to the north of ES07S13.

Site ES07S15

Once at a height of at least 256 m AMSL, this heavily eroded hill top is now a narrow ridge (~2 m maximum) made up of chert boulders with very little deposit. No lithics were seen on the ridge but the slopes directly below the ridge are covered with an enormous number of chert lithics which are clearly derived directly from the ridge above. The site overlooks a major wadi to the east; also there are long views to the south and east. The site is ~14 km north of ES(F)06A.

Site ES07S16

A chert lithic scatter was found at a height of ~242 m AMSL on a low rounded hill adjacent to, and south of, the ES07S15 hill and ~14 km north of ES(F)06A and ES(F)06D.

Site ES07S18

A discrete chert lithic scatter was found at a height of ~227 m AMSL on a rounded hill. The site is just south of the Sharjah/Fujairah border and ~20 km NNE of ES(F)06A.

Site ERM07A

In the south of Ras al Khaimah, UAE, a discrete chert lithic assemblage was found at a height of ~213 m AMSL on a rounded hill with red chert outcrops. Chert lithics also litter the heavily eroded sides of this isolated hill. Destruction of

the site has been exacerbated by military activity. This site, like those in Sharjah is situated on the western foothills of the Hajar Mountains. To the west of the site there are long views over the desert and to the south across a gravel plain. This site is directly N of site ES07S18 and ~40 km NNE of ES(F)06A.

The Lithic Assemblages: Sampling and Analysis

Nine lithic assemblages are included in this analysis, totalling 421 chipped stone artifacts. Only sites with sample sizes of greater than ten were examined. Because of the sheer density of the surface scatters from which the material was collected, there was an effort to select technologically and typologically diagnostic pieces in the field (e.g., Levallois debitage, retouched tools, bifacial elements, faceted platforms, prepared cores). Therefore, the assemblages represent deliberately skewed samples of artifacts with recognizable features, by no means a true representation of the full scope of chipped stone pieces found at each findspot (hence, tools comprise nearly 40% of the total assemblage).

Given that these are surface sites lacking stratigraphic context, we must assume that the lithic collections are palimpsests, deposited over the course of multiple occupational phases. Despite the inherent limitations posed by skewed samples and mixed industries, the data from these assemblages are useful for describing the variety of lithic technologies present throughout the region. While their nature as surface sites precludes any absolute or relative chronological attribution, it is still possible to note the range of technological features observed at each locality. Specific counts of artifact class (Table 1), blank type (Table 2), platform morphology (Table 3), dorsal scar patterns (Table 4), core types (Table 5), and tool types (Table 6) are presented and synthesized in a list of significant techno-typological features present within each assemblage (Table 7).

General Observations

Most of the artifacts were struck from chert slabs/nodules that outcrop in numerous seams found throughout the limestone foothills of the western Hajar Mountains. The material ranges in colour from dark maroon to red, and from fine to course-grained in texture. The artifacts are coated by a variably thick veneer of patina and/or desert varnish. Most of the knapped edges are relatively sharp, showing only some degree of aeolian abrasion but no rounding from fluvial activity.

Table 1 Artifact class

Artifact class	ERM07	ESF06	ESF06	ESF06	ESF06	ESF07	ESF07	ESF07	Gabbro
n (%)	A	A–C	D	E	F	S10	S14	S15	Hill
Debitage	26 (34.7)	48 (47.5)	9 (36.0)	20 (35.7)	35 (58.3)	23 (51.1)	6 (33.3)	2 (18.2)	4 (23.5)
Cores	2 (2.7)	3 (2.9)	2 (8.0)	15 (26.8)	10 (16.7)	2 (4.4)	3 (16.7)	3 (27.3)	6 (35.3)
Tools	46 (61.3)	50 (49.5)	9 (36.0)	5 (8.9)	12 (20.3)	18 (40.0)	7 (38.9)	4 (36.4)	6 (35.3)
Chips	–	–	–	11 (19.6)	–	–	–	1 (9.1)	1 (5.9)
Chunks	1 (1.3)	–	5 (20.0)	5 (8.9)	3 (5.0)	2 (4.4)	2 (11.1)	1 (9.1)	–
Total	75	101	25	56	60	45	18	11	17

Table 2 Blank type

Blank type	ERM07	ESF06	ESF06	ESF06	ESF06	ESF07	ESF07	ESF07	Gabbro
n (%)	A	A–C	D	E	F	S10	S14	S15	Hill
Flakes	**46 (76.7)**	**58 (67.4)**	**10 (71.4)**	**11 (45.8)**	**20 (46.5)**	**29 (85.3)**	**8 (80.0)**	**2 (50.0)**	**5 (71.4)**
Regular	33	42	8	10	13	22	7	2	4
Cortical	6	4	1	1	4	4	–	–	1
Debordant	2	1	–	–	–	–	–	–	–
Levallois	5	11	1	–	3	3	1	–	–
Blades	**6 (10.0)**	**22 (25.6)**	**1 (7.1)**	**9 (37.5)**	**22 (51.2)**	**3 (8.9)**	**1 (1.0)**	**2 (50.0)**	**1 (14.3)**
Regular	5	16	1	9	21	1	–	2	1
Ccortical	–	2	–	–	–	–	–	–	–
Debordant	1	3	–	–	–	2	–	–	–
Bladelets	–	1	–	–	1	–	–	–	–
Other	**8 (13.3)**	**6 (7.0)**	**3 (21.4)**	**4 (16.7)**	**1 (2.3)**	**2 (5.9)**	**1 (1.0)**	**0 (0.0)**	**1 (14.3)**
Kombewa	–	1	1	–	–	–	–	–	–
Biface thinning	3	3	2	3	1	–	1	–	1
Core trimming	5	2	–	1	–	2	–	–	–
Total	60	86	14	24	43	34	10	4	7

Table 3 Platform type

Platform type	ERM07	ESF06	ESF06	ESF06	ESF06	ESF07	ESF07	ESF07	Gabbro
n (%)	A	A–C	D	E	F	S10	S14	S15	Hill
Unmodified	**24 (57.1)**	**51 (63.8)**	**6 (50.0)**	**10 (55.6)**	**25 (75.8)**	**16 (59.3)**	**6 (100.0)**	**4 (100.0)**	**5 (71.4)**
Straight	20	33	4	9	23	10	6	2	4
Cortical straight	3	16	1	1	2	4	–	1	1
Cortical curved	1	2	1	–	–	2	–	1	–
Modified	**18 (42.3)**	**29 (36.3)**	**6 (50.0)**	**8 (44.4)**	**8 (24.2)**	**11 (40.7)**	**0 (0.0)**	**0 (0.0)**	**2 (28.6)**
Dihedral	4	5	2	2	3	1	–	–	1
Dihedral ½ cortex	–	1	–	2	–	2	–	–	–
Faceted straight	8	16	2	4	3	3	–	–	–
Faceted curved	6	6	2	–	2	3	–	–	1
Transverse	–	–	–	–	–	2	–	–	–
Chapeau de gendarme	–	1	–	–	–	–	–	–	–
Total	42	80	12	18	33	27	6	4	7

Table 4 Dorsal scar pattern

Scar pattern	ERM07	ESF06	ESF06	ESF06	ESF06	ESF07	ESF07	ESF07	Gabbro
n (%)	A	A–C	D	E	F	S10	S14	S15	Hill
Unidirectional	9 (18.4)	15 (18.8)	1 (10.0)	4 (17.4)	7 (18.4)	9 (30.0)	1 (11.1)	1 (25.0)	1 (16.7)
Unidirectional – crossed	18 (36.7)	19 (23.8)	5 (50.0)	10 (43.5)	10 (26.3)	10 (33.3)	2 (22.2)	–	2 (33.3)
Unidirectional – parallel	5 (10.2)	17 (21.3)	1 (10.0)	5 (21.7)	12 (31.6)	2 (6.7)	–	3 (75.0)	1 (16.7)
Convergent	3 (6.1)	20 (25.0)	1 (10.0)	–	5 (13.2)	1 (3.3)	2 (22.2)	–	–
Bidirectional	4 (8.2)	–	–	–	1 (2.6)	–	–	–	–
Radial	9 (18.4)	5 (6.3)	2 (20.0)	1 (4.3)	3 (7.9)	6 (20.0)	4 (44.4)	–	1 (16.7)
Transverse	–	2 (2.5)	–	1 (4.3)	–	–	–	–	1 (16.7)
Crested	–	–	–	1 (4.3)	–	1 (3.3)	–	–	–
Transverse – crested	1 (2.0)	2 (2.5)	–	1 (4.3)	–	1 (3.3)	–	–	–
Total	49	80	10	23	38	30	9	4	6

Table 5 Core type

Core type	ERM07	ESF06	ESF06	ESF06	ESF06	ESF07	ESF07	ESF07	Gabbro
n	A	A–C	D	E	F	S10	S14	S15	Hill
Simple unidirectional	1	–	5	8	–	–	1	2	2
90°	1	1	1	1	1	–	–	–	1
Globular	–	–	–	–	–	–	–	–	1
Kombewa	–	1	–	–	–	–	–	–	–
Discoid	–	–	1	–	2	–	–	–	1
Levallois, radial	–	–	1	1	–	1	2	–	–
Levallois, lateral	–	–	–	–	1	–	–	–	1
Levallois, distal	–	–	–	–	–	–	–	1	–
Levallois, unidirectional	–	–	–	4	–	1	–	–	–
Total	2	2	8	14	4	2	3	3	6

Table 6 Tool type

Tool type	ERM07	ESF06	ESF06	ESF06	ESF06	ESF07	ESF07	ESF07	Gabbro
n (%)	A	A–C	D	E	F	S10	S14	S15	Hill
Sidescrapers	**9 (19.6)**	**9 (18.4)**	**1 (11.1)**	**0 (0.0)**	**3 (25.0)**	**2 (11.1)**	**0 (0.0)**	**0 (0.0)**	**2 (33.3)**
Simple	5	4	–	–	3	–	–	–	–
Bilateral	2	–	–	–	–	1	–	–	–
Backed	1	2	–	–	–	–	–	–	2
Bifacial	–	2	1	–	–	–	–	–	–
Dejete	1	1	–	–	–	1	–	–	–
Endscrapers	**6 (13.0)**	**3 (6.1)**	**1 (11.1)**	**1 (20.0)**	**1 (8.3)**	**1 (5.6)**	**0 (0.0)**	**0 (0.0)**	**1 (16.7)**
Simple	2	1	–	–	–	–	–	–	–
Transverse	1	–	–	–	–	1	–	–	–
Ogival	1	1	–	–	–	–	–	–	–
Nosed	1	1	1	–	1	–	–	–	–
Thumbnail	1	–	–	1	–	–	–	–	1
Small bifaces	**4 (9.7)**	**7 (14.3)**	**3 (33.3)**	**0 (0.0)**	**3 (25.0)**	**3 (16.7)**	**2 (28.6)**	**1 (25.0)**	**1 (16.7)**
Foliate	1	4	1	–	–	1	–	–	–
Limande	–	1	–	–	–	2	1	–	1
Partially-retouched	1	1	–	–	1	–	–	–	–
Misc fragment	2	1	2	–	2	–	1	1	–
Heavy duty tools	**13 (28.3)**	**17 (34.5)**	**1 (11.1)**	**1 (20.0)**	**3 (25.0)**	**9 (50.0)**	**0 (0.0)**	**2 (50.0)**	**2 (33.3)**
Handaxe	–	–	–	1	–	–	–	1	–
Backed knife	8	13	1	–	2	5	–	1	1
Chopper	3	2	–	–	1	2	–	–	1
Backed biface	2	1	–	–	–	2	–	–	–
Cleaver	–	1	–	–	–	–	–	–	–
Other	**14 (30.4)**	**13 (26.5)**	**3 (33.3)**	**3 (60.0)**	**2 (16.7)**	**3 (16.7)**	**5 (71.4)**	**1 (25.0)**	**0 (0.0)**
Retouched flake	4	5	1	1	–	2	1	–	–
Retouched blade	–	1	–	–	–	1	1	–	–
Notch	1	2	–	–	–	–	1	–	–
Denticulate	3	2	–	–	1	–	1	–	–
Truncation	1	2	–	1	–	–	1	1	–
Perforator	5	1	1	1	1	–	–	–	–
Burin	–	–	1	–	–	–	–	–	–
Total	46	49	9	5	12	18	7	4	6

Table 7 Techno-typological indicators

	ERM07	ESF06	ESF06	ESF06	ESF06	ESF07	ESF07	ESF07	Gabbro
	A	A–C	D	E	F	S10	S14	S15	Hill
Technological features									
High blade frequency	+	++	+	++	+++	+	+	+	+
Edge preparation	+	+			+++				
Core maintenance	++	+	+			+			
Bidirectional scars	++				+				
Platform faceting	++	++	++	+	+	++	+		+
Levallois, radial	++	+	++	+	+	+	+		+
Levallois, unipolar	+	++							
Kombewa		+	+						
Biface thinning	+	+	+	++	+		+		+
Typological features									
Handaxes				+				+	
Points and foliates	+	++	+		++	++	+	+	+
Sidescrapers	++	++	+		++	+			++
Endscrapers	++	+	+	+	+	+			+
Group	A1	A2	A1	A3	B1	A1	A1?	A3?	A1?

There are several indicators that the primary means of blank removal was via hard hammer percussion. Most of the debitage, including blades and éclat de taille,[1] exhibit large striking platforms and prominent bulbs of percussion. Furthermore, few pieces are lipped and striking platforms tend to be quite large in relation to blank proportions, indicated by relative platform size (blank area divided by platform area).

While the sites exhibit a range of variability, there are technologically diagnostic features shared among every assemblage that allow for vague chronological attributions bracketed between the late Lower Paleolithic and early Upper Paleolithic. These common features include faceted platforms, Levallois[2] cores and flakes (Fig. 6), bifacial tool production (Figs. 7–9), and simple, unidirectional, hard hammer blades (Fig. 10q). Bifacial tools produced by façonnage reduction range from large handaxes (Fig. 9) to small, thin, leaf-shaped [Khasfian] points (Fig. 7g–j). Levallois cores exhibit both radial and unidirectional-convergent methods of convexity maintenance. In a few cases, the frequency of blade-proportionate blanks reaches over 50%.

Based on the combination of various features recorded at each findspot, the assemblages have been organized into four different groups: A1, A2, A3, and B1. These categories represent arbitrarily defined units that should *not* be considered true lithic industries. They are useful for articulating the range of technologies distributed across the landscape, but not for defining discrete analytical units. The samples from ES07S11 to S18 were too small to be included in these groupings.

---

[1]This term refers to an analytical unit that specifically describes shaping flakes produced during bifacial reduction.
[2]For the purposes of this paper, our definition of Levallois is sensu stricto: faceted striking platforms, flat longitudinal profile, and a preferentially prepared working surface to establish and maintain convexity.

Group A1 (ES(F)06D, ES07S10, ES07S14, ERM07A, Gabbro Hill)

Group A1 is most notably characterized by faceted striking platforms, which are found on over 30% of all blanks (Table 3). Radial forms of core reduction appear to be the most prominent strategy (Fig. 10o–p, r–s), indicated by the relatively high frequencies of Levallois cores with centripetally prepared working surfaces, discoids, and debitage bearing radial scar patterns (Table 4). Tools from these sites include scrapers (both sidescrapers and endscrapers) often showing heavy, invasive retouch on thick, flat blanks. There are diminutive bifacial tools classified as foliates and limandes (Fig. 7), as well as partially retouched points (Fig. 6b, Table 6). Heavy duty tools such as backed knives, choppers, and backed bifaces are also present. Many of these tool types belong to a continuum ranging from bilateral convergent sidescrapers to partially retouched points, and from backed sidescrapers to backed bifaces (Fig. 8).

Group A2 (ES(F)06A, ES(F)06B, ES(F)06C)

This group exhibits the basic features shared among all A-group assemblages; that is, faceted striking platforms (Table 3), Levallois reduction strategies (Table 6), sidescrapers with invasive retouch, and small bifacial points (Table 6). What distinguishes A2 is the predominance of unidirectional-convergent scar patterns resulting from the production of Levallois points (Fig. 6a, c, Table 4), the high percentage of blade-proportionate blanks (Table 2), and, in one case, a

Fig. 6 Levallois blanks (**a**: ESF07S19, **e**: ESF06C); retouched/partially-retouched points (**b**: ESF06S27, **d**: ESF06D); Unipolar Levallois cores (**c**: ESF07MAR, **f**: ESF07wadi)

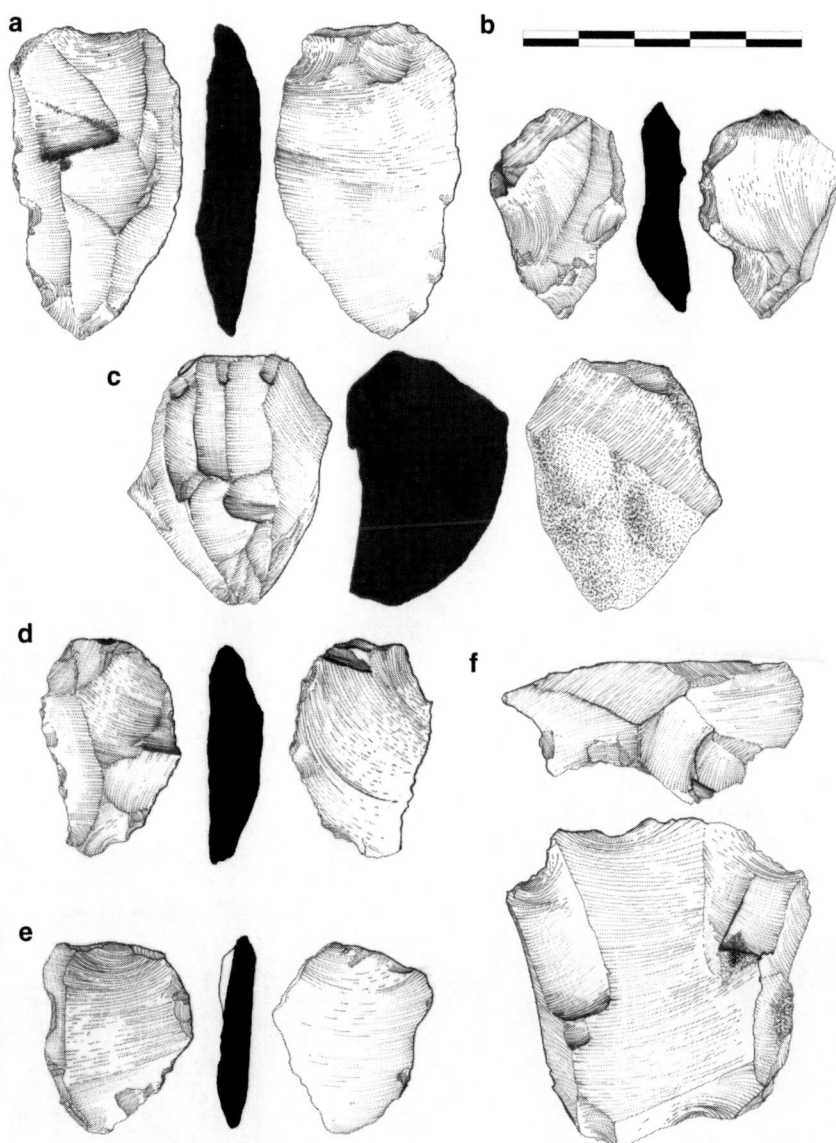

chapeau de gendarme striking platform (Table 3). The presence of a unipolar Levallois method may explain the high frequency of blade-proportionate pieces, which are merely the byproducts of convexity maintenance and not related to a true, prismatic blade industry.

Group A3 (ES(F)06E, ES07S15)

Group A3 exhibits a core reduction strategy resembling the A1 assemblages – radially prepared Levallois cores with platform faceting (Table 5). Differentiating the two different groups of assemblages is the additional component in A3 of large, flat bifacial handaxes (Fig. 9) with trimmed or

untrimmed butts (Table 6), as well as the higher frequency of biface thinning flakes and blade-proportionate debitage (Table 2). It is unclear whether the handaxes are associated with the radial Levallois pieces, blades, neither, or both.

Group B1 (ES(F)06F)

Group B1 has been given a separate letter designation to signify that it is the most distinct of all groups. Unlike the other assemblages, modified striking platforms are below 25% (Table 3), while blade-proportionate pieces comprise over 50% of the debitage (Table 2). There are a few indications that Levallois technology is still present within the

Fig. 7 Foliates (**g**: ESF07green, **h**: ESF06A,
i: ESF06A, **j**: ESF06A)

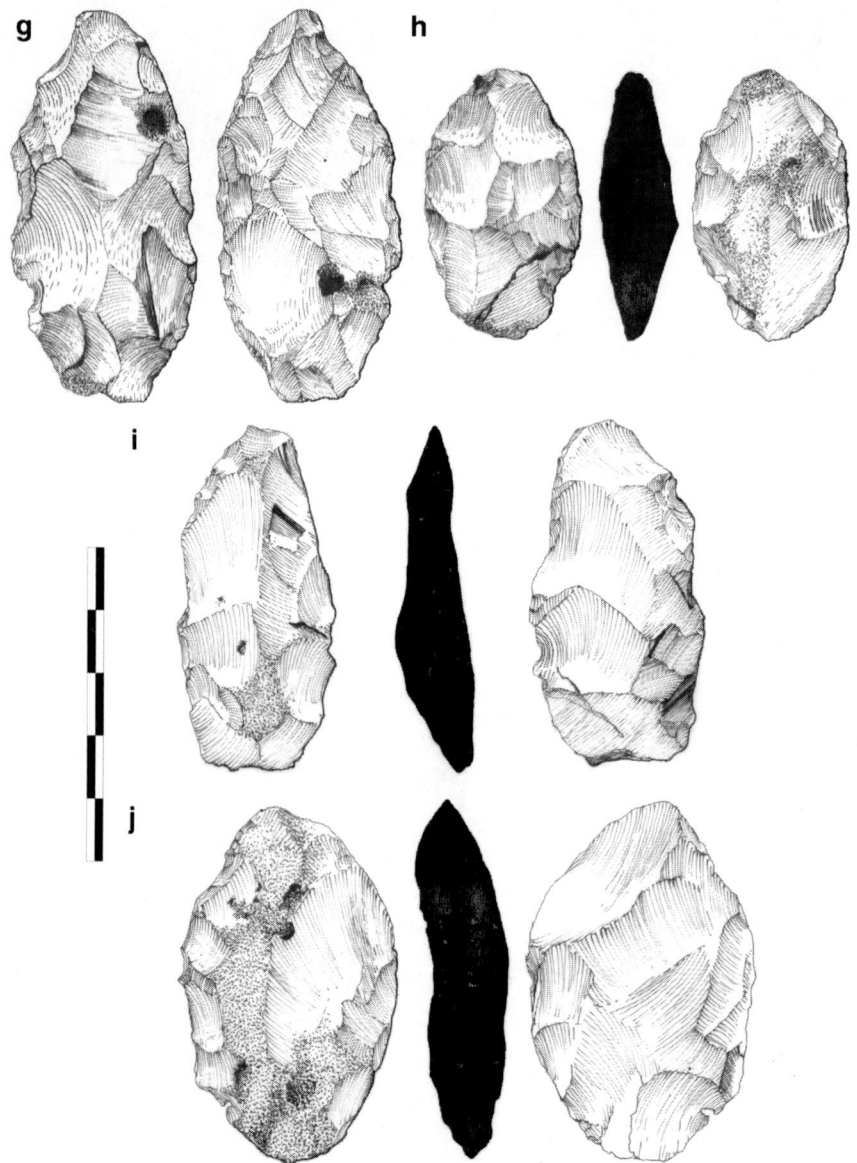

assemblage; however, the frequency is significantly less than at all other findspots. In contrast, the most prevalent mode of reduction is the removal of simple, unidirectional, blade-proportionate blanks from volumetric cores (Fig. 10q). Also distinguishing Group B1 is the frequency of edge preparation,[3] which, at 38%, is nearly double that of other sites. There are only 12 tools in this group (Table 6), so it is difficult to make any meaningful observations regarding typology. One noteworthy characteristic

is the absence of bifacial pieces; neither foliates, limandes, nor handaxes were found. Only one biface thinning flake was recorded, also indicating that there was minimal manufacture of such tools.

Regional Context

Since the assemblages described above were all collected from surface contexts, the following observations must be considered tentative. Taking into account the prevalence of faceted striking platforms, Levallois cores, retouched points, bifacial pieces, blades, handaxes, and discoids, the artifacts recovered in this collection can be attributed to several

---

[3]Edge preparation is indicated by grinding/abrasion at the interface between the dorsal face and the striking platform. This method of abrading the core is used to remove brittle overhanging bits along the edge, which allows for more accurate removal of flakes or blades.

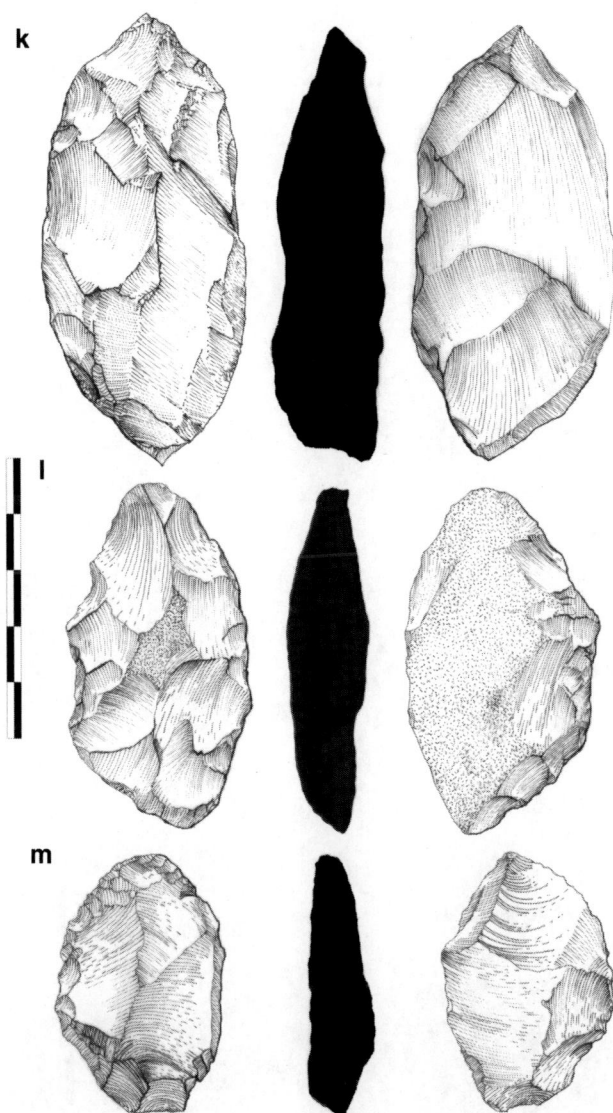

Fig. 8 Backed bifaces (**k**: ERG07A18, **l**: ESF07S44, **m**: ESF06a)

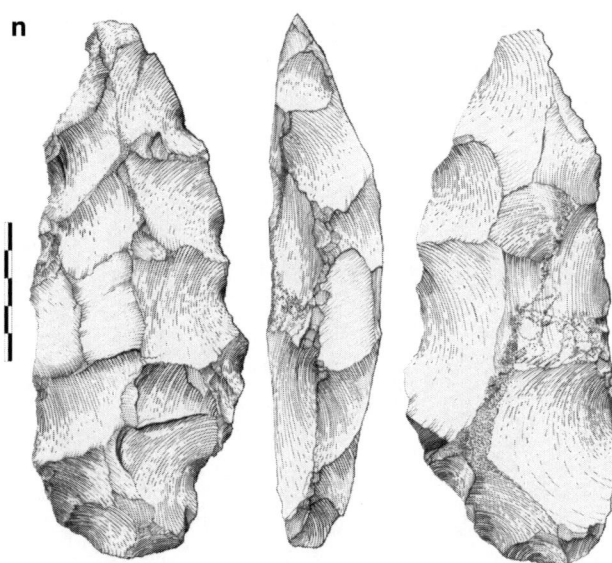

Fig. 9 Large biface (**n**: ESF07S15)

different phases between the late Lower Paleolithic and the early Upper Paleolithic. If the artifacts reported within Group A3 are indeed coeval, then it is likely this is the oldest group of assemblages. The same suite of technological and typological features observed within Group A3 – unipolar hard hammer blade cores, centripetal Levallois reduction, discoids and lanceolate bifacial handaxes – was recorded at the Wadi Qilfah 1–4 complex (Rose 2006, 2007) and other surface sites throughout central Oman (Jagher, 2009), suggesting affinities to the as yet undated Sibakhan Industry of southern Arabia (Rose, 2006). Given the evidence for multiple pluvial episodes during MIS 6 (Parker and Rose, 2008), it would not be surprising to discover hominin occupations associated with these wet periods.

Bifacial handaxe production accompanied by radial Levallois technique also commonly occurs in the late Middle Pleistocene of both East Africa (e.g., Leakey et al., 1969;

McBrearty, 2001; Van Peer et al., 2003; Tryon, 2006) and the Near East (Hours et al., 1973; Jelinek, 1990; Copeland, 2000). Less than 50 km away, a similar assemblage consisting of handaxes, foliates, radial cores, and blades was discovered at Jebel Faya Rockshelter, level C, with OSL dates indicating an age >85 ka (Uerpmann et al., 2007). Assemblages bearing a similar suite of technologies were also discovered in Dawadmi (Whalen et al., 1984) and Wadi Fatimah (Whalen et al., 1988) in central Saudi Arabia. Albeit problematic, U-series dates were obtained from calcium carbonate concretions on chipped stone artifacts found at these sites, placing them potentially as far back as 250 ka.

There is very little to discern Groups A1 and A2 from one another, the primary feature being radial (A1) versus convergent (A2) Levallois methods. In both cases, prepared core reduction is accompanied by the manufacture of bifacial foliates. These variable Levallois modalities are reminiscent of the Levantine Middle Paleolithic (Jelinek, 1992; Monigal, 2002), which is comprised of Levallois points (both short and elongated), as well as centripetal Levallois flakes. One specimen from ES(F)06A (A2) with unidirectional-convergent scars exhibited a classic 'chapeau de gendarme' striking platform, further suggesting a Middle Paleolithic attribution. However, it should be pointed out that bifaces are completely absent from Levantine MP assemblages; as such, this clearly distinguishes the Arabian MP. On the other hand, the repeated co-occurrence at these findspots of prepared core technologies with diminutive bifacial foliates is the hallmark of East African MSA assemblages (e.g., Wendorf and Schild, 1974; Pleurdeau, 2005). Given its position bridging East Africa and the Near East, shared elements from both adjacent regions is expected within the archaeological record and is a

Fig. 10 Unipolar cores (**o**: ESF07S18, **q**: ESF07S15) Centripetal Levallois cores/ discoids (**p**: ESF07massive, **r**: ESF07S46, **s**: ESF07S38)

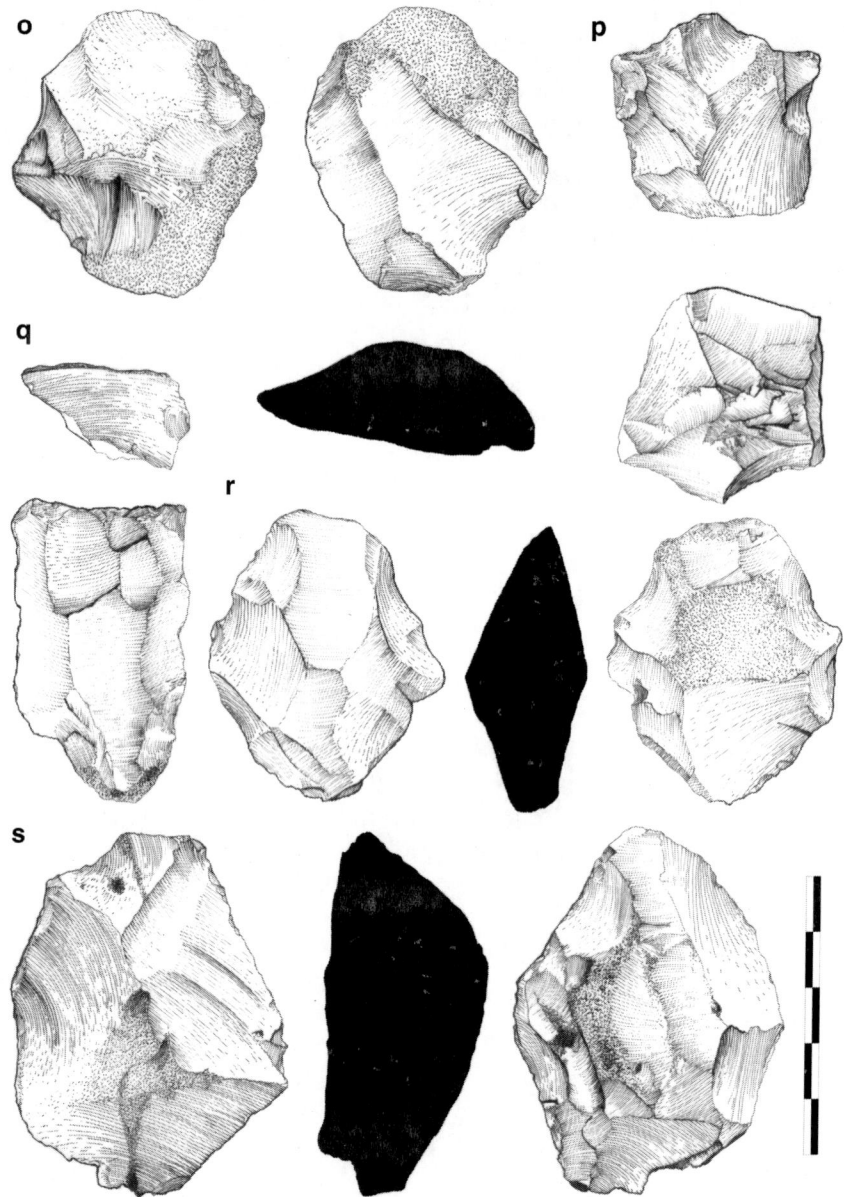

pattern also noted from "Upper Paleolithic" sites in South Arabia (Rose and Usik, 2009).

Based solely on technological indicators, Group B1 appears to be the most recent, which is apparent in the high percentage of blade-proportionate blanks as well straight, unmodified striking platforms. Although lacking tools, this particular form of hard hammer blade blank production resembles similar leptolithic (sensu Piette, 1880) scatters reported throughout southern Arabia (e.g., Amirkhanov, 1994; Zarins, 2001; Rose, 2002; Rose, 2006, 2007). There are too few data to determine a precise age for this group; it is reasonable to assume the artifacts were created sometime during MIS 3, in the episodic wet phase that lasted from ca. 50 to 20 ka.

Conclusion

The discovery of Paleolithic assemblages in Sharjah and Ras al Khaimah has opened up a new, extensive area of research in the UAE. The high-level sites in these limestone areas have clearly provided important evidence from which we can now begin to construct distribution patterns of Paleolithic occupation and land-use in these regions. Dating the sites presents many challenges, especially as the deposits are invariably decalcified, resulting in the loss of organic environmental evidence, but Optically Stimulated Luminescence (OSL) may prove useful. Also, although general principles apply for the retention of high-level sites in-situ, site specific investigations are required to understand a particular site formation and the context in which the Paleolithic artifacts were found. At this stage of our investigations it is impossible to calculate how many more high-level sites await discovery. Furthermore, during the field surveys, numerous rock shelters/caves of various sizes and at different altitudes were to be seen in close proximity to both the high and low-level surface-scatters. Whether or not these rock shelters/caves will provide further evidence of Paleolithic occupation has yet to be ascertained. Until such time as the high-level sites and/or the rock shelters/caves are excavated, the major source for determining dates (other than by technical and formal attributes of the Paleolithic artifacts) is by combining the results of the research described here with that of data derived from existing excavated Paleolithic sites. A good candidate for this is the ongoing excavation of the multi-period rock shelter (which includes the Paleolithic) at Jebel Faya, Sharjah (see Uerpmann et al., 2007). This rock shelter is ~18 km west (i.e., on the other side of the Al Madam Plain) from the ES(F)06 sites at Fili.

Due to the dearth of Paleolithic research in the Arabian peninsula over the past two decades it is perhaps not an exaggeration to say that these recently discovered, and still to be revealed, high-level Paleolithic sites are now under siege, with the greatest threats coming from hunting and military activities in the form of digging fox-holes; the relentless demand for building material (particularly, limestone for the manufacture of cement) to satisfy the requirements of the twenty-first century burgeoning housing complexes and perhaps that of climate change. The irony is however, that just as the window on the Paleolithic of the UAE opens – the opportunities to obtain access to potential areas of interest are becoming problematical, as more and more land is being fenced off for various reasons. We may indeed be running out of time.

Acknowledgments These investigations and research were carried out under the directions of His Highness Dr. Shaikh Sultan Bin Mohammad Al Qasimi, Member of the Supreme Council and Ruler of Sharjah, Emirate, and under the patronage of the Culture and Education Department in Sharjah. We are greatly indebted to Dr. Sabah Jasim, Director of Antiquities in Sharjah for his unfailing support. For permission to conduct field-investigations in Ras al Khaimah, we would like to thank Dr. Christian Velde of the Ras al Khaimah Department of Antiquities and Museums. Thanks are also due to Dr. Sarah Milliken (Institute of Archaeology, University of Oxford) for contributing to the initial artifact analysis and photographs. Our thanks to William Spring for producing the artifact illustrations.

References

Alsharekh A. The archaeology of Central Saudi Arabia: lithic artefacts and stone structures in Central Saudi Arabia. Saudi Arabia: Deputy Ministry of Antiquities and Museums; 2006.

Amirkhanov H. Research on the Paleolithic and Neolithic of Hadhramaut and Mahra. Arabian Archaeology and Epigraphy. 1994;5:217–28.

Boucharlat R, Dalongeville A, Hesse A, Sanlaville P. A survey in Sharjah Emirate, U.A.E. In: Boucharlat R, editor. Survey in Sharjah Emirate, U.A.E. on behalf of the Department of Culture, Sharjah, First Report (1984). Sharjah; 1984. p. 5–15.

Briand BR, Dalongeville R, Ploquin A. The industries of Mleiha: mineralogical and archaeological data recovered from the site and its environments in 1990. In: Boucharlat R, editor. Archaeological surveys and excavations in the Sharjah Emirate, 1990 and 1992: A Sixth Interim Report, Sharjah. 1992. p. 45–48.

Calley S, Santoni MA. Sounding at the prehistoric site al-Qassimiya. In: Boucharlat R, editor. Archaeological surveys and excavations in the Sharjah Emirate, 1986: A Third Preliminary Report, Sharjah. 1986. p. 13–15.

Cauvin MC, Calley S. Preliminary report on lithic material, Appendix 1. In: Boucharlat R, editor. Survey in Sharjah Emirate, U.A.E. on behalf of the Department of Culture, Sharjah, First Report (1984). Sharjah; 1984. p. 17–20.

Copeland L. Yabrudian and related industries: the state of research in 1996. In: Ronen A, Weinstein-Evron M, editors. Toward modern humans: the Yabrudian and Micoquian, Proceedings of a Congress Held at the University of Haifa, November 3–9, 1996. Oxford: BAR International Series; 2000. 850, p. 97–118.

Debenath A, Dibble H. Handbook of Palaeolithic typology: volume I, Lower and Middle Paleolithic of Europe. Philadelphia, PA: University of Pennsylvania; 1994.

Forster P, Matsumura S. Did early humans go north or south? Science. 2005;308:965–6.

Gopher A, Barkai R, Shimelmitz R, Khalaily M, Lemorini C, Heshkovitz I, et al. Qesem Cave: an Amudian site in central Israel. Journal of the Israel Prehistoric Society. 2005;35:65–92.

Hours F, Copeland L, Aurenche O. Les industries Paléolithiques du Proche-Orient, essai de corrélation. L'Anthropologie. 1973;77(3/4):229–80.

Inizan ML. Premiers résultats des fouilles préhistoriques de la région de Khor. In: Tixier J, editor. Mission Archéologique Française à Qatar. Doha: Dar al-Uloom; 1980. p. 51–97.

Jagher R. The Central Oman Palaeolithic Survey: recent research in Southern Arabia and reflection on the prehistoric evidence. In: Petraglia MD, Rose JI, editors. The evolution of human populations in Arabia: paleoenvironments, prehistory and genetics. The Netherlands: Springer; 2009. p. 139–50.

Jelinek AJ. The Amudian in the context of the Mugharan tradition at the Tabun Cave (Mt. Carmel), Israel. In: Mellars P, editor. The emergence of modern humans. Edinburgh: Edinburgh University Press; 1990. p. 81–90.

Jelinek AJ. Problems in chronology in the Middle Paleolithic and the first appearance of modern humans in Western Eurasia. In: Akazawa T, Aoki K, Kimura T, editors. The evolution and dispersal of modern humans in Asia. Tokyo: Hokusen-sha Press; 1992. p. 253–76.

Kapel H. Atlas of the Stone-Age cultures of Qatar. Denmark: Aarhus University Press; 1967.

Leakey M, Tobias PV, Martyn JE, Leakey R. An Acheulian industry with prepared core technique and the discovery of a contemporary hominin mandible at Lake Baringo, Kenya. Proceedings of the Prehistoric Society. 1969;3:48–76.

McBrearty S. The Middle Pleistocene of East Africa. In: Barham L, Robson-Brown K, editors. Human roots, Africa and Asia in the Middle Pleistocene. Bristol: Western Academic and Specialist Press; 2001. p. 81–98.

Millet M. The lithic industry. In: Boucharlet R, editor. Archaeological surveys and excavations in the Sharjah Emirate, 1988. Sharjah; 1997. p. 24–30.

Monigal K. The Levantine Leptolithic: blade technology from the Lower Paleolithic to the dawn of the Upper Paleolithic. Ph.D. dissertation, Southern Methodist University, Dallas; 2002.

Parker AG, Rose JI. Climate change and human origins in southern Arabia. Proceedings of the Seminar for Arabian Studies. 2008;38:25–42.

Petraglia MD, Alsharekh A. The Middle Paleolithic of Arabia: implications for modern human behaviour and dispersals. Antiquity. 2003;77(298):671–84.

Piette E. Nomenclature des temps anthropiques primitifs. Laon: Imprimerie Le Vasseur; 1880.

Pleurdeau D. Human technical behavior in the African Middle Stone Age: the lithic assemblage from Porc-Epic Dave (Dire Dawa, Ethiopia). African Archaeological Review. 2005;22(4):177–97.

Rose JI. The Question of Upper Pleistocene connections between East Africa and South Arabia. Current Anthropology. 2004;45(4): 551–5.

Rose JL. Among Arabian sands: defining the Paleolithic of Southern Arabia. Ph.D. dissertation, Southern Methodist University, Dallas; 2006.

Rose JI. The Arabian Corridor Migration Model: archaeological evidence for hominin dispersal into Oman during the Middle and Upper Pleistocene. Proceedings of the Seminar for Arabian Studies. 2007;37:219–37.

Rose JI, Usik VI. The "Upper Paleolithic" of South Arabia. In: Petraglia MD, Rose JI, editors. The evolution of human populations in Arabia: paleoenvironments, prehistory and genetics. The Netherlands: Springer; 2009. p. 169–85.

Scott-Jackson JE. Lower and Middle Paleolithic artefacts from deposits mapped as clay-with-flints – a new synthesis with significant implications for the earliest occupation of Britain. Oxford: Oxbow Books; 2000. p. 125–38.

Scott-Jackson JE, Scott-Jackson WB, Jasim SA. Middle Paleolithic – or what? New sites in Sharjah, UAE. Proceedings of the Seminar for Arabian Studies. 2007;37:277–9.

Scott-Jackson JE, Scott-Jackson WB, Rose JI. Upper Pleistocene stone-tools from Sharjah, UAE. Initial investigations: interim report. Proceedings of the Seminar for Arabian Studies. 2008;38.

Smith GH. New prehistoric sites in Oman. Journal of Oman Studies. 1977;3:71–81.

Tryon C. 'Early' Middle Stone Age lithic technology of the Kapthurin formation (Kenya). Current Anthropology. 2006;47(2):367–75.

Uerpmann H-P, Uerpmann M, Kuttered J, Handel M, Jasim SA, Marks A. The Stone Age sequence of Jebel Faya in the Emirate of Sharjah (UAE). Paper presented at the Seminar for Arabian Studies 2007; 2007.

Usik V. The variants of Levallois method of Middle Paleolithic industries of Ukraine. In: Kulakovska L, editor. The Middle Paleolithic variability on the territory of Ukraine. Kiev: Shliakh; 2003. p. 32–62.

Wendorf F, Schild R. A Middle Stone Age sequence from the Central Rift valley, Ethiopia. Warsaw: Polska Akademia Nauk; 1974.

Whalen N, Siraj-Ali J, Davis W (1984) Excavation of Acheulean sites near Saffaqah, Saudi Arabia. Atal 8:43–45

Whalen N, Siraj-Ali J, Sindi H, Pease D, and Badein M (1988) A complex of sites in the Jeddah-Wadi Fatima Area. Atal 11:77–87

Zarins, J 2001 The Land of Incense: archaeological work in the Governorate of Dhofar, Sultanate of Oman, 1990–1995. Muscat: Sultan Qaboos University Publications

Chapter 11
The Central Oman Paleolithic Survey: Recent Research in Southern Arabia and Reflection on the Prehistoric Evidence

Reto Jagher

Keywords Huqf • Oman • Prehistory • Survey • Technology

Introduction

The principal objective of the Central Oman Paleolithic Survey (COPS) program is the exploration of the earliest human occupation in the southern part of the Arabian peninsula. The COPS, organized by the Institute for Prehistory and Archaeological Science (IPAS) of the University of Basel (Switzerland), was carried out in 2007, in the Huqf area of the Sultanate of Oman. This project is complementary to the ongoing research of the IPAS in Syria which has been operating for more than 20 years (Le Tensorer and Hours, 1989; Le Tensorer, 1996, 2004; Le Tensorer et al., 1997, 2001; Le Tensorer et al., 2007). During 5 weeks of field work, a tremendous amount of information was collected shedding new light on Omani prehistory. A total of 623 sites were surveyed for information on prehistory and geology. In total, 369 archaeological sites were recorded. Even though the main goal of the survey was to discover Early Paleolithic sites, the preliminary results of the COPS survey are astonishing in the array of archaeological sites that were identified. The 2007 discoveries revealed a significant and diverse prehistoric legacy in Central Oman reflecting a proliferate population unexpected in an arid area.

The archaeology of hunter-gatherer societies in southern Arabia is still in its beginning, despite an increasing number of reported discoveries over the past few years. To date, the number of archaeological sites from the Pleistocene is still very limited. Although there are reported discoveries, no general framework has been possible, as well-dated sites and consistent typological and technological studies of representative assemblages are largely missing. The present

outlines of South Arabia's early history are mostly based on conventional, and in many cases outdated theoretical concepts. This lack of knowledge is surprising, given that the Arabian peninsula is located in the heart of the crossroads between Africa and Eurasia, and it occupies a privileged position for the spreading of hominins from Africa to the rest of the world (Fig. 1).

Based on current archaeological evidence, the "Out of Africa" route for early human migrations is assumed to have proceeded along River Nile, through the Levant and the Northern Middle East into Eurasia. Notwithstanding strong biogeographic, paleoclimatic and geologic evidence, the alternative route from Ethiopia over the narrow Strait of Bab al Mandab and along the southern margin of Arabia, i.e., through Yemen, Oman into Asia is rarely considered (Petraglia, 2003; Petraglia and Alsharekh, 2003; Rose, 2004, 2007; Whalen et al., 2002; Whalen and Fritz, 2004).

Geographic Setting

The geographical location for the COPS was selected based on the identification of a topographically structured environment offering a broad spectrum of ecological niches, access to fresh water and availability of raw material for the production of stone tools. As precipitation patterns were subject to considerable changes during the Paleolithic period, today's aridity was not considered an argument for ruling out specific areas. Based on the above criteria, the Huqf region in Central Oman, a territory embedded between the continental plains of the Jidat Al Harasis and the coast of the Indian Ocean, was identified as a promising research area.

The Huqf is characterized by low hills with a rich geological setting, offering a broad variety of different bedrock types, supporting a great diversity of ecological habitats, and allowing access to different landscapes permitting varied modes of subsistence. The situation at the intersection of different ecotypes and the many ecological niches within it must have been attractive for hunters and gatherers throughout time.

R. Jagher (✉)
Institute of Prehistory and Science in Archaeology,
University of Basel, Spalenring 145, CH-4055, Basel, Switzerland
e-mail: Reto.jagher@unibas.ch

M.D. Petraglia and J.I. Rose (eds.), *The Evolution of Human Populations in Arabia*, Vertebrate Paleobiology and Paleoanthropology,
DOI 10.1007/978-90-481-2719-1_11, © Springer Science+Business Media B.V. 2009

The entire area is presently only slightly covered by dunes (both fossil and active), thus permitting a perfect observation. Furthermore, paleoclimatic evidence from immediately adjoining areas points repeatedly at moderate weather conditions for extended periods. The presence of known archaeological sites within a particular region is encouraging, but not a decisive argument in choosing a survey area.

The chosen survey area is delimited by the towns of Al Ghaba and Filim in the North, Ad Duqm in the South, Al Ajaiz in the South and the Jiddat Al Harasis plateau comprising the Jaaluni Oryx-station to the West. The defined sector extends about 200 km from North to South and approximately 40–60 km from the coast into the interior (Fig. 2).

Despite their apparent uniformity, the wide expanses of the Rub' al Khali exhibit complex formations with an intri-

cate history (Warren and Allison, 1998; Preusser et al., 2005). Their emergence extends over a long period and they are in continuous and asynchronous shift (Goudie et al., 2000; Bray and Stokes, 2003; Radies et al., 2004). Rather than global climatic changes, local phenomena such as the supply of sand and persistence of winds are decisive for the accretion of massive dunes (Kocurek and Lancaster, 1999; O'Connor and Thomas, 1999).

Our perception of the Arabian Desert is biased by its current substantial constraints. Human pressure in the past destroyed much of its natural resources. Indigenous game was wiped out for food or as competition to livestock. Intensive use of pastures destroyed the plant cover beyond recovery, and people abandoned the exploited areas as the desert expanded. What is left today is only a meager image of the original potential, today providing few clues about the possibilities for a hunter and gatherer subsistence.

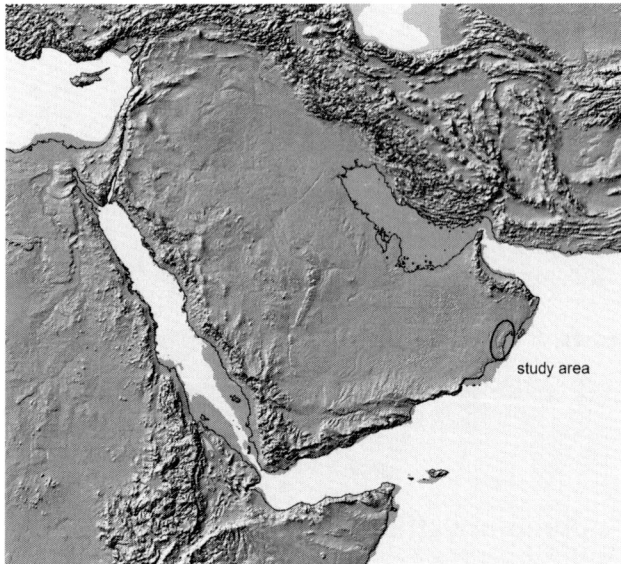

Fig. 1 Map of the Middle East showing exposed land bridges at −100 m below sea level during Pleistocene cold periods

Climate in the Late Pleistocene

A crucial point for evaluating the potential for human settlement in this part of the world is climatic change and the ensuing alterations of the environment. During the last million years, global climate was subject to significant and swift changes, due to celestial mechanics of earth's orbit and the activity of the sun (Leuschner and Sirocko, 2003; Ivanochko, 2004). Important shifts in temperature triggered a massive extension of the polar ice sheets, lowering considerably global sea levels, uncovering today submerged land bridges (Fig. 1). Consequences for the Arabian subcontinent were a drying up of the Persian Gulf and a narrowing of the Bab al Mandab (Rohling et al., 1998; Siddall

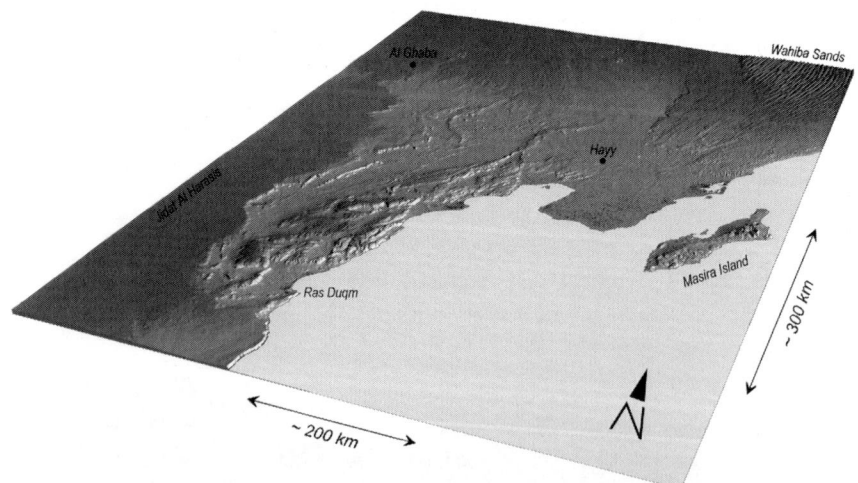

Fig. 2 Digital Elevation Model of the Huqf and adjacent areas (scale is varying in this perspective). For a better rendering of the topographical structures the vertical scale is exaggerated tenfold

et al., 2003). Cooler temperatures imply a reduced evaporation which is essential for the growth of plants (Haude, 1969). The most important motor for climate fluctuations in Arabia is the shifting monsoon belt, nearing at present its southern maximum. In a cycle of about 21,000 years BP, markedly humid phases are possible over the Arabian realm for a few millennia. However, these conditions were subject to local particularities, hindering a clear predictability of their effects.

For a reconstruction of the past climate in southeastern Arabia, one depends on a combination of comprehensive and regional archives to understand the local implications of global changes. These are preserved in deep-sea sediments of the Indian Ocean off the Arabian coast, in stalagmites from caves or continentals deposits of different origins, such as in lake deposits or sand dunes. The compiled data from deposits in the Rub' al Khali but especially from stalagmites from Omani Caves and sediments drilled from the Wahiba Sands allow a more or less detailed reconstruction of the past climate over the last 200 ka (Preusser et al., 2002; Fleitmann et al., 2003).

The most recent humid period lasted from about 10 to 6 ka (Lézine et al., 1998; Neff, 2001; Parker et al., 2006). During the maximum of the last ice age about 20 ka, arid conditions persisted (Hoelzmann et al., 2004). However, it is impossible to estimate how hostile this climate was to life and if human settlement completely ceased or whether people continued to occupy the former coasts now covered by more than 100 m of water. Further back in the past, more humid periods are indicated by sediments of perennial lakes at many places in the Rub' al Khali between 25 and 35 ka (McClure, 1976; Clark and Fontes, 1990; Wood and Imes, 1995). Another humid period between 78 and 82 ka conform to increased monsoon activity (Burns et al., 2001; Neff, 2001). Again, dry conditions prevailed until the time between about 120 and 135 ka, when a warm and humid climate existed during the Last Interglacial. Soil formation in the sediments of the Wahiba Sands indicate considerable vegetation. The preceding well-marked global cooling, with a sharp drop of sea levels, was not an extremely arid period in southern Arabia, as might be expected. Paleosols from the Wahiba Sands dated between 130 and 160 ka point to several short lived humid cycles, again with a substantial expansion of vegetation (Radies et al., 2004). The earliest clearly established climatic cycle is again a phase with increased rains, which occurred between 180 and 200 ka (Burns et al., 2001). Beyond this point, continental data about paleoclimatic changes are erratic and limited to isolated observations. In summary, climatic conditions throughout the last 200 ka in southern Arabia appear not to have been as harsh as prevalent schemes might suggest. Hence, autochthonous cultural traditions could have persisted over much of this time.

Prehistory in Oman

For many decades, the Fertile Crescent and adjoining regions in the Northern part of the Middle East have been the focus of intense Paleolithic research. In comparison, only a very limited number of discoveries relating to this ancient period have been reported from the southern part of the Arabian peninsula. Nevertheless, the few findings made in this area indicate a human presence throughout the Paleolithic, demonstrating the importance of the region (e.g., Petraglia, 2003; Petraglia and Alsharekh, 2003; Whalen and Fritz, 2004; Rose, 2006).

Since the early 1970s, archaeological investigations in the Sultanate of Oman experienced an important impetus in research. Pioneering surveys were carried out under the difficult conditions of a rough geographical terrain and a limited infrastructure. Archaeological activities focusing on prehistoric periods were initiated along the Oman Mountains and their southern foothills where a multitude of prehistoric sites were identified (Pullar, 1973; Doe, 1976; Smith, 1976, 1977; Pullar and Jäckli, 1978; Pullar, 1985; Edens, 1988b). To the south, the Dhofar region as a continuation of the Yemenite realm became a second centre of archaeological research in Oman (Whalen and Schatte, 1997; Zarins, 2001; Cremaschi and Negrino, 2002). Due to the poor preservation conditions of archaeological evidence in the interior of the Sultanate, research zoomed in on coastal archaeology (Biagi, 1988, 2004), and important excavations were initiated along the coast (e.g., Biagi et al., 1984; Maggi, 1990; Tosi and Usai, 2003; Uerpmann and Uerpmann, 2003; Biagi and Nisbet, 2006). Because of better conservation of sites, the investigation of Holocene sea-side settlements rapidly turned into a well-established research field in Oman.

In the meantime, the interior of the Sultanate mostly remained *terra incognita* from an archaeological point of view. Until recently, only a limited number of prehistorians ventured into the interior of the country reporting the discovery of Stone Age sites. All of these discoveries were surface finds, and in most cases only selective collections were recovered. The terminology referring to chronology, technology and typology of stone tools in Oman and the region has been inconsistent.

Concerning the Paleolithic period of the south-eastern part of the Arabian peninsula, there are only a few communications published, reporting the discovery of archaic stone tools (Inizan and Ortlieb, 1987; Biagi, 1994; McClure, 1994). In many cases, the age of the tools is not established or the short descriptions of the artifacts make it difficult to assess their value (Whalen et al., 1988; Amirkhanov, 1994; Whalen and Schatte, 1997). There is no doubt about the presence of early humans during the Pleistocene in south-eastern Arabia, but the modest database concerning the Paleolithic of South-Eastern Arabia makes it difficult to draw general conclusions (Petraglia, 2005).

Similarly, it is difficult to obtain a comprehensive overview of the Neolithic period. Although there is an impressive number of sites and published reports, the literature is contradictory in many aspects. As for the earlier periods, much of the information available consists of short communications or is dealing with selected problems from sites scattered over a huge geographic area. Except for a few basic types, often abused as type fossils, there is not much agreement about typology and technology of stone tools from the Neolithic (Kapel, 1967; Edens, 1982, 1988; Di Mario, 1989). Also, the chronological framework remains controversial, as for many sites no dates are available, and where existing, their interpretation is often complex due to methodological factors (Magnani et al., 2007). There is reasonable agreement in literature with respect to the general evolution of stone tools in the course of the Neolithic period, and to a certain degree different regional patterns can be discerned (Uerpmann, 1992; Edens and Wilkinson, 1998). However, the detailed relationship of these observations in space and time is still discussed (Kallweit, 2003; Uerpmann, 2003).

Problems of Pleistocene Archaeology in Arabia

Archaeological ethnicity is conceived basically on the physical legacy limited to objects mostly of imperishable materials. These reflect just a minimal fraction of the cultural wealth of prehistoric civilizations. The archaeological concept of ancient cultures is traditionally structured by typological concepts, attributing cultural and chronological importance to selected objects. Such index fossils can be an effective means to allocate new discoveries to known cultural entities. As a basic tool this approach is reliable in areas where the prehistoric legacy is soundly documented and statements are fully confirmed by multiple observations. In regions with limited and incomplete information, this approach is rather treacherous as the full spectrum of possibilities is only fragmentarily known and cursory appearances can mislead interpretation.

It is in the nature of archaeological nomenclature that observations like "Acheulean like" handaxes are readily attributed to the Acheulean period. However, such handaxes may occur in completely different cultural contexts of varying age. The same goes for many other terms or tool types. Most of this terminology has been minted in Europe and later adopted for better or worse all over the world. The original

implications and meanings of these expressions were lost, as these terms were dissociated from their primary context and adapted to a new one. Still their suggestive power remained implying relationships that are not established by the local archaeological evidence per se.

It is innate to the nature of stone tools that congruent forms and techniques were devised independently on repeated occasions in different areas and in various cultural contexts. So, similar looking objects can have a perfectly different background. Apparent resemblance is not compelling evidence, especially when dealing with huge geographic areas and covering a very long history. This danger is imminent when only selected tool forms or production concepts such as isolated technological aspects are considered for comparisons, without taking into account the complete context on both sides. This unidirectional approach is highly selective, leading to inconsistent reasoning.

When considering the archaeology of hunter and gatherer societies of the Arabian peninsula, the above mentioned aspects have to be kept in mind. Research over the past few decades was intermittent and dispersed over a huge geographical area with little chronologic control and was mostly limited to preliminary observations gathered by a multitude of archaeologists with different backgrounds. From this scattered information, no general framework has been devised summing up the Pleistocene prehistory in this part of the world. In many studies there is a strong urge for long-range comparisons due to the absence of comparable neighboring sites (because of a lack of discoveries). It is not the intention of this chapter to disqualify fellow archaeologists. To avoid pointless polemics I refrain purposefully from any quotations in this context. However, Arabia's early prehistory strongly needs a comprehensive review on a common base with a precise description of these discoveries, but also to return to basic reflections and to refrain from complex speculation overstretching the basic results in a disproportionate way.

Lower Paleolithic Sites in the Huqf

One of the starting points for the COPS project were the reported rich Early Paleolithic sites investigated by Norman Whalen in 2002 (Whalen, 2003). He published on a concentration of 61 localities situated along a narrow band about 0.8–1.5 km wide and less than 10 km long (area III in Fig. 3). From these sites a total of 4494 artifacts were described, of which 1676 classified tools were attributed to the Oldowan and Acheulean. Based on morphology and patination of the artifacts but also by analogies from East Africa, an age of

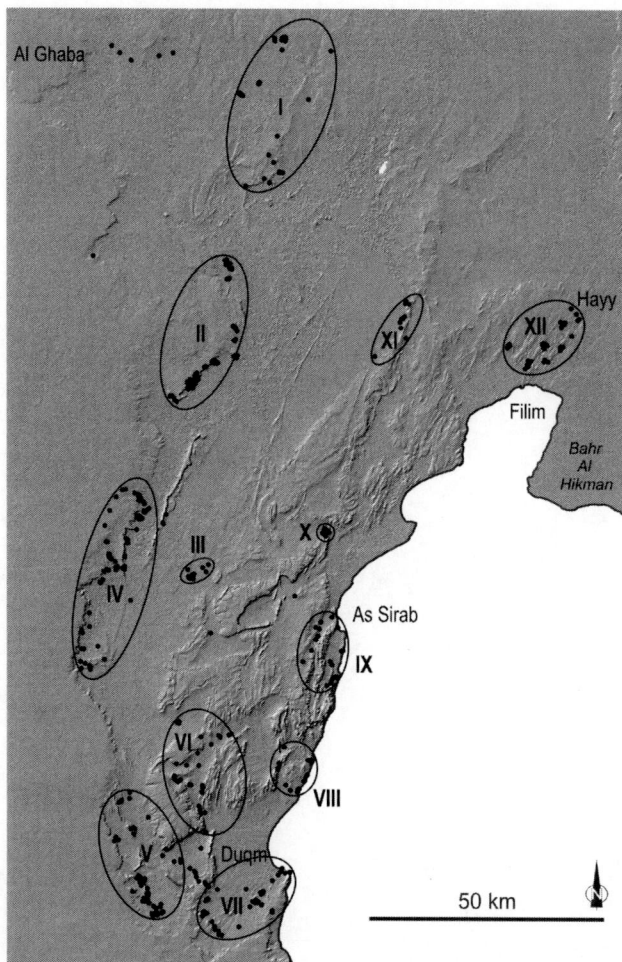

Fig. 3 Map of the Huqf with outlines of the survey areas, dots indicate places recorded during field work

about 1.5 million years for the Oldowan tools was suggested. The well demarcated spatial distribution and high concentration of Whalen's discoveries suggested a particular topographic situation or an exceptional geologic setting. The specific conditions of preservation at these places would have been of special interest, as it would have made possible, if understood, to recognize similar situations in other sectors of the Huqf. Despite intensive surveying within the core area, no artifacts were found. Covering a combined surface of more than 8 ha, not a single artifact was discovered. However, a number of naturally broken pebbles evoking archaic tools were found on occasion. A thorough re-examination of all available topographic data in every aspect perfectly matched with the geographic information provided by Whalen. A personal assessment of Whalen's discoveries was not possible. Despite the kind assistance of the staff, the original finds could not be traced in the repository of the Ministry of Heritage and Culture in Muscat.

The 2007 COPS-Survey

The principal intention of the 2007 field season was to obtain a basic overview of the situation in different sectors of the study area. Instead of a detailed investigation of a few sites, we aimed for a broader appreciation over a larger area. During survey, no artifacts were sampled from archaeological sites. Hence systematic counts of the different tool categories are lacking and only qualitative observations can be given for the time being (Table 1).

During the 2007 survey, 623 locations were mapped (Fig. 3), of which 369 were identified as archaeological sites. A total of 340 of these contained tools. In addition to flint artifacts, all evident archaeological structures were systematically mapped. Their significant numbers primarily reflect land use patterns during the Late Holocene period. These structures account for less than 10% of all places investigated.

All sites located were found on the natural floor and none of them was discovered in a stratified situation. However, preservation of many sites is surprisingly good. On several occasions flakes lying next to each other could be refitted, suggesting that disturbance and erosion at many places was limited. As these surfaces are uncovered, a superposition of different periods was possible. In fact, at a number of places a multitude of different time periods was noticed. However, in most sites where diagnostic artifacts were found, no major mixture of "early" and "late" material was observed. Only 12% (i.e., 20 out of 161 sites with chronologically diagnostic material) show a blending of different periods. Open air sites in the Huqf are far better preserved than may be expected under similar conditions. However, these statements are based on preliminary field observations and need further investigation for full confirmation.

Late Quaternary and Holocene sedimentation in the study area is mostly limited to colluvial deposits at the feet of cliffs and the adjacent debris fans (Fig. 4). At present, erosion is mainly confined to existing channels, cut during periods of increased rainfall earlier in the Holocene. In many cases the ongoing erosion events had cut deeper into the bedrock than previous ones. Contemporary erratic rainfalls may locally have some effect, but overall contribute little to the general evolution of the topography. As major river systems are missing in the Huqf, no extended fluviatile deposits are expected. As an alternative, rockshelters offer protected spaces for sediment accumulation. In many cliffs natural overhangs evolved along stratigraphic joints within the geologic formations. Instead of a flat platform beneath the overhang, there is a sharp drop to the exterior where no sizeable sedimentation can accumulate.

Table 1 Inventory of observations recorded during the 2007 survey, bear in mind that multiple observations at the same site are possible

	Survey	Archaeology	Artifacts	Structures	Geology
Number of observations	623	369	361	54	602
No finds	283				
Undiagnostic sites	179				
Diagnostic sites	161				
Structures only		29			
Flint artifacts		340			
Small débitage			24		
Light baldes			18		
Heavy blades			99		
Cores			81		
Heavy bifacials			59		
Small bifacials			13		
Large foliates			17		
Medium foliates			14		
Small foliates			13		
Arrow heads			8		
Retouched flake-tools			15		
Triliths				23	
Stone circles				14	
Tumili				10	
Shell midden				7	
No raw material					343
Good raw material					128
Bad raw material					83
Rock shelters					22
Water holes					26

Fig. 4 Typical landscape along Huqf escarpment (survey area II-3) showing heavily eroded alluvial terraces with factory sites (*black patches* in the *centre* of the picture). When the flint features a dark patina (such as here) sites are easily located. Clear streaks in the middle ground are the active wadis where possible artifacts are dislocated from their original context

Fig. 5 Factory site stretching over a vast surface next to a flint outcrop, nearly every piece of flint stone in the picture is a man made artifact

Outcome of the 2007 Season

The number of discoveries in the 2007 survey exceeded our expectations, not only in quantity but also in the size of sites. Sites, especially workshop areas comprising up to an estimated million artifacts or more, are quite familiar in the southern part of the Huqf (Fig. 5). All these particular observations, as well as a multitude of much smaller single work shop sites, reflect an extremely extensive production of stone tools throughout the surveyed area. Basically, two alternative hypotheses can be outlined. Either these sites reflect an unusual high production over a limited time or the same places have been in use over a very long time. In the first case, a high demand in stone tools over a short period would mean a considerable increase in population density or exchange over a substantial area. Both assumptions are quite unlikely for a subsistence of hunting and gathering. For the time being the use of these sites over long periods is much more likely.

To ensure an unbiased approach as possible, artifacts found during fieldwork were not classified according to cultural stages. The premature attribution to an alleged cultural entity can be a pitfall as classical archaeological concepts rarely consider variability but relay strongly on specific index types. To avoid entanglement in speculations, we deliberately refrained from adopting preconceived concepts, with the intention to evade premature conclusions.

An open system with clearly delimited elements was devised, allowing an impartial assembling of the observed components. In addition, data collection in the field had to be in a way that the growing experience in the course of the survey did not distort the integrity of information. Therefore our observations in the field were split into aspects of débitage and retouched tools. Based on experience from direct field observations, the débitage was subdivided into four categories: simple or undiagnostic flakes, 'heavy blades', 'light blades' and 'small débitage' for deliberately produced small flakes. The presence of cores was just noted without going into details. Retouched artifacts were categorized as 'heavy bifacials', 'small bifacials', foliates (subdivided into three classes of size), arrowheads and retouched flakes. Typological aspects were just generally noted as no comprehensive counts were carried out. This plain and rigid classification was basic in order to obtain a stringent database and to ensure the compatibility of the observations among each other. Specific observations were noted as a commentary to each site, describing local particularities or affinities to earlier results thus clearly separating fundamentals and interpretations.

Cores

Cores and the dependent technologies can be useful instruments to determine cultural entities. However, a quick assessment of cores in the field is rather difficult. The effort for a comprehensive study of the core technologies exceeds by far the possibilities of the ongoing phase of the COPS project. In a first approach the observation of the products, is much more rewarding, as the general outlines become visible in a more efficient way (Fig. 6).

Small Debitage

The small débitage facies is characterized by a predominant presence of conspicuously small flakes about 4 cm long. These diminutive blanks which are rather broad and rarely laminar, are a deliberate choice of the prehistoric flintknappers, as the size of local raw material would have permitted much larger

Fig. 6 Exceptionally large core for the production of flakes over 10 cm long

artifacts. Half of the 24 diagnostic sites (15% out of 161) where small débitage was found are classified as big sites. At these places the small débitage is the dominating group, indicating the importance of this category. A conclusive cultural attribution of the small débitage is difficult. Yet tentative field observations are consistent with foliates of Holocene age.

Light Blades

This group comprises elongated flakes of about 5–8 cm long with a thin cross-section. Preliminary field observations indicate a production using an elementary scheme including cores with only little preparation. These blades show quite regular but not strictly parallel edges. In most cases the dorsal ridges are sinuous. Formal as well as metric standardization of these products is quite low. Occasionally, individual blades are more regular but never achieve an outstanding level.

The remnant of the striking platform is plain and only in rare cases facetted. The maintenance of the crest is made by simple abrasion. The flaking angle is clearly acute-angled. Production of these blades was usually unidirectional with only rare exceptions having opposite striking platforms. In summary, the production scheme is rather basic and in many cases an opportunistic strategy can be noticed. Light blades were not very common during the 2007 survey. They were present on 18 (i.e., 11%) out of 161 sites with diagnostic artifacts, making them rather a rare phenomenon. The age of this facies is difficult to estimate. There is, for the time being no clear preference of any tool type associated with the light blades, giving little clue of their cultural affiliation. In any case the light blades show no affinity at all to the initial Neolithic or the Upper Paleolithic of the Levant.

Heavy Blades

This kind of blank can be considered a hallmark of the pre-history in the Huqf. Not only for their wide distribution throughout the territory but also for their distinctive characteristics. Heavy blades are easily recognized by their highly standardized shapes. The elongated blades, usually between 8 and 10 cm, but up to about 15 cm long, show well aligned parallel edges and dorsal ridges and the eponymous thick section. Despite their impressive dimensions, these blades present a surprisingly low longitudinal curvature, a feature seldom seen in blade technology of this size. The heavy blades are produced with hard hammer and a basic, but highly efficient procedure without much preparation of the core. Except of rare exceptions, the striking platform is unprepared. The maintenance of the core edge between the successive generations of blades is reduced to a summarily executed short flaking. Associated cores show predominantly a unipolar and only exceptionally bidirectional reduction. The same can be noticed on the blades themselves.

This method of blade production is in sharp contrast to Neolithic and Upper Paleolithic schemes known from the Arabian realm. These follow completely other techniques resulting in morphologically different products. For the moment the best fitting analogies to the heavy blades are found in the northern part of the Middle East in the Hummalian technocomplex, dating back to about 250 and 180 ka (Le Tensorer et al., 2005; Meignen, 2007). Indeed, this statement is no proof for a similar age of the Huqf discoveries. Nevertheless geomorphologic observations on some sites suggest quite an old age well before the Holocene.

The plainness of the heavy blades makes them a peculiarity of southern Arabian prehistory as such a technology has never been described. The importance of this discovery is highlighted by a wide distribution of these artifacts throughout the study area. As a matter of fact, heavy blades are the most common lithic group present in the Huqf. It was recorded at 99 places, i.e., by far more than half (59%) of all sites producing diagnostic artifacts (Fig. 7).

Flake Tools

Another striking observation was the rarity of retouched flake tools. Compared to the number of prehistoric sites discovered, the poverty in retouched flake tools is absolutely striking. As flake tools of manifold forms were related to a multitude of daily activities, they can be expected in every prehistoric settlement in notable numbers. Retouched flakes occur just at 15 out of a potential 340 sites. This scarcity eludes explanation for the time being.

Fig. 7 Dense cluster of production waste from the manufacturing of heavy blades (scale = 30 cm). Such heavy concentration spreading over hundreds of square meters were regularly observed where good raw material is plentiful

Heavy Bifacials

Heavy bifacials are tools of respectable size usually about 10–15 cm long that are worked all over the two faces (Fig. 8). They were made from suitable blocks or from massive flakes. Their proportions vary from broad, i.e., nearly circular contours, to quite slender and elongated shapes. With only rare exceptions, the outline of the heavy bifacials is bipolar with no clear base or tip. The cross-sections are consistently rather thick. Thin sections were only exceptionally observed and usually in tools of mediocre workmanship. The quality of manufacture of these tools is highly varied.

At a first glance, the heavy bifacials may recall classical handaxes of the Acheulean. However, there are profound inherent differences. The typical bipolarity of Lower Paleolithic handaxes with a clear base and an opposed tip, a basic concept of these tools, is missing among the Huqf artifacts. Despite a strong formal variability among Paleolithic handaxes this basic characteristic is thoroughly respected throughout time and space. The peculiarities of the Huqf bifacials clearly separate them from their counterparts of the Acheulean and in the subsequent Early Middle Paleolithic in the northern realm of the Middle East (Jagher, 2000).

Heavy bifacials were among the most common discoveries during our field work. A total of 59 places more than one third (i.e., 37%) of all diagnostic sites produced such artifacts. Their incidence is strongly correlated with the heavy blades. Although all these occurrences are surface sites, with the possibility of chronological overlapping, there is a very strong affinity between the two types of tools. The age of these tools is difficult to estimate. Clearly

Fig. 8 Selection of heavy bifacials from different sites in the Huqf area, Central Oman scale = 10 cm

comparable cultural groups are for the time unknown. Repeated geomorphologic observations however suggest a clearly pre-Holocene age.

Small Bifacials

This class comprises tools retouched also all over the two faces of the artifact. The size fluctuates between less than 6 and up to about 8 cm. The style of manufacture, but also the broader proportion as well a markedly thicker section, clearly distinguish them from foliates of the same size. Small bifacials are rather rare tools found only in 13 locations. In several sites, these tools occur in substantial numbers and show a surprisingly deliberate standardization in size. The limited database prevents a further interpretation, as no dominant association exists for the time being. Despite their rarity, small bifacials are a distinct tool type, well differentiated from analogous tools such as the heavy bifacials or foliates.

Foliates

Foliates are tools with retouch completely covering the two faces of the artifact. In proportion to the width, the thickness is thin. The slender section is usually biconvex and only on rare occasions D-shaped or plano-convex. The original proportion of the outline can rarely be determined with certainty as most of pieces were broken. However, the fragments indicate a length to width ratio ranging from less than 1.2 up to about 1.3. Elongated forms with very slender shapes of 1.4 and more were rarely observed (Fig. 9).

Fig. 9 Collection of foliates from a Holocene settlement (site 503, area-VII), as in this picture these tools are usually broken

Throughout the survey area, 33 sites with foliates were mapped, i.e., at 20% places with diagnostic artifacts. Usually they occur in good numbers and in many cases they were found in large quantities. In general, the quality of retouch rarely reaches the high perfection of elaboration observed in many cases of such projectiles in general attributed to the Arabian Bifacial Tradition.

As most of these tools were broken, the width was taken for classification. Three generic groups were defined through this argument: Small foliates about 2–3 cm wide, medium foliates with a width between 4 and 5 cm and large foliates, more than 5 cm wide. All three occur at a more or less equal frequency. The partial information about the length from complete or reassembled foliates, suggest dimensions of 5–6 cm for the small ones whereas the large foliates easily exceed 10 or 12 cm.

Arrowheads

The typical tanged arrowheads, one of the hallmarks of the Arabian Neolithic, were discovered at a mere eight places out of 340 potential sites and always in surprisingly small numbers. Of the eight sites with classical arrowhead types, two can clearly be attributed to the Fasad group, which is confirmed by just two typical Fasad points. The other six sites with tanged arrowheads show not only a high variability relating to the size and form of the points but also in the execution of the retouch, having little in common among each other. The highly elaborated arrow heads, typical for the Arabian Bifacial Tradition, until know, are missing in the Huqf sites.

The scant discoveries of Neolithic arrowheads were not expected. The numerous discoveries from this period reported by the pioneering surveys during the 1970s (Pullar, 1973, 1985; Pullar and Jäckli, 1978) and more recently from the many investigations in the Dhofar region (Zarins, 2001), suggested quite a continuous occupation throughout the interior of the Sultanate during the Neolithic. This assumption was also nourished by climate data indicating superior rainfall conditions in that period. The low density of Neolithic arrowheads in the Huqf remains enigmatic at the moment.

Another unexpected observation is the complete absence of trihedral rods among the Huqf sites. This characteristic tool is typical for the margins of the Rub' al Khali Neolithic (Crowfoot Payen and Hawkins, 1963; Pullar, 1973, 1985). Present also in the neighboring Dhofar, they are considered an index type for the final period of the Neolithic (Zarins, 2001).

Microliths

Microlithic tools of diminutive size less than 2 cm long are the hallmark of the Iron Age in southern Arabia (Zarins, 2001). During the 2007 survey, one isolated site in the southern Huqf was located. A closely delimited accumulation of small flakes, 2–3 cm long, was found adjacent to a stone circle of unknown age. Among these artifacts made of excellent non-local flint, two geometrical microliths ten and twelve millimeters long were discovered. Despite the small size of the flakes at this place, sites with "small débitage" in general cannot be attributed to the Iron Age, as the diagnostic tools of this period were not observed at these localities.

Chronology

At this stage of the project, it is difficult to propose a clear age for the different discoveries. Currently there are no absolute or relative dates for any of the Huqf sites. A relative chronology is possible just in its rough outlines. Due to the lack of precise diagnostics for stone tools inventories from South-Eastern Arabia, a comparison with the Huqf materials is almost impossible. In this chapter we limit ourselves to a basic model, as these problems need further investigation.

Patination of the flint tools is extremely uniform and differences of diverse artifact series on the same site are insignificant. Patination is not a gradual change developing indefinitely over time, but proceeds until equilibrium with its surrounding is reached. After that, basically no further change occurs. This process may happen in quite a short time. As far as our field observations demonstrate, this is what happened in the Huqf. Consequently, prehistoric artifacts of clearly different age show an indistinguishable state of weathering on the same site.

As an initial approach, a general age determination for the archaeological sites in the Huqf is proposed based on the morphology of the artifacts found. In a generic model, we distinguish between an older Pleistocene period and a younger Holocene phase. Sites considered of Pleistocene age comprise localities with heavy blades and/or heavy bifacials. Whereas the Holocene group consists of discoveries of light blades, small débitage, foliates and arrow heads. On this simplified scale however some interesting observations are possible. Sites with mixed materials of both periods are surprisingly rare. Only 12% out of 161 sites considered in this model belong to this category. So-called "Pleistocene" sites account for 64% of the localities, whereas "Holocene" series contribute 24%.

The old age of the so-called Pleistocene artifacts is sustained by a number of geomorphologic observations. Repeatedly at sites where heavy blades and bifacials occur, erosional gullies cut deep into the old surfaces on which these artifacts were found. The still ongoing erosion partially destroyed these sites. This observation is no absolute proof for a very high age. Nevertheless these sites clearly predate the current topographic situation.

Conclusions

The postulated high potential for prehistoric archaeology in the center of the Sultanate of Oman was entirely confirmed. A total of 369 archaeological sites located during the first leg of the COPS project clearly demonstrate the importance of the Huqf area for the prehistory of south-east Arabia. Many of these were factory sites of different periods. Some production-areas extend over several thousand square meters with millions of artifacts. Such extended concentrations were observed repeatedly and can be explained either by a high production over a short period for a sizable population or reflect a long and repeated exploitation of these places. As these people survived on a hunter and gatherer subsistence, a high population density can be ruled out. Therefore a production over a substantial period can be assumed.

For the time being, there is no evidence of an Early Paleolithic occupation as it is known from the Levant. Acheulean-like artifacts, such as true handaxes, were found only in very few exceptional specimens. The context in every case clearly demonstrated that these selected objects represent the extremes in a broader spectrum of forms and not a proper type. Furthermore, these pieces showed exactly the same state of preservation as the rest of the artifacts from the respective sites. Also the post-Acheulean periods widely known from the surrounding zones in Africa and the Levant, were not yet identified in the Huqf area. However, there is clear evidence of some Middle Paleolithic occupation in Yemen (Inizan and Ortlieb, 1987).

The outcome of the first field season of the COPS project allowed a remarkable new insight into the early prehistory of the south-eastern Arabian peninsula. Preliminary results indicate the clear presence of a lithic tradition from the Pleistocene. As far as can be said today, all the well known Paleolithic entities in the Levant seem to be absent in southern Arabia and no direct link between these neighboring areas can be established so far. Furthermore, the initial field work did not reveal any Early Paleolithic sites comparable to those found in the Fertile Crescent or Eastern Africa.

Acknowledgments We would like to express our gratitude to H.E. Sultan Bin Hamdoon Al-Harty, Undersecretary of the Ministry of Heritage and Culture, Ms. Biuwba Ali Al Sabri Director of Excavations and Archaeological Studies. We are also indebted to Salim Al Maskeri of Shuram L.L.C in Muscat, H.E. Youcef Fartas, Honorary Consul of Switzerland to the Sultanate of Oman and to Dr. Barbara Stäuble (Muscat), last but not least the participants of the COPS 2007 survey. Christine Pümpin, Fabio Wegmüller and Ines Winet (all from the University of Basel). Furthermore our gratitude goes to the main sponsors of the COPS project; Petroleum Development Oman (PDO) L.L.C. and Bank Muscat S.A.O.G.

References

Amirkhanov H. Research on the Palaeolithic and Neolithic of Hadramaut and Mahra. Arabian Archaeology and Epigraphy. 1994;5:217–28.

Bar-Yosef O. The Lower Palaeolithic of the Near East. Journal or World Prehistory. 1994;8(3):211–65.

Biagi P. Surveys along the Oman coast. Preliminary Report on the 1985–1988 campaigns. East and West. 1988;38:271–91.

Biagi P. An Early Palaeolithic site near Saiwan (Sultanate of Oman). Arabian Archaeology and Epigraphy. 1994;5:81–8.

Biagi P. Surveys along the Oman coast: a review of the prehistoric sites discovered between Dibab and Qalhat. Adumatu. 2004;10:29–50.

Biagi P, Nisbet R. The prehistoric fisher-gatherers of the western coast of the Arabian sea. A case of seasonal sedentarisation? World Archaeology. 2006;38(2):220–38.

Biagi P, Torke W, Tosi M, Uerpmann H-P. Qurum: a case study of coastal archaeology in northern Oman. World Archaeology. 1984;16(1):43–61.

Bray HE, Stokes S. Chronologies for Late Quaternary barchan dune reactivation in the southeastern Arabian peninsula. Quaternary Science Reviews. 2003;22:1027–33.

Burns SJ, Fleitmann D, Matter A, Neff U, Mangini A. Speleothem evidence from Oman for continental pluvial events during interglacial periods. Geology. 2001;29(7):623–6.

Clark ID, Fontes J-C. Palaeoclimatic reconstruction in northern Oman based on carbonates from hyperalkaline groundwaters. Quaternary Research. 1990;33:320–36.

Cremaschi M, Negrino F. The frankincense of Sumhuram: palaeoenvironmental and prehistorical background. In: Avanzini A, editor. Khor Rori Report 1. Pisa: Edizioni Plus, Università di Pisa; 2002. p. 325–63.

Crowfoot Payen J, Hawkins S. A surface collection of flints from Habarut in southern Arabia. Man. 1963;240:185–8.

De Cardi B, Doe DB, Roskams SP. Excavation and survey in the Sharqiyah, Oman, 1976. Journal of Oman Studies. 1977;3:17–33.

Di Mario F. The western ar-Rub al-Khali "Neolithic": new data from the Ramlat Sabatayn (Yemen Arab Republic). Annali dell'Istituto Universitario Orientale di Napoli. 1989;49(2):109–48.

Doe DB. Gazetteer of sites in Oman 1976. The Journal of Oman Studies. 1976;3:35–57.

Edens C. Towards a definition of the western ar-Rub al-Khali "Neolithic". Atlal. 1982;6:109–23.

Edens C. The Rub al-Khali 'Neolithic' revisited: the view from Naqdam. In: Potts DT, editor. Araby the Blest. Copenhagen: Museum Tusculanum Press; 1988a. p. 15–43.

Edens C. Archaeology of the sands and adjacent portions of the Sharqiyah. The Journal of Oman Studies. 1988b;3:113–30.

Edens C, Wilkinson TJ. Southwest Arabia during the Holocene: recent archaeological developments. Journal of World Prehistory. 1998;12(1):55–119.

Fleitmann D, Burns SJ, Neff U, Mangini A, Matter A. Changing moisture sources over the last 330000 years in Northern Oman from fluid-inclusion evidence in speleothems. Quaternary Research. 2003;60:223–32.

Glennie KW. The desert of Southeast Arabia. Bahrain: Gulf PertoLink; 2005.

Goudie A, Colls A, Stokes S, Parker A, White K, al-Farraj A. Latest Pleistocene and Holocene dune construction at the north-eastern edge of the Rub al Khali, United Arab Emirates. Sedimentology. 2000;47:1011–21.

Haude W. Erfordern die Hochstände des Toten Meeres die Annahme von Pluvial-Zeiten während des Pleistozäns? Meteorologische Rundschau. 1969;22(2):29–40.

Hoelzmann P, Gasse F, Dupont LM, Salzmann U, Staubwasser M, Leuschner DC, et al. Palaeoenvironmental changes in the arid and subarid belt (Sahara-Sahel-Arabian peninsula) from 150 kyr to present. In: Battarbee RW, Gasse F, Stickley CE, editors. Past climate variability through Europe and Africa, vol. 6. Dordrecht: Springer; 2004. p. 219–56.

Inizan M-L, Ortlieb L. Préhistoire dans la region de Shabwa au Yémen du Sud. Paléorient. 1987;13(1):5–22.

Ivanochko T. Sub-orbital variations in the intensity of the Arabian sea monsoon. Ph.D. dissertation, University of Edinburgh, Edinburgh; 2004.

Kallweit H. Remarks on the Late Stone Age in the U.A.E. In: Potts DT, Naboodah H, Hellyer P, editors. Proceedings of the First International Conference on the Archaeology of the United Arabian Emirates (Abu Dhabi 15–18 April). London: Trident Press; 2003. p. 55–64.

Kapel H. Atlas of the Stone-Age cultures of Qatar: Reports of the Danish archaeological expedition to the Arabian Gulf. Aarhus University Press; 1967.

Kocurek G, Lancaster N. Aeolian system sediment state: theory and Mojave Desert Kelso dune field example. Sedimentology. 1999;46:505–15.

Le Tensorer J-M. Les cultures Paléolithiques de la steppe syrienne. L'exemple d'El Kowm. Annales Archéologiques Arabes Syriennes. 1996;42:43–61.

Le Tensorer J-M. Nouvelles fouilles à Humal (El Kowm, Syrie centrale) premiers résultats. British Archaeological Reports, International Series 1263. 2004. p. 223–39.

Le Tensorer J-M, Hours F. L'occupation d'un terrritoire à la fin du Paléolithique moyen à partir de l'exemple d'El Kowm (Syrie). Etudes et Recherches Achéologiques de l'Université de liège (ERAUL). 1989;33:17–114.

Le Tensorer J-M, Muhesen S, Jagher R, Morel P, Renault-Miskovsky J, Schmid P. Les premiers hommes du désert syrien – fouilles syro-suisses à Nadaouiyeh Aïn Askar. Paris: Editions du Muséum d'Histoire Naturelle; 1997.

Le Tensorer J-M, Jagher R, Muhesen S. Paleolithic settlement dynamics in the El Kowm area (central Syria). In: Conard N, editor. Settlement dynamics of the Middle Paleolithic and Middle Stone Age. Tübingen: Kerns Verlag; 2001. p. 101–22.

Le Tensorer J-M, Hauck T, Wojtczak D. Le Paléolithique ancien et moyen d'Hummal (El Kowm, Syrie centrale). Swiatowit V (XLVI) Fasc. B. 2005;179–86.

Le Tensorer J-M, Jagher R, Rentzel P, Hauck T, Ismail-Meyer K, Pümpin C, et al. Long-term site formation processes at natural springs Nadaouiyeh and Hummal in the El Kowm Oasis, Central Syria. Geoarchaeology. 2007;22(6):621–39.

Leuschner DC, Sirocko F. Orbital insolation forcing of the Indian monsoon – a motor for global climate changes? Palaeogeography, Palaeoclimatology, Palaeooecology. 2003;197:83–95.

Lézine A-M, Saliège J-F, Robert C, Wertz F. Holocene lakes from Ramlat as-Sab'atayn (Yemen) illustrate the impact of monsoon activity in southern Arabia. Quaternary Research. 1998;50:290–9.

Maggi R. The chipped flint assemblage of RH6 (Muscat, Sultanate of Oman): some considerations on technological aspects. East and West. 1990;40:293–300.

Magnani G, Bartolomei P, Cavulli F, Esposito M, Marino EC, Neri M, et al. U-series and radiocarbon dates on mollusk shells from the uppermost layer of the archaeological site KHB-1, Ra's al Khabbah, Oman. Journal of Archaeological Science. 2007;34:749–55.

McClure HA. Radiocarbon chronology of late Quaternary lakes in the Arabian desert. Nature. 1976;263:755–6.

McClure HA. A new Arabian stone tool assemblage and notes on the Aterian industry of North Africa. Arabian Archaeology and Epigraphy. 1994;5:1–16.

Meignen L. Le phénomène laminaire au Proche-Orient, du Paléolithique inférieur aus débuts du Paléolithique supérieur. In: Evin J, editor. Un Siècle de Construction du Discours Scientifique en Préhistoire: XXVIe Congrès Préhistorique de France – Congrès du centenaire de la Société Préhistorique Française, Avignon, 21–25 Septembre. Paris: Société Préhistorique Française; 2007. p. 79–94.

Neff U. Massenspektrometrische Th/U-Datierung von Höhlensintern aus dem Oman. Klimaarchive des asiatischen Monsuns. Ph.D. dissertation, Ruprecht-Karls-Univeristät Heidelberg, Heidelberg; 2001.

O'Connor PW, Thomas DSG. The timing and environmental significance of Late Quaternary linear dune development in western Zambia. Quaternary Research. 1999;52:44–55.

Parker AG, Goudie AS, Stokes S, White K, Hodson MJ, Manning M, et al. A record of climate change from lake geochemical analyses in southeastern Arabia. Quaternary Research. 2006;66:465–76.

Petraglia MD. The Lower Paleolithic of the Arabian peninsula: occupations, adaptations, and dispersals. Journal of World Prehistory. 2003;17(2):141–79.

Petraglia MD. Hominin responses to Pleistocene environmental change in Arabia and South Asia. In: Head MJ, Gibbard PL, editors. Early–Middle Pleistocene transitions: the land–ocean evidence. London: The Geological Society of London; 2005. p. 305–19.

Petraglia MD, Alsharekh A. The Middle Palaeolithic of Arabia: implications for human origins, behaviour and dispersals. Antiquity. 2003;77:671–84.

Platel J-P, Bourdillon-De Grissac C, Babinot J-F, Roger J, Mercadier C. Modalités de la transgression campagienne sur le massif du Haushi-Huqf (Oman): stratigraphie. Bulletin de la Société Géologique Française. 1994;165:147–61.

Preusser F, Radies D, Matter A. A 160000 year record of dune development and atmospheric circulation in Southern Arabia. Science. 2002;296(June 14):2018–20.

Preusser F, Radies D, Driehorst F, Matter A. Late Quaternary history of the coastal Wahiba Sands, Sultanate of Oman. Journal of Quaternary Science. 2005;20(4):395–405.

Pullar J. Harvard archaeological survey in Oman, 1973. Flint sites in Oman. Proceedings of the Seminar for Arabian Studies. 1973;4:33–48.

Pullar J. A selection of aceramic sites in the Sultanate of Oman. The Journal of Oman Studies. 1985;7:49–87.

Pullar J, Jäckli B. Some aceramic sites in Oman. The Journal of Oman Studies. 1978;4:53–74.

Radies D, Preusser F, Matter A, Mange M. Eustatic and climatic controls of the development of the Wahiba Sand Sea, Sultanate of Oman. Sedimentology. 2004;51:1359–85.

Rohling EJ, Fenton M, Jorissen FJ, Bertrand P. Magnitudes of sea-level lowstands of the past 500000 years. Nature. 1998;9394(July 9):162–5.

Rose J. The question of Upper Pleistocene connections between East Africa and South Arabia. Current Anthropology. 2004;45:551–5.

Rose J. Among Arabian sands. Defining the Palaeolithic of southern Arabia. Ph.D. dissertation, Southern Methodist University, Texas; 2006.

Rose J. The Arabian Corridor Migration Model: archaeological evidence for hominin dispersals into Oman during the Middle and Upper Pleistocene. Proceedings of the Seminar for Arabian Studies. 2007;37:1–19.

Siddall M, Rohling EJ, Almogi-Labin A, Hemleben C, Melschner D, Schmelzer I, et al. Sea-level fluctuations during the last glacial cycle. Nature. 2003;423(June 19):853–8.

Smith GH. New Neolithic Sites in Oman. The Journal of Oman Studies. 1976;2:189–98.

Smith GH. New prehistoric sites in Oman. The Journal of Oman Studies. 1977;3:71–81.

Tosi M, Usai D. Preliminary report on the excavations at Wadi Shab, Area 1, Sultanate of Oman. Arabian Archaeology and Epigraphy. 2003;14(2003):8–23.

Uerpmann M. Structuring the Late Stone Age of southeastern Arabia. Arabian Archaeology and Epigraphy. 1992;3:65–109.

Uerpmann M. The dark millenium – remarks on the Final Stone Age in the Emirates and Oman. In: Potts DT, Naboodah H, Hellyer P, editors. Proceedings of the First International Conference on the Archaeology of the United Arabian Emirates (Abu Dhabi 15–18 April). London: Trident Press; 2003. p. 73–81.

Warren A, Allison D. The palaeoenvironmental significance of dune size hierarchies. PALAEO. 1998;137(1998):289–303.

Whalen N. Lower Palaeolithic Sites in the Huqf area of Central Oman. The Journal of Oman Studies. 2003;13:175–82.

Whalen NM, Fritz GA. The Oldowan in Arabia. Adumatu. 2004;9:7–18.

Whalen NM, Schatte KE. Pleistocene sites in southern Yemen. Arabian Archaeology and Epigraphy. 1997;8:1–10.

Whalen NM, Siraj-Ali J, Sindi HO, Pease DW, Badein MA. A complex of sites in the Jeddah-Wadi Fatimah area. Atlal. 1988;11:77–85.

Whalen NM, Zoboroski M, Schubert K. The Lower Palaeolithic in Southwestern Oman. Adumatu. 2002;5:27–33.

Wood WW, Imes JL. How wet is wet? Precipitation constraints on late Quaternary climate in the southern Arabian peninsula. Journal of Hydrology. 1995;164:263–8.

Zarins J. The land of incense – Archaeological work in the Governorate of Dhofar, Sultanate of Oman 1990–1995. Muscat: Sultan Qaboos University Publications; 2001.

Chapter 12
The Middle Paleolithic of Arabia:
The View from the Hadramawt Region, Yemen

Rémy Crassard

Keywords Dispersals • Hadramawt • Levallois • Middle Paleolithic • Yemen

Arabia: A New "El Dorado" for Evolutionary Scholars?

While prehistoric research in the Arabian peninsula is still in its primary stages of development, the very existence of this book is proof of a recent growing interest in the region. Yet, the interest in the prehistory of the region is outshined by the dearth and frailty of the available data. We must then ask ourselves, why such interest and enthusiasm? And is it really justified to theorize about the contribution of Arabia for human prehistory if the data remain scant? It can certainly be explained, as Petraglia (2007: 383) correctly states, by the progressive reorientation of research towards areas of the world where it is more "logical" to look in order to understand "the evolutionary history of geographically widespread populations". Consequently, this phenomenon is akin to a revolution in the small world of Arabian prehistoric research; a revolution that carries great aspirations for crucial questions such as the origin of the dispersion of anatomically modern humans out of Africa. While the data are scarce, the passion which one can have for the prehistory of a region such as Arabia is fully justified by the simple recognition of its being an area laden with enormous possibility.

Given the possibilities offered by this vast peninsula, what data can we rely on? There have been archaeological surveys carried out in Arabia for more than half a century. These surveys have shed light on the existence of an Arabian Paleolithic. Nonetheless, the presence of a Paleolithic in Arabia became problematic as it became necessary to find points of comparisons with well-established, or at least, better established, neighboring industries, such as those of East Africa and

R. Crassard (✉)
Leverhulme Centre for Human Evolutionary Studies,
University of Cambridge, The Henry Wellcome Building,
Fitzwilliam Street, Cambridge, CB2 1QH, UK
e-mail: rc461@cam.ac.uk

the Levant. The chrono-cultural framework was thus gradually modeled after the great phases of the Lower and Middle Paleolithic: Oldowan, Acheulean, and Mousterian or Middle Stone Age. All discoveries made in Arabia were adapted to this framework and not vice versa; simply put, the Arabian framework was not made by regional discoveries. This very fact is of great importance, as no site has ever been properly chronometrically dated to the Paleolithic period. Any pebble-tool was thus Oldowan, any biface was associated to the Acheulean, and any Levallois core to the Mousterian. All of these labels happen to be associated with dates that are in many cases of a low degree of accuracy which renders such designations highly debatable at the very least, especially in light of the fact that these sites are dated typologically and not absolutely. It appears then, that the lithic industries are the best available data to study, waiting for some better dated and archaeologically richer contexts.

The Arabian Middle Paleolithic Background

Questions concerning affinities between Arabia and its better-documented neighboring regions, such as the Levant and East Africa, during the Middle Paleolithic have long been the subject of debate. However, the Middle Paleolithic of Arabia suffers from numerous lacunae. From a paleoanthropological point of view, no hominin fossils have been discovered thus far. In addition, all of the artifacts which presume a Paleolithic age were collected from the surface of sites, which are undated. All that remains are human expansion or demographic models based on genetic data. Given the current state of the research which mainly relies on models and undated in situ artifacts, it is safe to say that the existence of an Arabian Middle Paleolithic that is more or less contemporary with geographically close and well-identified cultural complexes elsewhere, would present an occasion to consider and discuss such networks of diffusion and dispersal.

In Europe, the Levant and Africa, the Middle Paleolithic is generally characterized by Levallois debitage, which more or less comes to fruition in the Upper Acheulean period, and develops widely thereafter. A mental conceptualization that can be clearly seen with the Levallois concept, consequently dominates lithic production. This conceptualization is illustrated by a predetermination of the debitage products, which are obtained from more or less variable Levallois methods of debitage (Boëda, 1994). As the Levallois debitage is present in various regions of the world, the analyses based on this classificatory unit are then of significant value for making comparisons.

The presence of Levallois debitage in Arabia was identified rather early by the first pioneers of Yemeni archaeology. Caton-Thompson (1938, 1953) was the first archaeologist to have detected a potential Pleistocene human presence in Hadramawt. Her work was followed by Van Beek et al. (1963), Inizan and Ortlieb (1987), Inizan (1989), Amirkhanov (1991, 1994a) and Zimmerman (2000), to quote only the principal archaeologists. However, of the sites that produced a Levallois industry in the whole of the Arabian peninsula, not a single one was able to provide either intact stratigraphy or relative or absolute dates. A cumulative review of the state of Paleolithic research in Arabia was, nonetheless, carried out by Petraglia and Alsharekh (2003), and one, earlier, by Zarins (1998).

Levallois Industries and the Middle Paleolithic: How and Why Study Surface Material in Yemen?

Universal Dating of Sites with Levallois Technology?

Until very recently, not a single stratified site in Arabia had produced sufficient indications of a Levallois debitage for such a method to have been associated with a precise date or period. Despite the large numbers of Levallois artifacts collected from the surface of sites, and particularly on the plateaus of Hadramawt in Eastern Yemen, the poor contexts and lack of dated comparisons rendered them chronologically unidentifiable. Having said this, what dates can one allot temporarily to the use of the Levallois concept in Yemen?

Levallois debitage was in use for more than 400 thousand years. It appears from the Acheulean period in Africa and is attested in Western Europe from the end of the isotopic stage 10 alongside Middle Acheulean type assemblages and in particular in the Somme basin (Tuffreau, 2004: 81–82) of northern France (ca. 600–400 ka). It spreads throughout Eurasia in the Middle Paleolithic (from 300 ka) during the Mousterian period (300–30 ka), starting at isotopic stage 8. The presence of a Levallois technology is therefore quite ancient throughout Africa, the Near-East, Europe and Asia.

Nevertheless, the typical attribution of the use of the Levallois concept to periods of the Middle Paleolithic can be misplaced as this debitage modality has been proven to be present in more recent lithic sets throughout the world. It is for such reasons that the use of an old system of dating the Levallois technology in Yemen, especially in the absence of reliable relative or absolute dates held such reservations. Nonetheless the data presented hereafter lead us to allot a Pleistocene date to the Levallois methods from the Hadramawt area.

Some Elements Largely in Favor of Pleistocene Dating

The Arabian peninsula is located at the crossing between Africa, the Levant and Asia. It would thus seem, on the basis of a diffusionist theorization, that Arabia was in one way or another, in contact with populations that used this type of debitage. Even if one considers the possibility of a late contact with African MSA traditions or Levantine Middle Paleolithic ones, it still would have happened during the Pleistocene. Moreover, paleoenvironmental and geological data indicate periods when the sea level was very low and crossing the Red Sea would have been feasible. Other paleoclimatic data indicate periods during which aridity was less extreme and the peopling of Arabia would have been facilitated. These above-mentioned periods occur during the Pleistocene (Burns et al., 2003).

Patina as a relative dating method is fundamentally more important on Levallois pieces. Comparison of the patina acquired on Levallois pieces as opposed to typical Early Holocene artifacts systematically produced the same result. Although a comparison of the degree of patina is an unreliable method that the author partly denounced in a previous study (Crassard, 2007: 71–73), not a single non-heavily patinated Levallois element has ever been collected from a surface site. This is absolutely not the case for Holocene industries, which can be patinated, but are not in the majority of instances.

Finally, stratified Holocene assemblages have never provided components of a Levallois debitage. We can thus relatively date Levallois industries to a period of time preceding the typical Holocene industries found in Hadramawt (Crassard et al., 2006: 169; Crassard, 2007: 251–252).

Consequently, without more precision, the Levallois industries from Yemen favor a Pleistocene period date. What then is the significance of using a technological approach (study of techniques) for such imprecisely dated data?

Lithic industries can deliver technical, cultural and at times chronological information, but their study cannot be an end in and of itself. The numerous surface sites in Arabia cannot be dated precisely. However, the lithic evidence is impressive due to the enormous quantity of surface finds. From this point of view, the study of lithic technology constitutes a relevant tool to distinguish convergences, diffusions and autonomous inventions. While the technological approach alone is not sufficient, it is a important heuristic tool.

Fig. 1 The Arabian peninsula and Yemen
location

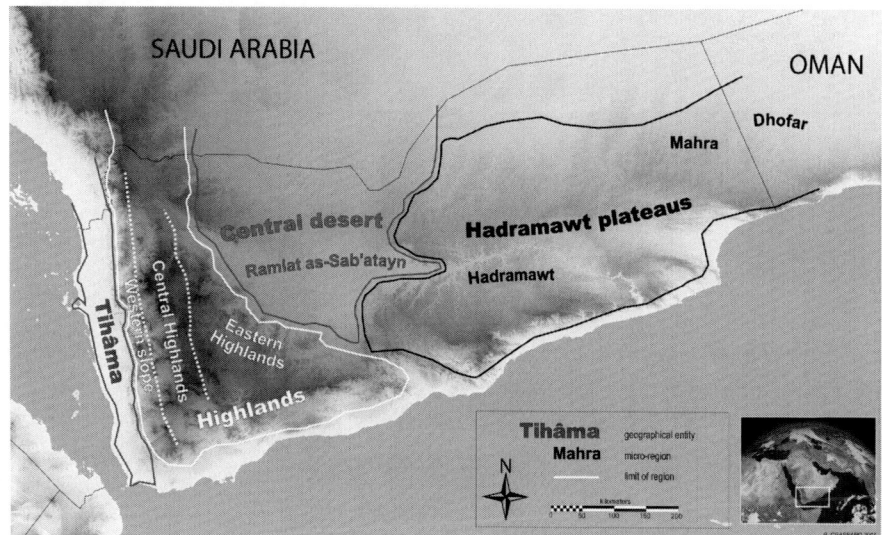

Fig. 1 The Arabian peninsula and Yemen
location

Levallois Assemblages from Hadramawt: The Contexts

The Hadramawt Region

In Southern Arabia (Fig. 1), the Aden Gulf rift opened approximately 34 million years ago during the Early Oligocene. This fissure was created earlier than the rifting of the Red Sea, and was accompanied by the rising of the Hadramawt (Eastern Yemen) and Dhofar plateaus (Western Oman; Sanlaville, 2000: Fig. 2). The Paleocene and Eocene limestone Hadramawt plateau (locally called *Jawl* meaning "plateau") reaches a maximum height of more than 1,000 m.

A vast network of abrupt valleys, which are often canyons, leads rainwater towards the Wâdî Hadramawt. This wadi consists of a gigantic gorge that crosses most of eastern Yemen (from west to east). The eastern half bears the name of Wâdî Masîla, until it reaches its delta near the Arabian Sea. Collapsing cones were formed by erosional processes at the base of the high limestone cliffs and in some cases provide access to the top of the plateaus. East of the Hadramawt plateaus is the modern Yemeni province of Mahra. In this region the geological limestone formations gradually descend to the limestone hills of Dhofar in Western Oman.

The Context of the Discoveries

A significant number of lithic industries from all periods of prehistory were recently collected as part of two archaeological research projects in Hadramawt (Fig. 2): The Roots of Agriculture in Southern Arabia Project (RASA) and the French Archaeological Mission in Jawf-Hadramawt (HDOR).

Fig. 2 The Hadramawt region and the project locations

The HDOR and RASA fieldwork projects produced 48 surface sites (21 for HDOR and 27 for RASA) that delivered characteristic Levallois elements (Tables 1 and 2). In most of these cases, the presence of this lithic type was discrete forming part of moderate surface assemblages, except some rare cases where remains of debitage clusters were still visible (remains of workshops). The state of preservation of the lithics collected ranged from average (heterogeneous assemblages, good readability of the scars) to very bad (heterogeneous assemblages, rare artifacts, very eroded and strong patina). Most of the samples collected were heavily patinated or eroded. Only a few sites were characterized by less patinated flint industries, on which knapping stigmata was clearly readable. A selective collecting strategy, which

Table 1 Levallois cores from HDOR and RASA projects, classified by sites

Sites	Core #	Length	Width	Thickness	Quadrangular/oval	Triangular	A1	A2	A3	B1	B2	B3	B4	C	Abandoned	Non-Levallois
HDOR 412	1	63	63	–	X?		X									
	2	69	54	–	X?			X								
	3	50	42	–		X				X						
	4	66	54	–	X					X						
	5	60	58	–	X?										X	
HDOR 417	1	82	67	–	X?				X							
	2	84	66	–	X?									X		
HDOR 526	1	45	41	17	X?		X									
	2	49	44	23		X						X				
RASA 2004 84-0	1	72	60	–	–	-										X
	2	53	51	–		X				X						
	3	55	51	–	X?			X								
RASA 2004 84-2	1	51	42	20		X				X						
	2	56	61	–	X?			X								
	3	57	47	–	X?				X							
RASA 2004 124-1	1	72	82	23		X						X				
	2	75	82	27	X?		X									
	3	79	71	37		X				X						
	4	80	59	35		X				X						
	5	58	64	39	X?			X								
	6	74	39	30		X				X						
	7	91	78	50		X			X							
RASA 2004 149-1	1	63	54	17	X?		X									
	2	54	57	12		X						X				
	3	44	33	25		X		X								
	4	45	35	26		X		X								
RASA 2004 149-2	1	59	52	20		X					X					
	2	60	42	22		X						X				
	3	62	57	25	X?		X									
	4	54	41	18		X						X				
	5	43	48	21	X?		X									
	6	62	43	25		X								X		
	7	60	48	28	X?				X							
	8	52	45	19	–	-										X
RASA 2004 153-1	1	60	46	18		X				X						
	2	69	52	20	X										X	
RASA 2004 165-1	1	66	38	22		X									X	
	2	50	42	23	–	-										X
RASA 2004 166-1	1	55	46	13		X				X						
	2	47	43	11	X?		X									
	3	62	53	23		X?				X						
	4	77	38	23		X?						X				
HDOR 500	1	69	44	24		X?			X							
HDOR 520	1	58	44	16		X						X				
HDOR 527	1	54	53	25		X						X				
HDOR 566	1	98	90	50		X					X					
HDOR 571	1	57	31	14		X					X					
HDOR 574	1	78	62	24		X?					X					
RASA 2004 135-1	1	68	55	14		X					X					
RASA 2004 136-1	1	84	59	43		X			X							
RASA 2004 141-1	1	73	65	29	X?									X		
RASA 2004 168-1	1	65	33	24		X					X					

Table 2 Synthesis of Levallois cores analysis, classified by schemes

Sites	Core #	Preparation phase	Production phase	Scheme
HDOR 412	1	Centripetal	Unique preferential flake debitage	A1
HDOR 526	1	Centripetal	Unique preferential flake debitage	
RASA 2004-124-1	2	Centripetal	Unique preferential flake debitage	
RASA 2004-149-1	1	Centripetal	Unique preferential flake debitage	
RASA 2004-149-2	3	Centripetal	Unique preferential flake debitage	
RASA 2004-149-2	5	Centripetal	Unique preferential flake debitage	
RASA 2004-166-1	2	Centripetal	Unique preferential flake debitage	
HDOR 412	2	Centripetal	Recurrent preferential flakes debitage	A2
RASA 2004-84-2	2	Centripetal	Recurrent preferential flakes debitage	
RASA 2004-124-1	5	Centripetal	Recurrent preferential flakes debitage	
HDOR 417	1	Opposed lateral	Unique preferential flake debitage	A3
RASA 2004-84-0	3	Opposed lateral	Unique preferential flake debitage	
RASA 2004-84-2	3	Opposed lateral	Unique preferential flake debitage	
HDOR 412	3	Convergent unipolar	"Classical" Levallois point debitage	B1
HDOR 412	4	Convergent unipolar	"Classical" Levallois point debitage	
RASA 2004-124-1	7	Convergent unipolar	"Classical" Levallois point debitage	
RASA 2004-149-1	3	Convergent unipolar	"Classical" Levallois point debitage	
RASA 2004-149-1	4	Convergent unipolar	"Classical" Levallois point debitage	
RASA 2004-149-2	7	Convergent unipolar	Recurrent "classical" Levallois points debitage	
RASA 2004-84-0	2	Convergent unipolar and lateral	"Constructed" point debitage	B2
RASA 2004-84-2	1	Convergent unipolar and distal	"Constructed" point debitage	
RASA 2004-124-1	3	Convergent unipolar and distal	"Constructed" point debitage	
RASA 2004-124-1	4	Convergent unipolar and distal	"Constructed" point debitage	
RASA 2004-124-1	6	Convergent unipolar and distal	"Constructed" point debitage	
RASA 2004-136-1	1	Convergent unipolar and lateral–distal (?)	"Constructed" point debitage	
RASA 2004-153-1	1	Distal convergent	"Constructed" point debitage	
RASA 2004-166-1	1	Convergent unipolar and distal	"Constructed" point debitage	
RASA 2004-166-1	3	Convergent unipolar and distal	"Constructed" point debitage	
HDOR 500	1	Bipolar and lateral–distal	"Constructed" point debitage	B3
HDOR 566	1	Bipolar and lateral–proximal	"Constructed" point debitage	
HDOR 571	1	Bipolar	"Constructed" point debitage	
HDOR 574	1	Bipolar and lateral–distal	"Constructed" point debitage	
RASA 2004-124-1	1	Bipolar and lateral	"Constructed" point debitage	
RASA 2004-135-1	1	Bipolar and lateral–proximal	"Constructed" point debitage	
RASA 2004-149-2	1	Bipolar	"Constructed" point debitage	
RASA 2004-168-1	1	Bipolar	"Constructed" point debitage	
RASA 2004-149-1	2	Proximal–lateral and opposed lateral	"Constructed" point debitage	B4
HDOR 520	1	Proximal–lateral and opposed lateral	"Constructed" point debitage	
HDOR 526	2	Proximal–lateral and opposed lateral	"Constructed" point debitage	
HDOR 527	1	Proximal–lateral and opposed lateral	"Constructed" point debitage	
RASA 2004-149-2	2	Proximal–lateral and opposed lateral	"Constructed" point debitage	
RASA 2004-149-2	4	Proximal–lateral and opposed lateral	"Constructed" point debitage	
RASA 2004-166-1	4	Proximal–lateral and lateral	"Constructed" point debitage	
HDOR 417	2	Centripetal	Recurrent centripetal debitage	C
RASA 2004-141-1	1	Centripetal	Recurrent centripetal debitage (?)	C
HDOR 412	5	Bipolar	Undetermined (abandoned)	Undet.
RASA 2004-149-2	6	Proximal–lateral and distal	Undetermined (abandoned)	Undet.
RASA 2004-153-1	2	Parallel unipolar and distal	Levallois debitage? (abandoned)	Undet.
RASA 2004-165-1	1	Proximal–lateral and opposed lateral	B4 Levallois debitage? (abandoned)	Undet.

consisted mainly of cores, was carried out for the majority of Levallois sites. These cores were later analyzed systematically and provided a significant study of the technical schemes utilized in the final debitage phases in the Hadramawt region. Nevertheless, it is important to keep in mind that the information obtained from these cores is incomplete as only the last stage of debitage is represented.

The majority of the elements that have led to a better comprehension of Levallois variability in South Arabia were discovered in the Wâdî Wa'shah and Wâdî Sanâ, tributaries of the Wadi Hadramawt. To avoid redundancy, the detailed characteristics of these sites will not be described here. It suffices to say that the majority of these sites consist of surface scatters located on the plateau tops. By contrast, the location of Holocene sites in the region suggests settlement variability (Crassard, 2007: 154–170, 348–354).

Technological Analysis and Terminology

The lithic material that originates from the Hadramawt surveys was analyzed on the basis of a grid established by E. Boëda (1994: 22, 35–39), which deals with the characteristics of predetermined removals. Such a process allows for the identification of technical schemes, also called methods, and which are equivalent to the knapper's Levallois conceptualization.

The preferential cores are grouped under several types of preferential removals (quadrangular, oval or trapezoidal flake and Levallois point), whereas the recurrent cores are sorted by modes of manipulation of the debitage surface: unipolar (parallel), bipolar, centripetal (Boëda, 1994: 257–258). This technological (study of techniques) approach aims to gain insight into the technical cultural tradition. Such insight can only be attained through a gradual deciphering of the technologies used and through an understanding of the constraints applied to the technical norm of a group (see Boëda, 1994: 263).

Technological Analysis of Levallois Cores

The technological analysis concerned 56 cores in total (15 for HDOR and 41 for RASA). A total of 10 cores was found isolated, whereas 46 came from 11 "homogeneous" assemblages. These assemblages are mainly made up of cores, as well as some products that resulted from a Levallois debitage modality. These assemblages originate from surface sites in the Hadramawt. The following analysis of these Levallois cores (final stages of debitage) focuses on the variability of the technical schemes involved in the knapping process.

Levallois Assemblages from Hadramawt: The Data

The First Synthesis on the Levallois Debitage of Hadramawt: Two Methods, Three Groups and Eight Modalities

Through the study of the methods of debitage made on Levallois cores from the HDOR and RASA project sites, it was possible to isolate three different groups (A, B and C):
Group A: Levallois debitage of one (sometimes two) oval, quadrangular or trapezoidal preferential flake(s)

Group B: Levallois point debitage
Group C: centripetal recurrent Levallois debitage

These three groups represent the technical schemes that allowed for a predetermined product to be obtained. A total of eight technical schemes, or debitage modalities, were identified (A1, A2, A3, B1, B2, B3, B4 and C). They reveal technical variability within groups A and B. A thorough statistical analysis was not undertaken as the cores do not originate from "closed" archaeological contexts.

Through the analysis of the final debitage phases of the cores from Hadramawt, two objectives can be distinguished:
A quantitative objective:

One (even two) products per flaking surface (preferential product Levallois debitage)
Several products per flaking surface (recurrent Levallois debitage)

A qualitative objective:

Debitage of Levallois flakes or points
In the case of the Hadramawt cores, these two abovementioned objectives (quantitative and qualitative) help us to isolate two methods of Levallois debitage (preferential product or centripetal recurrent) which are associated with eight different modalities: A1 to A3, B1 to B4 and C.

Group A

Group A is characterized by:

A debitage of preferential flakes with centripetal preparation (schemes A1 and A2)
A debitage of preferential flakes with "crossed" preparation (scheme A3)

Scheme A1

Scheme A1 (Fig. 3) is characterized by a Levallois debitage of a unique preferential flake with centripetal preparation. This scheme leads to the obtainment of a preferential flake by shaping the debitage surface centripetally. This particular scheme is represented by seven cores, including three from the same site (RASA 2004-149-1, cores 1, 3 and 5).

Scheme A2

Scheme A2 (Fig. 4) is characterized by a Levallois debitage of recurrent preferential flakes with centripetal preparation. This scheme is represented by four cores collected from four different sites. It is similar to the A1 scheme. However, a second removal is often intended after the first predeter-

Fig. 3 An example of one core showing Scheme A1 (lithic drawing J. Espagne)

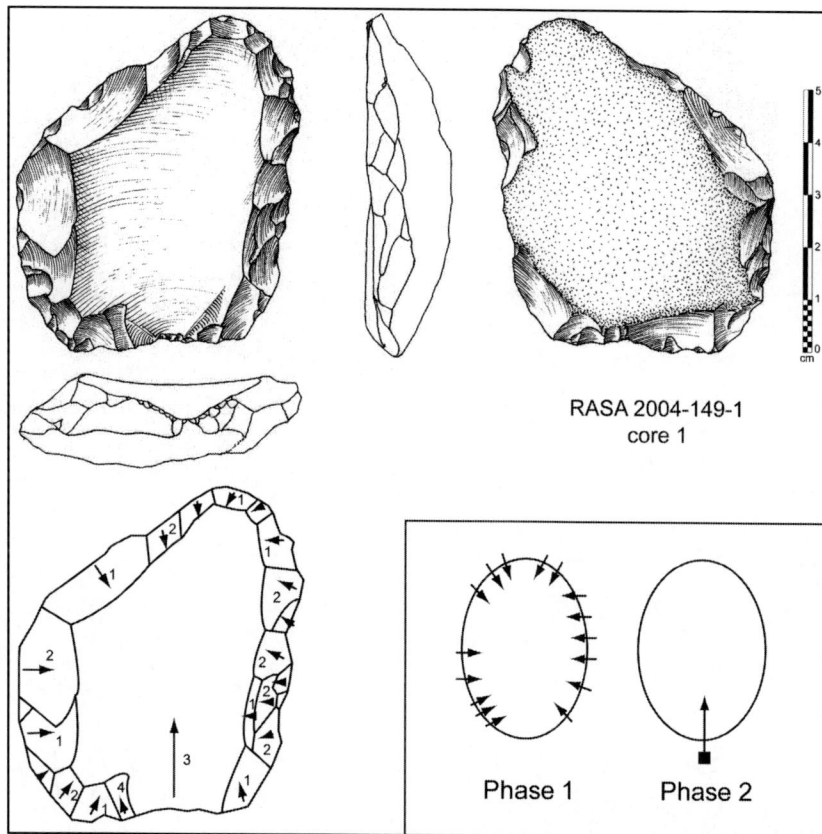

RASA 2004-149-1
core 1

Phase 1   Phase 2

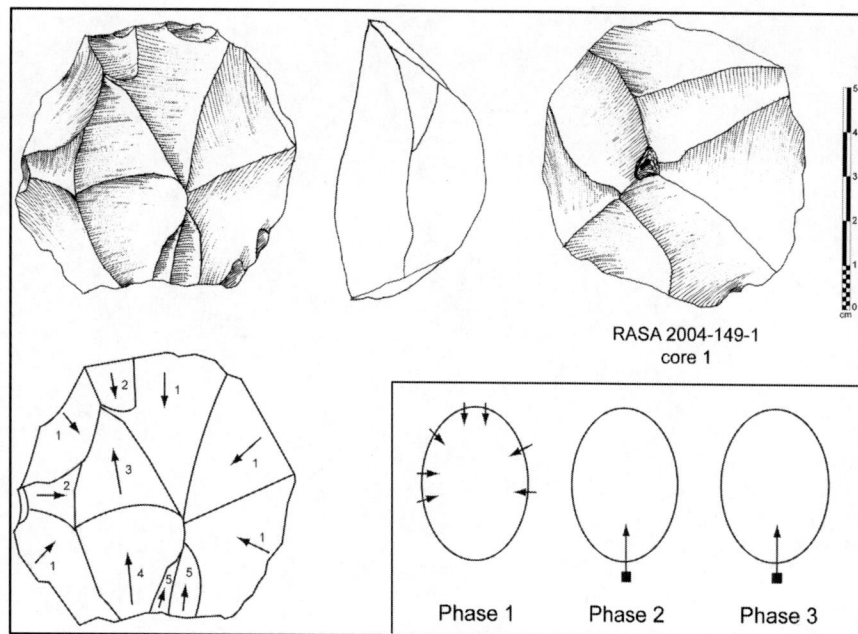

RASA 2004-149-1
core 1

Phase 1   Phase 2   Phase 3

Fig. 4 An example of one core showing Scheme A2 (lithic drawing R. Crassard)

mined removal. The second removal is knapped from the same debitage axis (proximal part of the core). The second "predetermined" removal occurs at stage 3 of the debitage. This occurs without new convex preparation or new Levallois arris. In fact, this scheme never follows a lateral arris from the negative of the first predetermined removal. One can see there the search for a second removal which seeks to extend in parallel of the first. It can also be interpreted as a resharpening removal. Consequently these pieces could be cores abandoned during unfinished debitage operations. Scheme A2 remains debatable.

Scheme A3

Scheme A3 is characterized by a Levallois debitage of unique preferential flake with opposed lateral preparation. This scheme is represented by two cores from two different sites. Scheme A3 is similar to scheme A1, except that the preparation of the lateral convexities from the sides of the cores, is carried out in an opposed way. This type of "crossed" preparation suggests a (voluntary?) intention to produce short and a priori wide Levallois flakes. The abandonment of unfinished debitage cores will be considered in the final interpretation of this scheme.

Group B

Group B is characterized by:

A debitage of preferential triangular flakes with convergent unipolar preparation (scheme B1)

A debitage of preferential triangular flakes with "crossed" preparation (unipolar + lateral, or bipolar + lateral) (schemes B2, B3 and B4)

Scheme B1

Scheme B1 (Fig. 5) is characterized by a Levallois debitage of preferential triangular flakes with convergent unipolar preparation (called "classical" Levallois point production). This scheme concerns six cores, of which two originate from the same site (RASA 2004-149-1, core 3 and 4). This type of method is described by Boëda as the only method (described as "type 3") along with two others (described as "type 7" or "type 3+7"), that characterize a Levallois point core (Boëda, 1994: 86). According to this remark, we will label the cores which present a B1 scheme "classical Levallois point cores", in contrast to schemes B2, B3 and B4 cores which have the same type of predetermined product, but acquired through distinct means.

Scheme B2

Scheme B2 (Fig. 6) is characterized by a Levallois debitage of preferential triangular flakes with convergent unidirec-

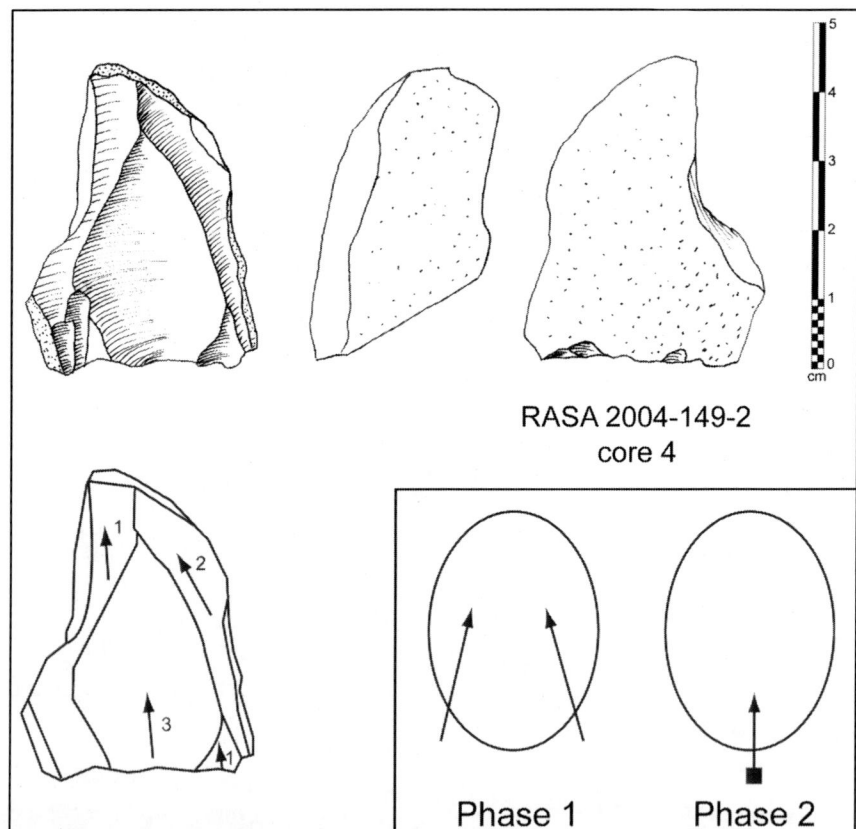

Fig. 5 An example of one core showing Scheme B1 (lithic drawing R. Crassard)

Fig. 6 An example of one core showing Scheme B2 (lithic drawing R. Crassard)

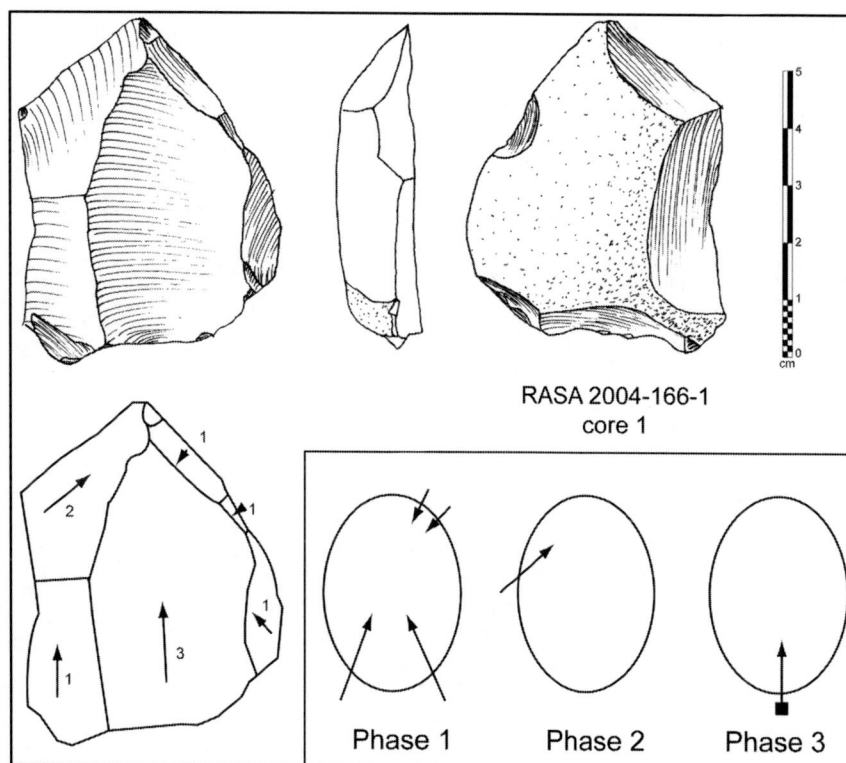

RASA 2004-166-1
core 1

Phase 1   Phase 2   Phase 3

tional preparation and lateral or distal convex reinstallation (called "constructed" Levallois point production). This scheme concerns nine cores, including three from the same site (RASA 2004-124-1, cores 3, 4 and 6), and two others from site (RASA 2004-166-1, cores 1 and 3).Scheme B2 resembles B1 one but can be differentiated by the production of lateral and/or distal convexity by *complementary* removals to the typical convergent removals of the B1 scheme. Such a scheme can be interpreted in two ways. The first interpretation of the method is systematic and aims to deliberately acquire a "constructed" Levallois point (This term is taken from E. Boëda who uses it for industries description from layer VI3 b' at Umm el-Tlel (Syria), Boëda et al., 1998: 249. It is opposed to the scheme that we call with "classical" point.). The second interpretation of the method is that it consists of a convexity reinstallation starting with convergent removals which would not have been sufficiently long and which would not have crossed (if convergent removals came initially); or of a predetermined preparation allowing a debitage of convergent removals which are not necessarily supposed to cross each other (if convergent removals followed).

Scheme B3

Scheme B3 (Fig. 7) is characterized by a Levallois debitage of preferential triangular flakes with bipolar preparation and

installation of lateral convexity (also included in the category of the "constructed" Levallois point production). This scheme concerns eight cores that originate from various surface sites. It is very similar to the B2 scheme because it consists of the production of a "constructed" Levallois point. In this case, the preparation is quasi-systematically bipolar, with convexity installation removals, which, like the B2 scheme, may have been produced in a *predetermined* way (first) or in a *repair* action (second).

Scheme B4

Scheme B4 (Fig. 8) is characterized by a Levallois debitage of preferential triangular flakes with proximo-lateral preparation and opposed lateral (and/or lateral) convexity installation (also included in the category of the "constructed" Levallois point production). This scheme concerns seven cores, of which two originate from the same site (RASA 2004-149-2, cores 2 and 4). It resembles the B1 scheme, but without debitage of two convergent removals. Instead there is only one proximo-lateral removal, which is the base of the predetermined operation. These are supplemented by removals of convexity reinstallation, thus completing the "construction" of the Levallois point. This scheme is rather heterogeneous in its implementation. However, it is homogeneous in its general conceptualization.

Fig. 7 An example of one core showing Scheme B3 (lithic drawing J. Espagne)

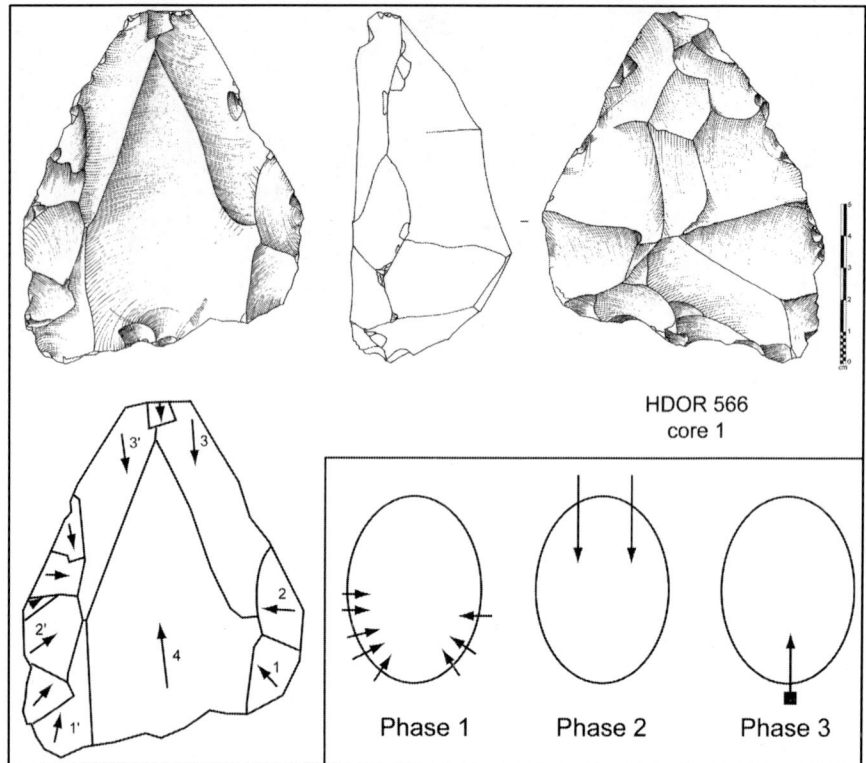

HDOR 566
core 1

Phase 1   Phase 2   Phase 3

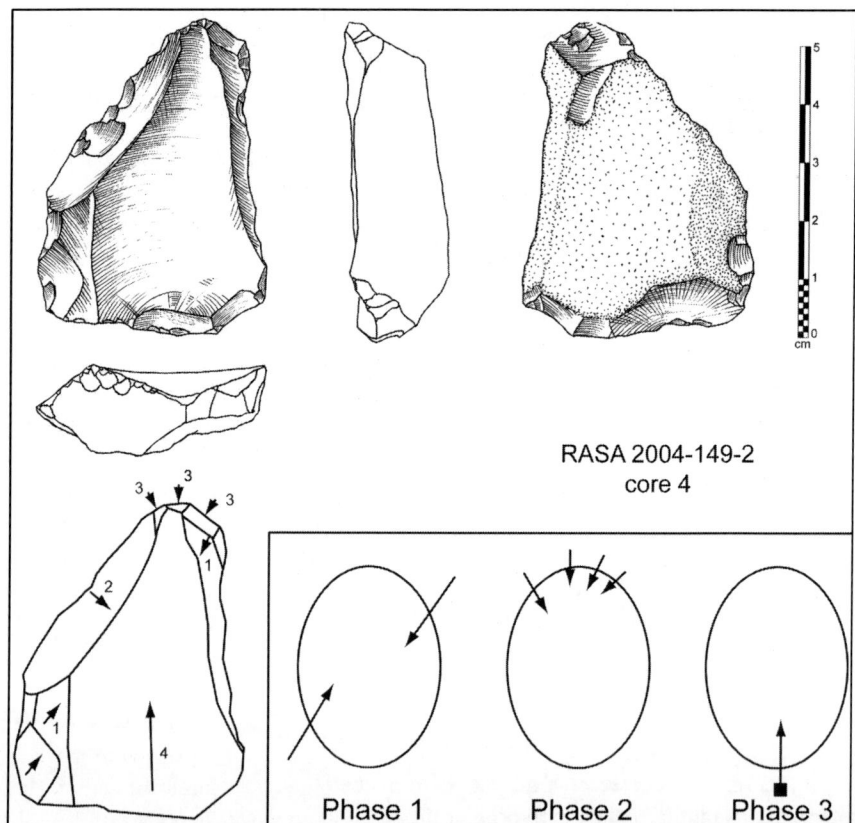

RASA 2004-149-2
core 4

Phase 1   Phase 2   Phase 3

Fig. 8 An example of one core showing Scheme B4 (lithic drawing J. Espagne)

Group C

Group C is characterized by:

A centripetal recurrent Levallois debitage (scheme C)

Scheme C

Scheme C is characterized by a centripetal recurrent Levallois debitage. No particular scheme was observed in this poorly represented group (only two cores). The relevance of this group is not clear, seeing as how it consists of a small number of cores that are severely eroded and patinated.

Four cores could not be linked to a particular group. It is probable that they are related to unfinished debitage and that they were abandoned in the process of knapping.

Interim Conclusions

The analysis of these cores demonstrates the futility of a simple typological analysis which would not have allowed us to differentiate different methods, even for the same production (of points for instance). By contrast, this study has shown that within the same Levallois method, variability can be observed.

The documentation of the variability of the methods that may result from the Levallois concept must be carried out in the rest of the Arabian peninsula. It is important, in the case of this study, to emphasize the technical schemes that are absent in our assemblages, but known elsewhere. The laminar Levallois debitage (of which core 1 from site RASA-168-1 could possibly be an example?) and the recurrent unipolar or bipolar Levallois debitage, are methods that were not identified and which do not appear to be used in Hadramawt.

Comparison with Other Levallois Industries from Hadramawt and Elsewhere in Yemen

Comparison with Industries of Hadramawt from Other Archaeological Projects

In order to extend the study of the Levallois cores from HDOR and RASA projects, some comparisons with industries from other archaeological operations in Hadramawt were carried out on the base of the available drawings in publications and according to observations of the material itself in some Yemeni museums.

In Hadramawt, taken as a geological feature, it was possible to observe lithic pieces at the museum of Say'ûn (Hadramawt region) which come from the Russian-Yemeni missions (directed by A. Sedov and H. Amirkhanov); other pieces were observed at the museum of 'Ataq (Shabwa region) which come from projects of M.-L. Inizan.

The Russian-Yemeni Mission

The Russian-Yemeni Mission to Yemen carried out a great part of its fieldwork in Hadramawt, mainly in the Wâdî Daw'an region surrounding the site of Raybûn. The prehistoric sites were discovered by Amirkhanov (Amirkhanov, 1991, 1994a, b, 1996a, b, 1997, 2006). A study of the lithics at the museum of Say'ûn that were collected by Amirkhanov, established that very few pieces had been collected. In addition all of the pieces have a strong patina which, as previously mentioned, makes the technological reading very difficult. A total of 190 of the 857 studied pieces turned out to be un-knapped natural stones. Eighteen Levallois cores from fourteen different sites were identified. Five fragments of Levallois flakes were also identified. The dominating technical scheme (10 out of 18) is one that aims to acquire unique preferential flakes through centripetal preparation (scheme A1). The Levallois point cores also represent a significant part (8 out of 18) of the Levallois production with prevalence of "classical" points (B1) or of "constructed" points (B2) schemes. The B3 and B4 schemes were not encountered in the study collection. Three other cores appear to have been recurrent centripetal Levallois debitage cores (diagram C) but their state of conservation is too poor to confirm their exact nature.

Sites from Shabwa Region

The majority of the prehistoric sites in the Shabwa region (Hayd Al-Ghalib, Wâdî Muqah) of Yemen were discovered by Inizan and Ortlieb. Certain Levallois pieces were sketched and published (Inizan and Ortlieb, 1985, 1987; Inizan, 1989). Their analysis had already differentiated certain technical schemes. Three schemes have already been identified by these pioneers of prehistoric Yemeni archaeology, namely the Levallois debitage of unique preferential flake with centripetal preparation, unipolar recurrent Levallois debitage of triangular flakes, and bipolar recurrent debitage of Levallois point and debordant flakes.

Some of these pieces were recently studied by the author at the Museum of 'Ataq. The previously described schemes correspond to those published by Inizan and consist of a search for preferential flakes using centripetal preparation (A1 scheme) or a search for "classical" Levallois points (B1 scheme). The presence of Levallois "constructed" point cores (schemes B2 to B4) is also attested. The recurrent debitage of flakes is also represented, but the low number of cores stored in the museum of 'Ataq cannot be viewed as representative of this scheme. Viewed in relation to the number of HDOR and RASA cores studied, which enabled us to have enough of a representative sample to be able to determine the dominant schemes used, the Shabwa collection is too small for such determinations.

Conclusions: Levallois Debitage in Hadramawt

It is rather clear that the technical schemes from groups A and B were employed throughout the Hadramawt region. The few Levallois pieces from other sites that were accessible for study and comparison with the HDOR and RASA assemblages indicated that there was in fact a relative homogeneity in the Hadramawt region.

The principal characteristics of this Levallois debitage in Hadramawt can be summarized by (Fig. 9): (1) the production of unique preferential Levallois flakes: recurrent debitage appears absent from observed pieces. The presence of workshops in proximity to the raw material can explain this phenomenon; (2) the prevalence of modalities that aim for

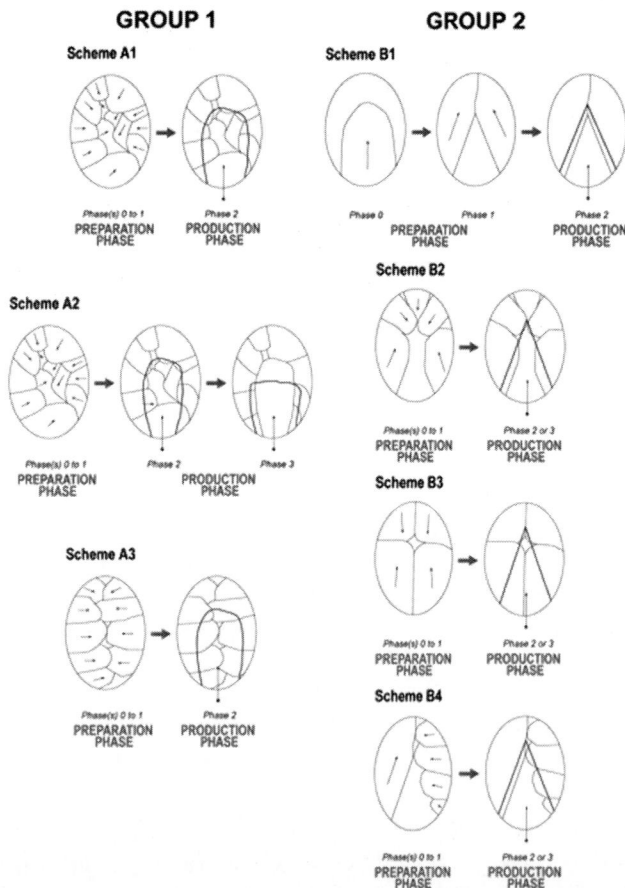

Fig. 9 Synthesis of the Levallois debitage schemes from Hadramawt

triangular flakes, if they are "classical" Levallois points or the so-called "constructed" points; and (3) a complexity in the knappers' technical behavior, especially in the production of the "constructed" points, when there are convex installations or reinstallation removals at various stages of the debitage.

What Are the Variable Methods of Levallois Debitage in Yemen?

According to Petraglia and Alsharekh (2003: 677), researchers working in Arabia have often noted that Levallois technology is not as well represented as in the Levantine Mousterian industries. The situation now appears quite different in Yemen. It is a lack of research that is at the origin of poor representation of Levallois technology in Yemen and not its absence or rarity in the archaeological record.

Very few Middle Paleolithic sites have been recorded in Yemen (save in Hadramawt). Until recently, Hadramawt industries were the only ones in South Arabia (even, in the Arabian peninsula) to have been analyzed technically. Nonetheless, the use of the Levallois concept has been documented in many regions in Yemen, including:

1. Aden region (Whalen and Pease, 1992; Whalen and Schatte, 1997): Wâdî Shahar and Wâdî Ghadin. It is worth noting that there is a presence of recurrent centripetal Levallois debitage of flakes in the Aden region (Whalen and Schatte, 1997, from Fig. 3: 3, 6, and 9. The three drawn cores are originally interpreted respectively as "polyhedron", "discoid" and "discoid").
2. Sâfer region (desert of Ramlat as-Sab'atayn): Wâdî Hirâb (Cleuziou et al., 1992: 9. Sites: HRB 7, HRB 20, HRB 21, HRB 25, HRB 26, HRB 27, HRB 30, HRB 31, HRB 33) and Wâdî Sadbâ (Cleuziou et al., 1992: 9. Sites: SDB 2, SDB 6).
3. Shabwa region: Khushm Tuhayfa in Wâdî Thib, Wâdî Muqqah and Hayd al'Ghalib (Inizan and Ortlieb, 1987; Inizan, 1989).
4. Western Hadramawt region: Wâdî Jirdân (YLNG-012 site; Crassard and Hitgen, 2006).
5. Say'ûn region (central Hadramawt): Wâdî al-Gabr (site al-Gabr 1) and Wâdî Hadjar (Amirkhanov, 1994a: 218); Wâdî bin 'Alî (Zimmerman, 2000).
6. Eastern Hadramawt region: Wâdî Wa'shah and Wâdî al-Khûn (Crassard and Bodu, 2004) region; Wâdî Sanâ and Wâdî Shumiliya (Crassard, 2004).
7. Khamis bani Saad region (Tihâma): Shi'bat Dihya sites (Macchiarelli and Peigné, 2007), including stratified site SD-1.

However, the absence of detailed technical studies for the greater majority of the discovered pieces does not allow us to establish comparisons with those of Hadramawt.

Discussion: Repercussion of the Results from Eastern Yemen

Anatomically Modern Humans' Dispersal Routes Out of Africa

Recent paleoanthropological debates have given place to discussions about the geographical origin of anatomically modern humans (AMH), and about human dispersal on Earth (e.g., Aiello, 1993; Klein, 1998; Stringer, 2000, 2002, 2003; Bräuer et al., 2004; Macaulay et al., 2005). One theory in particular considers that the ancestors of AMH originate from Africa alone, and appear between 200 and 100 ka (Cann et al., 1987; Stringer and McKie, 1996; White et al., 2003; McDougall et al., 2005). This theoretical model is commonly called the "Single Origin Model". It is based on the hypothesis that *Homo sapiens* initially appeared in a restricted zone of Africa, about 200 ka, and dispersed toward other areas of the globe, first in the Levant around 100 ka and after to Eurasia between 70 and 50 ka. *Homo sapiens* then gradually replaced ancestral species. For the supporters of this theory, the area where the first AMH would have first speciated would be in East or South Africa.

Although the paleontological, archaeological and genetic evidence is increasingly converging, and suggests an African origin for AMH, there is still disagreement regarding human dispersal routes out of the African continent. Different models of diffusion have been proposed (Kingdon, 1993; Lahr and Foley, 1998; Van Peer, 1998; Hublin, 2000; Stringer, 2000; Bar-Yosef and Belfer-Cohen, 2001; Ambrose, 2003). Nevertheless, such models remain hypothetical as there is little archaeological data to support them.

One of the proposed dispersal routes runs from Eastern Africa to the Levant, running along the Nile valley and crossing over through the Sinai (Tchernov, 1992; Bar-Yosef and Belfer-Cohen, 2001). This hypothesis holds ground as it has been substantiated by data showing a clear resemblance between Middle Paleolithic assemblages from the Nile region and from the Levant (McBurney, 1975; Clark, 1989; Van Peer, 1998).

A second human dispersal route has gained popularity in recent years. It has been proposed that humans crossed from Africa to Arabia via the strait of Bab al Mandab. This dispersal route is often referred to as the "Southern Dispersal Route" (Brandt, 1986; Nayeem, 1990; Kingdon, 1993; Lahr and Foley, 1994, 1998; Walter et al., 2000; Mithen and Reed, 2002; Ambrose, 2003; Petraglia, 2003; Rose, 2004a, b; Derricourt, 2005; Field and Lahr, 2005; Forster and Matsumura, 2005; James and Petraglia, 2005; Macaulay et al., 2005; Beyin, 2006) or the "Bab al Mandab connection" (Cleuziou, 2004: 126) but has never been confirmed due to a lack of archaeological evidence. The lithic industries

discovered by Amirkhanov in Hadramawt, which were published without a proper description of the pieces, and using purely typological labels, are regularly evoked by proponents of this theory in comparison with African industries. In addition the lithic industries discovered by Whalen in the Southern Yemeni Highlands are also used as references for African comparisons despite the fact that their analysis remains insufficient (Whalen and Pease, 1992; Whalen and Schatte, 1997).

The Levallois debitage methods and modalities that were recognized in Hadramawt (Wâdî Wa'shah and Wâdî Sanâ) thus constitute an important corpus of reference for a comparison with lithic industries from East Africa and the Levant.

What Are the Possible Comparisons with Neighboring Regions?

Since chronological data for the Levallois assemblages from Yemen do not yet exist, it is impossible to discuss affinities with industries from elsewhere. In contrast, typo-technological comparisons are justified as a means of comparison, as long as the final exploitation stages of the cores from Hadramawt (Wâdî Wa'shah and Wâdî Sanâ) fall into clear patterns. This first level of analysis, which involves the comparison of technical schemes is carried out as a technological exercise and with the full knowledge that contemporaneity is not necessarily a factor in the comparisons that may arise. Nonetheless, this does not prevent discussions of possible human dispersals and diffusions.

We propose to make a first assessment of the resemblances and differences that were observed between Middle Paleolithic assemblages from East Africa and the Levant, and the Levallois debitage characteristics from Hadramawt. The comparisons will be centered on the production modalities of Levallois points, which are more distinctive (than the "traditional" modalities of Levallois debitage of flakes) and whose characteristics indicate different debitage conceptions. This first comparative study, which is based on material from Yemen that was studied with technological accuracy, is nevertheless preliminary and will be developed in future studies.

Some Comparisons with Northeastern Africa and the "Nubian Mousterian"

In Northeast Africa, and especially in Nubia and the Nile Valley, several Levallois debitage methods have been recognized. Two principal methods were identified within the Nubian Levallois assemblages and from other areas in Northeast Africa (Guichard and Guichard, 1965; Vermeersch et al., 1990; Van Peer, 1991, 1992, 1998; Wurz et al., 2005), i.e., the "Nubian Method Type 1" and the "Nubian Method Type 2".

The Nubian Method Type 1 (Fig. 10) is well known from the Egyptian and Sudanese (especially the Lower Nile Valley) Paleolithic assemblages (Guichard and Guichard, 1965: 68–69; Van Peer, 1992: 40–41, Fig. 21/2). This technical scheme develops in this way:

1. Phase 1: Preliminary shaping of a narrow and oval core.
2. Phase 2: Removal of two long flakes from the pointed distal part of the core, with close bulb negatives. These removals create a central arris in the axis of symmetry of the core which will be used as a guiding arris for the predetermined removal.
3. Phase 3: Preparation retouches of the proximal part of the core (striking platform).
4. Phase 4: Removal of a Levallois point (predetermined triangular flake), from the proximal part of the core, which follows the central guiding arris.

This Nubian Method Type 1 is scarcely present in Hadramawt. It resembles what was observed on some cores of the bipolar preparation scheme B3, for example core HDOR 566-1 (Crassard, 2007: vol. 2, p. 7) and, more convincingly, core HDOR 571-1 (Crassard, 2007: vol. 2, p. 8).

The Nubian Method Type 2 (Fig. 10) is a Levallois debitage method that was also recognized in the Lower Nile Valley (Northern Sudan and Southern Egypt; Guichard and Guichard, 1965: 69; Van Peer, 1992: 41, Fig. 21/1). The shaping of the core resembles the preparation involved in the Nubian Method 1. The characteristic preparation of the Nubian method 2 takes place with the removal of flakes from lateral and distal segments of the core. This method can resemble a modality seeking the production of non-triangular preferential flakes. However, the preparation regularly creates a central guiding arris in the longitudinal symmetry axis of the core. The obtainment of a point is thus the final objective of this type of debitage.

The Nubian Method Type 2 is closely associated to a modality of debitage seeking "constructed points", found particularly in the scheme B2 and maybe B4 from Hadramawt.

In Nubia, the presence of "classical" Levallois point cores is mentioned (non Nubian methods; Guichard and Guichard, 1965: 85–86) and suggests a resemblance with the B1 scheme ("classical" points) described in Hadramawt. The preferential flake cores are also mentioned, and are comparable to the Hadramawt C group.

A "Nubian Mousterian" cultural group was proposed and was divided into the rather well-defined groups N and K (Van Peer, 1991: 111). These groups are differentiated on the basis of the Levallois methods involved in each. In the group N assemblages, the Nubian methods and the Levallois "classical" debitage of preferential flakes, are associated. Contrarily to this, the group K assemblages are not made using the Nubian method. The tools associated with these groups are very rare in Nubia.

There are similar components in Hadramawt that sometimes include the joint presence of schemes from groups A or B with that of scheme C and which correspond to the description of the Nubian Mousterian N group.

Some Comparisons with the Near-East and the Levantine Mousterian

Levallois modalities were studied on many mostly stratified sites in the Levant. They are associated with the Middle Paleolithic ("Levantine Mousterian"; Jelinek, 1982; Marks, 1992).

Close to Mount Carmel in Israel, the site of Kebara delivered an important corpus of Levallois material dating to between 60 and 48 ka. This assemblage was used to understand the technical variability of the Mousterian industries and the technical behavior of the Neanderthals in the Near-East (Meignen and Bar-Yosef, 1990). At the site of Kebara the Levallois debitage is present in all of the archaeological levels and the production of points, using predetermined convergent unipolar removals, dominates the assemblages. The obtained preferential products are especially short points with wide bases. This type of debitage in particular, presents a lesser degree of preparation of the debitage and dorsal

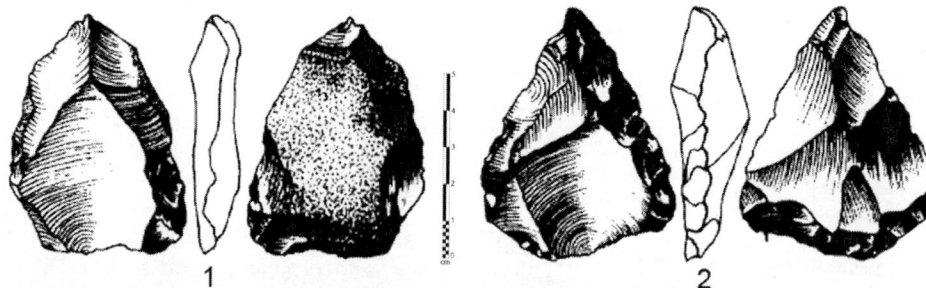

Fig. 10 Two cores showing the Nubian Method 1 (*left*) and 2 (*right*), after Van Peer (1992: Fig. 21/2)

surfaces, which indicates good control of the knapping operations.

These dominant characteristics are found on other sites in the Levant. In Tabun (Copeland, 1975; Jelinek, 1981), about fifteen kilometers North of Kebara, the convergent unipolar preparation modality is similar in layers B and D, with an even larger proportion of points than at Kebara. The presence of Levallois debitage of flakes with centripetal preparation in contemporary layers where Levallois debitage also aims towards the production of points, does not allow us to demonstrate the presence of a linear evolution for Mousterian industries. Given this, it becomes impossible to use the study of lithic industries as a dating criterion.

This modality of convergent unipolar production of Levallois points appears to be a characteristic element of the Near-East in certain periods (Meignen, 1995), but which remains less present in Northeast Africa. Crew (1975) highlighted a dominance of preparation/exploitation operations from the proximal zone of the cores in the Levant, which was clearly distinguished from the more diversified preparation/exploitation schemes in Northeast Africa (mainly Libya). Among these African schemes, removals coming from the lateral segments were more frequent.

This Levantine production of "classical" points is very similar to the B1 scheme identified in Hadramawt.

The convergent unipolar Levallois modality is also found in some layers of the Paleolithic site of Umm el-Tlel, in Syria, and in particular in the Mousterian layer VI3 b', dated to 65–50 ka (Boëda and Muhesen, 1993: 55–56, Figs. 19–20). At Umm el-Tlel, this scheme is called 'scheme A' and belongs to the set of points "à trois coups" ("with three hits"). 'Scheme B' (orthogonal preparation) and 'F' (on the ventral face of a flake) are also found in the same archaeological level at this site.

Other modalities from level 'VI3 b' at Umm el-Tlel consist of the production of "constructed" points (Fig. 11; Boëda et al., 1998: 249–250, Fig. 9). 'Scheme C' from this site (with lateral preparation) resembles scheme B2 from Hadramawt though it is not strictly identical. Scheme B2 from Hadramawt is based on preliminary convergent unipolar

removals. Lateral removals are a later addition and are meant to reinstall convexities and the central guiding arris.

'Scheme E' from Umm el-Tlel (with bipolar preparation) also resembles the debitage concept of schemes B2 and B3 from Hadramawt. Once again, the Hadramawt schemes differ slightly from those proposed by Boëda, but are part of the same knapping concept which seeks the production of points.

Finally, 'scheme B' from Umm el-Tlel (with orthogonal preparation) can be compared to scheme B4 ("constructed points") from Hadramawt. The first ('B') presents a preparation "à trois coups", whereas the second (B4) is less strict: several removals are made from the lateral part rather than only one. To conclude, no example of the schemes D and F from Umm el-Tlel were recognized in Hadramawt.

Besides the Levallois debitage of "classical" points, not a single Umm el-Tlel scheme perfectly matches those of the assemblages from Hadramawt. Nevertheless, it is possible to highlight resemblances between the various debitage modalities of "constructed" points. These resemblances provide a common objective (point production), achieved by "alternative" modalities, in comparison to the strict modality of convergent unipolar preparation. These types of modalities also exist elsewhere in the Near-East, particularly in Kebara, but are not dominant (L. Meignen, pers. comm., 2006).

Conclusions

The techniques from the Arabian Middle Paleolithic remain largely unknown and mostly ignored by scholars. Nevertheless, it would seem, thanks to the studies of Hadramawt's Levallois industries, that there are technical similarities with some Mousterian industries from the Levant. Also, from our first comparisons, no conclusive archaeological affinities between East Africa and Arabia sustain the Southern dispersal route model. Whether these resemblances with the Levant material are evidences of a specific link, and during which period(s), remains unknown. Where do the original

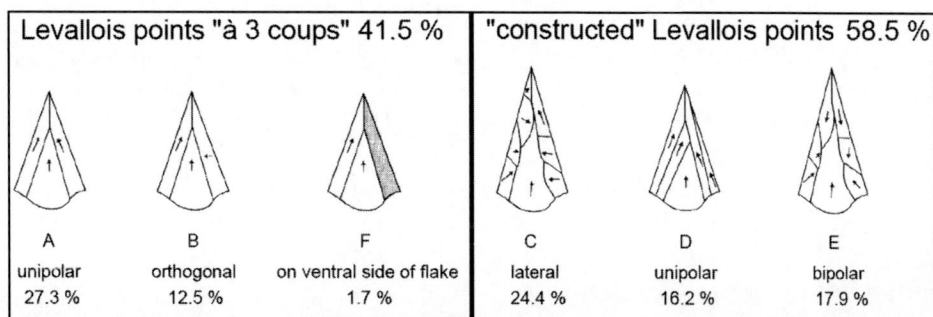

Fig. 11   Alternate flaking processes for the production of Levallois points from Umm el-Tlel, after Boëda et al. (1998: Fig. 9)

technical traditions come from? Did the Levant have a primary influence on Arabia, or is the contrary possible? These questions emphasize the enormous potential represented by South Arabia, in the still young studies concerning the peopling of Asia and the hominin expansion out of Africa.

Until now, no conclusion can be definitively proposed, as no Levallois site has been chronometrically dated in South Arabia (see Marks, 2009). Nevertheless, the existence of technological affinities is undeniable with some neighboring areas of Hadramawt where Levallois assemblages are found. It appears anyway that resemblances are more convincing with the Levant, considering the "constructed" debitage conception or the knapper's qualitative choices. If the chronological data could confirm a dating to ca. 50 ka for the Levallois industries from South Arabia, it would be then necessary to reconsider with attention the role of this area in the occupation and peopling modalities by Middle Paleolithic human groups. Relations, whose character remains to be defined, with the Levantine Mousterian would be then more probable than with an African Middle Stone Age (MSA) or Nubian Mousterian.

The variability of the Levallois debitage modalities we observe in South Arabia could then be explained by local evolution during the Upper Pleistocene, due to a possible situation of cultural isolation. A population, locally embedded, could have developed specific technological habits in the region. This is, for instance, something that happened in the Levantine Middle Paleolithic and in the Late MSA of Eastern Africa during the Oxygen Isotope Stages 5 and 4. A similar situation is largely attested during the Early/Mid-Holocene period (seventh to fifth million BC) in Yemen, like the existence of innovative and unique technical systems in the Old World such as fluting (Crassard et al., 2006; Crassard, 2007). This hypothesis of some regional locally emerged specificities developed as early as the Middle Paleolithic fits quite well with our preliminary analysis and the neighboring data. Later on, Southern Arabia seems to have been in a cultural isolation as suggested by a nearly absence of Upper Paleolithic phenomenon, correlated with a possible Middle Paleolithic-like complex until the first Holocene industries. Before us, Whalen (Whalen et al., 1981) had proposed to see in the South Arabian Paleolithic assemblages an endemic technological development, incorporating some technical and stylistic traditions which would have allowed optimizing human adaptation to the environmental conditions.

Furthermore, the absence of an Upper Paleolithic in Arabia is an additional problem to the regional prehistory definition. This period is quite simply unknown; perhaps even non-existent in Yemen's chronology and in its closest regions. The Upper Paleolithic problem, even if some insights seem to have been discovered in Eastern Yemen (Amirkhanov, 2006), is particularly important when one can see a possible technical continuity of the Levallois debitage all along the

Upper Pleistocene, until the "explosion" of the sophisticated industries during Early/Mid-Holocene (Crassard et al., 2006; Crassard, 2008), which means the absence of a clear long technical transition, the absence thus of an Upper Paleolithic. This remains to be proven, and it is another story.

Acknowledgments I gratefully acknowledge the assistance provided by Dr Abdullah Bawazir, President of the General Organization for Antiquities and Museums of the Republic of Yemen and his able staff, particularly Abdul Rahman Assaqaf, Hussein Alaydarus and Khairan Alzubaidi. I thank the RASA Projects directors: Joy McCorriston, Erich Oches and Abdulaziz bin Aqil, and the French Archaeological Mission in Jawf-Hadramawt directors: Michel Mouton, Frank Braemer, and Anne Benoist. For their comments on previous drafts and general discussion about South Arabian Prehistory, I gratefully acknowledge the help of Hizri Amirkhanov, Pierre Bodu, Anne Delagnes, Marie-Louise Inizan, Jacques Jaubert, Roberto Macchiarelli, Liliane Meignen, Jeffrey Rose and especially Michael Petraglia. I owe special thanks to Lamya Khalidi and Michael Haslam for comments on a draft of this chapter. I thank the *Fondation Fyssen* for providing financial support for my stay in Cambridge, and I thank the LCHES staff for their warm hospitality.

References

Aiello LC. The fossil evidence for modern human origins in Africa: a revised view. American Anthropologist. 1993;95:73–96.

Ambrose S. Population bottleneck. In: Robinson R, editor. Genetics. Farmington Hills: Macmillan References; 2003. p. 167–71.

Amirkhanov HA. Palaeolithic in South Arabia. Moscow (in Russian): Scientific World; 1991.

Amirkhanov HA. Research on the Palaeolithic and Neolithic of Hadramaut and Mahra. Arabian Archaeology and Epigraphy. 1994a;5:217–28.

Amirkhanov HA. Recherches sur le néolithique dans le Hadramaout en 1991. Chroniques Yéménites. 1994b;1994:54–5.

Amirkhanov HA. Notes on the stone tools from Raybun I settlement. In: Sedov A, Griaznevich P, editors. Raybun settlement (1983–1987 excavations). Moscow: Vostochnaya Litteratura; 1996a. p. 31–4.

Amirkhanov HA. Bilinear parallelism in the Arabian Early Neolithic. In: Afanasev G, Cleuziou S, Lukacs R, Tosi M, editors. The prehistory of Asia and Oceania, XIIIth International Congress of Prehistoric and Protohistoric Sciences. Abaco, Forli: Forli; 1996b. p. 135–9.

Amirkhanov HA. The Neolithic and PostNeolithic of the Hadramaut and Mahra. Moscow (in Russian): Scientific World; 1997.

Amirkhanov HA. Stone Age of South Arabia. Moscow (in Russian): Nauka; 2006.

Bar-Yosef O, Belfer-Cohen A. From Africa to Eurasia–early dispersals. Quaternary International. 2001;75:19–28.

Beyin A. The Bab al Mandab vs. the Nile-Levant: an appraisal of the two dispersal routes for early modern humans out of Africa. African Archaeological Review. 2006;23:5–30.

Boëda E. Le concept Levallois: variabilité des méthodes, Monographie du CRA 9. Paris: CNRS Editions; 1994.

Boëda E, Muhesen S. Umm el Tlel (El Kowm, Syrie): étude préliminaire des industries lithiques du Paléolithique moyen et supérieur 1991–1992. Cahiers de l'Euphrate (Editions Recherche sur les Civilisations, Paris). 1993;7:47–89.

Boëda E, Bourguignon L, Griggo C. Activités de subsistance au Paléolithique moyen: couche VI3 b' du gisement d'Umm El Tlel (Syrie). In: Brugal JP, Meignen L, Patou-Mathis M, editors. Economie

Préhistorique: Les Stratégies de Subsistance au Paléolithique. Sophia-Antipolis: XVIII Rencontres Internationales d'Histoire et d'Archéologie d'Antibes, Editions APDCA; 1998.

Brandt SA. The Upper Pleistocene and Early Holocene prehistory of the Horn of Africa. African Archaeological Review. 1986;4:41–82.

Bräuer G, Collard M, Stringer C. On the reliability of recent tests of the Out of Africa hypothesis for modern human origins. Anatomical Record. 2004;279A:701–7.

Burns SJ, Fleitmann D, Matter A, Kramers J, Al-Subbary A. Indian ocean climate and an absolute chronology over Dansgaard/Oeschger events 9 to 13. Science. 2003;301:1365–7.

Cann RL, Stoneking M, Wilson AC. Mitochondrial DNA and human evolution. Nature. 1987;325:31–6.

Caton-Thompson G. Geology and archaeology of the Hadramaut, Southern Arabia. Nature. 1938;142:139–42.

Caton-Thompson G. Some Palaeoliths from South Arabia. Proceedings of the Prehistoric Society. 1953;19:189–218.

Clark JD. The origin and spread of modern humans: a broad perspective on the African evidence. In: Mellars PA, Stringer C, editors. The human revolution: behavioural and biological perspectives on the origins of modern humans. Edinburgh: Edinburgh University Press; 1989. p. 565–88.

Cleuziou S. Pourquoi si tard? Nous avons pris un autre chemin. L'Arabie des chasseurs-cueilleurs de l'Holocène au début de l'Age du Bronze. In: Guilaine J, editor. Aux Marges des Grands Foyers du Néolithique, Périphéries Débitrices ou Créatrices? Paris: Errance Editions; 2004. p. 123–48.

Cleuziou S, Inizan M-L, Marcolongo B. Le peuplement pré- et proto-historique du système fluviatile fossile du Jawf-Hadramawt au Yémen (d'après l'interprétation d'images satellite, de photographies aériennes et de prospections). Paléorient. 1992;18(2):5–28.

Copeland L. The Middle and Upper Palaeolithic of Lebanon and Syria in the light of recent research. In: Wendorf F, Marks A, editors. Problems in prehistory: North Africa and the Levant. Dallas: SMU Press; 1975. p. 317–50.

Crassard R. Prehistory in Wadi Sana: the lithic industries, Roots of Agriculture in South Arabia Preliminary Report 2004, Unpublished preliminary report; 2004.

Crassard R. Apport de la technologie lithique à la définition de la préhistoire du Hadramawt, dans le contexte du Yémen et de l'Arabie du Sud. Ph.D. dissertation, Université Paris 1, Panthéon-Sorbonne; 2007.

Crassard R, Bodu P. Préhistoire du Hadramawt (Yémen): nouvelles perspectives. Proceedings of the Seminar for Arabian Studies. 2004;34:67–84.

Crassard R, Hitgen H. Final report: "Bronze Age Tombs" Project (YLNG-BAT 2006); Archaeological excavations. Unpublished internal report Yemen LNG, CEFAS/DAI, Sana'a (English and Arabic versions); 2006.

Crassard R, McCorriston J, Oches E, bin Aqil A, Espagne J, Sinnah M. Manayzah, early to mid-Holocene occupations in Wâdî Sanâ. Proceedings of the Seminar for Arabian Studies. 2006;36:151–73.

Crew H. An examination of the vatiability of the Levallois methods: its implication for the internal and external relationships of the Levantine Mousterian. Ph.D. dissertation, University of Columbia, Columbia; 1975.

Derricourt R. Getting "Out of Africa": sea crossings, land crossings and culture in the hominin migrations. Journal of World Prehistory. 2005;19:119–32.

Field JS, Lahr MM. Assessment of the southern dispersal: GIS-based analyses of potential routes at oxygen isotopic stage 4. Journal of World Prehistory. 2005;19:1–45.

Forster P, Matsumura S. Did early humans go north or south? Science. 2005;308:965–6.

Guichard J, Guichard G. The Early and Middle Palaeolithic of Nubia: a preliminary report. In: Wendorf F, editor. Contributions to the prehistory of Nubia. Dallas: SMU Press; 1965. p. 57–116.

Hublin J-J. Modern–nonmodern hominid interactions: a Mediterranean perspective. In: Bar-Yosef O, Pilbeam D, editors. The geography of Neanderthals and modern humans in Europe and the Greater Mediterranean, Bulletin 8. Cambridge, MA: Peabody Museum of Archaeology and Ethnology; 2000. p. 157–82.

Inizan M-L. Premiers éléments de Préhistoire dans la région de Shabwa (R.D.P. du Yémen). Raydan. 1989;5:71–7.

Inizan M-L, Ortlieb L. Yémen, R.C.P. 743, Préhistoire au Moyen-Orient, évolution des industries dans leur contexte paléo-écologique. Unpublished CNRS Activity Report; 1985. p. 9–21.

Inizan M-L, Ortlieb L. Préhistoire dans la region de Shabwa au Yémen du sud (R. D. P. Yémen). Paléorient 1987;13(1):5–22.

James HVA, Petraglia M. Modern human origins and the evolution of behavior in the Later Pleistocene record of South Asia. Current Anthropology. 2005;46(suppl):3–27.

Jelinek AJ. The Middle Palaeolithic in the Southern Levant from the perspective of the Tabun cave. In: Cauvin J, Sanlaville P, editors. Préhistoire du Levant. Paris: CNRS; 1981. p. 265–80.

Jelinek AJ. The Tabun Cave and Palaeolithic Man in the Levant. Science. 1982;216:1369–75.

Keita L. The "Africa and the rest of the world evolutionary hypotheses": an exercise in scientific epistemology. African Archaeological Review. 2004;21(1):1–6.

Kingdon J. Self-made man: human evolution from Eden to extinction. New York: Wiley; 1993.

Klein RG. Why anatomically modern people did not disperse from Africa 100,000 years ago. In: Akazawa T, Aoki K, Bar-Yosef O, editors. Neanderthals and modern humans in Western Asia. New York: Plenum; 1998. p. 509–21.

Lahr MM, Foley R. Multiple dispersals and modern human origins. Evolutionary Anthropology. 1994;3:48–60.

Lahr MM, Foley R. Towards a theory of modern human origins: geography, demography, and diversity in recent human evolution. Yearbook of Physical Anthropology. 1998;41:137–76.

Macaulay V, Hill C, Achilli A, Rengo C, Clarke D, Meehan W, et al. Single, rapid coastal settlement of Asia revealed by analysis of complete mitochondrial genomes. Science. 2005;308:1034–6.

Macchiarelli R, Peigné S. Le premier peuplement humain de l'Arabie méridionale: la perspective Tihama (Yémen). Bulletins et Mémoires de la Société d'Anthropologie de Paris, n.s. 2007;18:15–6.

Marks A. Upper Pleistocene archaeology and the origins of modern man: a view from the Levant and adjacent areas. In: Akazawa T, Aoki K, Kimura T, editors. The evolution and dispersal of modern humans in Asia. Tokyo: Hokusen-Sha; 1992. p. 229–51.

Marks AE. The Paleolithic of Arabia in an inter-regional context. In: Petraglia MD, Rose JI, editors. The evolution of human populations in Arabia: paleoenvironments, prehistory and genetics. The Netherlands: Springer; 2009. p. 295–308.

McBurney C. Current status of the Lower and Middle Paleolithic of the entire region from the Levant through North Africa. In: Wendorf F, Marks A, editors. Problems in prehistory: North Africa and the Levant. Dallas: SMU Press; 1975. p. 411–26.

McDougall I, Brown F, Fleagle J. Stratigraphic placement and age of modern humans from Kibish, Ethiopia. Nature. 2005;433(17): 733–6.

Meignen L. Levallois lithic production systems in the Middle Paleolithic of the Near East: the case of the unidirectional method. In: Dibble H, Bar-Yosef O, editors. The definition and interpretation of Levallois technology. Monographs in World Archaeology 23. Madison, WI: Prehistory Press; 1995. p. 361–79.

Meignen L, Bar-Yosef O. Kebara et le Paléolithique Moyen du Mont Carmel. Paléorient. 1990;14(2):123–30.

Mithen S, Reed M. Stepping out: a computer simulation of hominid dispersal from Africa. Journal of Human Evolution. 2002;43:433–62.

Nayeem MA. Prehistory and protohistory of the Arabian peninsula, vol.1: Saudi Arabia. Hyderabad: Hyderabad Publishers; 1990.

Petraglia MD. The Lower Paleolithic of the Arabian peninsula: occupations, adaptations, and dispersals. Journal of World Prehistory. 2003;17(2):141–79.

Petraglia MD. Mind the gap: factoring the Arabian peninsula and the Indian subcontinent into Out of Africa models. In: Mellars P, Boyle K, Bar-Yosef O, Stringer C, editors. Rethinking the human revolution. Cambridge: McDonald Institute Monographs; 2007. p. 383–94.

Petraglia MD, Alsharekh A. The Middle Palaeolithic of Arabia: implications for modern human origins, behaviour and dispersals. Antiquity. 2003;77:671–84.

Rose JI. The question of Upper Pleistocene connections between East Africa and South Arabia. Current Anthropology. 2004a;45(4):551–5.

Rose JI. New evidence for the expansion of an Upper Pleistocene population out of East Africa, from the site of Station One, northern Sudan. Cambridge Archaeological Journal. 2004b;14(2):205–16.

Sanlaville P. Le Moyen-Orient arabe: le milieu et l'homme. Paris: Armand Colin; 2000.

Stringer C. Coasting out of Africa. Nature. 2000;405:24–7.

Stringer C. Modern human origins: progress and prospects. Philosophical Transactions of the Royal Society London (B). 2002;357:563–79.

Stringer C. Human evolution: out of Ethiopia. Nature. 2003;423:692–5.

Stringer C, McKie R. African Exodus: the origins of modern humanity. London: Pimlico Press; 1996.

Tchernov E. Biochronology, paleoecology, and dispersal events of hominids in the southern Levant. In: Akazawa T, Aoki K, Kimura T, editors. The evolution and dispersal of modern humans in Asia. Tokyo: Hokusen-sha; 1992. p. 149–88.

Tuffreau A. L'Acheuléen de l'Homo erectus à l'Homme de Neandertal. Paris: La Maison des Roches Editeur; 2004.

Van Beek GW, Cole GH, Jamme Aluf GH. An archaeological reconnaissance in Hadramaut: a preliminary report. Annual Report of the Smithsonian Institute; 1963.

Van Peer P. Interassemblage variability and Levallois styles: the case of the northern African Middle Paleolithic. Journal of Anthropological Archaeology. 1991;10:107–51.

Van Peer P. The Levallois reduction strategy. Monographs in World Archaeology 13. Madison, WI: Prehistory Press; 1992.

Van Peer P. The Nile corridor and the Out-of-Africa model: an examination of the archaeological record. Current Anthropology. 1998;39:115–40.

Vermeersch PM, Paulissen E, Van Peer P. Le Paléolithique de la vallée du Nil égyptien. L'Anthropologie. 1990;94:435–58.

Walter RC, Buffler RT, Bruggemann JH, Guillaume MMM, Berhe SM, Negassi B, et al. Early human occupation of the Red Sea coast of Eritrea during the Last Interglacial. Nature. 2000;405:65–9.

Whalen NM, Pease DW. Archaeological survey in southwest Yemen. Paléorient. 1992;17(2):127–31.

Whalen NM, Schatte KE. Pleistocene sites in southern Yemen. Arabian Archaeology and Epigraphy. 1997;8:1–10.

Whalen NM, Killick A, James N, Morsi G, Kamal M. Saudi Arabian archaeological reconnaissance 1980, preliminary report on the Western Province survey. Atlal. 1981;5:43–58.

White TD, Asfaw B, DeGusta D, Gilbert H, Richards G, Suwa G, et al. Pleistocene Homo sapiens from Middle Awash, Ethiopia. Nature. 2003;423:742–7.

Wurz S, van Peer P, le Roux N, Gardner S, Deacon HJ. Continental patterns in stone tools: a technological and biplot-based comparison of early Late Pleistocene assemblages from northern and southern Africa. African Archaeological Review. 2005;22(1):1–24.

Zarins J. View from the south: the Greater Arabian peninsula. In: Henry D, editor. Prehistoric archaeology of Jordan. Oxford: BAR International Series 705; 1998. p. 179–94.

Zimmerman P. Middle Hadramawt archaeological survey: preliminary results, October 1999 season. Yemen Update. 2000;42:27–9.

Chapter 13
The "Upper Paleolithic" of South Arabia

Jeffrey I. Rose and Vitaly I. Usik

Keywords Demography • Dhofar • Paleoclimate • South Arabia

Introduction

The practice of assigning names to archaeological periods in Arabia is inherently problematic. Just as the Arabian subcontinent is the geographic bridge between Africa and Eurasia; similarly, it is wedged between the bifurcation of Eurasian and African taxonomic schema. This distinction represents separate evolutionary trajectories as expressed in the development of regional lithic technologies. For instance, if we refer to the Arabian "Middle Paleolithic" (MP), we are using a Eurasian name and insinuating closer affinities to this part of the world between 250 and 40 ka, whereas the Arabian "Middle Stone Age" (MSA) presumes a connection to sub-Saharan Africa during a similar interval. This distinction is critical for evaluating the origin and expansion of early modern humans, which predicts linked stone tool technologies on either side of the Red Sea during the Middle and/or Late Stone Age (LSA).

Hence, our use of the term Upper Paleolithic (UP) in reference to South Arabia is no accident. It is a deliberate attempt to highlight closer archaeological affinities with lithic industries found in North Africa and Southwest Asia, rather than sub-Saharan Africa. Indeed, a similar connection has already been made based upon Middle and Upper Paleolithic discoveries in Yemen (Delagnes et al., 2008; Crassard, 2009) and the United Arab Emirates (Marks, 2009). For the purposes of this chapter, "Upper Paleolithic" should be considered an archaeological phase, however, since there is so little evidence from this period in Arabia, we cannot presume a temporal range. The apparently wide range of blade technologies in South Arabia (Amirkhanov, 1994, 2006; Delagnes et al., 2008; Crassard, 2009; Marks, 2009) suggests a long-term tradition of linked laminar[1] technologies that spans at least MIS 4 through early MIS 1 (~75–8 ka).

The new data presented in this chapter comes from archaeological fieldwork conducted by the Central Oman Pleistocene Research (COPR) from 2002 to 2008. We include al-Hatab Rockshelter, an Arabian UP site with AMS and OSL ages placing it within the Terminal Pleistocene and Early Holocene, Ras Aïn Noor, an Arabian UP site buried in aeolian sands at the edge of an ancient spring, as well as a surface scatter sampled from Dhanaqr, situated on a rock outcrop overlooking the confluence of two drainage systems in the eastern Nejd Plateau (Rose, 2006).

Using observations from lithic assemblages collected at these three sites, as well as other reported occurrences from southern Arabia with similar technological features (e.g., Amirkhanov, 1994, 2006; Delagnes et al., 2008) we begin to define and articulate relevant features of the South Arabian UP. Broad technological trends are examined within the framework of the genetic and paleoenvironmental records. It is concluded that the current body of evidence does *not* support an 'Out of Africa' scenario via the Bab al Mandab Strait from MIS 4 onward.

J.I. Rose (✉)
Department of Anthropology and Geography,
Oxford Brookes University, Oxford, OX3 0BP, UK
e-mail: jeffrey.i.rose@gmail.com

V.I. Usik
Institute of Archaeology, National Academy of Science,
B. Khmelnitsky Street 15, 01030, Kiev-30, Ukraine
e-mail: vitaly_usik@yahoo.com

---

[1] For the purposes of this paper, we define "laminar" as a simple, unidirectional mode of core reduction utilizing one or more working surfaces, with unidirectional-convergent or unidirectional-parallel flakes often removed from an elongated longitudinal axis of the core. This is not necessarily a true prismatic blade technology in the sense of volumetric cores, crested blade production, core maintenance, rejuvenation, etc.

M.D. Petraglia and J.I. Rose (eds.), *The Evolution of Human Populations in Arabia*, Vertebrate Paleobiology and Paleoanthropology, DOI 10.1007/978-90-481-2719-1_13, © Springer Science+Business Media B.V. 2009

The Arabian Paleoclimate During the Latter Half of the Upper Pleistocene

There are meager climatic data from MIS 4 and early MIS 3 in southern Arabia. Indirect evidence can be gleaned from composite signals expressed in a summed probability curve (Parker and Rose, 2008) as well as the index of Indian Ocean Monsoon activity (Fleitmann et al., 2007), which suggest this period was characterized by increasingly hyperarid conditions throughout the interior culminating around 70 ka, followed by a return to a more humid regime by 50 ka (Fig. 1). Evidence for MIS 4 aridification is also inferred from geological profiles in the Rub' al Khali, which attest to a stage of aeolian deposition immediately below the MIS 3 lake marls. While most of central and southern Arabia was probably uninhabitable, bathymetric and hydrographical data suggest certain areas along the emerged coastal plain were ameliorated around this time (Bailey et al., 2007; Parker and Rose, 2008).

Geologists working in the Rub' al Khali sand sea have uncovered evidence of a landscape that was once marked by a network of rivers and small lakes (Fig. 2) spread across the interior (McClure, 1984). Radiocarbon measurements on freshwater mollusk shells and marls indicate the lakes reached their highest levels sometime prior to 37 ka (McClure, 1976, 1978). These playas ranged from ephemeral puddles to pools up to ten meters deep, and numbered well over a thousand. They are primarily distributed along an east–west axis across the centre of the Rub' al Khali basin, covering a distance of some 1,200 km (McClure, 1984). Similar lake

basins have been reported in the an-Nafud in northern Arabia (Garrard and Harvey, 1981; Schultz and Whitney, 1986).

In addition to interior paleolakes, other signals of the MIS 3 wet-phase include depositional terraces in the Wadi Dhaid, UAE; their stratigraphic position suggests an age between 35 and 22 ka (Sanlaville, 1992). Interdunal lake deposits (called *shuquq* in Arabic) recorded in the Liwa region of the UAE produced 31 OSL and C14 dates that cluster between 46.5 and 21.5 ka (Wood and Imes, 1995; Juyal et al., 1998; Glennie and Singhvi, 2002). Paleosols were recorded in the ad-Dahna desert, which are interstratified between MIS 4 and MIS 2 aeolian deposits (Anton, 1984). Clark and Fontes (1990) dated calcite formations from ancient hyperalkaline springs in northern Oman, producing radiocarbon ages between approximately 33 and 19 ka. Two soil horizons clustering around 26 and 19 ka were discovered around the central plateau of the Yemeni highlands, characterized as molissols – soils that form on landscapes covered by savannah vegetation (Brinkmann and Ghaleb, 1997).

The MIS 2 hyperarid phase was more extreme than the peninsula had experienced since the Penultimate Glaciation, if not earlier (Anton, 1984). Ages obtained from dune formations in the Rub' al Khali (McClure, 1984; Goudie et al., 2000; Parker and Goudie, 2007), an-Nafud (Anton, 1984), and the Wahiba Sands (Gardner, 1988; Glennie and Singhvi, 2002) all signal a major phase of aeolian accumulation between 17 and 9 ka. Calcite fractures in northern Oman corroborate the evidence for increasing aridity, indicating there was considerably less moisture in the environment starting around 19 ka (Clark and Fontes, 1990). Sometime around 13,500 years ago this period of environmental desiccation

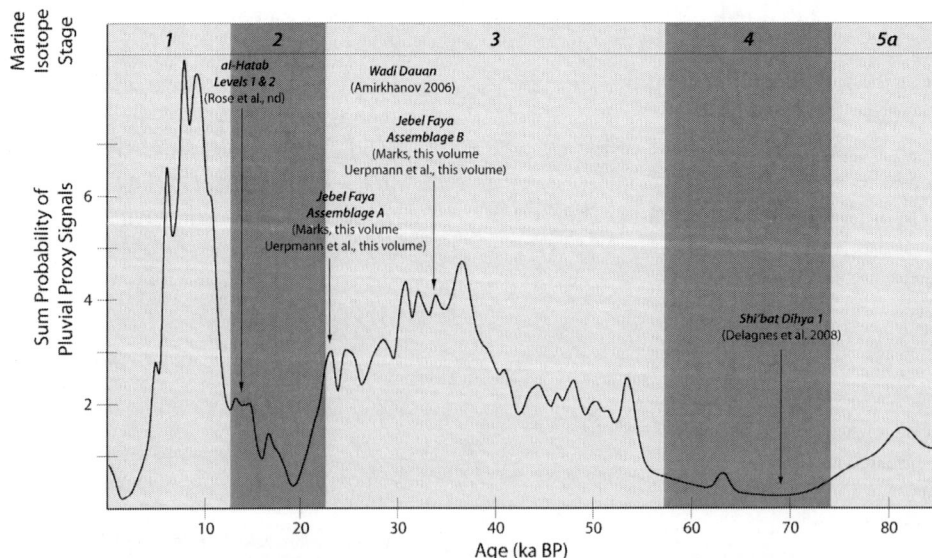

Fig. 1 Arabian paleoenvironmental curve adapted from Parker and Rose (2008, Fig. 4, pp. 31) displaying summed probability curve of pluvial proxy signals from MIS 5a–MIS 1. Dated Upper Pleistocene archaeological sites are also depicted to show their general chronological position in relation to paleoclimatic conditions

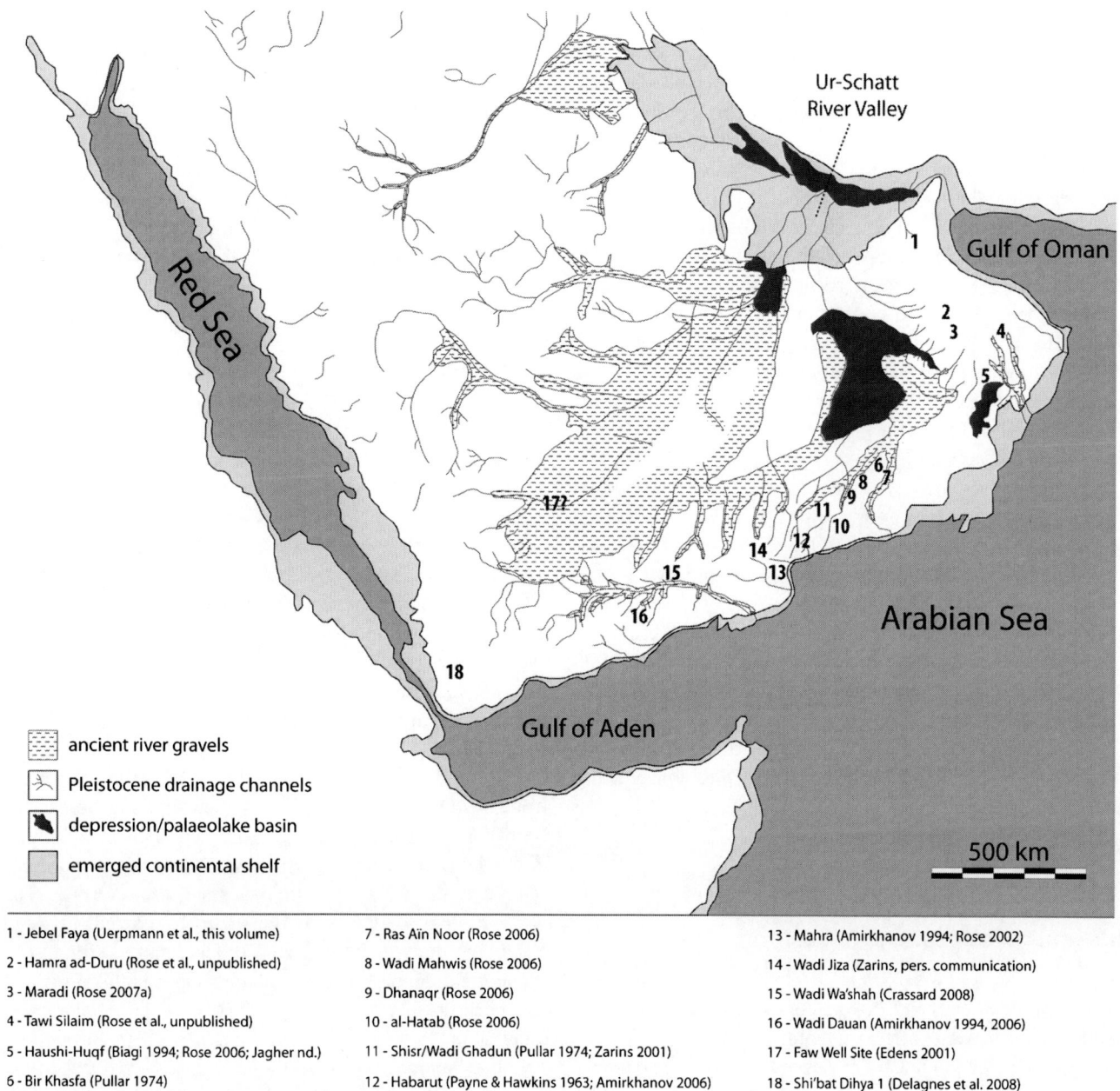

Fig. 2 Map of ancient drainage channels, alluvial deposits, and paleolake basins throughout the Arabian peninsula. The boundaries of the continental shelf indicate the extent of Arabia during periods of reduced sea levels (roughly between 75 and 8 ka). Sites mentioned in this chapter are also shown

1 - Jebel Faya (Uerpmann et al., this volume)

2 - Hamra ad-Duru (Rose et al., unpublished)

3 - Maradi (Rose 2007a)

4 - Tawi Silaim (Rose et al., unpublished)

5 - Haushi-Huqf (Biagi 1994; Rose 2006; Jagher nd.)

6 - Bir Khasfa (Pullar 1974)

7 - Ras Aïn Noor (Rose 2006)

8 - Wadi Mahwis (Rose 2006)

9 - Dhanaqr (Rose 2006)

10 - al-Hatab (Rose 2006)

11 - Shisr/Wadi Ghadun (Pullar 1974; Zarins 2001)

12 - Habarut (Payne & Hawkins 1963; Amirkhanov 2006)

13 - Mahra (Amirkhanov 1994; Rose 2002)

14 - Wadi Jiza (Zarins, pers. communication)

15 - Wadi Wa'shah (Crassard 2008)

16 - Wadi Dauan (Amirkhanov 1994, 2006)

17 - Faw Well Site (Edens 2001)

18 - Shi'bat Dihya 1 (Delagnes et al. 2008)

came to an end, as the Indian Ocean Monsoon again picked up in strength and again deposited rainfall across southern Arabia (Overpeck et al., 1996; Ivanochko et al., 2005).

The transformation of the South Arabian landscape throughout the latter half of the Upper Pleistocene had a profound effect upon the submerged continental shelf. Taking into account the shallow bathymetry of the Persian Gulf (Lambeck, 1996) and Red Sea basins (Bailey et al., 2007), nearly half a million square kilometers of contiguous land were repeatedly submerged and exposed by glacio-eustatic cycles of marine transgression and regression. The emergence

of the continental shelf around Arabia probably had direct implications for prehistoric occupation, since the exposed landmass provided abundant sources of freshwater juxtaposed to a severely desiccated landscape.

Faure et al. (2002) describe the formation of littoral freshwater upwelling they refer to as "coastal oases," highlighting the importance of such habitats for early humans groups. Depressed sea levels cause an increase of hydrostatic pressure on submarine rivers; consequently, greater amounts of freshwater flow through these aquifers. Eventually, this process leads to the creation of springs in favorable loci on the emerged shelf with lithology and topography conducive to upwelling. One extreme example of this phenomenon is the submerged seeps at the bottom of the Persian Gulf. The area around modern Qatar is the terminus of several submarine rivers that flow eastward beneath Arabia, creating a mass of upwelling in plumes scattered throughout the eroded karstic sea bed lining the Gulf basin (Church, 1996).

Throughout most of the Upper Pleistocene and Early Holocene, a considerable amount of runoff in southwest Asia was funneled into the Gulf basin via submarine aquifers flowing beneath Arabia, the Karun drainage network originating in the Zagros Mountains, and the Tigris and Euphrates Rivers flowing from the Anatolian Plateau. All of these systems converged in the centre of the Gulf basin, forming the Ur-Schatt River (Fig. 2), which ran through a deeply incised canyon that is still evident in the extant bathymetry (Seibold and Vollbrecht, 1969; Sarnthein, 1972). The most recent phase of Ur-Schatt River downcutting culminated during the Last Glacial Maximum, when global sea levels were reduced by 120 m and the basin was exposed in its entirety (Bernier et al., 1995; Lambeck, 1996; Williams and Walkden, 2002). Prior to the Early Holocene incursion into the Gulf basin, the floodplain was exposed to varying degrees for at least 75,000 years, when eustatic sea levels were more than 40 m lower (Siddall et al., 2002). Therefore, any discussion of human occupation in Arabia during this phase of prehistory must consider the demographic impact of this episodically exposed, large and favorable environmental niche.

Results of the Central Oman Pleistocene Research Program

The identification of mtDNA haplogroup M1 among living populations in East Africa (Quintana-Murci et al., 1999) provided the first glimmer of evidence for early human movement across the Arabian Corridor. Prompted by this discovery, the COPR project was initiated in 2002 to search for direct evidence of a modern human migration out of Africa. From 2002 to 2008, COPR conducted six seasons of archaeological survey and excavation in ad-Dakhliyah and Dhofar regions of Oman.

Ad-Dakhliyah is situated in north-central Oman and comprises the western Hajar Mountain range, accompanying foothills, and a sprawling alluvial plain that begins at the mountain piedmont and extends southward for two hundred kilometers. This plain is interlaced by a dense network of seasonally active widian weakly dipping into the Haushi-Huqf Depression. The bajada landscape displays little relief, declining from 230 m in the north to approximately 100 m in the south (Rogers et al., 1992). During the COPR campaign, archaeological sites were discovered on low terraces throughout the alluvial plain, associated with the low-energy widian (plural of wadi) that drain into the Haushi-Huqf Depression, and within the eroded limestone foothills situated between the Hajar Mountains and ad-Dakhliyah plain.

Following the geomorphic divisions proposed by Zarins (2001), the Dhofar governorate is divided into four zones: coastal plain, Dhofar Escarpment, Nejd Plateau, and Rub' al Khali Desert. The coastal plain of Dhofar stretches for about 50 km aligned southwest-northeast, and reaches a maximum of 15 km in width; it slopes gradually and steadily upward, some 200 m asl at the base of the escarpment. The plain is made up of early Quaternary travertine, ancient terraces, and alluvial fans overlying Tertiary limestone strata (Platel et al., 1992). These coastal deposits are cut by several drainage systems that are active during the summer monsoon season. Salalah, the second largest city in Oman, is situated along the shore in the centre of this plain. The city is surrounded by fields of date and coconut palms, banana trees, mangroves, as well as sorghum, millet, indigo, and cotton. Littoral South Arabia falls within the Sudano-Zambezian phytogeographic zone, which spans sub-tropical Africa into the western portion of the Indian subcontinent (Takhtajan, 1986). Paleobotanical investigations indicate the vegetation was considerably denser along the coastal plains and south-facing mountain slopes in antiquity (Radcliffe-Smith, 1980; Sale, 1980).

The Nejd Plateau is a dissected tableland stretching north from the Dhofar Escarpment. Widian draining across the Nejd into the Rub' al Khali Basin were active throughout the Pleistocene, with at least three distinct terrace systems spanning the last two million years (Zarins, 2001). The southern edge of the Nejd is marked by jagged hills and inselbergs derived from early Tertiary marine strata. The Rus Formation, particularly well developed in this region, is an Eocene bed with very high quality brown tabular chert found at the base and small, gray nodular cherts throughout the unit. As one travels north across the Nejd the vertical relief is reduced to a flat, undulating plain carpeted by Quaternary gravels overlying Late Tertiary limestone beds. The landscape is marked by occasional inselbergs and Rus Formation cherts still occur as lag deposits and in small outcrops exposed on the surface (Fig. 3). The Nejd gradually descends into a vast basin that houses the largest sand sea in the world – the Rub' al Khali.

Al-Hatab Rockshelter (OM.JA.TH.29)

Al-Hatab Rockshelter is a partially-collapsed rock overhang found at the southern end of the Nejd Plateau, just a few kilometers north of the present-day watershed divide along the Dhofar Escarpment. The rockshelter is situated inside a small tributary in the upper courses of Wadi Dawkhah, behind a wide terrace that is about 15m above the active channel (Fig. 4). Not only would this small tributary have provided an ample source of fresh, running water when it was active in the Terminal Pleistocene, there are also abundant fine-grained nodular and tabular chert deposits outcropping throughout the immediate landscape.

The tributary is roughly 15 m long and 10 m wide. There is a small limestone overhang perched approximately 5 m above the gully and oriented parallel to the drainage system. Most sediment inside the overhang has been scoured clean by erosion; however, scree slopes flank both sides of the gully and are comprised of slope waste, wind-borne sands, and eboulis from the collapsed portion of the limestone overhang. A shallow channel incised these sediments, which is how the site was initially recognized.

Nine square meters were excavated from an interstratified sequence of colluvial and aeolian deposits, yielding nearly 2,000 chipped stone artifacts. There are five sedimentary units labeled A to E (Fig. 5). Lithic artifacts were excavated from units A, B, and the upper portion of C. The assemblage was divided into archaeological Level 1 (unit A) and Level 2

Fig. 3 Chert hills comprising the lowest terrace at Ras Aïn Noor

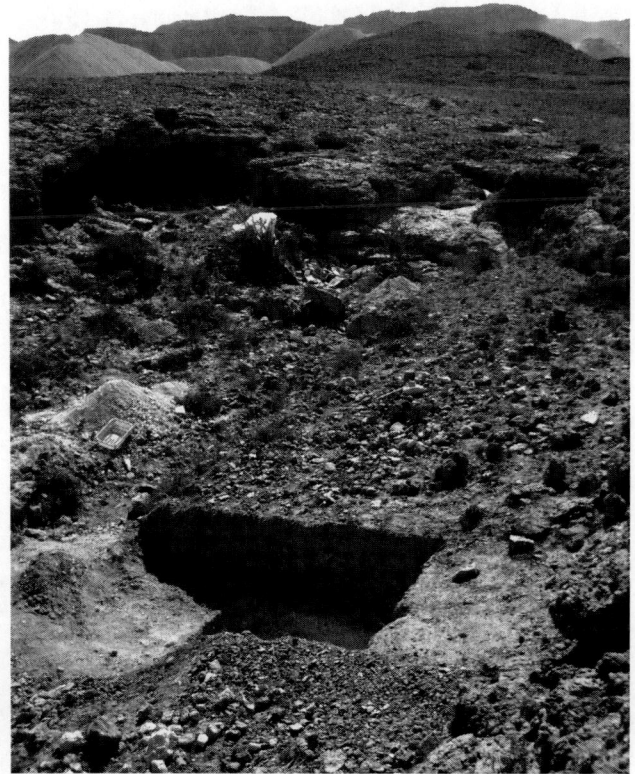

Fig. 4 Al-Hatab collapsed rockshelter with excavation unit in foreground

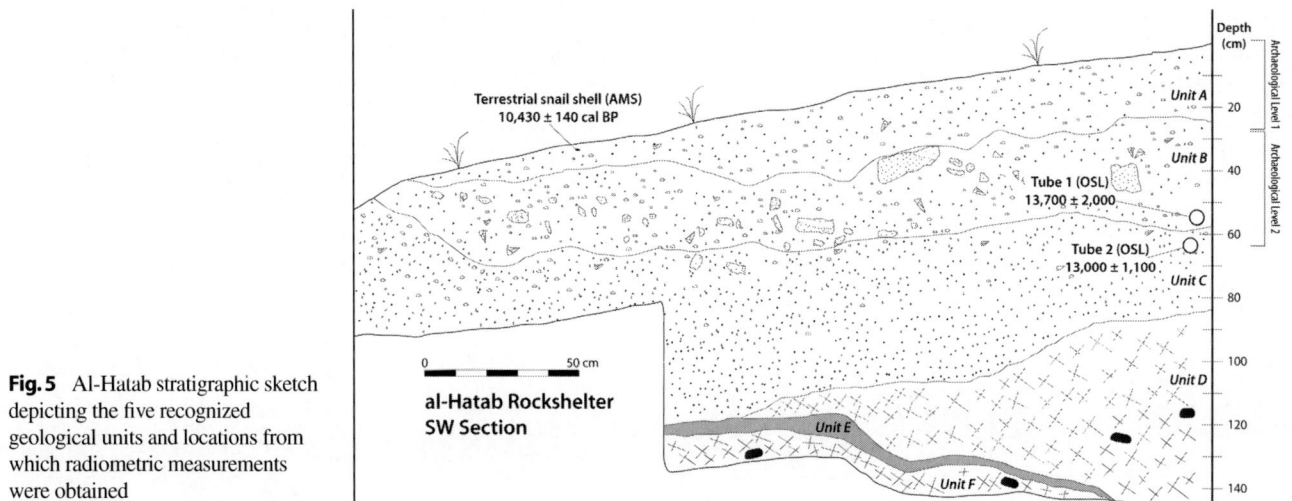

Fig. 5 Al-Hatab stratigraphic sketch depicting the five recognized geological units and locations from which radiometric measurements were obtained

(units B and the top of C). While this deposit is not a living surface, the frequency of chips and the artifacts' pristine state of preservation indicate they probably came from no more than a few meters up the low-gradient slope. Two OSL ages were obtained from the section: one from the top of unit C (13,000 ± 1,100) and the other from unit B (13,700 ± 2,000), producing a bracket date for the level 2 horizon between 14,100 and 11,700 BP (Rose et al., nd). A *terminus ante quem* for the archaeological material in unit A is determined by an AMS date of 10,430 ± 140 cal. BP (Beta-237899) on a terrestrial snail shell excavated from the top of this unit. The snail, *Euryptyxis latireflexa*, is non-borrowing species indicative of dense grass cover (Cremaschi and Negrino, 2005). As such, its presence in Level 1 is attributed to Early Holocene sedimentation, rather than post-depositional site formation processes. This relatively early age for the level 1 Fasad facies predates a similar tool assemblage excavated at KR213 Rockshelter some 30 km to the southeast (Cremaschi and Negrino, 2005), possibly explained by the 14C reservoir effect on shell, which has not been adjusted for on the al-Hatab measurement.

Contiguous one centimeter sediment samples were excavated from the southwest section of al-Hatab to a depth of 73 cm, from units A through C. To obtain a preliminary sketch of paleoenvironmental conditions at the time of deposition, organic and carbonate content was measured using the loss on ignition (LOI) technique (Rose et al., nd).

Unit A is comprised of fine unconsolidated silt that is relatively poor in carbonates and organics. There is a significant increase in both categories at the interface between units A and B, which steadily increases with depth through unit B. This is accompanied by the presence of large angular clasts in the unit B matrix, suggesting a period of alluvial deposition due to increased runoff through the local gully. The transition to unit C is marked by a spike in carbonate deposition and the disappearance of large angular clasts. Both of these trends indicate an abrupt shift from arid (unit C) to pluvial (unit B) conditions. The carbonates probably derive from dry wadi channels nearby; the reactivation of these channels would have significantly reduced the amount of carbonate material available for aeolian transport during the unit B depositional phase.

The characteristics and dates of the al-Hatab stratigraphic section fit comfortably with the regional paleoclimate record.

There is ample evidence for a sharp spike in Indian Ocean Monsoon activity during the Terminal Pleistocene (Overpeck et al., 1996; Ivanochko et al., 2005). Dates of 13,500–13,000 for this pluvial event correlate with OSL measurements at the al-Hatab unit B/unit C interface. Hence, overlying units A and B were deposited during the Terminal Pleistocene and Early Holocene wet-phase(s). The decrease in frequency of large clasts from units B to A may indicate a gradual reduction in runoff over the course of this period. Given the very fine, compacted sediments in unit C, the high carbonate content, and the absence of archaeological material, this stratum probably formed during the hyperarid phase associated with the LGM.

While lithic techno-typological features are fairly similar between Levels 1 and 2, some differences have been noted in the variety of raw material found between these two groups. In both cases, the tool manufacturers selected locally available fine-grained chert nodules derived from the Rus Formation, however, the Level 1 material is chocolate brown or yellow in color, while Level 2 cherts are more often shades of gray with banding. Tables 1 through 5 summarize technological features of the al-Hatab assemblage including (respectively) artifact classes, blank types, platform types, dorsal scar patterns, and tool types. The two predominant reduction strategies are simple unidirectional blades struck from volumetric/partial-volumetric cores (Fig. 6a–d) and the façonnage production of small bifacial foliates. This is followed, to a lesser degree, by the manufacture of twisted bladelets from unidirectional volumetric cores, a few carinated pieces, and a low percentage of Kombewa cores and flakes were also identified. On blade-proportionate debitage, the bulbs of percussion are prominent and lipped platforms are rare, implying the use of hard hammer percussion for blade production.

The al-Hatab toolkit is predominantly comprised of burins, endscrapers, notches (Fig. 7h), perforators, carinated pieces, and bifacial foliates. Many of the burins demonstrate multiple spalls struck from a truncated edge (Fig. 7e). Most sidescrapers were made on thick cortical flakes, suggesting that such blanks were deliberately chosen for this purpose. The manufacture of bifacial foliates is also significant since façonnage reduction is notably absent in the Near East during the Middle, Upper and Epi-Paleolithic periods. Given this fact, al-Hatab is probably *not* related to potentially coeval

Table 1 Artifact classes reported from Dhofar UP sites

Artifact class n (%)	Ras Aïn Noor, Level 1	Ras Aïn Noor, Level 2[a]	Al-Hatab, Level 1	Al-Hatab, Level 2	Dhanaqr
Debitage	76 (56.3)	11	569 (45.4)	348 (52.5)	239 (55.6)
Cores	2 (1.5)	2	62 (5.0)	35 (5.3)	24 (5.9)
Tools	7 (5.2)	1	178 (14.2)	80 (12.1)	26 (6.0)
Chips	39 (28.9)	3	299 (23.9)	148 (22.4)	33 (7.7)
Chunks/unident.	11 (8.1)	2	144 (11.5)	52 (7.9)	108 (25.1)
Total	135	19	1252	663	430

[a] Percentages not listed for sample sizes under 50.

Table 2   Blank types reported from Dhofar UP sites

Blank type n (%)	Ras Aïn Noor, Level 1	Ras Aïn Noor, Level 2[a]	Al-Hatab, Level 1	Al-Hatab, Level 2	Dhanaqr
Flakes	**41 (49.4)**	**7**	**438 (60.2)**	**254 (61.0)**	**179 (66.8)**
Flakes	35 (42.2)	5	348 (47.8)	217 (52.1)	161 (60.1)
Cortical flakes	6 (7.2)	2	90 (12.4)	37 (8.9)	14 (5.2)
Levallois flakes	–	–	–	–	4 (1.5)
Blades	**34 (41.0)**	**5**	**190 (26.1)**	**130 (31.3)**	**71 (26.5)**
Blades	19 (22.9)	2	107 (14.7)	86 (20.7)	39 (14.6)
Cortical blades	–	–	18 (2.5)	5 (1.2)	4 (1.5)
Debordant blades	3 (3.6)	2	25 (3.4)	13 (3.1)	25 (9.3)
Bladelets	12 (14.5)	1	40 (5.5)	26 (6.3)	3 (1.1)
Other	**8 (9.6)**	**–**	**100 (13.7)**	**32 (7.6)**	**18 (6.8)**
Kombewa flakes	–	–	5 (0.7)	1 (0.2)	1 (0.4)
Biface thinning flakes	4 (4.8)	–	74 (10.2)	20 (4.8)	12 (4.5)
Core trimming elements	4 (4.8)	1/N	10 (1.4)	5 (1.2)	2 (0.8)
Burin spalls	–	–	11 (1.5)	6 (1.4)	3 (1.1)
Total	83	12	728	416	268

[a]Percentages not listed for sample sizes under 50

Table 3   Platform types reported from Dhofar UP sites

Platform type n (%)	Ras Aïn Noor, Level 1[a]	Ras Aïn Noor, Level 2[a]	Al-Hatab, Level 1	Al-Hatab, Level 2	Dhanaqr
Unmodified	**44**	**8**	**477 (94.1)**	**293 (95.1)**	**205 (92.8)**
Straight	36	5	327 (64.5)	193 (62.7)	148 (67.0)
Cortical straight	7	3	107 (21.1)	70 (22.7)	39 (17.6)
Cortical curved	1	–	20 (3.9)	11 (3.6)	4 (1.8)
Dihedral	–	–	15 (3.0)	11 (3.6)	9 (4.1)
Dihedral ½ cortex	–	–	8 (1.6)	8 (2.6)	5 (2.3)
Modified	**–**	**–**	**30 (5.9)**	**15 (4.9)**	**16 (7.2)**
Faceted straight	–	–	19 (3.7)	10 (3.2)	10 (4.5)
Faceted curved	–	–	7 (1.4)	3 (1.0)	6 (2.7)
Faceted transverse	–	–	4 (0.8)	2 (0.6)	–
Total	44	8	507	308	221

[a]Percentages not listed for sample sizes under 50.

Table 4   Dorsal scar patterns reported from Dhofar UP sites

Scar pattern n (%)	Ras Aïn Noor, Level 1	Ras Aïn Noor, Level 2[a]	Al-Hatab, Level 1	Al-Hatab, Level 2	Dhanaqr
Unidirectional	24 (30.4)	4	269 (42.4)	154 (41.5)	88 (35.3)
Unidirectional-crossed	22 (27.8)	1	171 (27.0)	91 (24.5)	45 (18.1)
Unidirectional-parallel	13 (16.5)	3	81 (12.8)	69 (18.6)	48 (19.3)
Convergent	15 (19.0)	2	53 (8.4)	29 (7.8)	28 (11.2)
Bidirectional	1 (1.3)	–	11 (1.7)	4 (1.1)	13 (5.2)
Radial	1 (1.3)	1	25 (3.9)	16 (4.3)	13 (5.2)
Transverse	–	–	8 (1.3)	4 (1.1)	4 (1.6)
Crested	3 (3.8)	–	16 (2.5)	4 (1.1)	10 (4.0)
Total	79	11	634	371	249

[a]Percentages not listed for sample sizes under 50.

Near Eastern industries. Nor can the assemblage be said to resemble contemporary finds in East Africa, where the Late Stone Age exhibits markedly different features such as backed blades and bladelets, geometric microliths, discoids, and concave-base points. Thus, we suggest the Terminal Pleistocene lithic assemblage from al-Hatab represents a local, autochthonous population in South Arabia. The implications of this are discussed at the end of the chapter.

Ras Aïn Noor (OM.JA.SJ.32)

Ras Aïn Noor was discovered by COPR in 2004 and underwent formal investigation during fieldwork activities carried out in 2007. The relict spring, Aïn Noor, derives its name from the al-Noor oil camp approximately 20 km to the north. This findspot belongs to a large complex of lithic scatters situated on an outcrop of high-quality Rus Formation chert

Table 5 Tool types reported from Dhofar UP sites

Tool types n (%)	Ras Aïn Noor, Level 1[a]	Ras Aïn Noor, Level 2[a]	Al-Hatab, Level 1	Al-Hatab, Level 2	Dhanaqr[a]
Sidescrapers	–	–	32 (18.0)	18 (22.5)	3
Endscrapers	1	–	16 (5.6)	4 (5.0)	–
Burins	–	–	25 (14.1)	11 (13.8)	4
Notches	–	–	17 (9.6)	9 (11.3)	4
Denticulates	–	–	8 (4.5)	–	1
Perforators	–	–	8 (4.5)	3 (3.8)	–
Truncations	–	–	5 (2.8)	1 (1.3)	1
Carinated pieces	–	–	3 (1.7)	3 (3.8)	–
Retouched pieces	5	–	42 (23.6)	18 (22.5)	9
Levallois points	–	–	–	–	3
Bifacial foliates	–	–	1 (0.6)	2 (2.5)	–
Partially-retouched points	1	–	5 (2.8)	3 (3.8)	–
Fasad points	–	1	1 (0.6)	–	1
Misc bifacial elements	–	–	3 (1.7)	2 (2.5)	–
Heavy duty tools[b]	–	–	18 (10.1)	7 (8.8)	–
Total	7	1	178	80	26

[a]Percentages not listed for sample sizes under 50.

[b]Category includes naturally-backed knives, tranchets, and miscellaneous large chopping tools.

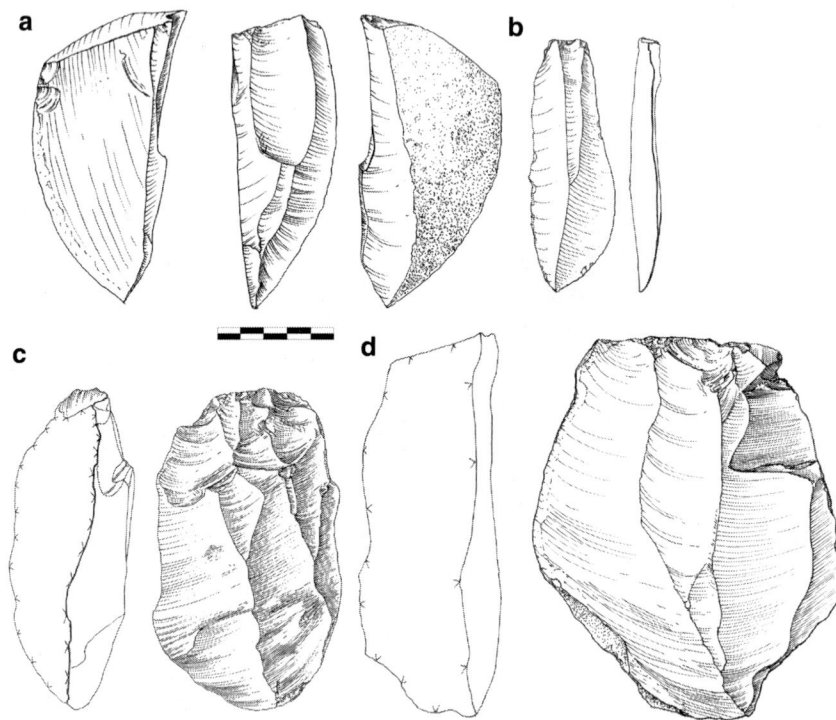

Fig. 6 Blade cores from al-Hatab with unidirectional-parallel and unidirectional-convergent scar patterns (**a,c,d**) as well an elongated, pointed blade (**b**)

at the edge of an ancient spring. The area investigated is located at the southeastern end of the spring, on the lowest of three terraces rising approximately 5, 10, and 15 m above the low-energy lacustrine basin. The crescent-shaped basin has a diameter of some ten kilometers and abuts a series of low

Tertiary limestone hills into which the terraces have been eroded.

Three 1 × 1 m test-pits were excavated at Ras Aïn Noor, two of them located inside the basin (Fig. 3) and one on the 5 m terrace. Pit 1 was sterile, while a moderate density

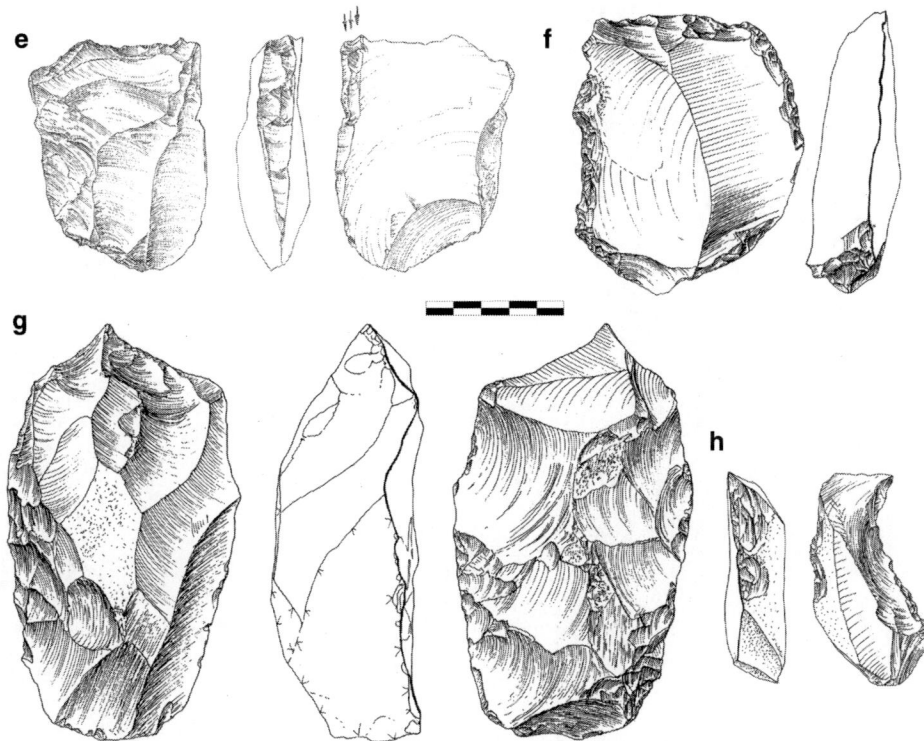

Fig. 7 Tools from al-Hatab including a burin on truncation (**e**), an atypical endscraper (**f**), a miscellaneous bifacial element (**g**), and a notch (**h**)

Fig. 8 Test pit 2 at Ras Aïn Noor. The surface of archaeological level 1 is shown, with flat-lying lithic artifacts immediately above tufa spring deposits

of lithic artifacts was collected from Pit 2 to a depth of 140 cm (Fig. 8). There was also chipped stone material recovered from the surface of Pit 3, although the subsurface strata were sterile. Given that the material from the top of Pit 3 was surface scatter and potentially mixed, we have not included it in this analysis.

Two archaeological levels were recorded in Pit 2. Artifacts from the more recent phase of occupation were excavated between 0 and 50 cm below the surface, while older material was recovered from depths ranging between 100 and 110 cm (Fig. 9). These two archaeological levels are separated by a thick layer of travertine spring deposits interstratified with lacustrine sediments. Level 1 material was excavated in a matrix of aeolian sands with some low-energy fluvial and colluvial input. Artifacts from Level 2 occur in the upper portion of a brecciated calcite layer bearing a coarse sandy matrix. Attempts to obtain OSL ages from geological unit 9 (depicted in Fig. 9) proved inconclusive, although technological parallels with the al-Hatab assemblage (specifically the presence of Wa'shah method core reduction *sensu* Crassard, 2008, as well as evidence for foliate production in the form of biface thinning flakes) suggest a Terminal Pleistocene/Early Holocene temporal attribution.

Level 1 yielded 135 chipped stone artifacts and 19 pieces were collected from Level 2; unfortunately, the low sample sizes preclude a detailed technological analysis. The artifacts were all manufactured from local, high-quality Rus chert. The technological features presented suggest both phases of archaeological occupation employed a similar mode of reduction, which was characterized almost entirely by the simple, unidirectional removal of blades and bladelets from volumetric cores. Even though no bifacial tools were recovered, it is noteworthy that four biface thinning flakes were found in

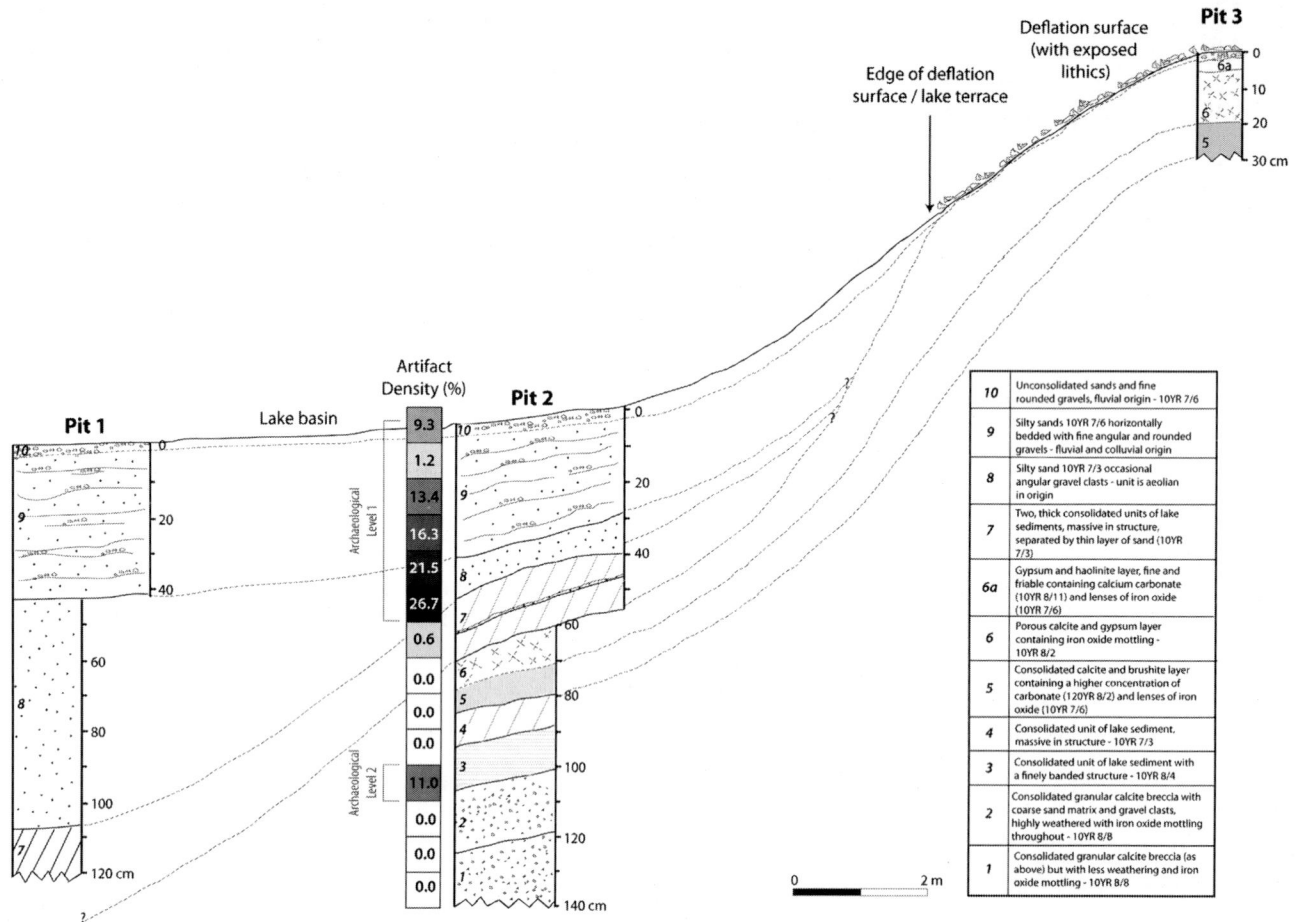

Fig. 9 Stratigraphic profiles of test pits 1, 2, and 3 at Ras Aïn Noor. Artifact distribution between 10-cm excavated spits is also included

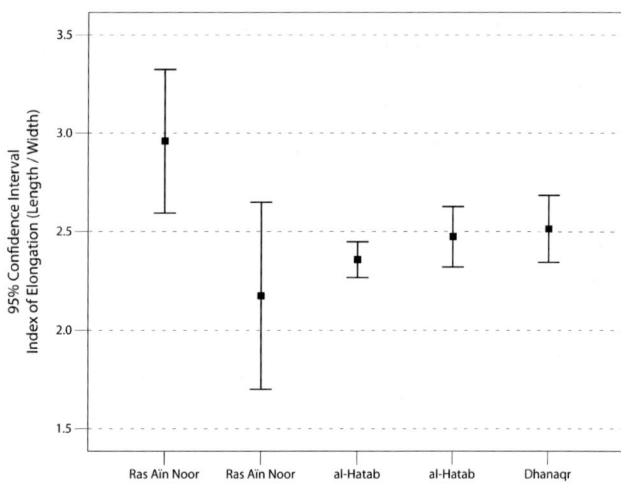

Fig. 10 Error bars comparing indices of elongation (length divided by width) between the Dhofar UP sites

Level 1. With over 40% blade-proportionate blanks, the material from Ras Aïn Noor is considerably more elongated than all other leptolithic assemblage considered in this chapter (Fig. 10). In every case, striking platforms are unmodified.

There is also scant evidence for edge preparation, only 13% of the artifacts exhibit grinding along the proximal-dorsal edge. Unidirectional-parallel and unidirectional-convergent scar patterns occur in the highest frequencies at Ras Aïn Noor, in conjunction with volumetric/partially-volumetric blade cores.

Dhanaqr (OM.JA.TH.21)

Dhanaqr was dubbed as such due to its proximity to a local well bearing the same name. The findspot is located at the confluence of Wadi Ribkhut and Wadi Dhahabun in the northern Nejd, in a location where the widian broaden out and drain across the plain on their way toward the Rub' al Khali Basin. The landscape is capped by Quaternary sediments composed of reworked fluvial sands, alluvial fans, depositional terraces, calcareous paleosols, and travertines. There is considerably less vertical relief than in the southern Nejd, though occasional hills and inselbergs rise up from the vast plain.

Dhanaqr is situated on a Tertiary rock outcrop belonging to the Andhur Member, a geological bed of yellowish orange shale with thin-bedded whitish bioclastic limestones and green marls (Platel et al., 1992). The rocky exposure measures roughly 5 km from east to west, 2 km from north to south, and rises about 30 m above the wadi channel. Though there is no chert naturally occurring in this geological unit, immediately to the south there is a deflated gravel plain with Rus Formation chert nodules and slabs outcropping in low density on the surface.

The lithic scatter at Dhanaqr was found on the interior flank of a small cluster of outcropping limestone hills; on a slope that dips gently toward the centre of the outcrop (Fig. 11). The location overlooks both widian, and would

Fig. 11 Dhanaqr surface scatter within a Tertiary outcrop at the confluence of Wadi Ribkhut and Wadi Dhahabun

have presented a tactical hunting advantage by providing both seclusion and elevation over the alluvial plain.

Chipped stone debris was collected both on the surface of the hill slope and in a subsurface layer of loose, unconsolidated sandy gypsum that carpets the hill that ranged in thickness from zero to ten centimeters. A recent nearby oil rig camp caused minor trampling and disturbance of the site's surface, although the high percentage of complete artifacts (70%) suggests this activity had minimal affect on the condition of the lithic assemblage. 27 m² were systematically sampled in 1 × 1 m units. Where present, the sandy-gypsum surface mantle was scraped clean and screened for artifacts.

The 325 lithic artifacts collected at Dhanaqr are derived from Rus cherts that range from extremely fine-grained and high-quality (15%), to poor, riddled with fracture planes and showing a high degree of shatter (24%), and everything between (60%). There are also traces of artifacts struck from limestone (0.5%). The higher quality material is glossy, orangish-yellow, while the more brittle chert tends more toward a dull brown. For the most part, techno-typological features are fairly homogenous and suggest a single phase of occupation, although given that this site is primarily a surface scatter we cannot rule out the possibility of some intrusive elements.

A few distinct technologies were noted at Dhanaqr. By far the most frequent was the production of blades struck from the narrow working surface of volumetric cores (Fig. 12),

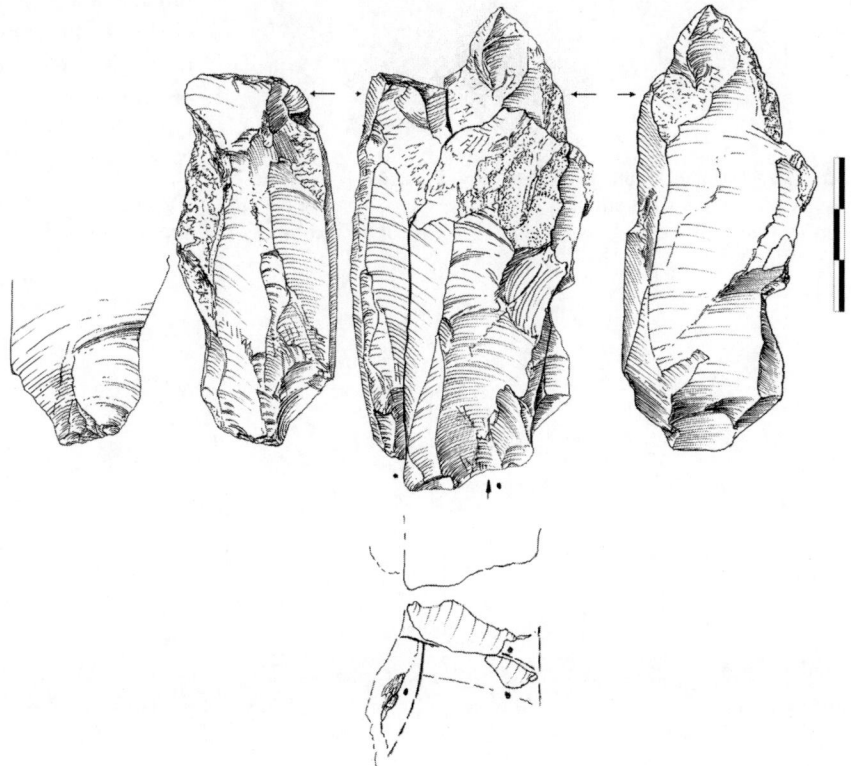

Fig. 12 Blade refits from Dhanaqr showing simple, unidirectional-parallel blade removals. A debordant blade was removed adjacent to the working surface as a means of re-establishing convexity across the working surface

with blade-proportionate pieces representing 27% of the total debitage. In contrast to the predominant use of a simple unidirectional method at al-Hatab, Dhanaqr exhibits a somewhat higher percentage of pieces with bidirectional reduction: 29% of cores are opposed platform and 5% of the debitage have bidirectional scar patterns. In most cases, the distal platforms are supplementary (i.e., short, non-invasive distal removals are used for establishing convexity, rather than to obtain substantial blanks for tools). The bidirectional cores belong to a continuum that, in a few cases, exhibit characteristics of the Levallois technique: there are three flat, prepared cores with evidence of both platform faceting and distal convexity maintenance across the working surface.

While not nearly as prominent as blade production, there is evidence for the façonnage manufacture of bifacial pieces. Biface thinning flakes are present (5%), and one miscellaneous bifacial preform was collected (Fig. 13). All of the façonnage pieces were made from high-quality Rus chert, suggesting the preferential treatment of raw material for specific modes of reduction. There is evidence that the higher

quality material was more intensively exploited. One Kombewa flake-core was found, which was made on a thick biface thinning flake that had been longitudinally split during reduction. The lateral-steep edge was used as a striking platform for subsequent blanks (Fig. 14).

There are just 26 tools in this assemblage, none of which are particularly diagnostic of any industry, region, or time period. The most frequent types are irregularly retouched pieces (made on an array of blank types), followed by burins, notches, denticulates, sidescrapers, and a bifacial preform.

Discussion

The South Arabian UP

Given the paucity of sites and dearth of absolute sites, it is not yet possible to construct a reliable or comprehensive synthesis of Upper Paleolithic archaeology in southern Arabia. Al-Hatab Rockshelter described in this chapter, Shi'bat Dihya 1 near the Red Sea coast in Yemen (Delagnes et al., 2008) both provide evidence for human occupation in southern Arabia in proximity to coastal environments during MIS 4 and MIS 2, respectively. The site of Jebel Faya 1, with assemblages dated to MIS 5 and MIS 3 (Marks, 2009; Uerpmann et al., 2009) and Upper Paleolithic sites recorded in the Wadi Hadramaut (Amirkhanov, 2006) are examples of Arabian UP findspots found in more marginal environments, with periods of occupation corresponding to paleoclimatic wet phases.

Two distinct modes of reduction are present in South Arabia during the latter half of the Upper Pleistocene: the façonnage creation of bifacial leaf-shaped points and the simple, unidirectional manufacture of blade-proportionate blanks struck from single platform, volumetric cores. Described as diminutive, thin, biconvex tools formed by soft hammer percussion, specimens have been reported at Bir Khasfa (Pullar, 1974;

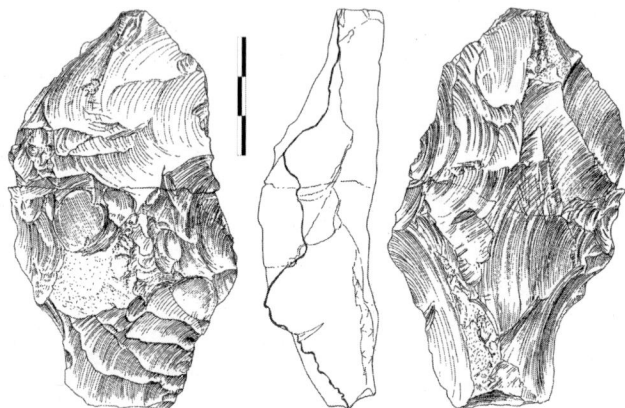

Fig. 13 Miscellaneous bifacial preform from Dhanaqr. The two halves were refit together from different parts of the site, suggesting the preform was discarded during manufacture as a result of the breakage

Fig. 14 Refit Kombewa flake made from a biface thinning flake

Rose, 2004a, 2006), Fahud (Pullar, 1974), and at scattered surface findspots published by Smith (1977) and Villiers-Petocz (1989) from museum collections stored in Muscat. Leaf-shaped points made via this former technology have been designated 'Type 5' of the Rub' al Khali Neolithic (Edens, 1982, 1988). Other scholars include pieces categorized as bifacial foliates with the Middle Holocene Saruq facies (Uerpmann, 1992); there are even foliates from earlier deposits dated to MIS 5 or older (Marks, 2009). Clearly, the presence of bifacially shaped tools is not a useful chronological marker to differentiate South Arabian Pleistocene and Holocene lithic assemblages. On the other hand, some regional geographic patterning may be emerging. The presence of bifacially-manufactured tools is significantly greater from sites around the Gulf basin refugium and adjacent areas of southeastern Arabia (Jagher, 2009; Marks, 2009; Scott-Jackson et al., 2009; Wahida et al., 2009), as opposed to the predominant laminar technologies discussed in this chapter found throughout southern and southwestern Arabia.

Some researchers have pointed out similarities between South Arabian foliates and similar specimens found in Magosian, Doian, and Aterian assemblages from Africa (Caton-Thompson, 1939, 1954, 1957; Van Beek et al., 1963; Gramly, 1971; Pullar, 1974; and Villiers-Petocz, 1989; Rose, 2004a). In East Africa, however, bifacial foliates are found in association with backed blades, microliths, outils écaillés, Levallois cores, microlithic cores, discoids, and thumbnail scrapers (Clark, 1954; Graziosi, 1954; Gresham, 1984; Merrick, 1975; Anthony, 1978; Clark et al., 1984; Brandt, 1986; Ambrose, 1998; Rose, 2004b; Pleurdeau, 2005). Conversely, the Khasfian foliates are most often found in association with single platform, volumetric blade cores, Kombewa cores, burins, nosed endscrapers, perforators, carinated pieces, and naturally-backed knives. While the tools are morphologically similar, they appear to belong to separate techno-typological traditions (contra Rose, 2004a). More likely, the presence of Khasfian foliates in South Arabia is associated with an Upper Pleistocene and Holocene tradition of bifacial tool production reported throughout the eastern portions of Arabia as early as MIS 5 (Marks, 2009), and as late as the Middle Holocene (Charpentier, 2008).

Throughout this paper, we have emphasized a very general definition of the South Arabian "Upper Paleolithic" that carries no temporal connotation. This is because a possible continuum of laminar technologies are known throughout the region between approximately 75 (Delagnes et al., 2008) and 8 ka (e.g., Charpentier, 2008). Al-Hatab provides two points on the timeline between roughly 13 and 10 ka. Based on a TL date from the base of a Late Pleistocene sandy loam in Wadi Hadramaut, the Yemeni-Soviet Expedition bracketed a stratified UP assemblage found there between approximately 30 and 18 ka (Amirkhanov, 1994, 2006). Undated lithic occurrences bearing a suite of potentially-related technological characteristics were also documented in the 'Asir Highlands

(Zarins et al., 1980), Habarut (Amirkhanov, 1994), Wadi Ghadun (Zarins, 2001), the Faw Well Site (Edens, 2001) and Wadi Jiza (Rose, 2002).

Given this wide temporal range and the ambiguity of techno-typological features, we recognize the UP in the most general sense by the ubiquitous presence of simple, hard and soft hammer laminar technologies. There is a considerable degree of variability within this leptolithic complex: (1) flat, unidirectional and bidirectional core reduction (e.g., the Dhanaqr and Wadi Dauan assemblages [Amirkhanov, 1994, 2006]), (2) prismatic blades and bladelets, crest-preparation, double-backed bladelets, endscrapers, and burins (e.g., the Faw Well assemblage [Edens, 2001]), (3) Wa'shah elongated point production (sensu Crassard, 2008) (e.g., al-Hatab, Wadi Wa'shah, and Ras Aïn Noor), and (4) Fasad facies found all throughout southern Arabia (sensu Charpentier, 2008). It is significant that there are no reported instances of microlithic assemblages, lunates, backed blades, bipolar core reduction, or other such features of the East African LSA, nor geometric microliths such as those found in the Levant during MIS 2. This suggests that the South Arabian laminar tradition developed independently in Arabia, with minimal external influence from MIS 3 onward.

The material from Jebel Faya Assemblages A and B, with OSL dates placing them roughly coeval in MIS 3, exhibit a markedly different array of characteristics (Marks, 2009; Uerpmann et al., 2009) and may potentially represent a separate and concurrent stone tool tradition around the Gulf basin refugium. The Jebel Faya material comprises multiple platform cores with flat converging and flat 90-degree flaking surfaces, sometimes from a faceted circumference. To a lesser extent, there is blade production from unidirectional-parallel cores. While technologically quite different, the tool assemblages bear an array of types similar to the UP findings in Dhofar and Hadramaut: burins, endscrapers, denticulates, and sidescrapers.

Given the propensity for laminar reduction in southern Arabia over the course of the Upper Pleistocene, we question why blade manufacture was so frequent. Are these blanks the product of a specialized reduction strategy designed to remove specifically-proportioned blanks or the unintentional byproduct of a simple unidirectional reduction strategy? We argue the latter is the case in Dhofar. The assemblages analyzed from Dhofar most often contain single platform, volumetric cores with unidirectional-parallel, unidirectional-convergent, and bidirectional scar patterns across the working surface. This core reduction technique is organized by recurrent unidirectional blanks struck from the long axis of the core. We do not consider this South Arabian UP laminar reduction strategy to be formal blade technology in the sense of those found in the Levantine UP (e.g., volumetric cores, prismatic blades, crest preparation etc.), rather, they belong to a possibly related, albeit separate techno-typological family.

So, we are able to make a few general observations regarding the Upper Paleolithic found in the southern portions of the peninsula: (1) there are multiple phases of human occupation in South Arabia throughout the latter half of the Upper Pleistocene, (2) there are elements loosely related to the Levantine sequence, however, the South Arabian Upper Paleolithic probably belongs to a unique and locally-derived lithic tradition, (3) there do not appear to be any links with East Africa (with the exception of the Hargeisan) from MIS 4-onward, and (4) assemblages from southern and south-western Arabia are dominated by different laminar-based technologies between 75 and 8 ka.

Demographic Implications

Merging these archaeological and paleoclimatic data with recent evidence from the burgeoning field of genetics, we address the role of southern Arabia in the emergence of modern humans, in light of the observation that there appears to be minimal exchange with East Africa between MIS 4 and MIS 2.

Analyses of mitochondrial DNA (mtDNA) (Kivisild et al., 2004; Metspalu et al., 2004), Y-Chromosome DNA (yDNA) (Ke et al., 2001; Cadenas et al., 2008), and X-Chromosome DNA (xDNA) (Garrigan et al., 2005; Yotova et al., 2007) suggest *Homo sapiens* initially developed in sub-Saharan Africa between 300 and 50 ka, the timing of this coalescence showing a wide range of variability depending upon the specific marker one examines. During this process of expansion out of Africa, some geneticists argue that early humans did not always replace local archaic groups encountered in their travels; there may have been varying degrees of admixture (e.g., Eswaran et al., 2005; Garrigan et al., 2005; Plagnol and Wall, 2006). Upper Pleistocene demographic pulses through the 'Arabian Corridor' probably resulted from early human groups tracking the growth of ecosystems to which they were already adapted (Lahr and Foley, 1998), whether it be colonization along the exposed continental shelf during MIS 4 (Stringer, 2000; Field et al., 2007), the range expansion of big-game hunters into the ameliorated interior (Rose, 2007), or a more complex combination of different dynamics.

Some scholars speculate the modern human demographic expansion, represented by the branching of L3 into M and N lineages, began in East Africa (e.g., Lahr and Foley, 1994, 1998; Ambrose, 1998; Kivisild et al., 2004). Coalescence dates from the earliest detectable mtDNA bottleneck release are 70,600±21,000 BP, represented by the M2 subclade in India (Metspalu et al., 2004), while the M1 subclade in Ethiopia coalesces at 48,000±15,000 BP (Quintana-Murci et al., 1999). Considering this temporal overlap of M coalescence between the two regions, there is no reason to assume

that the founder M population originated in East Africa rather than South Asia (or any number of locations within this broadly defined area). These modern geographic designations were not relevant to the early humans under discussion; the Red Sea flanking the western side of Arabia was vastly constricted, while the Persian Gulf basin was more or less dry land between 75 and 8 ka. Thus, it is not surprising that a number of genetic studies point to early human migration *into* Africa (e.g., Altheide and Hammer, 1997; Hammer et al., 1998; Cruciani et al., 2002; Coia et al., 2005; Olivieri et al., 2006). González et al. (2007) report that the most ancient M1 lineages are concentrated in Northwest Africa and the Near East.

From an archaeological perspective, Straus and Bar-Yosef (2001: 2) entertain the same possibility: "there is, however, no reason a priori to exclude the possibility that intercontinental contacts occurred on a two-way street, especially at Suez, via Sinai, or across the shallow Bab al Mandab, so close to that corridor to sub-Saharan Africa, the Nile." Marks (2005) and Otte et al. (2007) envisage similar scenarios during the MP/UP transitions in the Near East and Zagros regions. Both scholars argue that the archaeological evidence from Eastern Europe and Western Asia indicate the expansion of European UP technologies radiated from these areas, rather than Africa, during early MIS 3. Echoing this proposition from a biological perspective, Schillaci (2008) proposes the spread of Levantine-derived peoples into Australasia between 60 and 40 ka based on fossil evidence and phylogenetic relationships between populations.

One potentially additional piece of evidence for this hypothesized Near Eastern/Arabian-derived human expansion is the anomalous Hargeisan Industry found in the Horn of Africa. Known from a small number of findspots around Hargeisa (Clark, 1954), Boosasso (Graziosi, 1954) and Midhishi Cave in the Golis Mountains of northern Somalia (Gresham, 1984; Brandt, 1986), the Hargeisan has been found overlying MSA material and beneath LSA occupation layers. The industry is characterized by the presence of end-scrapers and burins produced by a volumetric blade technology, sometimes found in conjunction with bifacial foliates. The Hargeisan is incongruous with other roughly contemporary material in East Africa, leading to the conclusion that it represents:

> a local and probably hybrid form…at first glance it would seem that these northern Somaliland industries are but a local form of the Magosian, but a detailed study shows that in the angle-burins and end-scrapers…are forms which are entirely foreign to the Magosian, and clearly demonstrate that we are dealing with a distinct cultural complex (Clark, 1954: 218-9).

On the western side of the Red Sea, Hargeisan sites are limited to the coast of the Horn of Africa, in proximity to the Bab al Mandab Strait. Taking into account: (1) the geographic distribution of Hargeisan-like findspots, (2) their MIS 3/MIS

4 age range, and (3) genetic signals for a back migration into Africa around this time, a tentative correlation is proposed between the bottleneck release of mtDNA haplogroup M1 into Africa with archaeological data that attests to the appearance of a "foreign" and "hybrid" lithic industry in the Horn of Africa at the MSA/LSA boundary. It is germane to consider the possibility that the Hargeisan is a fingerprint of early human groups expanding westward across the southern route of dispersal, back into Africa.

That is not to say this industry necessarily represents a single expansion event, but perhaps the southern extent of a relatively homogenous population spread throughout the Horn, Arabia, North Africa, and the Levant, marked by their widespread use of laminar technologies during MIS 4–MIS 2. Indeed, Amirkhanov (2006) notes parallels between his UP assemblages from Hadramaut and roughly coeval late MIS 3 industries in Northeast Africa.

This proposition raises several points that must be addressed to evaluate its efficacy. In addition to analysis and dating of more Arabian UP sites, we must better establish the timing of the Hargeisan and its relationship to Arabian UP assemblages, particularly those found in Yemen. Its geographic extent must be articulated; considering the Horn of Africa its western boundary, how far north and east does it extend? What is the relationship between Arabian UP assemblages and those flat, unidirectional-parallel blades cores reported from the Thar Desert in Rajasthan (James and Petraglia, 2005), the Aterian in the Sahara, and Nilotic UP assemblages?

We maintain that the evidence from Arabia indicates the post-MIS 4 human expansion did *not* originate in sub-Saharan Africa; rather, early modern humans have emerged from a geographic range encompassing areas of northeast Africa, Western Asia, Arabia, and South Asia. These populations would have been forced to contract into environmentally stable refugia around Arabia such as the Ur-Schatt River Valley, coastal oases, Yemeni Highlands, and/or the Dhofar Mountains during climatic downturns. As such, the fluctuating dynamic between landscape carrying capacity and population density may have been a critical mechanism driving early human dispersals from the region. Episodes of climate change caused large portions of the Arabian peninsula to become uninhabitable due to such calamities as the inundation of the emerged continental shelf and desertification throughout the interior. Given the potential importance of these once favorable, now uninhabitable zones, future investigations in and around Arabia should endeavor to explore the heart of the desert and bottom of the sea.

Acknowledgments Financial and logistical support for the COPR project comes from the US National Science Foundation and Ministry of Heritage and Culture in Oman. We are particularly grateful to Biubwa al-Sabri, Khamis al-Asmi, and Ali al-Mahrouqi for their invaluable assistance during our fieldwork. Finally, we wish to thank the reviewers of this chapter for their thoughtful and insightful comments.

References

Abu-Amero KK, Gonzalez AM, Larruga JM, Bosley TM, Cabrera VM. Eurasian and African mitochondrial DNA influences in the Saudi Arabian population. BMC Evolutionary Biology. 2007;7:32.

Altheide TK, Hammer MF. Evidence for a possible Asian origin of YAP+ Y chromosomes. American Journal of Human Genetics. 1997;61:462–6.

Ambrose S. Late Pleistocene human population bottlenecks, volcanic winter, and differentiation of modern humans. Journal of Human Evolution. 1998;34:623–51.

Ambrose S. Population bottleneck. In: Robinson R, editor. Genetics, vol. 3. New York: Macmillan Reference; 2003. p. 167–71.

Amirkhanov HA. The Paleolithic of South Arabia. Moscow (in Russian): Nauka; 1991.

Amirkhanov HA. Research on the Paleolithic and Neolithic of Hadramaut and Mahra. Arabian Archaeology and Epigraphy. 1994;5:217–28.

Amirkhanov HA. Stone Age of South Arabia. Moscow (in Russian): Nauka; 2006.

Anthony B. The prospect industry – a definition. Ph.D. dissertation, Harvard University, Cambridge, MA; 1978.

Anton D. Aspects of geomorphological evolution: paleosols and dunes in Saudi Arabia. In: Jado AR, Zötl JG, editors. Quaternary period in Saudi Arabia. Vol. 2: sedimentological, hydrogeological, hydrochemical, geomorphological, and climatological investigations of Western Saudi Arabia. Wien: Springer; 1984.

Bailey GN, Flemming NC, King GCP, Lambeck K, Momber G, Moran LJ, et al. Coastlines, submerged landscapes, and human evolution: the Red Sea Basin and the Farasan Islands. Journal of Island and Coastal Archaeology. 2007;2:127–60.

Bernier P, Dalongeville R, Dupuis B, de Medwecki V. Holocene shoreline variations in the Persian Gulf: example of the Umm al-Qowayn lagoon (UAE). Quaternary International. 1995;29(30):95–103.

Biagi P. An early Paleolithic site near Saiwan (Sultanate of Oman). Arabian Archaeology and Epigraphy. 1994;5:81–8.

Bordes F. Typologie du Paléolithique Ancient et Moyen. France: Université de Bordeaux; 1961.

Brandt SA. The Upper Pleistocene and Early Holocene prehistory of the Horn of Africa. The African Archaeological Review. 1986;4:41–82.

Brinkmann R, Ghaleb AO. Late Pleistocene mollisol and cumulic fluvents near Ibb, Yemen Arab Republic. In: Grolier MJ, Brinkmann R, Blakely JA, editors. The Wadi al-Jubah Archaeological Project: Volume 5 environmental research in support of archaeological investigations in the Yemen Arab Republic, 1982–1987; 1997. p. 251–8.

Cadenas AM, Zhivotovsky LA, Cavalli-Sforza L, Underhill PA, Herrera PJ. Y-chromosome diversity characterizes the Gulf of Oman. European Journal of Human Genetics; 2007. Online only.

Caton-Thompson G. Climate, irrigation, and early man in the Hadhramaut. Geographical Journal. 1939;93(1):18–35.

Caton-Thompson G. Some Paleoliths from South Arabia. Proceedings of the Prehistoric Society. 1954;29:189–218.

Caton-Thompson G. The evidence of South Arabian Paleoliths in the question of Pleistocene land connections with Africa. Pan-African Congress on Prehistory. 1957;3:380–4.

Charpentier V. Industries bifaciales Holocènes d'Arabie orientale, un exemple: Ra's al-Jinz. Proceedings of the Seminar for Arabian Studies. 1999;29:29–44.

Charpentier V. Hunter-gatherers of the "Empty Quarter of the Early Holocene" to the last Neolithic societies: chronology of the late prehistory of south-eastern Arabia (8000–3100 BC). Proceedings of the Seminar for Arabian Studies. 2008;38:59–82.

Church TM. An underground route for the water cycle. Nature. 1996;380:579–80.

Clark JD. The prehistoric cultures of the Horn of Africa. Cambridge: Cambridge University Press; 1954.

Clark I, Fontes J-C. Paleoclimatic reconstruction of northern Oman based on carbonates from hyperalkaline groundwaters. Quaternary Research. 1990;33:320–36.

Clark JD, Williamson KW, Michels MJ, Marean C. A Middle Stone Age occupation site at Porc Epic Cave, Dira Dawa (East-Central Ethiopia). Part I. African Archaeological Review. 1984;2:37–71.

Coia V, Wallace DC, Oefner PJ, Torroni A, Cavalli-Sforza LL, Scozzari R, et al. A back migration from Asia to sub-Saharan Africa is supported by high-resolution analysis of human Y-chromosome haplotypes. American Journal of Human Genetics. 2005;70:1197–214.

Crassard R. The Middle Paleolithic of Arabia: the view from the Hadramawt Region, Yemen. In: Petraglia MD, Rose JI, editors. The evolution of human populations in Arabia: paleoenvironments, prehistory and genetics. The Netherlands: Springer; 2009. p. 151–680.

Cremaschi M, Negrino F. Evidence for an abrupt climatic change at 8700 14C yr B.P. in rockshelters and caves of Gebel Qara (Dhofar-Oman): paleoenvironmental implications. Geoarchaeology. 2005;20:559–79.

Edens C. Towards a definition of the Rub al Khali 'Neolithic'. Atlal. 1982;6:109–24.

Edens C. The Rub' al-Khali 'Neolithic' revisited: the view from Nadqan. In: Potts D, editor. Araby the Blest: studies in Arabian archaeology. Copenhagen: Tusculanum Press; 1988. p. 15–43.

Edens C. A bladelet industry in southwestern Saudi Arabia. Arabian Archaeology and Epigraphy. 2001;12(2):137–42.

Eswaran V, Harpending H, Rogers AR. Genomics refutes an exclusively African origin of humans. Journal of Human Evolution. 2005;49:1–18.

Faure H, Walter RC, Grant DR. The coastal oasis: ice age springs on emerged continental shelves. Global and Planetary Change. 2002;33:47–56.

Field H. The cradle of Homo sapiens. American Journal of Archaeology. 1932;36(4):426–30.

Field J, Petraglia M, Lahr MM. The southern dispersal hypothesis and the South Asian archaeological record: examination of dispersal routes through GIS analysis. Journal of Anthropological Archaeology. 2007;26:88–108.

Fleitmann D, Burns SJ, Mangini A, Mudelsee M, Kramers J, Villa I, et al. Holocene ITCZ and Indian monsoon dynamics recorded in stalagmites from Oman and Yemen (Socotra). Quaternary Science Reviews. 2007;26:170–88.

Gardner RAM. Aeolianites and marine deposits of the Wahiba sands: character and paleoenvironments. Journal of Oman Studies Special Report. 1988;3:75–94.

Garrard A, Harvey CPD. Environment and settlement during the Upper Pleistocene and Holocene at Jubbah in the Great Nafud, northern Arabia. Atlal 1981;5:137–48.

Garrigan D, Mobasher Z, Kingan SB, Wilder JA, Hammer MF. Deep haplotype divergence and long-range linkage disequilibrium at Xp21.1 provide evidence that humans descend from a structured ancestral population. Genetics. 2005;170:1849–56.

Glennie KW, Singhvi AK. Event stratigraphy, paleoenvironment and chronology of SE Arabian deserts. Quaternary Science Reviews. 2002;21:853–69.

González AM, Larruga JM, Abu-Amero KK, Shi Y, Pestano J, Cabrera VM. Mitochondrial lineage M1 traces an early human backflow to Africa. BMC Genetics. 2007;8:223.

Goudie AS, Colls A, Stokes S, Parker AG, White K, Al-Farraj A. Latest Pleistocene dune construction at the north-eastern edge of the Rub' al-Khali, United Arab Emirates. Sedimentology. 2000;47(5):1011–21.

Gramly RM. Neolithic flint implement assemblages from Saudi Arabia. Journal of Near Eastern Studies. 1971;30(3):177–85.

Graziosi P. Missione preistorica Italiana in Somalia (estate 1953). Revista di Science Preistoriche. 1954;9:121–3. (in Italian).

Gresham T. An investigation of an Upper Pleistocene archaeological site in northern Somalia. M.A. thesis, University of Georgia, Athens; 1984.

Hammer MF, Karafet T, Rasanayagam A, Wood ET, Altheide TK, Jenkins T, et al. Out of Africa and back again: nested cladistic analysis of human Y chromosome variation. Molecular Biology and Evolution. 1998;15:427–41.

Hannah S, al-Belushi M. Introduction to the caves of Oman. Muscat: Sultan Qaboos University; 1996.

Ivanochko TS, Ganeshram RS, Brummer G-J, Ganssen G, Jung S, Moreton SG, et al. Variations in tropical convection as an amplifier of global climate change at the millennial scale. Earth and Planetary Science Letters 2005;235:302–14.

Jagher R. The Central Oman Palaeolithic survey: recent research in Southern Arabia and reflection on the prehistoric evidence. In: Petraglia MD, Rose JI, editors. The evolution of human populations in Arabia: paleoenvironments, prehistory and genetics. The Netherlands: Springer; 2009. p. 139–50.

James HVA, Petraglia MD. Modern human origins and the evolution of behavior in the later Pleistocene record of South Asia. Current Anthropology. 2005;46:S3–S27.

Juyal N, Singhvi AK, Glennie KW. Chronology and paleoenvironmental significance of Quaternary desert sediment in southeastern Arabia. In: Alsharhan AS, Glennie KW, Whittle GL, Kendall CGStC, editors. Quaternary deserts and climatic change. Rotterdam: Balkema; 1998. p. 315–25.

Kapel H. Atlas of the Stone-Age cultures of Qatar. Denmark: Aarhus University Press; 1967.

Ke Y, Su B, Song X, Lu D, Chen L, Li H, et al. African origin of modern humans in East Asia: a tale of 12,000 Y chromosomes. Science. 2001;292:1151–3.

Kivisild T, Reidla M, Metspalu E, Rosa A, Brehm A, Pennarun E, et al. Ethiopian mitochondrial DNA heritage: tracking gene flow across and around the Gate of Tears. American Journal of Human Genetics. 2004;75:752–70.

Lahr M, Foley R. Multiple dispersals and modern human origins. Evolutionary Anthropology. 1994;3:48–60.

Lahr M, Foley R. Towards a theory of modern human origins: geography, demography, and diversity in recent human evolution. Yearbook of Physical Anthropology. 1998;41:137–76.

Lambeck K. Shoreline reconstructions for the Persian Gulf since the Last Glacial Maximum. Earth and Planetary Science Letters. 1996;142:43–57.

Lézine A, Saliège J, Robert C, Wertz F, Inizan M. Holocene lakes from Ramlat as-Sab'atayn (Yemen) illustrate the impact of monsoon activity in southern Arabia. Quaternary Research. 1998;50:290–9.

Lézine AJ-J, Tiercelin C, Robert J-F, Saliège D, Cleuziou S, Inizan M-L, et al. Centennial to millennial-scale variability of the Indian monsoon during the Early Holocene from a sediment, pollen and isotope record from the desert of Yemen. Paleogeography, Paleoclimatology, Paleoecology 2007;243:235–49.

Marks A. Comments after four decades of research on the Middle to Upper Paleolithic transition. Mitteilungen der Gesellschaft für Urgeschichte. 2005;14:81–6.

Marks AE. The Paleolithic of Arabia in an inter-regional context. In: Petraglia MD, Rose JI, editors. The evolution of human populations in Arabia: paleoenvironments, prehistory and genetics. The Netherlands: Springer; 2009. p. 295–308.

McClure HA. Radiocarbon chronology of Late Quaternary Lakes in the Arabian Desert. Nature. 1976;263:755–6.

McClure HA. Ar Rub' Al-Khali. In: Al-Sayari SS, Zötl JG, editors. Quaternary period in Saudi Arabia. Vol. 1: Sedimentological, hydrogeological, hydrochemical, geomorphological, and climatological investigations in Central and Eastern Saudi Arabia. Wien: Springer; 1978.

McClure HA (1984) Late Quaternary paleoenvironments of the Rub' al-Khali. Ph.D. dissertation, University of London, London

Merrick HV Change in later Pleistocene lithic industries in Eastern Africa. Ph.D. dissertation, University of California, Berkeley, 1975.

Metspalu M, Kivisild T, Metspalu E, Parik J, Hudjashov G, Kaldma K, et al. Most of the extant mtDNA boundaries in South and Southwest Asia were likely shaped during the initial settlement of Eurasia by anatomically modern humans. BMC Genetics. 2004;5:26.

Miller AG, Morris M. The plants of Dhofar. Muscat: Royal Diwan, Government of Oman; 1988.

Monigal K The Levantine Leptolithic: blade technology from the Lower Paleolithic to the dawn of the Upper Paleolithic. Ph.D. dissertation, Southern Methodist University, Dallas, 2002.

Olivieri A, Achilli A, Pala M, Battaglia V, Fornarino S, Al-Zahery N, et al. The mtDNA legacy of the Levantine early Upper Paleolithic in Africa. Science. 2006;314:1767–70.

Otte M, Biglari F, Flas D, Shidrang S, Zwyns N, Mashkour M, et al. The Aurignacian in the Zagros region: new research at Yafteh Cave, Lorestan, Iran. Antiquity. 2007;81:82–96.

Overpeck J, Anderson D, Trumbore S, Prell W. The southwest Indian Monsoon over the last 18000 years. Climate Dynamics. 1996;12:213–25.

Parker AG, Goudie AS. Development of the Bronze Age landscape in the southeastern Arabian Gulf: new evidence from a buried shell midden in the eastern extremity of the Rub' al-Khali desert, Emirate of Ras al-Khaimah, UAE. Arabian Archaeology and Epigraphy. 2007;18:232–8.

Parker A, Rose J. Climate change and human origins in southern Arabia. Proceedings of the Seminar for Arabian Studies. 2008;38:25–42.

Plagnol V, Wall JD. Possible ancestral structure in human populations. PLoS Genetics 2006;2(7):105.

Platel JP, Roger J, Peters TJ, Mercolli I, Kramers JD, Le-Métour J. Geological map of Salalah, explanatory notes. Sultanate of Oman: Ministry of Petroleum and Minerals; 1992.

Pleurdeau D. Human technical behavior in the African Middle Stone Age: the lithic assemblage from Porc-Epic Dave (Dire Dawa, Ethiopia). African Archaeological Review. 2005;22(4):177–97.

Pullar J. Harvard archaeological survey in Oman, 1973: flint sites in Oman. Arabian Seminar. 1974;4:33–48.

Pullar J. A selection of aceramic sites in the Sultanate of Oman. Journal of Oman Studies. 1985;7:49–88.

Pullar J, Jäckli B. Some aceramic sites in Oman. Journal of Oman Studies. 1978;4:53–71.

Quintana-Murci L, Semino O, Bandelt H, Passarino G, McElreavey K, Santachiara-Benerecetti AS. Genetic evidence of an early exit of Homo sapiens from Africa through eastern Africa. Nature Genetics. 1999;23:437–41.

Radcliffe-Smith A. The vegetation of Dhofar. The Journal of Oman Studies, Special Report. 1980;2:59–86.

Rogers J, Platel JP, Dubreuilh J, Wyns R. Geological Map of Mafraq, Explanatory Notes. Sultanate of Oman: Ministry of Petroleum and Minerals; 1992.

Rose JI. Survey of prehistoric sites in Mahra, Eastern Yemen. Adumatu. 2002;6:7–20.

Rose JI. The question of Upper Pleistocene connections between East Africa and South Arabia. Current Anthropology. 2004a;45(4):551–5.

Rose JI. New evidence for the expansion of an Upper Pleistocene population out of East Africa, from the site of Station One, Northern Sudan. Cambridge Archaeological Journal. 2004b;14(2):205–16.

Rose JI. Among Arabian sands: defining the Paleolithic of southern Arabia. Ph.D. dissertation, Southern Methodist University, Dallas; 2006.

Rose JI. The Arabian Corridor Migration Model: archaeological evidence for hominin dispersals into Oman during the Middle and Upper Pleistocene. Proceedings of the Seminar for Arabian Studies. 2007;37:219–37.

Rose JI, Petraglia M, Foley R, Lahr MM, n.d. Ras Aïn Noor, in preparation

Sale JB. The environment of the mountain region of Dhofar. The Journal of Oman Studies, Special Report. 1980;2:17–54.

Sanlaville P. Changements climatiques dans la péninsule Arabique durant le Pléistocène Supérieur et l'Holocène. Paleorient. 1992;18:5–25.

Sarnthein M. Sediments and history of the postglacial transgression in the Persian Gulf and northwestern Gulf of Oman. Marine Geology. 1972;12:245–66.

Schillaci MA. Human cranial diversity and evidence for an ancient lineage of modern humans. Journal of Human Evolution. 2008;54:814–26.

Schultz E, Whitney JW. Upper Pleistocene and Holocene lakes in the An Nafud, Saudi Arabia. Hydrobiologia. 1986;143:175–90.

Seibold E, Vollbrecht K. Die Bodengestalt des Persischen Golfs. In: Seibold E, Closs H, editors. Meteor Forschungsergebnisse: Herausgegeben von der Deutschen Forschungsgesellschaft. Berlin: Gebrüder Borntraeger; 1969. p. 31–56 (in German).

Shepard EM, Herrera RJ. Genetic encapsulation among Near Eastern populations. Journal of Human Genetics. 2006;51:467–76.

Siddall M, Smeed DA, Mathiessen S, Rohling EJ. Exchange flow between the Red Sea and the Gulf of Aden. The 2nd Meeting on the Physical Oceanography of Sea Straits, Villefranche; 2002. p. 203–6.

Straus L, Bar-Yosef O. Out of Africa in the Pleistocene: an introduction. Quaternary International. 2001;75:19–28.

Stringer C. Paleoanthropology: coasting out of Africa. Nature. 2000;405:24–6.

Takhtajan AL. Floristic regions of the World. Berkeley, CA: University of California Press; 1986.

Uerpmann M. Structuring the Late Stone Age of southeastern Arabia. Arabian Archaeology and Epigraphy. 1992;3:65–109.

Uerpmann H-P, Potts DT, Uerpmann M. Holocene (Re-) Occupation of Eastern Arabia. In: Petraglia MD, Rose JI, editors. The evolution of human populations in Arabia: paleoenvironments, prehistory and genetics. The Netherlands: Springer; 2009. p. 205–14.

Van Beek G, Cole G, Jamme A. An archaeological reconnaissance in Hadramaut, south Arabia: a preliminary report. Annual Report of the Smithsonian Institution. 1963;1963:521–45.

Villiers-Petocz LE. Some notes on the lithic collections of the Oman Department of Antiquities. Journal of Oman Studies. 1989;10:51–9.

Wahida G, Al-Tikriti Y, Beech MJ, Al Meqbali A. A Middle Paleolithic Assemblage from Jebel Barakah, Coastal Abu Dhabi Emirate. In: Petraglia MD, Rose JI, editors. The evolution of human populations in Arabia: paleoenvironments, prehistory and genetics. The Netherlands: Springer; 2009. p. 117–24.

Whalen N, Pease DW. Archaeological survey in Southwest Yemen, 1990. Paléorient. 1991;17(2):127–31.

Williams AH, Walkden GM. Late Quaternary highstand deposits of the southern Arabian Gulf: a record of sea-level and climate change. In: Clift PD, Kroon D, Gaedicke C, Craig J, editors. The tectonic and climatic evolution of the Arabian Sea region. London: Geological Society, Special Publications 195; 2002. p. 371–86.

Wood WW, Imes JL. How wet is wet? Precipitation constraints on late Quaternary climate in the southern Arabian peninsula. Journal of Hydrology. 1995;164:263–8.

Crassard R. The "Wa'shah method": an original laminar debitage from Hadramaut, Yemen. Proceedings of the Seminar for Arabian Studies. 38:3–14

Rose JI, Usik VI, as-Sabri B, Schwenninger J-L, Clark-Balzan L, Parton A, et al. Archaeological evidence for indigenous human occupation in southern Arabia at the end of the Pleistocene, in review; n.d.

Yotova V, Lefebvre J-F, Kohany O, Jurka J, Michalski R, Modiano D, et al. Tracing genetic history of modern humans using X-chromosome lineages. Human Genetics. 2007;122(5):13.

Zarins J. The Land of Incense: archaeological work in the Governorate of Dhofar, Sultanate of Oman, 1990–1995. Muscat: Sultan Qaboos University Publications; 2001.

Zarins J, Whalen N, Ibrahim M, Mursi A, Khan M. Comprehensive Archaeological Survey Program: preliminary report on the Central and Southwestern Province Survey: 1979. Atlal. 1980;4:9–36.

Chapter 14
The Late Pleistocene of Arabia in Relation to the Levant

Lisa A. Maher

Keywords Faw Well • Late Pleistocene • Microlithic

Introduction

Our understanding of the Late Pleistocene of Arabia lags far behind that of the Levant, where decades of research have provided a highly refined cultural-chronological framework. Part of the reason for this is a difference in research intensity between the two regions, with the Levant much more intensively studied. More than that, these sites are elusive in Arabia. Very few have been documented and those that have been are small in size, deflated, and contain only lithics. This makes them difficult to categorize temporally and typologically and as a result, the Late Pleistocene remains poorly understood. Alongside issues relating to how to identify the Late Pleistocene in Arabia are those questions regarding potential connections with the Levant and North Africa.

From 25 to 10 ka, hunter-gatherer groups in the Levant, Arabia and Africa underwent compelling social, technological, and economic changes, while also experiencing dramatic fluctuations in climate and ecology. These three regions are connected to each other geographically and environmentally, but much less attention has been placed on potential cultural connections (although see Tosi, 1986; Lahr and Foley, 1994; Petraglia, 2003; Petraglia and Alsharekh, 2003; Rose, 2004a, for example). This chapter attempts to summarize and evaluate our current evidence for Late Pleistocene hunter-gatherer occupation of Arabia within the context of its much better known neighbors in the Levant and, to a lesser degree, North and East Africa.

In the Levant, between 23 and 12 ka, diverse Epipaleolithic (EP) hunter-gatherers occupied encampments that varied geographically and temporally. Representing a continuation of local Upper Paleolithic (UP) traditions, some of the defining features of the Levantine EP include the production of blade/lets from pyramidal-shaped cores and subsequently formed into microlithic tools (Belfer-Cohen and Goring-Morris, 2002). However, this microlithic toolkit is now known to be accompanied by a variety of other features, including bone tools, ground stone, art, personal ornamentation, burials, and sites showing internal spatial organization (e.g., Belfer-Cohen, 1991; Bar-Yosef, 1998; Nadel and Werker, 1999; Nadel, 2002; Goring-Morris, 2003). Different EP cultural entities are identified on the basis of the varying frequencies of microlithic tool classes, such as obliquely truncated and backed bladelets or micropoints (Kebaran), trapeze/rectangles (Geometric Kebaran), or lunates (Natufian). Many cultural complexes have been identified in the Mediterranean core, Negev and Sinai, and elsewhere on the basis of different toolkits and other associated site features (Bar-Yosef, 1970; Goring-Morris, 1987; Henry, 1995; Olszewski, 2001, 2006). By the end of this period some groups began to congregate in base camps, construct more permanent dwellings, intensify plant and animal exploitation, and make more mobile art and personal ornamentation. Complex transitions occurred within a small geographical area and narrow chronological time frame and so the long history of research here has produced a large number of well-excavated and well-dated sites.

In Africa, microlith technologies have historically been associated with the Late Stone Age (LSA); however, recent work has demonstrated a much earlier date for their appearance (ca. 70 ka, Bar-Yosef and Kuhn, 1999; McBrearty and Brooks, 2000), a possibly non-linear shift from the flake-based Middle Stone Age (MSA) to blade/microlith-based LSA, and a somewhat different trajectory for microlith production compared to the Levant (e.g., Ambrose, 2002). For example, some of these early microlith assemblages are not made from a clear blade/bladelet technology and the final tools can occur within an otherwise MSA assemblage (Ambrose, 2002). Microlithic industries resembling those of the Levant in their technology of production (blade/bladelet), frequency, ubiquity, and general form do appear in North and East Africa in the LSA (from ca. 20–10 ka) as the Ibero-maurasian, Capsian, and other industries

L.A. Maher (✉)
Leverhulme Centre for Human Evolutionary Studies,
University of Cambridge, Cambridge, CB2 1QH
e-mail: l.maher@human-evol.cam.ac.uk

M.D. Petraglia and J.I. Rose (eds.), *The Evolution of Human Populations in Arabia*, Vertebrate Paleobiology and Paleoanthropology,   187
DOI 10.1007/978-90-481-2719-1_14, © Springer Science+Business Media B.V. 2009

(McBurney, 1967; Wendorf and Schild, 1980; Close, 1987; Wendorf et al., 1993; Phillipson, 2005). In the Levantine EP, similarities in tool type (La Mouillah point) and techniques for retouch (Helwan) are hypothesized to result from close connections between North Africa and the Levant (Bar-Yosef, 1987).

Arabia, on the other hand, covers a significantly larger area than North Africa or the Levant yet in comparison remains *terra incognita*. Paleolithic research in Arabia is still in its infancy, but currently seems to focus on two issues. First, research is concentrated on exploring the increasingly strong evidence for several Out of Africa migrations through the Bab al Mandab straights into southern Arabia, and beyond (Lahr and Foley, 1994, 1998; Petraglia and Alsharekh, 2003; Rose, 2004a). Second, and more important to this discussion, the paucity of clearly Late Pleistocene sites throughout the region begs us to question the nature of Late Pleistocene occupation in Arabia. What do these sites look like? Where are they? How do they relate to contemporary sites in nearby regions? Are we missing these sites because geomorphological or other processes have erased their traces? Or, is it because this region was hyper-arid and unsuitable for occupation in the Late Pleistocene? Or, are we missing these sites because they look different than what we expect in comparison to nearby, better known, regions such as the Levant and North/East Africa? In this case, we are not actually 'missing' them, but may be misidentifying them as belonging to earlier or later periods. Arabian researchers tended to favor comparisons with Levantine assemblages and used the well-defined Upper Paleolithic and Epipaleolithic to guide the search for contemporary sites in Arabia (Zarins et al., 1980). But, perhaps we should expect them to be more similar to LSA sites in Africa, where the distinction between the MSA and LSA is not as clear-cut (Phillipson, 2005)?

This chapter assesses our current state of research for the Late Pleistocene of Arabia, beginning with a brief summary of paleoclimatic conditions and known Late Pleistocene sites. It also explores how these sites compare with contemporary ones in the Levant and Africa. An examination of how the Late Pleistocene archaeological record in Arabia compares with other areas inevitably leads to a discussion of current research issues and future directions of research. The focus here is on the Late Pleistocene of Arabia from the perspective of the well-defined microlithic industries of the southern Levant. However, this chapter also attempts to explore similar potential connections with North and East Africa. The goal of this chapter then is to pull together what we do know about the Late Pleistocene in Arabia and attempt to place this Arabian material (or lack thereof) into a broader perspective that considers how hunter-gatherer groups in Arabia articulated with those from adjacent areas and assesses how our relatively detailed knowledge of these other regions impacts our understanding of the Arabian Late Pleistocene.

It is tempting to draw comparisons between Arabia and its closest neighbors when attempting to characterize the poorly understood Late Pleistocene of Arabia. In fact, connections can be made between both the Levant, with the UP or EP-like bladelet assemblage from the Faw Well site (Edens, 2001), and Africa with the early LSA-like material from eastern Yemen (Amirkhanov, 1991). However, these can only be loose connections at present. They help us refine our research questions and direct our future efforts in Arabia. But, until we better understand the nature of Late Pleistocene sites from the excavation of stratified, well-dated sites and the typo-technological sequence, drawing meaningful connections between people and their movements in these regions can only be preliminary in nature.

In this light, the title of this chapter 'The Late Pleistocene of Arabia in relation to the Levant' is perhaps a bit misleading. One of the biggest questions in Late Pleistocene research in Arabia is the nature of the Late Paleolithic in Arabia and how do we identify it? Preliminarily, several decades of comprehensive survey, covering most of the Arabian peninsula, suggest that the lack of Late Pleistocene sites is to some extent a real gap in the archaeological record. However, recent studies (Crassard and Khalidi, 2005; Crassard, 2008; Edens, 2001; Inizan and Francaviglia, 2002; Khalidi, 2005, 2006, 2009) also point out that there are indeed blade and microlith assemblages in Arabia, albeit rare. So, we begin here with tantalizing hints of the Late Pleistocene in the Arabian record.

The Arabian Paleoenvironment

The Arabian peninsula encompasses the modern-day countries of Saudi Arabia, Yemen, Oman, United Arab Emirates, Qatar, Kuwait, and Bahrain. Covering an area of ~3 million km^2, it includes a vast expanse of land with a wide variety of ecological habitats, including deserts, steppes, coasts, ancient and modern river and lake systems, and mountains. Obviously it is impossible to summarize such a large and diverse area as the Arabian peninsula in just a few words. Indeed several chapters of this volume cover the paleoenvironment record of Arabia. Here is a only a brief summary that should be considered as no more than a very general statement focused on Late Pleistocene climatic trends, with selective reference to the more detailed and localized geomorphological and environmental studies necessary to fully appreciate prehistoric site distributions and settlement patterns within each area.

Zarins (1992) describes Arabia as belonging to a larger Greater Southwest Asian Arid Zone encompassing most of the Near East, including eastern Jordan and the Negev and Sinai from which some of our best evidence for UP and EP occupations in the Levant derives (e.g., Betts, 1998; Garrard and Stanley-Price, 1977; Garrard et al., 1988a, b, 1994; Goring-Morris, 1987, 1988; Henry, 1995; Marks, 1976, 1977; Phillips, 1994). This arid zone exhibits a diversity of topography, geology, soils and vegetation, all of which influence the preservation of archaeological sites, as well as our detection of them. In eastern Jordan and northern Arabia a number of freshwater springs, lakes and playas form part of the extensive drainage system that includes the Azraq Basin and Wadi Sirhan, extending for several hundred kilometers and draining towards to the Great Nafud. In this large area there is evidence for a Pleistocene lake covering about 4,500 km^2 (Garrard and Stanley-Price, 1975–1977: 110) that had important bearing for Late Pleistocene occupation in Jordan and northern Arabia.

The Paleoenvironment Between 30 and 10 ka

Detailed paleoclimatic studies of the Arabian peninsula provide a fairly refined reconstruction of climate and landscape change over the last several thousand years (e.g., McClure, 1976, 1978, 1988; Sanlaville, 1992, 2000; Wilkinson, 2009). These studies document the oscillating Pleistocene climate, with shifts from favorable to hyper-arid conditions (and back again).

For the Late Pleistocene and Early Holocene, McClure (1976) documents two significant wet phases. The first occurs between 37 and 17 ka, when lake levels were high and Arabia was occupied by hippopotami, water buffalo, elephants, gazelle, and oryx (McClure, 1994). Several radiocarbon dates from fluvio-lacustrine deposits or shells (McClure, 1976) place the wettest phase and paleosol formation (Anton, 1984) between 30 and 21 ka. The lakes of the first wet phase extended over parts of the central Rub' al Khali (McClure, 1978), the Nefud (Schulz and Whitney, 1987), and the Ramlat as-Sabatayn in Yemen (Cleuziou et al., 1992; Lézine et al., 1998). Contemporary paleosols have been documented in the ad-Dahna desert (Anton, 1984) Hadramaut in Yemen (Edens and Wilkinson, 1998), and the Yemeni highlands (Brinkmann and Ghaleb, 1987). The accumulation of river terrace sediments in the Late Pleistocene is documented in the Wadi Dhaid area (Sanlaville, 1992). Finally, calcites from alkaline spring deposits in Oman also date to between 33 and 19 ka (Clark and Fontes, 1990). However, as of the writing of this chapter, no dates currently exist from archaeological sites within this time period (McCorriston et al., 2002). Another wet phase occurred during the Early Holocene (~10–6 ka) and was accompanied again by high lake levels, dense vegetation (Lézine et al., Lézine et al., 2007; McCorriston et al., 2002), and even a period of peat formation in the Yemeni highlands (Davies, 2006).

In between these two wet phases, the Arabian peninsula appears to have experienced a period of extreme aridity, with only short humid phases around ~17, 14 and 12 ka (Sanlaville, 2000: 78). This arid phase set in after ~17 ka and is documented in several locations by extensive aeolian deposition and dune formation in Rub' al Khali, the Nefud, and Wahiba areas (Anton, 1984; Gardner, 1988; McClure, 1978). One of the causes of this Late Pleistocene hyper-aridity was shifts in monsoonal intensity. Decreased precipitation by the Indian Ocean Monsoon caused by lower solar radiation led to more desiccated climate prior to the Early Holocene (DeMenocal et al., 2000; Gasse and van Campo, 1994). Cold and dry conditions are documented at several localities in Arabia, coinciding partly with the cool and dry Younger Dryas (12.8–11.5 ka. BP). Wilkinson et al. (1997) document a dry and cold Late Pleistocene in Yemen followed in the Early Holocene by the accumulation of terrace sediments and wadi gravels and a phase of land surface stability during the formation of the Jahran soil complex ca. 7.2–4.35 ka.

Geomorphological studies in northern Arabia reinforce McClure's (1976) interpretations. Lacustrine deposits in the Jubba area (western Arabia) date to 25,630 ± 430 BP and coincide with similar deposits in the adjacent al-Jafr Basin of southern Jordan (Garrard et al., 1981: 141). Subsequently, evaporites document a phase of increased aridity and reduced precipitation after 20 ka. This pattern of moist followed by dry conditions is noted at several localities in the region, including the Lake Lisan (Neev and Emery, 1967) and the Sahara, where lake levels drop after ca. 15 ka (Garrard et al., 1981: 141). In the Azraq Basin (and likely northern Arabia) a sequence of humid conditions, followed by arid conditions, followed by a return to humid conditions by the end of the Pleistocene and beginning of the Holocene is documented (Garrard and Stanley-Price, 1977: 112).

The climatic shifts extending over the Late Pleistocene and Early Holocene are central to this discussion. Certainly, these shifts impacted prehistoric populations, perhaps even leading to an abandonment of parts of the peninsula coinciding with the hyper-arid phase. If true, this may explain the rareness of Late Pleistocene sites. Although it may be deterministic to presume that the lack of Late Pleistocene sites indicates a regional abandonment as the climate forced people out of a land now uninhabitable, the impact of such widespread and dramatic fluctuations should not be discounted.

Late Pleistocene Sites in Arabia

Terminology and Nomenclature

In an attempt to avoid or limit the use of terms that define highly-circumscribed cultural-chronological complexes of the Levant or Africa in Arabian contexts without critical evaluation of whether these Arabian assemblages carry the same connotations, chronological divisions based on paleoenvironmental data, such as Pleistocene and Holocene, rather than the cultural labels of Paleolithic, Epipaleolithic or Neolithic, are used here. A scarcity of data for the Arabian record precludes clear discernment of cultural complexes, phases or industries for this time span.

Therefore, Late Pleistocene refers to the entire period from the first phase of high lake levels (beginning ~37 ka) until the Early Holocene, and more specifically, Terminal Pleistocene, is used for the very end of this sequence, from about 20 to 10 ka. Although these terms are used in reference to the archaeological record between the comparatively better-defined Middle Paleolithic (Petraglia and Alsharekh, 2003) and Neolithic (e.g., Edens, 1982; Charpentier, 2004; Crassard et al., 2006) periods (arguably the former suffers from many of the same issues), it is premature to assign culturally qualitative terms to these assemblages until we have a better understanding of them from well-dated or stratified sites.

In describing these sites below, focus is placed on the record from Saudi Arabia as it is extremely detailed in nature as a result of the Comprehensive Archaeological Survey Programs initiated by the Department of Antiquities of Saudi Arabia and each survey's publication in *Atlal: The Journal of Saudi Arabian Archaeology*. The record for Yemen is similarly detailed, where several national and international research projects have been operating, and publishing, for several decades (e.g., Caton-Thompson, 1939; Garbini, 1970; de Maigret, 1986; Whalen and Pease, 1991; McCorriston, 2002). Unfortunately, data from Oman, and elsewhere in Arabia remained sparse for this time period, until relatively recently (Rose, 2004a, 2006).

Identifying and Defining the Late Pleistocene

The big questions of the Late Pleistocene revolve around identifying and characterizing these sites and attempting to reconstruct hunter-gatherer activities in Arabia. From 1976 to 1981, the Comprehensive Archaeology Survey Program initiated by the Department of Antiquities of Saudi Arabia covered large areas of previously under-examined territory within the country and did much for our understanding of both the geomorphological and archaeological record of the Pleistocene (Adams et al., 1977; Masry, 1977). A close association between the few discovered UP sites and ancient lacustrine deposits, extant wadi systems, and raised terraces in the southern Nejd, eastern 'Asir, central plateaus of northern Najd and Ha'il and the Wadi Sirhan drainage (McClure, 1976; Masry, 1977: 10) highlight that high resolution geomorphological data is key to discovering and understanding these site distributions.

Most of our dataset on Late Pleistocene sites derives from surface scatters of lithics found during survey and a very few soundings at sites (Zarins, 1998). Almost no clearly in situ, stratified sites have been documented. The ubiquity of Middle Paleolithic sites and lack of later Paleolithic sites may suggest that Arabia was occupied by local hunter-gatherer groups during the Middle Pleistocene, yet largely abandoned afterwards with the onset of arid conditions. Or, as has been suggested by certain researchers (Bailey, 2009; Rose and Usik, 2009), perhaps the absence of Late Pleistocene sites is due, in part, to populations living out on the now submerged continental shelf, until the Early Holocene when sea levels rose.

So, we are left with a lack of sites, especially stratified ones, no chronological records, and almost no diagnostic lithic assemblages for the Arabian Late Pleistocene. As a result, we have problems defining what sites or assemblages constitute the Late Pleistocene record, if there even is one. There is, of course, the possibility that Late Pleistocene lithics are not being recognized during survey and unstratified material is being mixed with earlier or later material. Some researchers (e.g., Cleuziou et al., 1992; Rose, 2004a; Uerpmann, 1992; Zarins, 1998) have pointed out that perhaps the Late Pleistocene in Arabia is represented by a continuation of Middle Paleolithic or MSA traditions, with only minor modifications. Therefore, these sites look much like those of the Middle Paleolithic, with the addition of a few new tool types.

Early attempts to identify and define the character of the Arabian record during this time period have tended to focus on the adjacent UP and EP of the Levant (Masry, 1977; Zarins et al., 1978; Zarins, 1998). While these connections no doubt existed in parts of Arabia, for example between the Azraq Basin in eastern Jordan and Wadi Sirhan in northern Arabia (see Garrard et al., 1981), one cannot assume they extended throughout Arabia or were exclusive. Connections are readily drawn between East Africa and Arabia in terms of floral and faunal trends (i.e., the Saharo-Arabian phytogeographic zone, Zohary, 1962) and so one might also consider interaction between these regions just as valid when discussing cultural connections (e.g., Bar-Yosef, 1987; Rose, 2004a). Of course, one should not think of the entire Arabian peninsula as belonging to the same unified cultural tradition during any time period. It seems equally possible that northern Arabia might exhibit stronger similarities with sites in the Levant, while the record of southern Arabia might more closely resemble that of East Africa. In these terms, there may be multiple UP, EP, LSA, or other traditions represented throughout Arabia, some of which might be influenced by the Levant, others by East Africa, and still others indigenous.

Alternatively, if there was abandonment of large parts of the Arabian peninsula during the Late Pleistocene, then tracing a continuous cultural trajectory from the Middle

Paleolithic through into the Holocene would be impossible. The current lack of typological, technological and chronological resolution towards the end of the Paleolithic means that, at present, it remains uncertain how these sites relate to any contemporary sites in other adjacent regions, or whether they should be considered solely within their own context (see below).

Late Pleistocene Sites

Despite our best efforts, compiling the exact number of potential Late Pleistocene sites reported throughout Arabia is extremely difficult (Fig. 1, Table 1). Many sites are not attributable to any particular part of the Paleolithic and, thus, must be assigned more generally to the Paleolithic (e.g., Parr et al., 1978; Zarins et al., 1979). A few multi-component Paleolithic sites have been excavated in eastern Yemen, which include UP horizons (Amirkhanov, 1994), but largely, our data is restricted to a few, small, deflated lithic scatters.

During the first Comprehensive Archaeological Survey of Saudi Arabia in 1976, Adams et al. (1977) surveyed over 20,000 km^2 in the Jawf and Wadi Sirhan areas of the eastern

and northern provinces. Focusing on wadis, springs, and sabkha margins, they made several surface collections and small soundings. Although geomorphological evidence suggested wetter Late Pleistocene conditions (McClure, 1976) and they concentrated collection on diagnostic artifacts useful for typological and chronological assignment, they report no definitive Late Pleistocene sites. Several surface scatters were assigned more generally to the Paleolithic, including a site with Mousterian tools accompanied by scrapers made on blades (Adams et al., 1977: 30). In the northern province, Adams et al. (1977: Plate 13:1–4) note a few sites (including site 200-32 near Jabal Umm Wu`il) containing large blades and flakes that could belong to the Paleolithic (Fig. 1). Several sites (sites 200-30, 200-31, 200-33a, and 200-4) with well-defined blade tools, some backed or denticulated, and burins and endscrapers that could be UP, but are all assigned to the Early Holocene on the basis of comparisons with the Pre-Pottery Neolithic B site of Beidha (Adams et al., 1977: Plate 13: 5–15, 16–38, Plate 14: 7–22). Finally, Zarins et al. (1981: 19) described two UP sites in the Wadi Sirhan identified during a photographic survey of the northern province in 1975 that exhibited "bipolar cores, double-struck burins, small fine blade technique (Tixier, oral communication)".

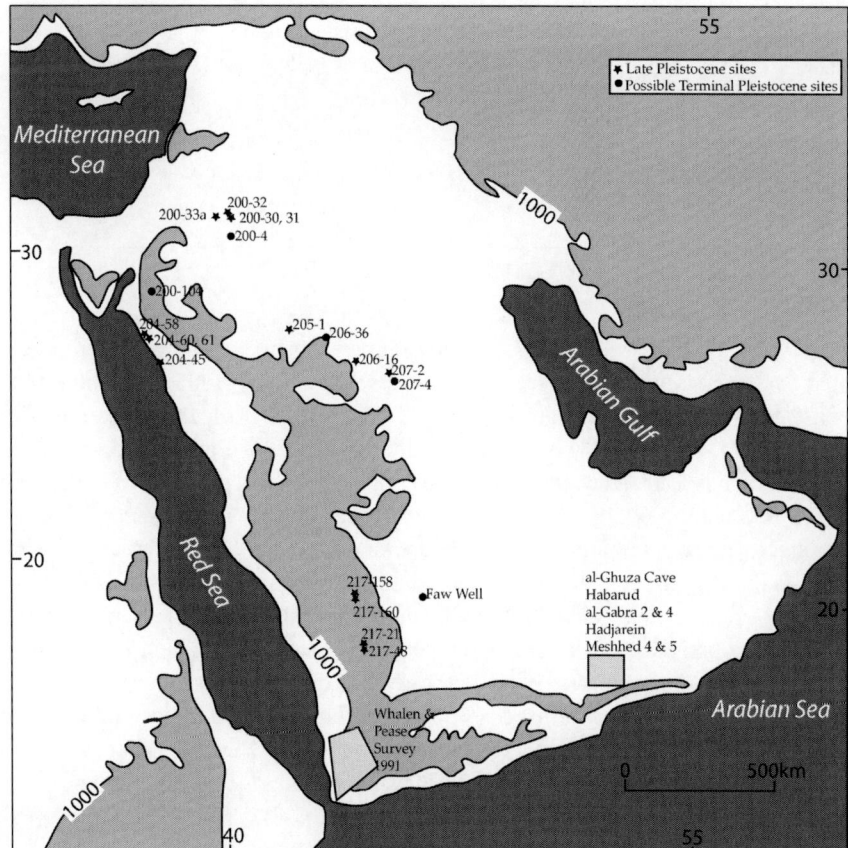

Fig. 1 Map of the Arabian peninsula showing the locations of selected Late Pleistocene and Terminal Pleistocene sites discussed in the text

Table 1 Late Pleistocene sites of the Arabian peninsula discussed in the text

Site designation	Cultural complex	Reference
200-32, Barduwil	Paleolithic, general	Adams et al. (1977)
200-30, 31, Umm Wu`il	Contains Paleolithic pieces?	Adams et al. (1977)
200-33, Dawqira	Contains Paleolithic pieces?	Adams et al. (1977)
Surface material from Jawf and Wadi Sirhan areas (not mapped)	Late Pleistocene	Adams et al. (1977)
Two possible unnamed sites from Wadi Sirhan (not mapped)	Late Pleistocene	Parr et al. (1978)
213-38, Ihsa/Umm ar-Rasas	Late Pleistocene	Anonymous (1989)
203-207, Kharj	Late Pleistocene	Anonymous (1989)
205-1, northern province	Late Pleistocene	Parr et al. (1978)
206-16, northern province	Late Pleistocene	Parr et al. (1978)
207-2, northern province	Late Pleistocene	Parr et al. (1978)
Surface material (not mapped) from the northern province	Paleolithic, general	Parr et al. (1978)
Six sites in central and Southwestern Province	Multi-component sites with UP horizon	Zarins et al. (1980)
Nine sites in central and Southwestern Province	Paleolithic, general (MP and UP)	Zarins et al. (1980)
200-45, Wadi Efthala	Late Pleistocene	Ingraham et al. (1981)
204-58, northwestern province	Paleolithic, general	Ingraham et al. (1981)
204-60, 61, northwestern province	Paleolithic, general	Ingraham et al. (1981)
204-78, northwestern province	Paleolithic, general	Ingraham et al. (1981)
Surface material, western province	Paleolithic, general	Whalen et al. (1981)
217-158, Wadi Tathlith	Late Pleistocene	Zarins et al. (1981)
217-160, Wadi Tathlith	Late Pleistocene	Zarins et al. (1981)
217-21, Bi`ra Hima	MP with Late Pleistocene pieces	Zarins et al. (1981)
217-48, Bi`ra Hima	MP with Late Pleistocene pieces	Zarins et al. (1981)
Surface scatters, central province	Late Pleistocene lithics	Gilmore et al. (1982)
200-134, central province	Post-Pleistocene site with Late Pleistocene pieces?	Gilmore et al. (1982)
Several sites, southwest Yemen	Multi-component Paleolithic sites with UP horizons	Whalen and Pease (1991)
Isolated surface material, southwest Yemen	Late Pleistocene	Whalen and Pease (1991)
Surface material in Hadramaut	Late Pleistocene	Caton-Thompson (1939)
22 sites with both stratigraphy or surface scatters, including Al-Ghuza Cave, Habarud, Al-Gabra 2 and 4, Meshhed 4 and 5, Hadjarein	Late Pleistocene (Upper Paleolithic)	Amirkhanov (1991, 1994)
200-4, ar-Raslaniyah	Terminal Pleistocene	Adams et al. (1977)
207-4	Terminal Pleistocene	Parr et al. (1978)
Faw Well Site	Terminal Pleistocene	Edens (2001)
200-104, al-Aynah	Terminal Pleistocene	Ingraham et al. (1981)

The 1977 Comprehensive Archaeological Survey documented two regions with Late Pleistocene sites with radiocarbon dates – a site in Ihsa/Umm ar-Rasas (site 213-38) is dated to 20,475 ± 750 BP, while another in the Kharj area (site 203-207) has four radiocarbon determinations between 18,535 ± 850 and 25,695 ± 820 BP – however; the lithics belonging to these sites are not described (Anonymous, 1991: 61). In their survey of the north, Parr et al. (1978: 31) recorded several sites that could not be assigned to particular part of the Paleolithic due to a lack of typologically-distinctive material and radiocarbon dates. Nine of these were quarry sites, which could have been re-used several times throughout the Paleolithic period, and none were clear habitation sites. While there were no definitive Late Pleistocene sites discovered in 1977, a few sites (sites 205-1, 206-16, and 207-2) had cores, flakes, blades, and multi-facetted burins which

could belong to the Late Pleistocene (Parr et al., 1978: 35). Both Parr et al. (1978) and Garrard et al. (1981) reported no clear occupation of the Jubba Region throughout the Late Pleistocene, despite favorable climatic conditions. Notably, parallels were drawn by Garrard et al. (1981) to a similar gap in the Azraq Basin record. However, subsequent examination of this area by Garrard and colleagues (Betts, 1988, 1998; Garrard et al., 1988a, b, 1994; Byrd, 1994) documented an extremely rich Late Pleistocene record.

The 1979 season of the Comprehensive Archaeological Survey Program, which was the second season of survey in the central and Southwestern Provinces, noted six multi-component Paleolithic sites with Upper Paleolithic occupations, and nine other sites that could not be securely assigned as Middle or Upper Paleolithic in age based on typological grounds (Zarins et al., 1980). The Upper Paleolithic occupations exhibited lithics made on local andesite, greenstone and quartzite, and included flakes and blades, a variety of scrapers, burins, notches, and knives, all with little patina and fresh edges (Zarins et al., 1980: 12, Plate 17). Zarins et al. (1980: 12) argue that based on these findings, the Arabian UP "…reflects a continuation of Mousterian types but emphasizes enhanced skill in flaking, especially pressure flaking; the production of finely made points … and refinement in blade tool technology". However, it should be kept in mind that UP sites are poorly represented and usually lack classic Upper Paleolithic tool types. Zarins et al. (1980: 16) acknowledge that assignment of these sites to the UP is based entirely on the presence of burins, borers, blades, notches, and scraper varieties – tools that could all be seen as remnants of the Mousterian, especially considering the absence of classic UP point types, carinated pieces, and other blade tools common to the Levantine Aurignacian. In addition, fine, pressure-flaked points could also be Holocene in date.

Ingraham et al. (1981) surveyed the northwestern province and reported no definitive UP sites, but did assign seven sites to a general Paleolithic age. These were located on the tops of plateaus and on wadi terraces and were identified by comparing the lithic assemblages with known Levantine UP sites (Ingraham et al., 1981: 65). No Paleolithic sites were noted in the Hijaz or Hisma regions; however, in Wadi Efthala site 200-45 was assigned as general Paleolithic because it contained both flakes and blades. In addition, four general Paleolithic sites were noted on the wadi terraces along the coastal plains (sites 204-58, 60, 61, 78) with circular enclosures and fragmentary walls of basalt and limestone, but unfortunately their assemblages are not described in detail.

In their 1980 survey of the western province, Whalen et al. (1981) report 20 of 37 sites as belonging to the Paleolithic in general, but specify that they did not find any definitive UP sites. They do report 16 Post-Paleolithic sites with small, retouched lithics differing from those of the Neolithic (Whalen et al., 1981: 46), but give little other details. These sites were found on wadi terraces, gravel-covered slopes or plateaus. Scraper varieties dominate these assemblages, but are accompanied by knives, burins, borers, and denticulates (Whalen et al., 1981: Plate 50/2, 5, and 8). None of these look distinctly UP and there seems to be several pieces reminiscent of the Middle Paleolithic.

In their 1980 survey of the Southwestern Province, Zarins et al. (1981) reported no clear Late Pleistocene sites. Possible UP sites in this region were identified in 1979 on the basis of steeply retouched double endscrapers, flakes, and blades (Zarins et al., 1980). Two sites discovered in 1980 in the Wadi Tathlith, sites 217-158 and 217-160, are likely UP, with tools described above along with strangulated blades. Zarins et al. (1981: 19) explain the rarity of classic (Levantine) UP sites as the persistent use of Mousterian tools, perhaps with a few new additions, until the onset of extreme aridity after

20 ka. A similar trend may be documented in the Bi`ra region of southern Arabia at sites 217-21 and 217-48, where Mousterian assemblages also contain subsequently reworked pieces (Zarins et al., 1981: Plate 17b).

In their 1981 survey of the Riyadh area, Zarins et al. (1982) noted that Paleolithic sites were in abundance, likely a result of a wetter Pleistocene climate. However, again the many undiagnostic lithics can only be assigned to a general Paleolithic age. Zarins et al. (1982: 30) re-iterate that perhaps "Mousterian industries dominated the peninsula well into the Late Paleolithic and traditional Upper Paleolithic industries associated with Levantine types are not present".

In their 1981 survey of the interior of Saudi Arabia, Gilmore et al. (1982) report evidence for extensive occupation of the margins of interior wadi systems throughout the Paleolithic, although the survey covered a large area and only general statements are made about the surface collections. Several sites in this area have been tentatively assigned to the Paleolithic period (Parr et al., 1978: 35; Zarins et al., 1980: 16; Gilmore et al., 1982). There is a rarity of microlithic assemblages, but 16 sites exhibit distinct lithic assemblages whose "[p]atination, workmanship, and material appear intermediate between the Middle Paleolithic and Neolithic tools of the area, and are similar to those labeled as possible Upper Paleolithic by Zarins et al. (1980: 16) from the western Nejd and eastern Hijaz" (Gilmore et al., 1982: 12–13, Plate 30). They do not mention the names of these sites or where they are found. The lithics are predominantly large flakes or blades with steep unifacial retouch struck from amorphous cores. Tools include side-scrapers and endscrapers, notches, gravers, burins, and roughly-shaped points. They suggest that these sites may have structures comprised of "low, interconnecting loops of piled rubble which form compound enclosures, probably habitations" (Gilmore et al., 1982: 13, Plates 5 and 7a). Four sites (204-116, 204-119a, 204-127 and 204-128) also contain small upright stone slabs arranged in irregular clusters (Gilmore et al., 1982: Plate 7b). "At site 204-116a they are arranged within regular circles, six to seven meters in diameter, formed of larger, broader slabs" (Gilmore et al., 1982: Plate 6b). There are several other sites labeled as Post-Paleolithic or Neolithic that may include pieces belonging to the Late Pleistocene (e.g., site 200-134).

In a survey of southwestern Yemen, Whalen and Pease (1991) report Paleolithic sites that are mainly small, multi-component sites. Those that contain Late Pleistocene occupation exhibit small blades formed into scrapers or truncated, with little patina. They also discovered 329 lithics as surface scatters or isolated finds that they assign to the UP, presumably on the basis of typology. The exact locations, or other details, of these sites are not given.

In her survey of the Hadramawt in Yemen, G. Caton-Thompson (1939) noted UP material occurring in thick sandy loams overlying Middle Pleistocene gravels. However, this material lacks pieces such as bladelets, backed blades, and microliths. Amirkhanov (1991: 615) reports 22 UP sites in the western Hadramawt that fall into two categories: four sites with in situ deposits and 18 deflated surface scatters. However, all have exhibited some degree of post-depositional alteration and little further information is provided. Most of the surface material derives from find spots recorded as 'workshop camps' (Amirkhanov, 1991: 615) and are useful primarily for the technological information provided. Of the stratified sites, Meshhed 4 and 5 are located in the Wadi Dauan, while Wadi Khurut 1 and 3 are within the Mahra region. Although he does not specify site function for these latter sites, he does report that most of the base camps are located on plateau settings (Amirkhanov, 1991: 616). The sites are all described as UP on the basis of flat cores with parallel striking surfaces (single or opposed platform) that he argues are much more standardized in size and shape than preceding MP cores (Amirkhanov, 1994). The tools all exhibit moderate to fine retouch. Endscrapers are the most common tools, but are accompanied by points, awls, and knives made from modified blade blanks (Amirkhanov, 1994: 220). The only available date is a radiocarbon date of 31,000 ± 2,000 BP obtained from a nearby geological section, only stratigraphically correlated to the lower part of one of these UP sites (Amirkhanov, 1994). Comparisons with Africa and the Levant are based on typo-technological grounds and Amirkhanov (1994: 223) draws strong parallels between UP sites in southern Arabia and the Nile Valley and Libyan Sahara (rather than the Levant).

Some details are provided for a few of the non-stratified sites in the Hadramawt by Amirkhanov (1991, 1994). Again, most appear to be situated on plateau settings, but two were noted as occurring in the bottom parts of wadi slopes (al-Gabr 2b and Hadjarein). Both contain a collection of heterogeneous tools formed on large blanks and include side-scrapers, carinated scrapers, and other large scrapers with wide symmetrical marginal notches. The site of al-Gabr 4 is reported only as exhibiting 'typical' UP lithics (Amirkhanov, 1994: 224, Fig. 8).

According to Amirkhanov (1991, 1994), UP tools can be defined by a shift from Mousterian points and side-scrapers on flakes to a new reduction sequence focused on blade blanks that are subsequently formed into a variety of tools, including scrapers, points, borers, and knives. He noted that UP horizons were missing Mousterian points and side-scrapers, demonstrating that they are not simply a continuation of the MP, with an increasing trend towards microlithization (see Phillipson, 2005). According to Amirkhanov (1991: 617), there is a distinctive UP in Yemen and he draws comparisons to the Kharga and late Khormusian of the Nile Valley on the basis of retouch and tool morphology.

The Terminal Pleistocene

The Terminal Pleistocene, from about 20 to 10 ka, is discussed separately here for two reasons. First, this chapter began with the intention of focusing on the very end of the Pleistocene period, namely, examining the evidence for a Pleistocene microlithic tradition in Arabia similar to that of the Levantine EP. Given the rarity of sites assigned to this period, it was more useful to view them within the context of other Late Pleistocene sites. Despite the futility of separating out the 'Epipaleolithic' from other Late Pleistocene sites on the basis of current data, this section is retained for a second reason. Several surveys have made a distinction between assemblages tentatively assigned to the UP (or LSA) and those that exhibit a microlithic technology and are referred to as 'Epipaleolithic' (e.g., Zarins et al., 1980; Tosi, 1986). It should be stressed that in Late Pleistocene research in Arabia, we do not yet have the resolution to definitively discern UP/LSA from EP/LSA sites until our picture of the entire Pleistocene, especially the MP/MSA to UP/LSA transition (if there is one), becomes clearer. Further, it seems difficult to reconcile assigning sites to the EP period, given the connotations of a Levantine connection, when we are still not sure how to differentiate UP and EP sites in Arabia, and what their potential relationships to the Levant or Africa may have been.

Not surprisingly, considering a hyper-arid climate during the Terminal Pleistocene, there are even fewer sites assigned to this time period than the 'Upper Paleolithic' (Fig. 1, Table 1). Those that are assigned to the Terminal Pleistocene are generally done so on the occurrence of a blade/bladelet-based microlithic technology, and described as being Epipaleolithic with reference to Levantine EP lithic typologies and behaviors (e.g., Edens, 2001; Zarins, 1998). Although he admits that sites are scarce, Zarins (1998: 185–186) gives the Epipaleolithic (including Natufian) distinct consideration while discussing the prehistory of Arabia in relation to the Levant and Africa where, as in these other regions, their position as Neolithic predecessors is highlighted. Difficulty in identifying these sites maybe even greater than the UP as it rests largely on the discovery of very small tools. Surveys must cover large areas, increasing the chance of missing these unobtrusive sites, particularly if buried. Otherwise undiagnostic surface finds, such as scrapers or burins can be assigned only to many different periods.

Terminal Pleistocene sites are well-documented throughout the Levant in a variety of different environmental settings. They vary considerably in their assemblages, size and nature of occupation, both across space and over time (Garrard et al., 1988a, 1994; Goring-Morris, 1987; Henry, 1995; Edwards, 2001; Olszewski, 2001; Belfer-Cohen and Goring-Morris, 2002; Nadel, 2002; Maher, 2007). Similarly, in Africa these sites may be absent from some areas, but are well-represented elsewhere (Tixier, 1963; McBurney, 1967; Wendorf and Schild, 1980; Close, 1987; Phillips, 1987; Wendorf et al., 1993; Zarins, 1998; Phillipson, 2005).

The 1976 comprehensive survey in Saudi Arabia discovered the site of ar-Raslaniyah (200–4) in the northern province. Here, shell, bone, ash, and lithics made from white, non-local flint were discovered on a small sand spit overlooking the Wadi Sirhan (Adams et al., 1977). The lithic assemblage is dominated by blades, burins made on blades, and pieces truncated on both ends (Adams et al., 1977: 34, Plate 14). Based on comparisons with Epipaleolithic sites in the Azraq Basin, Adams et al. (1977) attribute this site to the early Epipaleolithic, possibly to the Kebaran. Aside from this site, the authors report only surface scatters containing microliths, but no other clearly Terminal Pleistocene sites.

The only evidence for Terminal Pleistocene occupation of the northern province discovered during the 1977 survey was a single isolated lunate/crescent microlith found from site 207-4 in the al-Majma region (Parr et al., 1978), which could be Pleistocene or later. Site 206-36, located east of Hail near al-Kuhayfiyah, exhibits stone walls with a large re-used mortar built-in and several bedrock mortars nearby (Parr et al., 1978: 39, Plate 27:b–c). The researchers attribute it the Natufian period, but with no detailed explanation as to why since there are no other remains, including lithics, reported from the site. Although several individual pieces from the 1977 survey resemble UP or EP material from the Levant (e.g., Parr et al., 1978: Plate 40: 55–56 single-platform cores, Plate 39: 46–52, 54 blade/bladelets), no detailed description of their location or other features are given. Parr et al. (1978: 35) point out that the virtual absence of Terminal Pleistocene sites here is in accordance with their absence from Qatar, Oman, Yemen, and the Rub' al Khali.

The site of Al-Aynah (200-104) north of Tabuk, discovered during the 1980 comprehensive survey, is primarily a Neolithic site, but may have an earlier occupation attested by the presence of bladelets and lunate-like microliths (Ingraham et al., 1981: 67–68). If so, it is the only known stratified site. In addition, the presence of lunates suggests possible connections to the Levant during this time. No clearly attributable sites belonging to the Terminal Pleistocene have been documented in southern Arabia.

Garrard et al. (1981) report no apparent occupation of the Jubba region during the Terminal Pleistocene, although as mentioned above, several sites have since been documented in nearby eastern and southern Jordan. Zarins (1998) suggests that Epipaleolithic sites like al-Aynah and other surface material along the margins of the Hamad and Nefud deserts resemble Levantine Natufian sites. Similar to the Azraq Basin, future work in this region designed specifically for detecting UP/EP sites will likely be very fruitful.

On the other hand, potential connections between Africa and southern Arabia should not be ignored. This is especially apparent as early Levantine EP research makes this same connection with North and East African when considering the appearance of Helwan (bifacial) retouch on microliths at Mushabian and Natufian sites in the Levant. Indeed, several sites in the southern Negev and Sinai have similarities to contemporary ones in Africa (Phillips and Mintz, 1977; Bar-Yosef, 1987; Phillips, 1987, 1994). Again, the nature of these connections remains speculative until further data is collected from Terminal Pleistocene sites in Arabia.

The presence of a microlithic technology, with cores, blade/bladelets and microliths can indeed be found in Arabia (e.g., in the Northern Province), albeit rare; however, geographical and chronological affiliations remain unclear. It does seem possible that the entire region was not simply abandoned during hyper-arid conditions of the Terminal Pleistocene. The possibility of occupation throughout the Pleistocene, although given the limits of current data, it may have been episodic or geographically restricted, has important implications for our understanding of hunter-gatherer adaptability and migrations, as well as interaction with neighboring regions. It may make more sense to see these movements between the Levant, Africa, both, or elsewhere, as fluid, despite (rather than caused by) changing climatic conditions.

The Faw Well Material

During a study of collections at the Saudi Arabian Department of Antiquities, Edens (2001) analyzed a bladelet assemblage from Qaryat al-Faw (Fig. 1). The assemblage comes from two sites (212-32, 212-34) in the al-Faw area of southern Arabia. It is now the only detailed analysis of a likely Late Pleistocene assemblage and highlights that bladelet industries did indeed exist in Arabia. The tentative assignment of a Late Pleistocene age is based on 216 flakes and blades, although attribute analysis was conducted on only 34 of 80 unbroken blades, all made on fine-grained, grey-brown flint.

Both core preparation and on-site production is documented by the presence of crested blades (one complete and one broken), six blades struck immediately after removal of the crested blade (one plunging), and core preparation flakes (Edens, 2001: 138). Later stage removals include narrow bladelets, usually less than 12 mm in width and, thus, fitting Tixier's (1974) definition of bladelets. Small platforms, sometimes pointed or broken, are noted from the proximal ends of blades.

Three cores were analyzed and all show removal scars consistent with blade and bladelet production. One core is a wedge-shaped, single platform core with blades removed from a narrow face (Edens, 2001: Fig. 2.4). One core is a flat, opposed platform core with faceted striking platforms and a relatively wide area of removal scars (Edens, 2001: Fig. 2.3). The third is a single platform core prepared by flakes struck from either side and with one blade removal.

Retouched pieces include backed blades, double backed blades or point varieties, endscrapers on blades, and denticulated or notched blades (Edens, 2001: Fig. 2.1). The cores and tools reported by Edens (2001) all fit clearly within a bladelet-focused technology comparable to those of the UP/EP of the Levant (e.g., Bar-Yosef, 2002; Belfer-Cohen and Goring-Morris, 2002) although, of course, blades are not restricted to these periods. Edens (2001: 141) also describes bladelets backed on both edges from the Faw Well site that, according to him, typologically resemble late Upper Paleolithic (Ahmarian) assemblages from the Levant dating to ~20 ka. Although a more detailed technological and typological analysis, along with absolute dates from this site are required for confirmation, it does present the possibility that UP/EP sites are indeed present in southern Arabia.

Summary discussions of spatial and temporal patterns during the Late Paleolithic (or UP/EP) of Arabia are, for the present time, premature. General discussions of lithic, particularly microlithic, form, size, typology, technology or *chaîne opératoire* remain extremely limited in scope. From the little data that we have so far, what can be at best tentatively assigned to the UP/EP, includes single-platform and opposed cores that can be flat or sub-pyramidal in shape. Both flake and blade/bladelet removals have been noted on these cores. Blade and bladelet tools include retouched blades, rare points, burins, a variety of scraper forms and, sometimes, microliths. However, many sites do not exhibit assemblages that fall neatly into these categories and we have to be cautious about generalizing about the Late Pleistocene in Arabia. There is a danger of putting the cart before the horse where its identification and origins are defined on the basis of known UP/EP or LSA industries from elsewhere when the nature and timing of connections with these regions is still unknown.

Other Microlithic Sites

Here, I would be remiss not to mention the presence of a distinctive and clear geometric microlith industry found at a few sites in southern Arabia, including at Hureidha, Wadi Jubba, and elsewhere in Yemen (e.g., Rahimi, 1987). At first glance, these trapezes, rectangles and crescents appear very similar to the geometric microliths of the middle and later Levantine EP (see Rahimi, 1987: 143, Fig. 8.2). However, there are several important technological differences that

distinguish them as not belonging to the Terminal Pleistocene. Microlith production here seems to focus on trapezoids and lunates made on small obsidian blades and flakes (Rahimi, 1987: 139–143). The microliths are generally not made from clear bladelet-based blanks, but rather from flakes of various shapes and sizes. They are not accompanied by any bladelet-based debitage or cores. There is no evidence for the use of the micro-burin technique. And, there are no small, single-platform, pyramidal cores with evidence of bladelet removals. Although they may resemble some Levantine EP tools in final shape or some East African, non-bladelet LSA industries (Phillipson, 2005), these geometric microliths were also excavated from stratified sites which clearly indicate their later age, usually Iron Age (Rahimi, 1987).

Zarins and Zahrani (1985) noted site 217-175 in the Tihama area as exhibiting a large concentration of lithics that included snapped blades and backed lunates. The following year, Zarins and al-Badr (1986) discovered more sites of a similar type, including one extending over 6000 m² and with several small shell middens. The lithics are similar to 217-175, with choppers, scrapers, blades, flakes, awls, groovers, burins, and sickles, as well as a distinct transverse projectile point. They also note several unbacked lunates and a microlith core made from obsidian. The complete tool assemblage and accompanying other artifacts, namely polished greenstone, stone bracelets, and basalt tools, seem much more similar to Neolithic or later sites (e.g., Henry, 1982; Gilead, 1989; Rosen, 1997).

The phenomenon of microliths appearing in several time periods is not unique to Arabia. The earliest microlithic technologies date to the MSA in Africa (e.g., McBrearty and Brooks, 2000; Ambrose, 2002), but continue to be used until relatively recent times (Phillipson, 2005). Chalcolithic sites in the Levant exhibit geometric microliths, mostly lunates, made on flakes (Henry, 1982; Gilead, 1989; Rosen, 1997). It seems that they are not the straightforward chronological indicators we once thought.

Summary of the Late Pleistocene

In sum, it seems that virtually none of the sites assigned to the Late Pleistocene throughout Arabia are clearly typical of the late UP/EP as we know it from the Levant – very generally, a blade-focused technology resulting in the production of long, parallel-sided blanks, some further modified into blade/bladelet tools such as endscrapers, burins, variously retouched blades, points, and microliths, all produced from prismatic cores (e.g., Bar-Yosef, 1994, 2002; Bar-Yosef and Kuhn, 1999; Belfer-Cohen and Goring-Morris, 2002; Goring-Morris and Belfer-Cohen, 2003). Similarly, none of these sites are clearly assignable to any particular LSA industry in

East or North Africa. However, it is obvious that some similarities exist between these sites and contemporary assemblages from both the Levant and Africa (see below).

So, why are Late Pleistocene sites so rare? It seems that there are a few likely possibilities. First, no humans occupied the region for much of the Late Pleistocene. Although the peninsula experienced favorable conditions from 37 to 17 ka, and occupation may have continued from the Middle Paleolithic, the period between 17 and 12 ka was hyper-arid and perhaps we should not be surprised to find very little dating to the latest UP and EP. Geomorphological deposits dating to the Late Pleistocene come from several areas throughout the Arabian peninsula (see above) and demonstrate that the lack of sites does not correspond to some large-scale scouring or removal of sediments of the appropriate age. It seems that none of these deposits contain any archaeological material and suggest the possibility that the lack of sites is a 'real' pattern, at least in some areas.

Second, Late Pleistocene sites are in Arabia, but have not yet been found. Perhaps as a result of the large geographical area encompassed by the Arabian peninsula necessitating less detailed coverage of each area during preliminary surveys, the demonstrated importance of allying geomorphological studies to these surveys in order to locate and target the appropriately-aged deposits, or the general temporal foci of current projects, these sites may simply be awaiting discovery.

Third, UP occupations have been found but there is no consensus on the validity of this material (it is not being recognized) as UP because it is based solely on typological comparisons with the Levant or, more recently, East Africa, rather than considering the Arabian material on its own terms. Amirkhanov (1991) reports UP blank blades, points, awls, and knives from al-Ghuza and other sites all dating to around 31 ka. Whalen and Pease (1991) report small pressure-flaked and truncated blades in southwestern Yemen. Further north, work at the Rub' al Khali lakes indicates no evidence of Late Pleistocene occupation, even though reconstructions indicate a favorable climate for most of the Pleistocene. In the Dhofar and Mahra areas, Zarins (1998) reports sites with typical UP blades, endscrapers, side-scrapers and points. However, their deep patination and other features suggest to him they could be a continuation of the local MSA tradition.

Inizan and Ortlieb (1987) argue for an absence of a true blade technology – the traditional succession of a flake-based Middle Paleolithic followed by a UP blade industry as we know it in Europe and the Levant does not occur in Arabia. Zarins et al. (1980: 16–17) suggest a number of reasons for this. Differing environmental conditions may have necessitated different tool kits. If the Arabian paleoenvironment can be considered as a continuation of that from East Africa, then perhaps the Arabian Late Pleistocene is as well. Or, it is possible that the use of Mousterian tools persisted until the Neolithic, and the few Late Pleistocene assemblages thus far represent a brief transitional period (Zarins et al., 1980: 16). In this hypothesis, the UP exists, but spans a very brief time period, while the Mousterian is lengthened and accounts for most of the sites from 30 ka until the Neolithic. The fact that there are so few UP sites in other areas of Arabia may support this idea. In this scenario, Arabia was not abandoned in the Late Pleistocene, but rather we must broaden our expectations of what the Late Pleistocene looks like.

Of those few Late Pleistocene sites that have been reported, they are generally identified on the basis of a blade-based technology with scrapers dominating, or as exhibiting some combination of or transition between MP flake tools and UP blade tools. So, is there a well-defined Late Pleistocene in Arabia, identifiable and distinct from both the preceding MP/MSA and the following Neolithic? And, do we know enough about this time period yet to identify what may be simple variation versus trends over time (especially when these sites are largely undated)? If the Late Pleistocene is similar to Africa or the Levant, can we even identify it in Arabia based on current evidence or without comparison with Levant? If not, what existed there instead? It seems that "the late occurrence of Middle Paleolithic technology overshadows the rarer blade bearing industries of the Upper Paleolithic. The Arabian situation may show a different Middle to Upper Paleolithic trajectory from other regions, supporting the notion that a uniform and directional change from flake- to blade-based industries should not be expected" (Petraglia and Alsharekh, 2003: 679).

Discussion

Are We Missing the Arabian Late Pleistocene?

As a Levantine prehistorian, the first thing that strikes me in regards to the Arabian record is the obvious lack of Late Pleistocene sites. In the Levant, several decades of intensive survey and systematic, rigorous excavation have provided a detailed techno-typological record of material culture set within an excellent chronological framework. In particular, stone tools, as the most numerous artifacts by far, have shed innumerable insights into hunter-gatherer behavior and activities over time and space.

In contrast, the geographical area encompassed by Arabia is enormous compared to the Levant. In the Levant, projects are largely bounded by geographical barriers (usually wadi systems) or modern political boundaries and the practicalities of working around these. For example, surveys in southern and eastern Jordan cover a geographical and ecological zone that extends into western and northern Saudi Arabia (e.g., the Wadi Sirhan), yet for practical reasons research is confined to Jordan's borders. Arabian projects account for the vastly larger area and the artificial separation of continuous environmental zones in numerous ways. The Comprehensive Archaeological Survey Program initiated by the Department of Antiquities of Saudi Arabia divided the country into several provinces, surveyed in different years, in an attempt to cover the entire country to some degree. The logistics of a project of this scope are complex and Adams et al. (1977) and Parr et al. (1978) discuss these in their survey reports, along with the inevitable result that some time periods and areas were more intensively covered than others. In other countries, such as Yemen and Oman, the geographical areas are more manageable and many projects seem to take an approach similar to that of Jordan. That is, carving the country into smaller, manageable research units defined by geographical features, such as the Hadramawt in Yemen (e.g., McCorriston, 2000). As in Jordan, these projects operate in well-defined and mutually-exclusive areas, covering sites from many time periods. The result is a highly refined reconstruction of local occupation, but with gaps for those areas where project boundaries are not contiguous. Covering large areas obviously entails consideration of how to do this in a manner that is representative of the entire archaeological record, including those sites that are less obtrusive (as the Late Pleistocene ones inevitably are).

Another problem is that there are so few radiometric or other dates from Late Paleolithic sites anywhere in the peninsula, particularly from secure archaeological contexts for the period covering 30 to 10 ka. These sites are represented by surface scatters of lithics, with no remaining evidence for subsurface stratigraphy or traces of any other features, if there ever were any. This lack of dates means that if Late Pleistocene sites do exist, but appear different from what we are used to in Africa and the Levant, then not being able to date them means we may simply be overlooking them.

The problem of identification is central to another related issue of the Arabian Late Pleistocene – that is, how to define sites belonging to this time period. The Late Pleistocene could be expected to look different in different parts of Arabia. For example, the north may have experienced Levantine influences (or vice-versa) and resemble nearby sites in Jordan or Iraq while southern Arabia experienced ties to Africa and more closely resembles LSA sites there. Rather than which is a more valid comparison – the Levant or Africa – perhaps they both are. Or, perhaps neither is valid and the Late Pleistocene of Arabia should be seen on its own terms (e.g., Zarins et al., 1980: 16).

It is also possible that we are 'missing' Late Pleistocene sites in Arabia because this time period is less focused-on than other periods. Like elsewhere, the history of research reflects both the current interests of the discipline and those of the researchers. In Arabia, research is focused on the earlier portions of the Pleistocene or Neolithic and later sites. In part, this stems from the more numerous (and more easily recognizable) findings from these periods. But, it may also reflect a discipline-wide interest in these periods for their contributions to big-picture debates regarding human dispersals Out of Africa (e.g., Lahr and Foley, 1994, 1998; Petraglia, 2003; Petraglia and Alsharekh, 2003; Rose 2004a, b) and the origins of agriculture (e.g., McCorriston et al., 2002, 2006). For example, a southern dispersal route through southern Arabia has very important implications for our understanding of human migrations and patterns of adaptation in Arabia. Not only would early humans be able to occupy southern Arabia from East Africa, but also northern Arabia through the Levantine corridor. This scenario places Arabia, geographically and culturally, as a potential hub of Paleolithic activity.

In this light, it seems that the activities of Late Pleistocene groups, or even questions of where they might be moving from or to, during the shift from wet to arid conditions when occupation of parts of Arabia may have been extremely challenging at best, is quite complex. Paleoclimatic reconstructions suggest extreme aridity at the end of the Pleistocene and some researchers have drawn attention to the possibility of abandonment of parts of this region during this time (e.g., Zarins et al., 1980). Alternatively, Flemming et al. (2003) point out the possibility that these missing sites, if corresponding to hyper-arid conditions in the central part of the peninsula, may have been coastal sites now underwater. This may be particularly likely if the interior regions were too desiccated for occupation, but coastal areas were not. Although it has been suggested here that Late Pleistocene sites can be found in Arabia, the little available data necessarily side-lines debates about human activities in Arabia during this time.

Connections to the Levant (and Africa)

Perhaps part of the problem in identifying Late Pleistocene sites in Arabia stems from our proclivity to draw parallels with the Levant where this period is well-represented. We know exactly what to expect of the UP/EP in the Levant. When we do not find it in Arabia, particularly in the north, we wonder why this period is 'missing' from the archaeological record. As pointed out by Tosi (1986: 462), we describe the ecology and vegetation of Arabia (and the southern Levant) as a continuation or westward migration of plants and animals of the Saharo-Arabian phytogeographic zone during favorable climatic conditions (Zohary, 1962). Yet, we do not view Arabia this way culturally – as connected to

African traditions influencing Arabia (and vice-versa) through the continued interaction of people (although see Rose, 2004a, b). To take this further, perhaps we consider Arabia too much as a place to be colonized, either from Africa or the Levant? While this may be true for some earlier and later periods, it is not necessarily so for the Late Pleistocene.

Tosi (1986) insists we should abandon a perspective that views Arabia in relation to the Levant and instead investigate Arabia critically and on its own terms. We cannot simply impose a core-periphery model to the region, with the better-known Levant influencing the interpretation of the lesser-known Arabian peninsula. However, I argue that maintaining some acknowledgement of the fluid and dynamic interaction of prehistoric people between the Levant and Arabia is logical and necessary in order to fully appreciate both the Levantine and Arabian archaeological record. Although we cannot at this point directly correlate occupations in the Levant and Arabia during the Late Pleistocene, for example, tracing an Upper Paleolithic to Epipaleolithic trajectory or discerning various EP complexes, this does not preclude their existence. Historically, Jordan has been linked with the southern Levant and enjoyed a comparatively intensive degree of research. Yet, we must also remember its connections with Arabia. Much of eastern and southern Jordan's geography, past and present, relates closely to that of Arabia and we must consider the likelihood that people's activities here likely did too.

Furthermore, the possibility of Levantine connections must be seen within the context of possible connections to Africa. In fact, in this respect Arabia could be seen as a geographical axis that hints at its potential importance as a centre of prehistoric activity – a place of occupation and migration, interaction and exchange – to and from multiple regions.

Late Stone Age occupations are documented at several sites in North and East Africa, most notably at the stratified cave site of Haua Fteah (McBurney, 1967). Clark (1954) constructed a chronological sequence for the Middle Stone Age (MSA) and Late Stone Age (LSA) of East Africa consisting of three MSA complexes (Acheulo-Levalloisian, Levalloisian, and Somaliland Stillbay) and four LSA complexes (Magosian, Hargeisan, Doian, and Wilton). The first two LSA complexes are transitional from the MSA, while the latter two LSA entities are equated with the Terminal Pleistocene. Clark (1954) argues that the MSA traditions continue into the early LSA, although this transition is poorly dated and lithic data is sparse. During this MSA/LSA transition, tools decrease in size generally and there is a shift from discoidal cores and Levallois technique to backed blades, burins, enscrapers, and microliths. Therefore, the Magosian and Hargesian industries are transitional, exhibiting characteristics of both the MSA and LSA with unifacial and bifacial points and side-scrapers from the MSA and backed blades and bladelets and microliths from the LSA. This blurs the transition from MSA to LSA and makes assigning sites to one or the other far from clear. If connections with Africa are demonstrable during the Late Pleistocene, this has important consequences for identifying these sites in Arabia. Rather than a clear blade or microlith industry, these sites may document a 'mixture' of typical MP/MSA and UP/LSA features and, indeed, similar sites have been found (e.g., Amirkhanov, 1994; Zarins, 1998: 184).

Conclusions

The presence of Late Pleistocene sites in Arabia, albeit rare, hint at some very interesting potential future research directions. Given the interest in tracing African–Arabian migration patterns throughout the Paleolithic and dispersals of anatomically modern humans, it seems that a better understanding of the Arabian Late Pleistocene, especially in relation to the Levant and Africa, can provide numerous insights. Movement and migration between all of these areas, and further east (although not discussed here) likely continued throughout this period. Therefore, there is no reason to expect direct connections with only one region, or that these connections didn't fluctuate in intensity or direction over time.

Although there is little data to-date and our interpretations of it must be considered preliminary; the presence of these sites provides tantalizing clues to what else may be there, given the opportunity to explore further. I hope that the summary presented here provokes an interest in this little-known period and contributes something useful to existing debates regarding the timing, geography, and nature hunter-gatherers in this region.

As pointed out by Petraglia and Alsharekh (2003) researchers have tended to use a tripartite division of the Paleolithic in Arabia, dividing it into Early, Middle, and Upper phases. However, for the latest phase, we still do not have a clear understanding of how it relates to the original Levantine industries from which these names derive. In the end, the big questions remain open to debate: What is the Arabian Late Pleistocene? Does it exhibit its own distinct or indigenous cultural or stylistic tradition and trajectory of development or change? Or, does it draw its closest parallels with the Levant, Africa, or both?

This chapter cannot help but highlight how little we know of this time period in Arabia and, as a result, an attempt is made to stress the need for refining our research questions about the Arabian Late Pleistocene, particularly in comparison to the Levant and Africa. The goal of this chapter was to present a synthesis of the Late Pleistocene as we currently know it, and examine some of the key research issues from the perspective of a Levantine prehistorian, although I have tried to balance this view in regards to Africa. So far, it seems the Late Pleistocene cannot be assigned as 'Levantine' in character, but there are some interesting hints at Levantine connections in the north, and African connections in the south, and, recent analyses (Edens, 2001) strongly suggest that there is a Late Paleolithic blade/bladelet or microlithic industry to be found. It is not that the entire peninsula was abandoned during the Late Pleistocene. Rather, something else may have been going on here that we don't yet fully recognize because we are focused on looking for blades and microliths that we expect from other areas.

Acknowledgments I would like to thank Mike Petraglia for getting me interested in the Arabian Paleolithic and giving me the opportunity to participate in this volume. I would also like to thank Danielle Macdonald for many useful discussions on the Late Pleistocene of Arabia, North Africa, and the Levant. I also thank Mike Harrower for providing me with several hard-to-find publications and for several useful discussions about Arabia. Remy Crassard and Dan Rahimi are thanked for discussions and their comments on the data. I would also like to thank Marta M. Lahr and Rob Foley for introducing me to the North African Epipaleolithic and for several discussions on the Paleolithic. Finally, I would like to thank Kevin Gibbs for reading over and commenting on an earlier version of this chapter.

References

Adams R, Parr P, Ibrahim M, al-Mughannum AS. Preliminary report on the first phase of the Comprehensive Survey Program. Atlal. 1977;1:21–40.

Ambrose S. Small things remembered: origins of early microlithic industries in Sub-Saharan Africa. In: Elston RG, Kuhn S, editors. Thinking small: global perspectives on microlithization. Arlington, VA: Archaeological Papers of the American Anthropological Association Number 12; 2002. p. 10–29.

Amirkhanov HA. The Paleolithic of South Arabia (English Summary). The Paleolithic of South Arabia. Moscow: Academy of Sciences; 1991. p. 599–632.

Amirkhanov HA. Research on the Palaeolithic and Neolithic of Hadramaut and Mahra. Arabian Archaeology and Epigraphy. 1994;3:217–28.

Anonymous. Radio carbon dating. Atlal 1986;13:61–4.

Anton D. Aspects of geomorphological evolution: paleosols and dunes in Saudi Arabia. In: Jado AR, Zötl JG, editors. Quaternary period in Saudi Arabia, vol 2: sedimentological, hydrogeological, hydrochemical, geomorphological, and climatological investigations of western Saudi Arabia. Wien: Springer; 1984.

Bailey G. The Red Sea, coastal landscapes, and hominin dispersals. In: Petraglia MD, Rose JI, editors. The evolution of human populations in Arabia: paleoenvironments, prehistory and genetics. The Netherlands: Springer; 2009. p. 15–37.

Bar-Yosef O. The Epipalaeolithic cultures of Palestine. Ph.D. dissertation, Hebrew University, Jerusalem; 1970.

Bar-Yosef O. Pleistocene connexions between Africa and Southwest Asia: an archaeological perspective. African Archaeological Review. 1987;5:29–38.

Bar-Yosef O. The Natufian culture in the Levant: threshold to the origins of agriculture. Evolutionary Anthropology. 1998;159–177

Bar-Yosef O. The Upper Palaeolithic revolution. Annual Review of Anthropology. 2002;31:363–93.

Bar-Yosef O, Kuhn S. The big deal about blades: laminar technologies and human evolution. American Anthropologist. 1999;101:322–38.

Belfer-Cohen A. The Natufian in the Levant. Annual Review of Anthropology. 1991;20:167–86.

Belfer-Cohen A, Goring-Morris AN. Why microliths? Microlithization in the Levant. In: Elston RG, Kuhn SL, editors. Thinking small: global perspectives on microlithic technologies, vol 12. Arlington, VA: American Anthropological Association; 2002. p. 57–68.

Betts AVG. The Harra and the Hamad. Excavations and surveys in Eastern Jordan, vol 1. Sheffield: Sheffield Academic; 1998.

Brinkmann R, Ghaleb AO. Late Pleistocene mollisol and cumulic fluvents near Ibb, Yemen Arab Republic. In: Grolier, MJ, Brinkmann R, Blakely JA, editors. The Wadi al-Jubah Archaeological Project, vol 5: environmental research in support of archaeological investigations in the Yemen Arab Republic, 1982–1987; 1987. p. 251–8.

Byrd B. Late Quaternary hunter-gatherer complexes in the Levant between 20,000 and 10,000 BP. In: Bar-Yosef O, Kra R, editors. Late Quaternary chronology and paleoclimates of the Eastern Mediterranean. Tuscon: Radiocarbon; 1994. p. 205–26.

Byrd BF, Garrard AN. The Last Glacial Maximum in the Jordanian desert. In: Soffer O, Gamble C, editors. The world at 18,000 BP. London: Unwin Hyman; 1989. p. 78–96.

Caton-Thompson G. Climate, irrigation, and early man in the Hadhramaut. Geographical Journal. 1939;93(1):18–35.

Charpentier V. Trihedral points: a new facet to the "Arabian Bifacial Tradition". Proceedings of the Seminar for Arabian Studies. 2004;34:53–66.

Clark JD. The prehistoric cultures of the Horn of Africa. Cambridge: Cambridge University Press; 1954.

Clark I, Fontes J-C. Paleoclimatic reconstruction of northern Oman based on carbonates from hyperalkaline groundwaters. Quaternary Research. 1990;33:320–36.

Cleuziou S, Inizan M-L, Marcolongo B. Le peuplement pré- et proto-historique du système fluviatile fossile du Jawf-Hadramawt au Yémen (d'après l'interprétation d'images satellite, de photographies aériennes et de prospections). Paléorient. 1992;18(2):5–28.

Close A, editor. Prehistory of arid North Africa: essays in honor of Fred Wendorf. Dallas: Southern Methodist University Press; 1987.

Crassard R. The "Wa'sha method": an original laminar debitage from Hadramawt, Yemen. Proceedings of the Seminar for Arabian Studies 2008;38(July):3–14.

Crassard R, Khalidi L. De la pré-Histoire à la Préhistoire au Yémen, des données anciennes aux nouvelles expériences méthodologiques. Chroniques Yéménites. 2005;12:1–18.

Crassard R, McCorriston J, Oches E, Bin 'Aqil A, Espagne J, Sinnah M. Manayzah, early to mid-Holocene occupations in Wâdî Sanâ. Proceedings of the Seminar for Arabian Studies. 2006;36:151–73.

Davies C. Holocene paleoclimates of southern Arabia from lacustrine deposits of the Dhamar highlands, Yemen. Quaternary Research. 2006;66:454–64.

de Maigret A. Archaeological activities in the Yemen Arab Republic, 1986. East and West. 1986;36:376–470.

DeMenocal P, Ortiz J, Guilderson T, Sarnthein M, Baker L, Yarunsinsky M. Abrupt onset and termination of the African Humid period: rapid climate responses to gradual insolation forcing. Quaternary Science Reviews. 2000;19:347–61.

Edens C. Towards a definition of the Rub al Khali 'Neolithic'. Atlal. 1982;6:109–24.

Edens C. A bladelet industry in southwestern Saudi Arabia. Arabian Archaeology and Epigraphy. 2001;12(2):137–42.

Edwards PC. Nine millennia by Lake Lisan: the Epipalaeolithic in the East Jordan Valley between 20,000 and 11,000 years ago. In: Bisheh G, editor. Studies in the history and archaeology of Jordan VII. Amman: Department of Antiquity of Jordan; 2001. p. 85–93.

Flemming N, Bailey G, Courtillot V, King G, Lambeck K, Ryerson F, et al. Coastal and marine palaeo-environments and human dispersal points across the Africa–Eurasia boundary. In: Brebbia CA, Gambin T, editors. Maritime and underwater heritage. Southampton: WIT Press; 2003. p. 61–74.

Garbini G. Antichità Yemenite. Annali Instituto Orientale di Napoli. 1970;30:537–48.

Gardner RAM. Aeolianites and marine deposits of the Wahiba Sands: character and palaeoenvironments. Journal of Oman Studies Special Report. 1988;3:75–94.

Garrard A, Stanley-Price N. A survey of prehistoric sites in the Azraq Basin, Eastern Jordan. Paléorient. 1977;3:109–26.

Garrard AN, Harvey CPD, Switsur VR. Environment and settlement during the Upper Pleistocene and Holocene at Jubba in the Great Nefud, Northern Arabia. Atlal. 1981;5:137–48.

Garrard A, Betts A, Byrd B, Colledge S, Hunt C. Summary of palaeoenvironmental and prehistoric investigations in the Azraq basin. In: Garrard A, Gebel, H, editors. The prehistory of Jordan: the state of research in 1986. Oxford: BAR International Series 396; 1988a. p. 311–37.

Garrard A, Colledge S, Hunt C, Montague R. Environment and subsistence during the Late Pleistocene and Early Holocene in the Azraq Basin. Paléorient. 1988b;14:40–9.

Garrard A, Baird D, Colledge S, Martin L, Wright K. Prehistoric environment and settlement in the Azraq Basin: an interim report on the 1987 and 1988 excavation seasons. Levant. 1994;26:73–109.

Gasse F, van Campo E. Abrupt post-glacial climate events in West Asian and North Africa monsoon domains. Earth and Planetary Science Letters. 1994;126:435–56.

Gilead I. Grar: a Chalcolithic Site in the Northern Negev, Israel. Journal of Field Archaeology. 1989;16(4):377–94.

Gilmore M, al-Ibrahim M, Murad AS. Preliminary report on the northwestern and northern region survey program 1981(1401). Atlal. 1982;6:9–25.

Goring-Morris AN. At the edge: terminal pleistocene hunter-gatherers in the Negev and Sinai. Oxford: BAR International Series; 1987.

Goring-Morris AN. Trends in spatial organization of terminal Pleistocene hunter-gatherer occupations as viewed from the Negev and Sinai. Paléorient. 1988;14:231–44.

Goring-Morris AN. Structures and dwellings in the Upper and Epi-Palaeolithic (ca. 42-10 ka BP) Levant: profane and symbolic uses. In: Vasil`ev SA, Soffer O, Kozlowski J, editors. Perceived landscapes and build environments: the cultural geography of Late Paleolithic Eurasia. Oxford: BAR International Series, 1122; 2003. p. 65–81.

Goring-Morris AN, Belfer-Cohen A, editors. More than meets the eye: studies on Upper Palaeolithic diversity in the Near East. Oxford: Oxbow Books; 2003.

Henry DO. The prehistory of Southern Jordan and relationships with the Levant. Journal of Field Archaeology. 1982;9:417–44.

Henry DO. Prehistoric cultural ecology and evolution: insights from Southern Jordan. New York: Plenum; 1995.

Ingraham M, Johnson T, Rihani B, Shatla I. Preliminary report on a reconnaissance survey of the Northwestern Province (with a note on a brief survey of the Northern Province). Atlal. 1981;5:59–84.

Inizan M-L, Francaviglia VM. Les périples de l'obsidienne à travers la mer Rouge. Journal des Africanistes: Afrique-Arabie, d'une rive à l'autre en mer d'Erythrée. 2002;72(2):11–9.

Inizan M-L, Ortlieb L. Prehistoire dans la region de Shabwa au Yemen du sud. Paléorient. 1987;13:5–22.

Khalidi L. The prehistoric and early historic settlement patterns on the Tihâmah coastal plain (Yemen): preliminary findings of the Tihâmah Coastal Survey 2003. Proceedings of the Seminar for Arabian Studies. 2005;35:115–27.

Khalidi L. Settlement, culture-contact and interaction along the Red Sea coastal plain, Yemen: the Tihamah cultural landscape in the late

prehistoric period, 3000–900 BC. Ph.D. dissertation, University of Cambridge; 2006.

Khalidi L. Holocene obsidian exchange in the Red Sea region. In: Petraglia MD, Rose JI, editors. The evolution of human populations in Arabia: paleoenvironments, prehistory and genetics. The Netherlands: Springer; 2009. p. 279–291.

Lahr M, Foley R. Multiple dispersals and modern human origins. Evolutionary Anthropology. 1994;3:48–60.

Lahr M, Foley R. Towards a theory of modern human origins: geography, demography, and diversity in recent human evolution. Yearbook of Physical Anthropology. 1998;41:137–75.

Lézine A-M, Saliège J-P, Wertz CR, Wertz F, Inizan M-L. Holocene lakes from Ramlat as-Sab'atayn (Yemen) illustrate the impact of monsoon activity in southern Arabia. Quaternary Research. 1998;50:290–9.

Lézine A-M, Tiercelin J-J, Robert C, Saliège J-F, Cleuziou S, Inizan M-L, et al. Centennial to millennial-scale variability of the Indian monsoon during the Early Holocene from a sediment, pollen and isotope record from the desert of Yemen. Palaeogeography, Palaeoclimatology, Palaeoecology. 2007;243:235–49.

Maher LA. Microliths and mortuary practices: new perspectives on the epipalaeolithic in northern and eastern Jordan. In: Levy TE, Daviau PM, Younker RW, Shaer M, editors. Crossing Jordan: North American contributions to the archaeology of Jordan. London: Equinox; 2007. p. 195–202.

Marks A. Prehistory and paleoenvironments in the Central Negev, Israel, vol 1. Dallas: Southern Methodist University Press; 1976.

Marks A. Prehistory and paleoenvironments in the Central Negev, vol 2. Dallas: Southern Methodist University Press; 1977.

Masry A. The historic legacy of Saudi Arabia. Atlal. 1977;1:9–19.

McBrearty S, Brooks A. The revolution that wasn't: a new interpretation of the origin of modern human behavior. Journal of Human Evolution. 2000;39:453–63.

McBurney CBM. The Haua Fteah (Cyrenaica) and the Stone Age of the South-East Mediterranean. Cambridge: University Press; 1967.

McClure HA. Radiocarbon chronology of Late Quaternary lakes in the Arabian Desert. Nature. 1976;263:755–6.

McClure HA. Ar Rub' al Khali. In: Al-Sayari SS, Zötl JG, editors. Quaternary period in Saudi Arabia, vol 1: sedimentological, hydrogeological, hydrochemical, geomorphological, and climatological investigations in central and eastern Saudi Arabia. New York: Springer; 1978. p. 252–63.

McClure HA. Late Quaternary palaeogeography and landscape evolution: the Rub' al-Khali. In: Potts D, editor. Araby the Blest. Copenhagen: CNIP; 1988. p. 7–11.

McCorriston J. Early settlement in Hadramawt: preliminary report on prehistoric occupation at Shi'b Munayder. Arabian Archaeology and Epigraphy. 2000;11:129–53.

McCorriston J, Oches EA, Walter DE, Cole KL. Holocene paleoecology and prehistory in highland southern Arabia. Paléorient. 2002;28(1):61–88.

McCorriston J, Heyne C, Harrower M, Patel N, Steimer-Herbet T, al-Amary I, et al. Roots of Agriculture (RASA) Project 2005: a season of excavation and survey in Wadi Sana, Hadramawt. Bulletin of the American Institute for Yemeni Studies. 2006;47:23–8.

Nadel D, editor. Ohalo II, A 23,000-year-old Fisher-Hunter-Gatherers' Camp on the shore of the Sea of Galilee. Haifa: Hecht Museum; 2002.

Nadel D, Werker E. The oldest ever brush hut plant remains from Ohalo II, Jordan Valley, Israel (19 000 BP). Antiquity. 1999;73:755–64.

Neev D, Emery KO. The Dead Sea: depositional processes and environments of evaporites. Geological Survey of Israel Bulletin. 1967;41:1–147.

Olszewski D. My "backed and trucated bladelet", your "point": terminology and interpretation in Levantine Epipalaeolithic assemblages. In: Beyond tools: redefining the PPN lithic assemblages of the Levant.

Third Workshop on PPN Chipped Lithic Industries, Ca'Foscari University of Venice, 2001, Studies in Early Near Eastern Production, Subsistence and Environment 9; 2001. p. 303–18.

Olszeweksi D. Issues in the Levantine Epipaleolithic: The Madamaghan, Nebekian and Qalkhan (Levant Epipaleolithic). Paléorient. 2006;32:19–26.

Parr P, Zarins J, Ibrahim M, Waechter J, Garrard A, Clark C, et al. Preliminary report on the second phase of the Northern Province survey 1397/1977. Atlal. 1978;2:29–50.

Petraglia M. The Lower Palaeolithic of the Arabian peninsula: occupations, adaptations, and dispersals. Journal of World Prehistory. 2003;17:141–79.

Petraglia M, Alsharekh A. The Middle Palaeolithic of Arabia: implications for modern human origins, behaviour and dispersals. Antiquity. 2003;77(298):671–84.

Phillips JL. Sinai during the Paleolithic: the early periods. In: Close A, editor. Prehistory of arid North Africa. Essays in honor of Fred Wendorf. Dallas: Southern Methodist University Press; 1987. p. 105–21.

Phillips JL. The Upper Palaeolithic Chronology of the Levant and the Nile Valley. In: Bar-Yosef O, Kra R, editors. Late Quaternary chronology and paleoclimates of the Eastern Mediterranean. Tuscon: Radiocaron; 1994. p. 169–76.

Phillips J, Gladfelter B. A survey in the Upper Wadi Feiran Basin, Southern Sinai. Paléorient. 1989;15:113–22.

Phillips J, Mintz E. The Mushabian. In: Bar-Yosf O, Phillips JL, editors. Prehistoric investigations in Gebel Maghara, Northern Sinai, Qedem 7. Monographs of the Institute of Archaeology. Jerusalem: Hebrew University; 1977. p. 149–83.

Phillipson D. African archaeology. 3rd ed. Cambridge: Cambridge University Press; 2005.

Rahimi D. Chipped stone assemblage. In: Glanzman WD, Ghaleb AO, editors. Site reconnaissance in the Yemen Arab Republic, 1984: the stratigraphic probe at Hajar ar-Rayani. Washington, DC: American Foundation for the Study of Man; 1987. p. 139–43.

Rose J. The question of Upper Pleistocene connections between East Africa and South Arabia. Current Anthropology. 2004a;54(4):551–5.

Rose J. New evidence for the expansion of an Upper Pleistocene population out of East Africa, from the site of Station One, Northern Sudan. Cambridge Archaeological Journal. 2004b;14(2):205–16.

Rose J. Among Arabian Sands: The Palaeolithic of southern Arabia. Ph.D. dissertation, Southern Methodist University; 2006.

Rose JI, Usik VI. The "Upper Paleolithic" of South Arabia. In: Petraglia MD, Rose JI, editors. The evolution of human populations in Arabia: paleoenvironments, prehistory and genetics. The Netherlands: Springer; 2009. p. 169–85.

Rosen SA. Lithics after the Stone Age: a handbook of stone tools from the Levant. London: Altamira; 1997.

Sanlaville P. Changements climatiques dans la péninsule Arabique durant le Pléistocène supérieur et l'Holocène. Paléorient. 1992;18:5–25.

Sanlaville P. Le Moyen-Orient arabe: Le milieu et l'homme. Collection U – Série Géographie. Paris: Armand Colin; 2000.

Schulz E, Whitney J. Upper Pleistocene and Holocene paleoenvironments in the An Nafud, Saudi Arabia. In: Schandelmeier H, editor. Current research in African Earth Sciences. Rotterdam: A.A. Balkema; 1987.

Tixier J. Glossary for the description of stone tools: with special reference to the Epipalaeolithic of the Maghreb. Pullman, WA: Newsletter of lithic technology: special publication no. 1; 1974.

Tosi M. Survey and excavations on the Coastal Plain (Tihamah). East and West. 1986a;36:400–14.

Tosi M. The emerging picture of prehistoric Arabia. Annual Review of Anthropology. 1986b;15:461–90.

Uerpmann M. Structuring the Late Stone Age of southern Arabia. Arabian archaeology and epigraphy. 1992;3:65–109.

Wendorf F, Schild R. Prehistory of the Eastern Sahara. New York: Academic; 1980.

Wendorf F, Schild R, Close A, et al. Egypt during the Last Interglacial. Plenum, New York; 1993.

Whalen N, Pease D. Archaeological survey in southwest Yemen, 1990. Paléorient. 1991;17:127–31.

Whalen N, Killick A, James N, Morsi G, Kamal M. Preliminary Report on the Western Province Survey. Atlal. 1981;5:43–65.

Wilkinson TJ. Environment and long-term population trends in southwest Arabia. In: Petraglia MD, Rose JI, editors. The evolution of human populations in Arabia: paleoenvironments, prehistory and genetics. The Netherlands: Springer; 2009. p. 51–66.

Wilkinson T, Gibson M, Edens C. The archaeology of the Yemen high plains: a preliminary chronology. Arabian Archaeology and Epigraphy. 1997;8:99–142.

Zarins J. Archaeological and chronological problems within the greater southwest Asian arid zone, 8500–1850 BC. In: Ehrich RW, editor. Chronologies in Old World archaeology, vol. 1. Chicago: University of Chicago Press; 1992. p. 42–76.

Zarins J. View from the South: the greater Arabian peninsula. In: Henry DO, editor. The prehistoric archaeology of Jordan. Oxford: BAR International Series 705; 1998. p. 179–194.

Zarins J, al-Badr H. Archaeological investigation in the southern Tihama Plain, part II (including Sihi, 217-107 and Sharja, 217-172) 1405/1985. Atlal. 1986;10:36–57.

Zarins J, Whalen N, Ibrahim M, Mursi A, Khan M. Comprehensive Archaeological Survey Program: preliminary report on the central and Southwestern Province survey: 1979. Atlal. 1980;4:9–36.

Zarins J, Murad A, al-Yish K. Comprehensive Archaeological Survey Program: the second preliminary report on the Southwestern Province. Atlal. 1981;5:8–42.

Zarins J, Rahbini A, Kamal M. Preliminary report on the archaeological survey of the Riyadh area. Atlal. 1982;6:25–38.

Zohary M. Plant life of Palestine. New York: The Rouald Press; 1962.

Chapter 15
Holocene (Re-)Occupation of Eastern Arabia

Hans-Peter Uerpmann, D.T. Potts, and Margarethe Uerpmann

Keywords Fasad • Herders • Hunters • Neolithic • Pastoralists • PPNB • Sharjah

Introduction

Population discontinuities on a micro-scale are familiar phenomena in the archaeological record of many parts of the world, and Western Asia is no exception. Multi-period sites often display stratigraphic features, gaps in ceramic sequences and distances between radiocarbon dates implying breaks in the history of settlement. However, there is often a presumption that if settlement evidence from one period is missing in one trench or set of associated trenches, it may be present elsewhere since not all areas necessarily contain the full stratigraphic record of occupation at any given site. Population discontinuities at a macro-scale, such as a valley system or drainage zone, are equally common in settlement pattern studies, and de-population for periods ranging from centuries to millennia is familiar to most archaeologists who have worked at this scale. There is, however, another aspect of discontinuity which is rarely addressed directly by archaeologists working in Western Asia, even when it is observed, namely the issue of population continuity or discontinuity between the Pleistocene and the Holocene.

Specialization in archaeology has had the unintended and unfortunate effect of compartmentalizing Paleolithic archaeology (and cognate fields like Pleistocene climatic and geological studies), turning it into a stand-alone field of study with little or no relationship to the study of later periods (Neolithic, Chalcolithic, Bronze Age, etc.). Similarly, the perspective of scholars who work on the later periods of human history often fails to reach back in time beyond the 'great' Pleistocene–Holocene divide. In the present chapter we shall consider the specific case of eastern Arabia, where opinions on the matter of occupational continuity or discontinuity between the Pleistocene and the Holocene have been evolving rapidly in recent years.

The Arabian Paleolithic and the 'Paleolithic' of Eastern Arabia

Beginning in the 1930s reports began to circulate of stone tools picked up at sites in Kuwait, eastern Saudi Arabia, Qatar, Bahrain and Oman (Potts, 1990). Thus, by the 1960s, when Danish archaeologists investigated lithic scatters in Qatar and eastern Saudi Arabia, they needed little convincing of the reality of Paleolithic occupation in the region (Kapel, 1967) and in this they were followed by other scholars working nearby (e.g., Masry, 1974; de Cardi, 1978). Terms like 'handaxe', 'Levallois-Mousterian' and 'possibly Acheulean' were all used in these early reports, but in 1976 the situation changed dramatically when the French prehistorians J. Tixier, who had years of experience working on North African assemblages, and M.-L. Inizan, who was familiar with the lithic industries of the Levant, initiated a research project in Qatar, one objective of which was to verify the claims made by Kapel. In their first season of work the French archaeologists concluded that the so-called Qatar A Group, considered Middle Paleolithic and Mousterian-related by Kapel, was in reality a Holocene industry that was deceptively similar to Pleistocene material (Tixier, 1986). As noted many years ago, this work had implications far beyond the boundaries of Qatar itself, throwing into question all of those finds from Oman, Bahrain, eastern Saudi Arabia and Kuwait that had previously been attributed to the Paleolithic (Potts, 1990). In effect, while superficially similar to Paleolithic material, yet Holocene in date, these assemblages exposed an underlying conundrum: that Paleolithic sites of Acheulean (Petraglia, 2003) and Mousterian

H.-P. Uerpmann (✉) and M. Uerpmann
Institut für Ur- und Frühgeschichte und Archäologie des Mittelalters, Naturwissenschaftliche Archäologie,
Eberhard-Karls-Universität Tübingen, Rümelinstr.,
19-23, D-72070, Tübingen, Germany
e-mail: hans-peter.uerpmann@uni-tuebingen.de;
margarethe.uerpmann@uni-tuebingen.de

D.T. Potts
Department of Archaeology, The University of Sydney,
NSW 2006, Australia
e-mail: dpot3385@usyd.edu.au

M.D. Petraglia and J.I. Rose (eds.), *The Evolution of Human Populations in Arabia*, Vertebrate Paleobiology and Paleoanthropology,
DOI 10.1007/978-90-481-2719-1_15, © Springer Science+Business Media B.V. 2009

(Petraglia and Alsharekh, 2003) affinity, so widespread on
the Arabian Shield the area of western Arabia composed of
'Precambrian crystalline and metamorphosed sedimentary
rock and volcanics' (al-Juaidi et al., 2003: 118), were completely absent on the Arabian Shelf ('an exposed sequence of
continental and shallow marine sedimentary rocks overlying
the rocks of the Arabian Shield' in eastern Arabia (al-Juaidi
et al., 2003: 118); on the absence of stone tools there, see
Zarins et al., 1982: 28). To distinguish dated from unconfirmed Paleolithic, we shall refer to the latter as 'Paleolithic'.

The situation began to change again, however, after lithic
scatters discovered at Shuwaihat, Hamra, Ras al-Aysh and
Jabal Barakah in western Abu Dhabi during the late 1980s and
early 1990s were published. As McBrearty noted, "The Barakah
radial cores and biface tip might fall within the range of artifacts
expected at … Neolithic biface tradition sites … On the other
hand, it is quite possible that the lithic artifacts from Jabal
Barakah are very ancient, perhaps dating to the Middle
Pleistocene. Nothing in their technical execution or state of
patination would exclude them from the Acheulian or Middle
Stone Age" (McBrearty, 1999: 383). Moreover, further south,
at Saiwan in the Sultanate of Oman, Biagi announced the
discovery of a Late Acheulean or Early Paleolithic surface site
in association with a Pleistocene lake bed, where large numbers
of sidescrapers as well as bifaces belonging to Bordes' Band
IV were recovered (Biagi, 1994).

Other ongoing studies have been adding new data to the
Paleolithic debate. Recent work at Jabal Barakah highlights
the absence of blade tools as compared to the presence of
Mousterian points, a handaxe and Levallois-type flakes
(Wahida et al., 2007, 2009) supporting the Middle Pleistocene
attribution tentatively suggested by McBrearty. Surface collections from sites ESF06A-C on a limestone ridge on the
western side of the Hajar Mountains in Sharjah include
highly patinated flakes and flake blades, some of which were
made from Levallois and discoidal cores and, as at Barakah,
an absence of blades or pressure-flaked, Holocene points of
the Arabian bifacial tradition (Scott-Jackson et al., 2007,
2009). More recently Rose (2004), at Wadi Qilfah 1-4 (findspots A7-10) on the ad-Dakhliyah plain of central Oman,
discovered débitage, cores and tools (over 80% unifacial)
exhibiting technological features "diagnostic of the late
Middle/Upper Pleistocene (i.e., unidirectional-convergent
Levallois method)" which he has termed the Sibakhan industry (Rose, 2006: 243, 286). Based on its techno-typological
characteristics (cf. the East African Kapthurin Formation
and the Levantine Mugharan Complex), he has tentatively
dated the Sibakhan to the OIS 9-7 (400–180 ka) or, less probably, the OIS 5e pluvial (128–120 ka) (Rose, 2006: 288,
297). Additionally, in light of comparison with Middle and
Late Stone Age assemblages from the Horn of Africa, Rose
has suggested that the 'small, soft hammer foliates, Khasfian
Foliates' from Bir Khasfa in Oman, discovered by Pullar in

the 1970s, may be Upper Pleistocene, dating to the pluvial
ca. 60–24 ka (Rose, 2006: 282, 298).

These new data suggest that, contrary to the position
adopted after the French expedition to Qatar, there may in
fact have been a hominin presence in eastern Arabia during
parts of the Pleistocene. The recent excavations by H.-P.
Uerpmann and M. Uerpmann at Jebel Faya (Sharjah) clearly
confirm this for the Upper Pleistocene. One problem
remains, however, to determine whether there was population continuity from the Pleistocene into the Holocene.
Since the majority of the recently studied 'Paleolithic' sites
are unstratified, or if stratified, do not also contain assemblages extending into the Holocene, their presence does not
necessarily provide us with evidence of the original population stock from which the Holocene inhabitants of eastern
Arabia were descended. As Rose has stressed, there were
several periods of hyper-aridity which could have acted as
tabula rasa 'events', wiping out existing populations. For
example, OIS 9-7 (400–180 ka) was followed by a period
of aridity in OIS 6, just as the pluvial phase in OIS 5e (ca.,
128–120 ka) and the sub-pluvial during OIS 5a (ca. 85–73
ka) were brought to an end by a severe episode of aridity in
OIS 4 (ca. 73–60 ka), followed by another pluvial from ca.
60–24 ka in OIS 3, and again by a hyper-arid period in OIS
2 (ca. 24–12 ka). While 'out of Africa' hominin dispersals
involving *Homo helmei* and/or *Homo sapiens* in the earlier
periods may have colonized Arabia (Petraglia and
Alsharekh, 2003: 680), it is highly unlikely that such populations could have survived the aridity of OSI 6 (Rose,
2004: 553) and/or OIS 4. As Glennie et al. (1994: 2–3)
observed, during the glacial periods "large areas of very
high atmospheric pressure associated with each of the ice
caps had the effect of squeezing all other air-pressure zones
towards the equator, resulting in an increase in global wind-
velocity … In desert areas, the wind probably blew at sand-
transporting speeds for much of each glacial winter …
During glacial extremes, therefore, the strong winds would
cause severe desiccation, even at reduced air temperatures,
producing conditions that were probably too severe for man
to tolerate".

Rose has suggested that the date of the posited Upper
Pleistocene industries of OIS 3 perhaps best reflects 'the
genetically-predicted time frame for the modern human
expansion across the Arabian Corridor' (Rose, 2006: 282,
298). The issue for scholars interested in the Holocene,
however, is not whether the mtDNA haplogroup M dispersed
into Arabia in OIS 3 (Rose, 2006: 306), but rather whether
this group survived OIS 2 and was still extant in the Early
Holocene.

Addressing this question properly means that we have to
look at environmental deterioration during the cold phases
of the Pleistocene in more detail. First of all it is important
not to look at the Arabian peninsula as a whole, but rather

treat its geographical subunits separately. There are marked differences in population density today, and the same will have been the case in the past. Generally speaking, the availability of water and food determines how many humans can exist in a given area. Large parts of eastern Arabia – in particular the Rub' al Khali – are not permanently inhabited today, but stone artifacts indicate that they were populated during the Stone Ages. Climatic changes led to an expansion or shrinking of the areas inhabitable with a given subsistence strategy.

Some Thoughts on Ecology and Subsistence

Before the 'Neolithic Revolution', which occurred along the northwestern edge of Arabia at the transition from the Pleistocene to the Holocene, hunting-and-gathering was the only available subsistence strategy. Unfortunately, man-made conditions make it impossible to determine where in Arabia subsistence hunters could survive today. It would be helpful if natural densities of wild animal and plant resources could be correlated with present patterns of precipitation and temperatures. However, this is impossible due to human over-exploitation of most Arabian landscapes. Only 'experienced ecological guesswork' based on knowledge from other arid zones can help to generate a more detailed approach to estimating potential regional population densities of hunter-gatherers in the past.

Based on such considerations aggregations of game animals, large enough to guarantee sustainable hunting success, might be possible along the edges of the higher mountains, where the natural relief would concentrate sufficient water for human existence and for the vegetation to be dense enough for sustaining plant-eating animals in numbers high enough to support a viable population of predators. In addition, the coastlines would have constituted a habitat where hunters and gatherers, who had learned to fish as well, were able to exploit reliable food resources. Given present conditions one could therefore assume hunters and gatherers lived, in variable densities, in the foothills along the Red Sea coast from Aqaba in the northwest to Aden in the southeast and from there along the east coast up to Dhofar. Penetration of the inland would have depended on the relief as well, mountains always forming more variable and therefore more favorable habitats than plains. The inland of Arabia will always have been less populated than the coastal areas with their additional sources of marine food at the shore-lines.

Southeastern Arabia, the so-called Oman peninsula, is isolated from the zone just described by the Rub' al Khali, the Wahiba Sands and the flat area between these two sand seas. Under present conditions a regular supply of surface water is not guaranteed in this vast intermediate area and wild game populations would be scarce there, highly mobile and unpredictable. The Hajar Mountains of Oman and the United Arab Emirates are more favorable again, and towards the coast of the Gulf of Oman conditions similar to those along the southern coasts could be expected.

Under present climatic conditions one might therefore expect a thin but fairly continuous population of hunter-fisher-gatherers in the coastal strip and its mountainous hinterland from Aqaba via Aden to Salalah. Small groups of this population would probably also exist in the mountainous parts of the Arabian Shield. Another population – isolated from the first one – would be able to exist along the northern Omani coast and in the Hajar Mountains. This division of Arabia into two separate regions appropriate for hunter-gatherers will also have existed during most of the Pleistocene. This ecological separation of the Oman peninsula from the rest of Arabia is at present the best available explanation for the lack of unambiguous Acheulean sites in northern Oman, the Emirates and along the Persian Gulf. That separation became reduced during periods of high monsoon activity, when inter-dunal lakes formed regularly, when there was more vegetation cover everywhere and when the southward runoff from the Hajar Mountains created a green belt along the western edge of the Wahiba Sands.

During intervals of moister climate plant biomass would generally have increased, and with it the potential populations of herbivores and their predators, including prehistoric hunters. Geographic patterns would have remained similar, though, except for a gradual expansion into the formerly uninhabited desert margins. Climatic deterioration on the other hand, i.e. decreasing precipitation, would cause the areas inhabitable by hunter-gatherers to shrink towards the parts where relief and air-currents concentrate the highest amounts of water. Judging from maps of modern rainfall distribution the Yemen highlands and their run-off areas would be the last part of the Arabian peninsula to become uninhabitable. At the present state of research one should therefore not discount the possibility that hunter-gatherers survived OIS 2 around the southern tip of Arabia, even if only along a narrow strip of the now flooded coast.

There is a possibility of a similar niche existing along the northeast coast of Oman, but the general lack of indications for human presence in southeastern Arabia during the LGM casts strong doubt on the uninterrupted presence of humans in this area.

Obviously, the key to understanding many of the questions surrounding continuity or population replacement is finding a stratified site that straddles the Late Pleistocene and Early Holocene. Recent excavations at Jabal Faya and Nad al-Thamam in Sharjah have contributed new data to the discussion of this critical period of transition.

Bridging the Pleistocene–Holocene Divide in Southeastern Arabia: Wadi Wutayya, the Jabal Faya Sites (FAY-NE01 and 10) and Nad al-Thamam (NTH) (Fig. 1)

In 1983, H.-P. Uerpmann and M. Uerpmann excavated a series of fireplaces in the Wadi Wutayya (Oman) which, among others, produced a late tenth to early ninth millennium BC radiocarbon date on charcoal (Table 1; see also Uerpmann, 1992: 69; Uerpmann and Uerpmann, 2003: 40), potentially pushing post-Paleolithic human occupation in the region into the early ninth or even late tenth

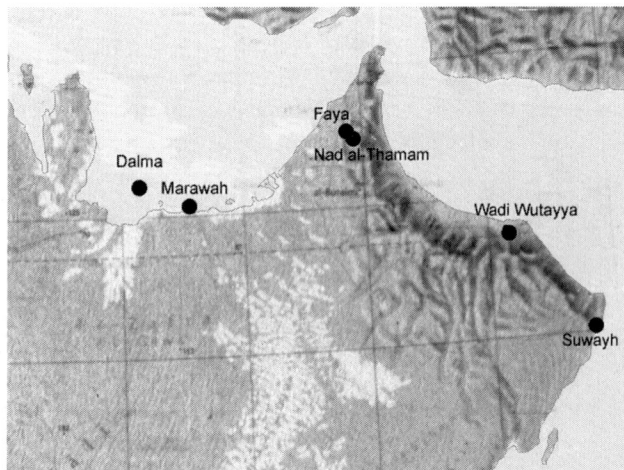

Fig. 1 Map of southeastern Arabia showing the principal sites discussed

millennium BC. For many years, however, this date remained anomalous. An ever lengthening list of sites came to light located in Kuwait, eastern Saudi Arabia, Qatar and the United Arab Emirates, which dated to the fifth millennium BC yielding lithic material of the 'Arabian Bifacial Tradition', a term of questionable meaning (e.g., Charpentier, 2004) but preferred by many authors as a common denominator for the widespread, mid-Holocene industries characterized by the presence of bifacially retouched artifacts. At coastal sites they often also yielded imported 'Ubaid-type ceramics from southern Mesopotamia. This floruit of occupation (cf. Biagi, 2006) seems to correspond to a mid-Holocene 'climatic optimum' (Parker et al., 2006a,b) that is widely attested across the Near and Middle East. More recently, slightly older dates were obtained on carbonized date stones from the island of Dalma, off the coast of Abu Dhabi (Beech and Shepherd, 2001: 86) and on shells at SWY 11 on the east coast of Oman (Berger et al., 2005: Fig. 4). This pushed the history of occupation on the coast back into the sixth millennium BC. Even older dates from the early or mid-sixth millennium BC were obtained on charcoal from Neolithic sites on Marawah island (Beech et al., 2005: 50; see also ADIAS Radiocarbon Archive: http://www.adias-uae.com/radiocarbon.html#marawah). Older dates cannot be expected from coastal sites, because any sites pre-dating the sixth millennium BC would now be under water as a result of the rise of global sea levels after the LGM. This explains why all of the recently discovered earlier sites are found in the interior.

Table 1 Early Holocene radiocarbon dates from southeastern Arabia, ninth–sixth millennium cal BC (calibrated with CALIB 5.1)

Site	Lab no.	Material	date BP	2σ cal BC Intcal04	2σ cal BC Marine04 R = 300 ± 100[a]	Contribution to probabilities
Wadi Wutayya	Hv12964	Ash	9,615 ± 65	9,230–8,800		1.0
FAY-NE01	Hd26089	Shell	9,583 ± 66	(9,217–8,773)	8,454–7,761	1.0
Nad al-Thamam (NTH)	Hd-24,356	Shell	8,434 ± 40	(7,581–7,379)	6,997–6,444	1.0
FAY-NE10	Hd-26,062	Shell	7,714 ± 59	(6,645–6,457)	6,176–5,698	1.0
	Hd-25,793	Shell	6,710 ± 45	(5,713–5,552)	5,210–4,687	1.0
	Hd-25,820	Shell	6,665 ± 45	(5,662–5,495)	5,190–4,643	1.0
	Hd-26,117	Charcoal	6,249 ± 47	5,180–5,061		1.0
	Hd-26,118	Charcoal	6,145 ± 49	5,217–4,952		1.0
SWY11	Pa-1,716	Shell	7,275 ± 60	(6,240–6,020)	5,722–5,307	1.0
Marawah 1	Hd-20,756	Ash	7,036 ± 30	5,809–5,989		1.0
	Hd-20,758	Ash	6,446 ± 56	5,506–5,304		1.0
	Hd-20,755	Ash	6,314 ± 74	5,469–5,071		1.0
Marawah 11 context 58	SUERC-3,612	Charcoal	6,750 ± 40	5,724–5,618		0.952
				5,578–5,563		0.048
Marawah 11 context 55	SUERC-3,608	Charcoal	6,675 ± 40	5,663–5,647		0.052
				5,644–5,512		0.927
				5,497–5,485		0.021
Dalma 11, context 15	AA-32,032	Carbonised Date stone	6,165 ± 55	5,228–4,961		0.959
				5,292–5,252		0.041

[a]Additional reservoir-correction.

Recent Discoveries in the Interior of Sharjah (UAE)

Since 2003 a joint Tübingen-Sharjah team has conducted excavations at FAY-NE01, a rock shelter located near the northeastern end of Jabal Faya. Two wadis that converge in front of the rock shelter have been responsible for a slow but steady accumulation of sediments over a long time. Flint sources, with traces of extraction, are located close to the site. To date, 150 m² have been exposed. The uppermost levels contain few copper or bronze artifacts, Iron Age ceramics, and fireplaces dating to the early historic Maleiha-Period. Below this were strata with Neolithic artifacts, including several so-called Fasad points on blades (Fig. 2d–i) in a trench located in front of the rock shelter.

Fasad points received their name (Charpentier, 1996) from that of the site where they were first reported by Pullar (1974). According to Charpentier (1996) they are a group of flint artifacts consisting of blades or flakes which are pointed at their distal end. The stem at the proximal end is usually the only (unifacially or bifacially) retouched part of the artifact. They represent a simple technological concept for producing projectile armatures with little effort. The first step is knapping a tipped blade or flake from a more or less prepared core. Then a stem is formed at the basal end, usually by steep direct retouch.

A number of such points were found in the early Neolithic layers of Faya NE01. They are similar to the points depicted from Fasad, but as Pullar (1974) already noted for her finds, they also resemble the typical Qatar-B points described earlier by Kappel (1967) and beyond those they remind one of the simple blade-points of the PPNB usually termed Byblos points. Beneath the strata in which these characteristic points were found at FAY-NE01 was a sterile layer of sand, below

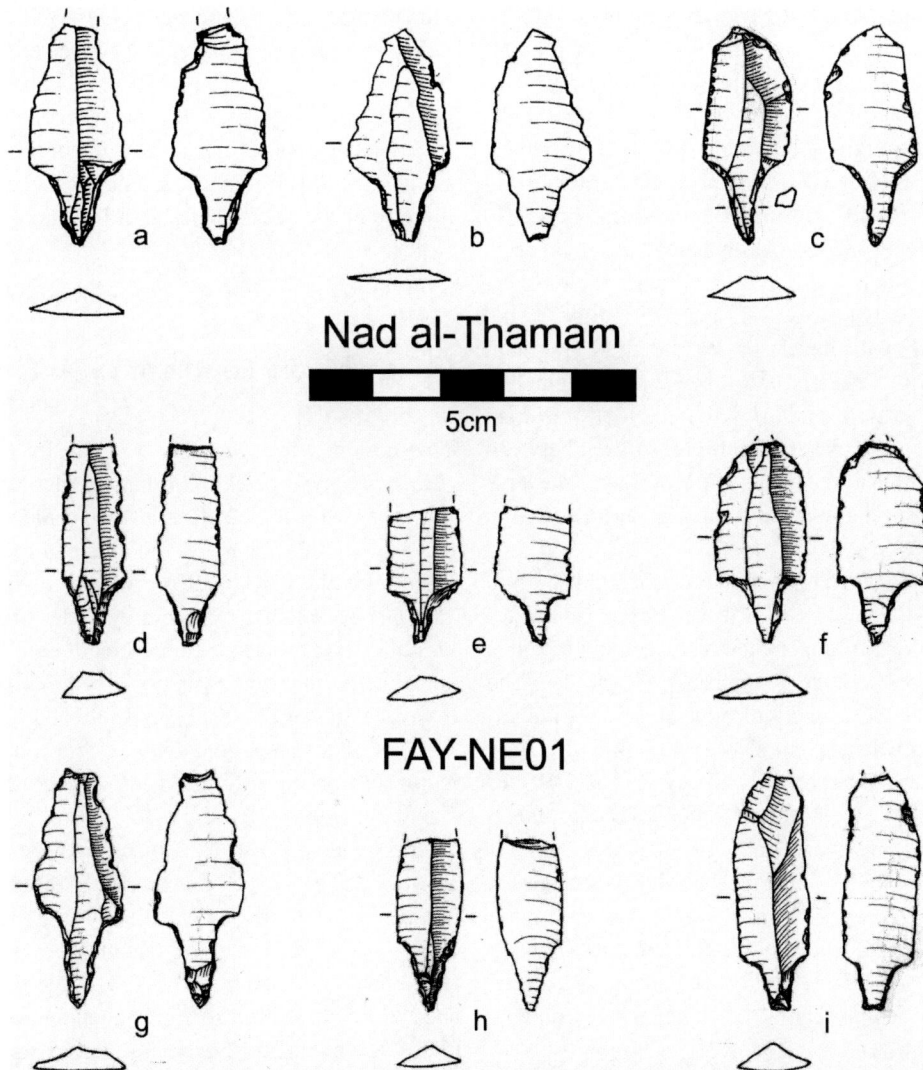

Fig. 2 Selection of Fasad points from Nad al-Thamam and FAY-NE01

which more flint-bearing strata were encountered. Preliminary results of OSL dating suggest an Upper Pleistocene date of these lower strata.

The dating of the Neolithic strata containing Fasad points on blades is problematic, mainly because a number of fire places of early historic date had been dug down into them from above. Nevertheless, a marine shell fragment found in association with the Fasad points yielded a date of 9,583 ± 66 BP, or 8,454–7,761 (2σ cal BC) if an additional marine reservoir effect between 200 and 400 (300 ± 100) years is taken into consideration. It is certain that this shell could only have arrived through human agency at FAY-NE01, which is ca. 60 km away from the nearest modern coastline. This suggests a very Early Holocene date for the first post-Paleolithic occupation which, as its most typical element, left the Fasad points behind.

Excavations within a nearby cave, FAY-NE10, revealed pre-Islamic and Iron Age material from the surface to a depth of ca. 60–80 cm. Below these layers were strata with half-finished, broken and discarded fragments of bifacial foliates and hard- and soft-hammer chips of their production, as well as groundstone fragments. Several radiocarbon dates (Table 1) indicate formation of this deposit in the final quarter of the sixth millennium BC. An arrowhead with trihedral cross section similar to a trihedral rod but only dorsally retouched was found approximately 15–20 cm below this level, and at 35 cm below these layers a marine shell fragment was recovered which yielded a radiocarbon date of 7,714 ± 59 BP (6,176–5,698 2σ cal BC; Table 1). From the levels below the shell fragment there are more flint artifacts and a few bone fragments. Thus, FAY-NE10 provides additional evidence for human presence in the area and use of local flint resources between about 5,000 and at least 6,500 BC. Most probably the cave served as a workshop for a considerable period of time (Uerpmann, in press).

A third site at Nad al-Thamam (NTH), ca. 15 km south of the Jabal Faya sites, consists of a dense surface scatter of lithics on a sand dune. Cultural material was found in a series of soundings to a depth of ca. 50 cm below the surface. The diagnostic lithics consist mainly of bifacial artifacts including foliates and arrowheads, but also Fasad points (Fig. 2a–c). Serrated and lanceolate arrowheads, typical of BHS 18 which is well dated to the fifth millennium BC (Uerpmann et al., 2000, 2006), were absent, suggesting that NTH is older. This is a possibility confirmed by the presence of a fragmentary shell (*Fasciolaria trapezium*) that can again only have reached this inland site through human activity. A radiocarbon date of 8,434 ± 40 BP yielded a 2σ date of 6,997–6,444 cal BC using Calib Rev. 5.1 and a reservoir correction of 300 ± 100 year.

Although the sample of excavated assemblages is small, it is significant that the Wadi Wutayya, Jabal Faya and Nad al-Thamam sites have now produced a series of calibrated dates extending from the late tenth through the late sixth millennium BC (Table 1). Furthermore, for the first time Fasad points may have been given chronological determinations at FAY-NE01 and NTH falling in the time-span from the late ninth to the early seventh millennium BC. Such points, often larger and heavier and made on flakes, are known from a number of sites in southeast Arabia (e.g., Charpentier, 1996), and are tentatively dated at Al-Haddah (BJD-1) to the late fifth or seventh to sixth millennium BC (Usai, 2000: 7). FAY-NE01 and NTH suggest that the ones made on blades or thin flakes are older still.

Excavations at FAY-NE01 have, at least locally, documented a population discontinuity or *tabula rasa* event (cf. Rose, 2006), represented by a sand horizon between the putative Paleolithic horizon and the early Neolithic levels. Nevertheless, the data described above indicates that human occupation was widespread throughout the Oman peninsula in the Early Holocene. However, the evidence at hand does not yet enable us to confirm or deny the hypothesis of population continuity or replacement between the Pleistocene and Holocene in eastern Arabia. It does raise a few questions, though, about another thesis that has been commonly enunciated in the archaeological literature of this area, namely the Levantine, PPNB-related origins of the earliest Holocene lithic industry found there, and, by extension, of the Early Holocene inhabitants as well.

Links with the Southern Levant

This notion, which we shall refer to for simplicity as the 'Levantine hypothesis', is founded upon two observations: first, that what were long considered the earliest lithics in the region, the Qatar B-type blade arrowheads, showed affinities with PPNB blade arrowheads in the Levant (Kapel, 1967: 18 based on a comment by Peder Mortensen), which was of course prior to the identification of the Fasad points; and second, that wherever preserved animal remains were excavated at Neolithic sites in South and Southeast-Arabia, domesticated sheep, goat and/or cattle were present, which had to have been introduced into eastern Arabia from outside the region.

It is of course difficult to exclude local domestication. However, at least the wild precursor of the sheep never seems to have inhabited the Arabian peninsula. The wild goat once existed in a small part of the northern Hajar Mountains (Uerpmann and Uerpmann, 2008), while the aurochs as the ancestor of domestic cattle existed along both shores of the Gulf (Uerpmann and Uerpmann, 2008), and also along the Red Sea, probably from southern Jordan all the way to the Yemeni highlands (e.g., Cattani and Bökönyi, 2002). Local domestication in Arabia is, nevertheless, quite unlikely. A cattle bone from BHS18 in the Emirate of Sharjah yielded an ancient MtDNA-sample which indicates an origin of the fifth millennium BC cattle of southeastern Arabia in the Fertile Crescent (Uerpmann and Uerpmann, 2008). For the areas on both sides of the Red Sea such results are still lacking, but there is no compelling genetic reason to assume a separate centre of domestication for the Neolithic cattle in those regions. Indeed, if independent domestication were assumed for southern Arabia, this part of the wider Near and Middle East would have been the only area outside the Fertile Crescent where ungulates were domesticated by hunter-gathers who had not gone through the previous adaptations to sedentism. Such adaptations, as a precondition to animal domestication, had happened *beforehand* in the Fertile Crescent at the transition from the Late Pleistocene (OIS 2) to the Holocene (OIS 1) (for more detailed considerations of theoretical aspects of animal domestication also including other parts of the world see Uerpmann, 2008).

Putting the observations of blade arrowheads and domestic herds together, many scholars working in eastern Arabia have long supported the tacit assumption that the earliest occupation of the region, which to this day has no demonstrated relationship with earlier Paleolithic populations in the Arabian peninsula, may have had its roots in the stockbreeding, blade arrowhead-using, PPNB cultures of the southern Levant (e.g., Uerpmann and Uerpmann, 1996: 132ff). Just how or why such a migration might have occurred has never been explicitly discussed. Several hypotheses suggest themselves.

Hypothesis 1: *The peopling of eastern Arabia by PPNB-related settlers was the result of widespread climatic deterioration to the north of the Arabian peninsula around 6,200 cal BC.*

The PPNB – including its terminal phase "PPNC" – lasted from about 8,700 to 6,200 cal BC (Drechsler, 2007a, b). Its end correlates closely with the global 8.2 ka BP cold event, which caused widespread climatic deterioration in the Near and Middle East as well (e.g., Lézine et al., 2007). One might therefore have postulated that the beginnings of settlement in eastern Arabia were prompted by a southern flight of PPNB villagers and herders away from the deteriorating environment of the southern Levant as a result of the 8.2 ka event. In view of the earlier evidence of settlement that has been accumulating during the past few years, in particular at Jabal Faya and Nad al-Thamam, such an hypothesis is no longer tenable because the dates for these potentially PPNB-related sites pre-date the 8.2 ka event. Nevertheless, this does not at all mean that the links with the PPNB must necessarily be severed. At least one other hypothesis suggests itself:

Hypothesis 2: *The peopling of eastern Arabia by PPNB-related settlers was the result of widespread population dispersal during the Early Holocene.*

Rossignol-Strick (1999: 528) has suggested that the "clement climatic conditions at the onset of the Holocene account for the explosion of the Early Pre-pottery Neolithic culture and the demographic human expansion." This period was "very wet with mild winters, as evidenced by the high abundances of oak and *Pistacia*" in the Mediterranean zones (Robinson et al., 2006: 1525). One could therefore suggest that population growth due to a favorable situation, as opposed to population pressure caused by shrinking resources, may have prompted social fission with some groups moving south into the Arabian peninsula with their herds of domestic sheep, goats and cattle.

What would such groups have found? Obviously, although we have no idea what sorts of communication mechanisms may have existed in the Early Holocene between discrete social groups, it is unlikely that anyone in the southern Levant had any idea what kind of environmental niches they might be heading towards as they made their way southward into the Arabian peninsula and eventually to the coast of the Persian Gulf or even the Arabian Sea. Nevertheless, enough studies have now been done on fossil lakebeds in the Rub' al Khali, Yemen and the UAE to confirm beyond any doubt that variations in the Earth's orbital pattern led to a northward displacement of the summer monsoon with maximum monsoon activity in the period around 7,000–6,000 cal BC (Lézine et al., 2007: 247). The al-Hawa lake sequence from central Yemen provides a fine-grained chronology of the periods with increased precipitation and the dryer episodes attested around 10,400, 6,200, 4,600 and 2000 cal BC (Lézine et al., 2007: 247).

It is interesting to note that the cool and dry event around 8.2 ka cal. BP or 6,200 cal BC is visible in the lake sequence from central Yemen as well. With regard to hypothesis 1 discussed above this would have meant that the PPNB herders would have left a bad (or deteriorating) situation for an even worse one, because the general climatic relations between the Levant and Arabia remained the same during the climatic changes. Therefore, expansion from the Levant must have happened before this time, and it may on the contrary be assumed that the PPNB-related herders in Arabia also suffered severely from the same period of drought.

Ecological Considerations Regarding Herders Versus Hunters

With regard to the accessibility of the Arabian peninsula for herders some remarks are necessary which are based on the respective considerations made above for hunter-gatherers. While the last depend on the availability of fugitive wild game, the herders always have their 'prey' with them, and they are very careful not to over-use this resource. Generally speaking, finding pasture is the main pre-occupation of the herder. Pastoralists and their animals live in a real symbiosis: the herders use their mental capacity to find pasture; the animals feed the herders once their own reproduction is secured by the latter. From the same landscape herders can extract much more biomass than hunters. They protect and care for their stock, using their intelligence to optimize its density at the highest possible level. Hunters, on the other hand, only diminish the stock of wild fauna in order to obtain food. The density of the wild animals is anyway regulated by the long-term *minimum* carrying capacity of the respective landscape, while under human control the density of domestic animals can be adjusted to the maximum, or at least the optimum, carrying capacity of the available territory.

With regard to the lakes which formed in the desert areas of Arabia during the Early Holocene, another important ecological fact must be noted. The formation of such lakes during moister climatic phases is typical for *desert* areas. They are not found in areas where the climate is always moist, because accumulations of water in natural depressions of the earth's relief will soon reach the sill around the depression and flow over it, thus causing erosion and the incision of the sill by a valley, which will drain the whole depression, creating a river system as opposed to standing water. The formation of lakes in southern Arabia in the Early Holocene indicates more moisture, but still not enough to make these lakes flow over their rims and create rivers. Evaporation from these lakes was always strong enough to prevent overflow. Thus, these lakes provided flourishing micro-environments, but do not indicate a general change of the desert ecology. It is obvious that herders could make good use of these special environments by monitoring which lakes had received high inflow during the rainy season in contrast to those which might be more likely to dry out early because the rains had missed their catchment areas. Hunters, of course, would have been able to do the same, but not for the sake of their prey, which still would have suffered from hunting in both environments, some being killed by hunters wherever they could reach them, others dying of starvation in the lake-basins that were drying out.

Ecologically speaking, hypothesis 2 thus makes a lot of sense: The density of PPNB settlements in southern Jordan is very high, obviously extending out into the desert to the southeast from an early date (Fujii, 2006, 2007). With slightly more precipitation during the Early Holocene than today, large areas in Arabia would have been good pasture lands, perhaps even better than in the moister mountain strip along the Mediterranean and the Rift valley, where increased precipitation would have led to an expansion of woody plants at the expense of grasses and herbs. Thus, moving into the desert may have been triggered by a pull-factor as much as by a push-factor from increased settlement density, although in reality both factors would probably have worked together. Nevertheless, apart from and in addition to a spread

of Neolithic herders from the Levant into Arabia a third hypothesis must not be neglected, based either on the potential presence of a relict population in the south of Yemen or on another 'out of Africa' movement at the onset of the Holocene.

Hypothesis 3: *The earliest settlement in southeastern Arabia reflects repopulation from South Arabia and/or northeastern Africa.*

For the time being this hypothesis can neither be tested nor discussed beyond a very theoretical level because of a lack of good chronological evidence for the earliest Holocene in South Arabia. The reason why it is even raised at all lies in the typology of a widespread facies of Neolithic industries in southern and southeastern Arabia. What is called the 'Arabian Bifacial Tradition' in the widest sense includes a number of elements, among them bifacial foliates and trihedral rods, which do not seem to originate in the Levant, but rather locally in South Arabia (Amirkhanov, 1996) or in the Late Paleolithic of the Horn of Africa (Clark, 1954; Rose, 2006). On a number of surface sites these elements are sometimes found together with blade arrowheads, but more often they appear alone at sites in many parts of southern and southeastern Arabia. Do they indicate the presence or arrival of another population unrelated to the assumed PPNB herders? Although up to now there does not seem to be well-dated evidence for the presence or arrival of such an independent Early Holocene population in Yemen, the typological similarities between Late Pleistocene/Early Holocene industries from the Horn of Africa and South Arabian industries are strong enough to consider this hypothesis as a realistic possibility.

Conclusion

This review suggests that at present there is no evidence to demonstrate population continuity between the Pleistocene inhabitants of southeastern Arabia and those of the Holocene. FAY-NE01 is the only archaeological site in southeastern Arabia known at present where a stratigraphic sequence straddles the transition from the Upper Pleistocene to the Holocene. The earliest known Neolithic deposits there are separated from the latest Paleolithic artifacts by a thick, sterile, sandy layer. This indicates population discontinuity at least for this site. The eagerly anticipated OSL dates on this sandy horizon below the Neolithic and on the deeper Paleolithic layers may provide answers as to the time-depth of this discontinuity. But for the time being these dates are not yet available.

As noted above, a 'Levantine hypothesis', broadly associated with PPNB colonization in the Early Holocene has some advantages, but it cannot account for the earlier evidence of occupation from Wadi Wutayya. This suggests that the situation may be far more complex than previously assumed, and that it is wrong to speak of 'colonization' in the singular. It is unlikely to have been an 'event' either, and much likelier to have been a process which may initially have involved hunters and gatherers coming from the south, soon followed by aceramic herders from the northwest using some variant of PPNB-related lithic technology. How and where they met and mixed and how they sorted out their subsistence economies is a fascinating topic for future research.

Yet it must also be emphasized that, even if one core of the Neolithic population was broadly southern Levantine PPNB-related, it remains true to say that we see in eastern Arabia very few of the typical traits associated with PPNB sites in Jordan (e.g., Banning, 1998, 2003). Apart from lithics, and here we must emphasize that the Fasad and other blade arrowheads of Qatar B type are similar to but certainly not identical to PPNB types, the only other 'typological' link between eastern Arabia and the Levant seems to be the elongated or apsidal room (Room 1 at site MR11) built of dry stone on Marawah island, off the coast of Abu Dhabi (Beech et al., 2005: Figs. 6–8). This building bears a superficial similarity to one from the late PPNB period at 'Ain Ghazal in Jordan (Banning, 1998: 205). Significantly, however, an almost complete ceramic jar from the same context has features with clear parallels to early 'Ubaid ('Ubaid 0) and Susiana pottery in Mesopotamia and Khuzestan (Beech et al., 2005: 46–47 and Fig. 10). The two axes of influence, southern Iraq/Khuzestan and southern Jordan, are certainly not incompatible, but probably serve to emphasize that the population history of eastern Arabia was more complex than we can presently tell. Ultimately, however, it will not be possible to further explore the possibility of even earlier links with southern Mesopotamia or the other side of the Persian Gulf, links as early as the radiocarbon dates from Jabal Faya and Wadi Wutayya would imply. Subsequent changes in sea-level have obliterated any trail of evidence linking southeastern Arabia and the north or northeast that might once have existed. In the case of the southern Levant, no such impediments exist to continued research. But there is much work to be done through the Arabian arid zone before we can truly say that we understand the dynamics of population expansion into eastern Arabia in the terminal Pleistocene or Early Holocene, and before we can relate the population there to any local Paleolithic forerunners, or discount the possibility of such ties once and for all.

References

Al-Juaidi F, Millington AC, McLaren SJ. Merged remotely sensed data for geomorphological investigations in deserts: examples from central Saudi Arabia. The Geographical Journal. 2003;169:117–30.

Amirkhanov H. Bilinear cultural parallelism in the Arabian Early Neolithic. In: Afanas'ev G, Cleuziou S, Lukacs R, Tosi M, editors. The prehistory of Asia and Oceania. Forli: Abaco; 1996. p. 135–40.

Banning EB. The Neolithic period: triumphs of architecture, agriculture, and art. Near Eastern Archaeology. 1998;61:188–237.

Banning EB. Housing Neolithic farmers. Near East Archaeology. 2003;66:4–21.

Beech M, Shepherd E. Archaeobotanical evidence for early date consumption on Dalma Island, United Arab Emirates. Antiquity. 2001;75:83–9.

Beech M, Cuttler R, Moscrop D, Kallweit H, Martin J. New evidence for the Neolithic settlement of Marawah Island, Abu Dhabi, United Arab Emirates. Proceedings of the Seminar for Arabian Studies. 2005;35:37–56.

Berger J-F, Cleuziou S, Davtian G, Cattani M, Cavulli F, Charpentier V, et al. Évolution paléographique du Ja'alan (Oman) à l'holocène moyen: impact sur l'évolution des paléomilieux littoraux et les stratégies d'adaption des communautés humaines. Paléorient. 2005;31(1): 46–63.

Biagi P. An Early Paleolithic site near Saiwan (Sultanate of Oman). Arabian Archaeology and Epigraphy. 1994;5:81–8.

Biagi P. The shell-middens of the Arabian Sea and Persian Gulf: maritime connections in the seventh millennium BP? Adumatu. 2006;14:7–22.

Charpentier V. Entre sables du Rub' al Khali et mer d'Arabie, Préhistoire récente du Dhofar et d'Oman: les industries à pointes de 'Fasad'. Proceedings of the Seminar for Arabian Studies. 1996;26:1–12.

Charpentier V. Trihedral points: a new facet to the 'Arabian Bifacial Tradition'? Proceedings of the Seminar for Arabian Studies. 2004;34:53–66.

de Cardi B. Qatar archaeological report: excavations 1973. Oxford: Clarendon; 1978.

Drechsler P. The Neolithic dispersal into Arabia. Proceedings of the Seminar for Arabian Studies. 2007a;37:93–109.

Drechsler P. The dispersal of the Neolithic over the Arabian peninsula. Ph.D. dissertation, Tübingen University, Germany; 2007b.

Edwards PC, Meadows J, Sayej G, Metzger MC. Zahrat Adh-Dhra' 2: a new pre-pottery Neolithic A site on the Dead Sea Plain in Jordan. Bulletin of the American Schools of Oriental Research. 2002; 327:1–15.

Fujii S. Wadi Abu Tulayha: a PPNB pastoral station in the al-Jafr basin, Southern Jordan. 13th Annual Meeting on Excavations in West Asia; 2006. p. 35–47.

Fujii S. Wadi Abu Tulayha: a PPNB Agropastoral outpost in the al-Jafr basin, Southern Jordan. 14th Annual Meeting on Excavations in West Asia; 2007. p. 45–51.

Glennie KW, Pugh JM, Goodall TM. Late Quaternary Arabian desert models of Permian Rotliegend reservoirs. Exploration Bulletin. 1994;274:1–19.

Kapel H. Atlas of the Stone Age cultures of Qatar. Aarhus: Jutland Archaeological Society Publications 6; 1967.

Lézine A-M, Tiercelin J-J, Robert C, Saliège J-F, Cleuziou S, Inizan M-L, et al. Centennial to millennial-scale variability of the Indian monsoon during the Early Holocene from a sediment, pollen and isotope record from the desert of Yemen. Palaeogeography, Palaeoclimatology, Palaeoecology. 2007;243:235–49.

Masry AH. Prehistory in northeastern Arabia: the problem of interregional interaction. Coconut Grove: Field Research Projects; 1974.

McBrearty S. Earliest stone tools from the Emirate of Abu Dhabi, United Arab Emirates. In: Whybrow PJ, Hill A, editors. Fossil vertebrates of Arabia. New Haven and London: Yale University Press; 1999. p. 373–88.

Parker AG, Preston G, Walkington H, Hodson MJ. Developing a framework of Holocene climate change and landscape archaeology for the lower Gulf region, southeastern Arabia. Arabian Archaeology and Epigraphy. 2006a;17:125–30.

Parker A, Davies C, Wilkinson T. The early to mid-Holocene moist period in Arabia: some recent evidence from lacustrine sequences in eastern and south-western Arabia. Proceedings of the Seminar for Arabian Studies. 2006b;36:243–55.

Petraglia MD. The Lower Paleolithic of the Arabian peninsula: occupations, adaptations, and dispersals. Journal of World Prehistory. 2003;17:141–79.

Petraglia MD, Alsharekh A. The Middle Paleolithic of Arabia: implications for modern human origins, behaviour and dispersals. Antiquity. 2003;77:671–84.

Potts DT. The Arabian Gulf in antiquity. vol 1st ed. Oxford: Clarendon; 1990.

Pullar J. Harvard archaeological survey in Oman, 1973, I flint sites in Oman. PSAS 1974;(4):33–48.

Robinson SA, Black S, Sellwood BW, Valdes PJ. A review of palaeoclimates and palaeoenvironments in the Levant and Eastern Mediterranean from 25,000 to 5,000 years BP: setting the environmental background for the evolution of human civilisation. Quaternary Science Reviews. 2006;25:1517–41.

Rose JI. The question of Upper Pleistocene connections between East Africa and South Arabia. Current Anthropology. 2004;45: 551–5.

Rose JI. Among Arabian Sands: defining the Paleolithic of Southern Arabia. Ph.D. dissertation, Southern Methodist University, Dallas; 2006.

Rossignol-Strick M. Sea-land correlation of pollen records in the eastern Mediterranean for the glacial–interglacial transition: biostratigraphy versus radiometric time-scale. Quaternary Science Reviews. 1995;14:893–915.

Rossignol-Strick M. The Holocene climatic optimum and pollen records of sapropel 1 in the eastern Mediterranean, 9,000–6,000 BP. Quaternary Science Reviews. 1999;18:515–30.

Scott-Jackson J, Milliken S, Scott-Jackson W, Jasim SA. Upper Pleistocene stone-tools from Sharjah, UAE, Initial investigations: Interim report. Unpublished paper presented at the Seminar for Arabian Studies; 2007.

Scott-Jackson J, Scott-Jackson W, Rose JI. Paleolithic stone tool assemblages from Sharjah and Ras al Khaimah in the United Arab Emirates. In: Petraglia MD, Rose JI, editors. The evolution of human populations in Arabia: paleoenvironments, prehistory and genetics. The Netherlands: Springer; 2009. p. 125–38.

Tixier J. The prehistory of the Gulf, recent finds. In: Al Khalifa HA, Rice M, editors. Bahrain through the ages, the archaeology. London: Kegan Paul International; 1986. p. 76–8.

Uerpmann M. Structuring the Late Stone Age of southeastern Arabia. Arabian Archaeology and Epigraphy. 1992;3:65–109.

Uerpmann H-P. Animal domestication. In: Pearsall DM, editor. Encyclopedia of archaeology. New York: Academic; 2008. p. 434–45.

Uerpmann M. In press. The Holocene Stone Age in Southeastern Arabia – A reconsideration

Uerpmann M, Uerpmann H-P. 'Ubaid pottery in the eastern Gulf – new evidence from Umm a-Qaiwain (U.A.E.). Arabian Archaeology and Epigraphy. 1996;7:125–39.

Uerpmann H-P, Uerpmann M. Stone Age sites and their natural environment. The Capital Area of Northern Oman, Part III. Beihefte zum Tübinger Atlas des Vorderen Orients Reihe A (Naturwissenschaften) Nr. 31/3. Dr. Ludwig Reichert Verlag, Wiesbaden; 2003.

Uerpmann M, Uerpmann H-P. Neolithic faunal remains from al-Buhais 18 (Sharjah, UAE). In: Uerpmann H-P, Uerpmann M, Jasim SA, editors. The natural environment of Jebel al-Buhais: past and present. The archaeology of Jebel al-Buhais. vol 2, Tübingen: Kerns Verlag, 2008. p. 97–132.

Uerpmann M, Uerpmann H-P, Jasim AS. Stone Age nomadism in SE-Arabia – palaeo-economic considerations on the Neolithic site of Al-Buhais 18 in the Emirate of Sharjah, U.A.E. Proceedings of the Seminar for Arabian Studies. 2000;30:229–34.

Uerpmann M, Uerpmann H-P, Jasim AS. Früher Wüstennomadismus auf der Arabischen Halbinsel. In: Hauser S, editor. Die Sichtbarkeit von Nomaden und saisonaler Besiedlung in der Archäologie. Mitteilungen des SFB, Differenz und Integration 9, Orientwissenschaftliches Zentrum der Martin-Luther-Universität Halle-Wittenberg. Heft 21/2006; 2006. p. 87–103.

Usai D. New prehistoric sites along the Omani coast from Ra's al-Hadd to Ra's al-Jins. Arabian Archaeology and Epigraphy. 2000;11:1–8.

Wahida G, Yasin W, Beech M. Barakah: a Middle Paleolithic site in Abu Dhabi. Unpublished paper presented at the Seminar for Arabian Studies; 2007.

Wahida G, Yasin Al-Tikriti W, Beech MJ, Al Meqbali A. A Middle Paleolithic assemblage from Jebel Barakah, Coastal Abu Dhabi Emirate. In: Petraglia MD, Rose JI, editors. The evolution of human populations in Arabia: paleoenvironments, prehistory and genetics. The Netherlands: Springer; 2009. p. 117–24.

Zarins J, Rahbini A-A, Kamal M. Preliminary report on the archaeological survey of the Riyadh area. Atlal. 1982;6:25–38.

Chapter 16
Early Holocene in the Highlands: Data on the Peopling of the Eastern Yemen Plateau, with a Note on the Pleistocene Evidence

Francesco G. Fedele

Keywords Early Holocene • Pre-Neolithic • Yemen Plateau

Introduction

The uplands of the western Arabian peninsula have featured negligibly in discussions about the Pleistocene and Early Holocene occupation of Southwest Asia. Paleolithic, or presumed Paleolithic implements, have only been reported occasionally, and these are often without context. In most instances such findings have not been approached with critical scrutiny. As far as human occupation is concerned, the whole chronological period between the Last Glacial Maximum and the beginning of the Holocene is relatively unknown. Whether there is reason to think of an actual void in human presence can not be assessed. In particular, no lithic assemblage resembling an "Upper Paleolithic" industry has been reported. An attempt in the 1980s to develop a Paleolithic archaeology on the eastern and central Yemen Plateau met with limited success (Bulgarelli, 1988) and was soon discontinued. The Early Holocene itself, here defined as the period earlier than the "Mid-Holocene Pluvial", has remained archaeologically unknown. Against this background, even modest information obtained from systematic archaeological fieldwork should be of interest.

Although we focus here on a small part of Yemen, a broader geographic perspective is essential. Here we are concerned with the western Arabian uplands, the fairly extensive mountainous "backbone" of the peninsula, which originated as a cordillera by the rifting of East Africa and Arabia along the Red Sea (Fig. 1). An often used and comprehensive name for these uplands is Yemen Mountains. However, an Arabic term is strangely lacking, perhaps as a result of historical contingencies that have emphasized divisions over geographic unity. The middle sector of the Yemen Mountains is rather loosely identified with the historical

region of 'Asīr, a part of "Greater Yemen" in a geographic sense; as a name, 'Asīr is presently connected with the mountainous province of southwestern Saudi Arabia whose core relief is the Sarāt massif. Steep gullies intersect the eastern border of the Yemen Mountains as the highlands fall away more or less gradually towards the Arabian interior, this latter occupied by lowlands and deserts. Reaching higher in altitude (up to 3,500–3,600 m a.s.l.) the mountain sector in present-day Yemen tends to have a more precipitous and dissected border onto the eastern interior.

In this chapter, I wish to report observations from the eastern Yemen Plateau which suggests the potential of the Yemen highlands for an understanding of the Early Holocene occupation of the southern Arabian peninsula. Additional information points to the environmental and archaeological potential of the region for the Late Pleistocene peopling as well, although on the basis of limited data. My own observations derive from excavations and surveys carried out between 1984 and 1990 in the region of Khawlān at-Tiyāl, which together with Al-Hadā to the south formed the core study area of the Italian Archaeological Mission to Yemen (cf. de Maigret, 2002). Two smaller areas will be of interest in particular (Fig. 1): the Wādī at-Tayyilah basin, primarily from the standpoint of a Neolithic excavation, and Wādī Khamar in the Jihānah district, made the object of a survey programme. These areas are located 40–60 km east-southeast of San'ā'.

The first intimation of a "pre-Neolithic" occupation of Early Holocene date was obtained in October 1984 during the initial testing at a Neolithic settlement on middle Wādī at-Tayyilah, site WTH3. An intentional search for earlier levels than the Neolithic was attempted through soundings, in the absence of any recognizable material on the surface. Further results were obtained in the two following seasons. At the same time, the environmental background to human activity in the region throughout the Holocene was established. Only research priorities and practical constraints prevented from documenting the pre-Neolithic evidence in greater detail and pursuing the investigation further. However, in 1987 and 1990, the examination of Wādī Khamar reinforced the opinion that during the earlier part of the Holocene human peopling may have been widespread, in the eastern

F.G. Fedele (✉)
Laboratory of Anthropology, University of Naples "Federico II",
via Mezzocannone, 8, 80134 Naples, Italy
e-mail: ffedele01@yahoo.it

M.D. Petraglia and J.I. Rose (eds.), *The Evolution of Human Populations in Arabia*, Vertebrate Paleobiology and Paleoanthropology,
DOI 10.1007/978-90-481-2719-1_16, © Springer Science+Business Media B.V. 2009

Fig. 1 Map of the eastern Yemen Plateau, central sector. *Boxed areas* include KHM, the Wādī Khamar basin (cf. Fig. 13), and WTH, the Thayyilah-NAB area (cf. Fig. 2). Other prehistoric localities: DA, Dulāʿ al-Aʿmās; GSH, Jabal Shaʿīr. *Below*: simplified geological map of Khawlān (Kohlan Sandstone and Amran Limestone unified; see Kruck et al., 1996, Sheet 5 Sanʿaʾ, for updating and detail); the *asterisk* on the Suhmān Plateau is Jabal al-ʿUrqūb

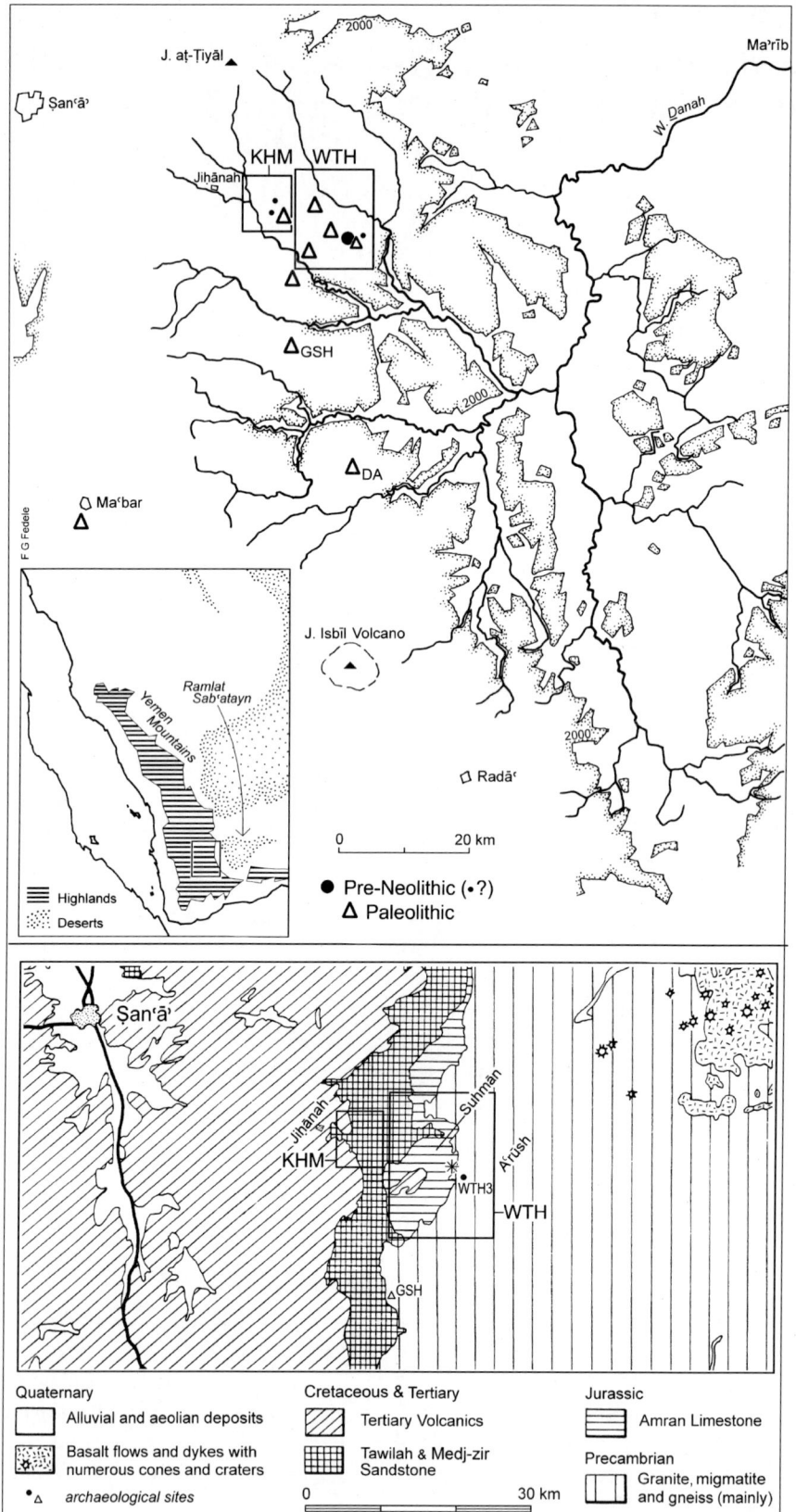

highlands at least. The Khamar survey has remained unpublished until now. In this context, this chapter summarizes primary evidence as documented and studied at the time of fieldwork. Where appropriate, the evidence has been reconsidered in the light of subsequent experience.[1]

Admittedly, due to a lack of exploration in Yemen, it is difficult to determine to what extent our study area and sampling results in the Khawlān at-Tiyāl are representative of the highlands at large. Such windows of preservation of Early and Mid-Holocene landscapes may, in fact, best occur in the least modified parts of the eastern Yemen Plateau, and be rather rare elsewhere in the western Arabian uplands. In areas of the uplands populated more densely from early historic times to the present the landscape appears to have been highly modified by human activity (Wilkinson, 2003, chapter 9), which leaves little opportunity for intact Holocene landscapes to be preserved on the surface, with or without archaeological evidence. Concurrently, earlier Holocene surfaces may be deeply buried as a result of widespread sedimentation. However, only systematic research in the future can provide firm data on the presence and distribution of relevant occurrences throughout the highlands.

Elsewhere on the Yemen Plateau, sparse but valuable data (e.g., de Bayle des Hermens, 1976; Garcia et al., 1991; Edens and Wilkinson, 1998: 55–65) suggest that identification of the pre-Neolithic archaeological record may largely depend on the amount of research and focus. Apposite research design is obviously needed to place any occurrences within a geoarchaeological and chronological framework, particularly if pre-Neolithic (including Paleolithic) sites turn out to be rare and scattered. Recently in the Hadhramawt, a re-orientation of strategy in prehistoric survey has allowed stratified Early Holocene contexts to be revealed in a region otherwise characterized by heavy erosion and human disturbance (Crassard and Khalidi, 2004; Crassard et al., 2006). Such observations might help cast in a wider context our data, which derive from a small study area explored for a relatively short time. The subject will be taken up again in the last part of this chapter.

Pre-Neolithic Evidence from the Wādī at-Tayyilah Basin (Al-A'rūsh)

Area and General Stratigraphy

The region reviewed here is the Khawlān at-Tiyāl, comprising a large portion of the mountainous territory east and south-east of Sanʿāʾ, between the Yemen Plateau's central basins and the outermost edge of the highlands, this latter bordering the interior lowlands and desert.

The area of particular interest is represented by the middle sector of Wādī at-Tayyilah[2] and a syncline furrow nearby, An-Najd al-Abyad (NAB) or 'white valley', which together will be called the Thayyilah-NAB area (Fig. 2). This area is located about 60 km east-southeast of Sanʿāʾ.

Overall, the Thayyilah-NAB area constitutes a well-defined, medium-sized basin of about 18 km², characterized by a uniform environmental record. It presents active drainage along the Tayyilah, coupled with a surrounding cluster of largely "fossil" catchment remnants, notably in the NAB furrow. To some extent the latter are relics of a mid-Holocene landscape, fossilized by widespread relief rejuvenation due to recent tectonic movements, still probably in progress (de Maigret et al., 1989; Fedele, 1990a). The name itself of An-Najd al-Abyad hints to the gray hue of fine-grained mid-Holocene sediments blanketing the valley, testimony to a milder environment than today. The interest of the Thayyilah-NAB area results from the large number of prehistoric sites associated with this peculiar, highly informative geologic framework.

The Khawlān at-Tiyāl includes a mosaic of mountains and small intermontane plains, with average annual precipitation of about 200 mm, as well as more dissected and barren fringes nearer to the edge of the Yemen Plateau, above 2,000 m in altitude. These uplands are scarred by seasonal streams or wadis (*widiān*) which eventually cut through the margin of the highlands and disappear from escarpments into the vast stretches of semidesert and desert to the east. The Thayyilah-NAB area belongs to the Wādī Danah drainage, the largest wadi system of inland northern Yemen. This is the rivercourse that flows down from the eastern highlands to Mārib and was responsible for the florescence of this ancient Sabaean capital.

In 1984–1985 a generalized sedimentary sequence spanning the terminal Pleistocene and Holocene was recognized over the entire mountainous part of Khawlān. Some of the most complete occurrences were studied in the Thayyilah-NAB area. The sequence may in fact be common throughout the eastern Yemen Plateau, at least within the 1,800–2,000-m altitude belt, although with slightly different local variants. The standard litho- and pedostratigraphy is summarized in Fig. 3, where relevant details are included. In the Thayyilah-

---

[1] The documentation presented in this chapter has several limits that need be explained. The excavations at WTH3 came to a halt after 1986, and that very season was curtailed, because of mounting tensions in the area, making any further exploration of the deeper horizons impossible. It is also unfortunate that well preserved charcoal was very rare at the site and, alas, several charcoal samples for radiocarbon were misplaced in Rome after preparation for shipping to the dating laboratory. The only measurement (Beta-23, 583) cannot be associated with the prehistoric occupation; a dating programme would require a return to the site. The Khamar survey was conducted at the very end of what eventually became my last season in Yemen, in February 1990, and time for post-survey work was very limited. Furthermore, because of expected continuation in the near future, artifacts from surface scatters or sections were rather observed than collected, in general, and priority was given to the Neolithic samples. The materials from Wādī Khamar and the deeper levels of WTH3, housed at the National Museum in Sanʿāʾ, could not be re-examined for the present publication.

[2] Thayyilah is an anglicized spelling of the original name, at-Tayyilah; several Arabic placenames in this chapter will be written in a simplified transliteration with diacritic signs omitted.

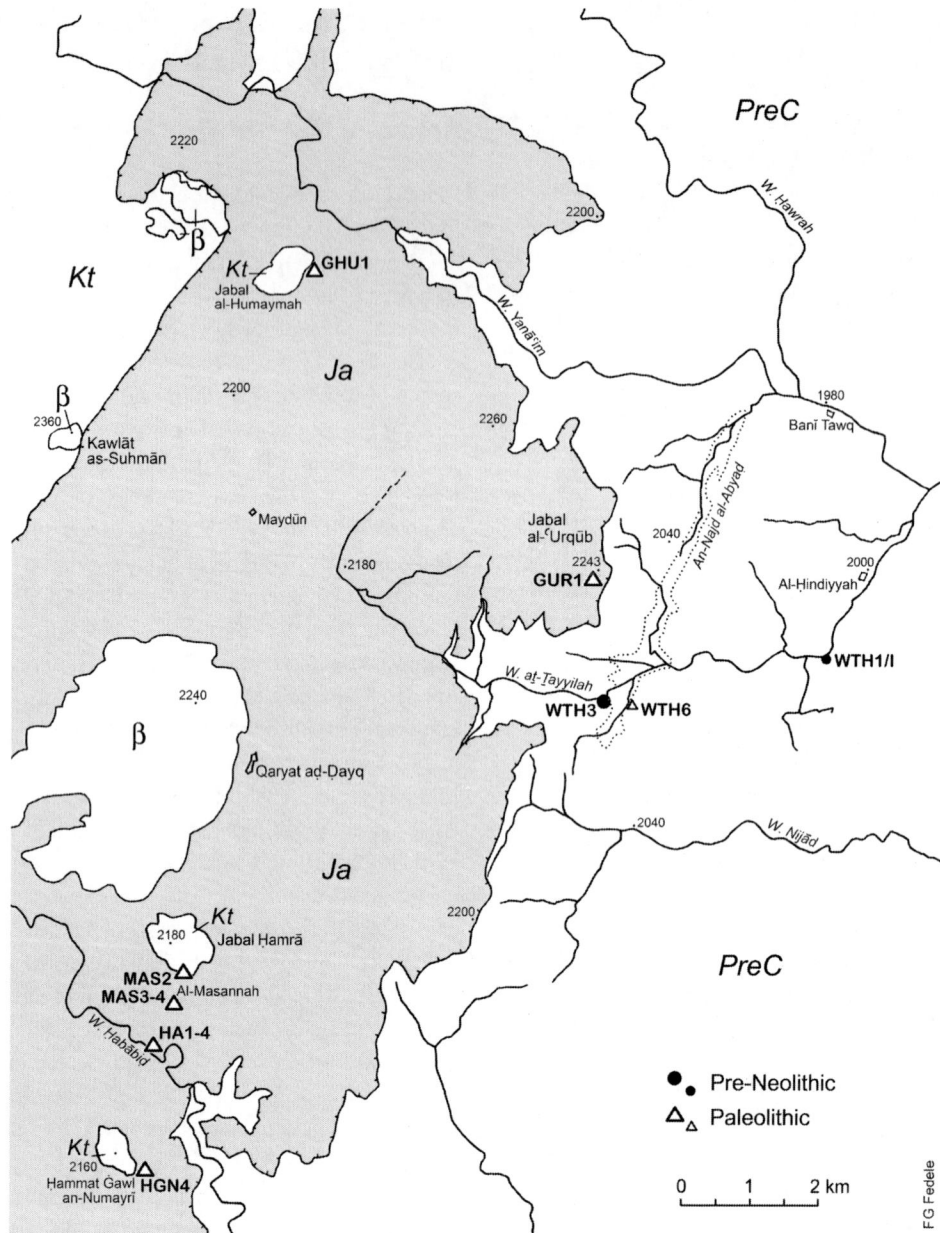

Fig. 2 Map of Suhmān and the Thayyilah-NAB area, this latter comprising the Wādī at-Tayyilah and Wādī an-Najd al-Abyad basins (center-east). Prehistoric sites of interest are indicated. Geology based on Marcolongo and Palmieri (1990), with updating from Kruck et al. (1996): PreC, Precambrian basement; Ja (*shaded*), Jurassic Amran Limestone and underlying Kohlan Sandstone; Kt, Cretaceous Tawilah Sandstone; β, Tertiary Volcanics, basalt

NAB area, the Holocene deposits on the lower slopes are mainly composed of redeposited aeolian silts, intergrading with alluvial or sometimes lacustrine sediments towards valley bottoms. The typical, widespread Holocene sequence forms wadi terraces varying in thickness from 2 to 6 m, but it can often be found upslope on the gentler sides of depressions and valleys. From the migmatite or granite bedrock to the top the succession is the following (numbering begins with "Stratum 2" in the light of the subsequent discovery of an earlier unit in Wādī Khamar, see below):

Stratum 2 ("Early gravels"): more or less cemented alluvial gravels, often a conglomerate, indicating high-energy transport of a torrential type; intercalated, indurated sandy units may be locally common;

Stratum 3 ("Light lower silts"): a complex of horizontally deposited silty-clayey sands, alluvial in origin but more colluvial towards the top, sometimes intercalated to gravel lenses; locally these units include, or show lateral variation to, laminated lacustrine sediments (3λ) or evaporitic deposits near former springs (calcareous sinter or travertine, 3t); they seem to indicate an alternating wet-dry regime and can be attributed to the earlier half of the Holocene;

Stratum 4 ("Gray paleosol"): a thick silty-clayey layer, sandier towards the top; it typically includes a gray to dark

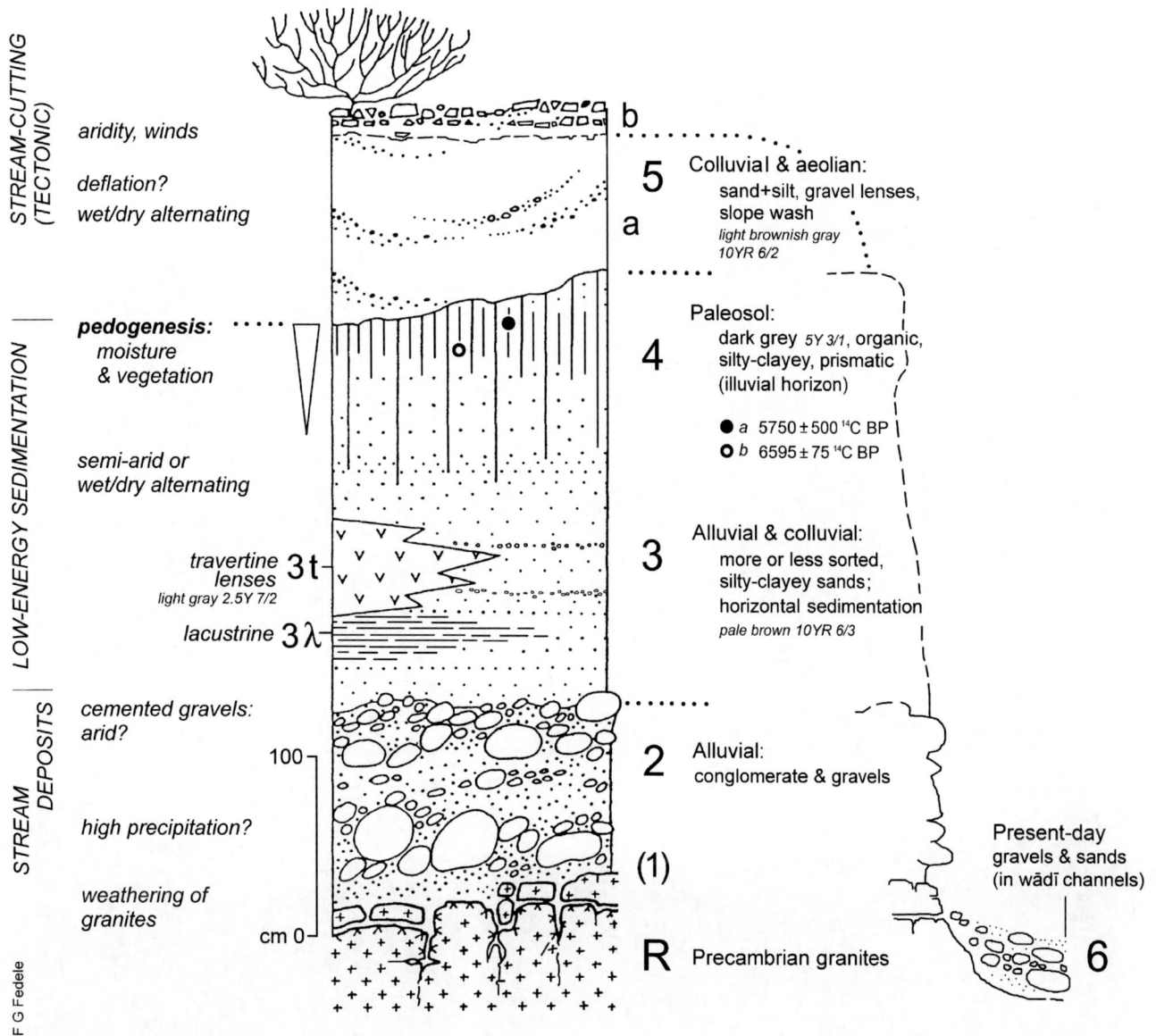

Fig. 3 Holocene–Pleistocene depositional series on the eastern Yemen Plateau, Thayyilah-NAB area: a generalized lithostratigraphy and paleoenvironmental sequence (based on de Maigret et al., 1989, and personal observations). R, granite/migmatite bedrock. Radiocarbon dates on soil's organic fraction by the Rome laboratory, unpublished (**a**, Middle Wādī at-Tayyilah; **b**, Wādī Swayhāt; as reported in Marcolongo and Palmieri, 1986; de Maigret et al., 1989)

gray band, rich in organic matter and accumulated calcium carbonate, usually bounded by a well-defined upper limit due to truncation (by deflation or erosion); the gray band is a humic-accumulation horizon of a paleosol clearly to be correlated with mid-Holocene conditions, as explained below;

Stratum 5 ("Light upper silts"): aeolian silt and colluvial lenses of sand and gravel, largely affected by aeolian deflation and normally topped (5b) by *hammada* stony surfaces with angular clasts, linked to recent aridity; upwards these units are increasingly associated with exfoliation of desert varnish from rock faces, a process that is at least partly coeval with Stratum 6;

Stratum 6 ("Modern gravels"), only to be found in furrows with active discharge: present-day wadi gravels and

sands, regularly associated with stream or river channels cut not only into the pre-existing Strata 2–5, or 1–5 elsewhere, but often also the bedrock (examples in Fig. 6).

The Early gravels may be related to a supposed phase of massive discharge around the Pleistocene-Holocene transition, and therefore have an age of ca. 12 to 9 ka (cal. BP); further comments are given in the section on Wādī Khamar below. Stratum 3 represents the rapid establishment of mild, moist, soil-forming conditions during the earlier part of the Holocene in the northern tropics, equally known across the Red Sea (e.g., Barnett, 1999, chapter 4, with references). The best instance of a limnic series in our area was observed at the bend of Wādī at-Tayyilah between site WTH1 and the hamlet of Al-Hindiyya, where two suites of laminated units

Fig. 4 *Above*: middle Wādī at-Tayyilah and overlooking Jabal al-ʿUrqūb; part of prehistoric site WTH3 in the foreground. *Below*: exposure of an Early Holocene sequence at WTH1/I

are present and contain gastropods and charcoal (site WTH1/I; Figs. 4, below, and 6). They are separated by several thin lenses of alluvial sands and gravels. Where present, travertines show plant remains and poorly developed vacuolar structure. At WTH1/I the exposures of Stratum 3 are no less than 4.5 m thick.

Stratum 4 is only of interest here as an easily recognizable marker and the upper limit of the Early Holocene as defined in this paper. In connection with this stratum, the depositional history was punctuated by one major phase of soil formation, simultaneously identified at WTH3 (de Maigret et al., 1984: 431–437; Fedele, 1985) and by the Italian geologists in the Thayyilah-NAB area (Marcolongo and Palmieri, 1986). On qualitative data this fossil soil was designated the "Thayyilah Paleosol" (Fedele, 1986, 1987, 1988, 1990a); quantitative analyses have subsequently improved its identification (Marcolongo et al., 1988; de Maigret et al., 1989; Marcolongo and Palmieri, 1990) while our brief survey of Wādī Khamar allowed to trace its presence further north and west. The Thayyilah Paleosol can be dated to the sixth–fifth millennia cal. BC on the basis of two radiocarbon determinations (Fig. 3). It is a local expression of a mid-Holocene soil

which represents a useful pedostratigraphic marker over a wide area of southwestern Arabia, given that similar pedogenetic bodies of the same general age have been reported from a number of locations at different altitudes (e.g., Overstreet et al., 1988; Overstreet and Grolier, 1996; Wilkinson, 1997; Lézine et al., 1998; McCorriston, 2000; French, 2003: 224–234; , Parker et al., 2006).

The connection of widespread soil formation with a period of milder and moister oscillations, plausibly resulting from higher rainfall (e.g., Wilkinson, 2005), is generally accepted, hence the frequent designation of Mid-Holocene Pluvial (see Fleitmann et al., 2007, for a detailed climatic framework). The period is also well documented in tropical eastern Africa including the Ethiopian highlands (e.g., Barnett, 1999). In Khawlān this soil's environmental significance is clear: pedo-sedimentary evidence, topography and a palynological test (Marcolongo et al., 1988, palynology by A. Lentini; Fedele, 1990a, Fig. 4) (Fig. 5) suggest the presence of high watertable, scattered ponds and some tree cover in upland basins. Well watered conditions can equally be inferred from the incidence of bovine husbandry in the Neolithic (Fedele, 2008) and by

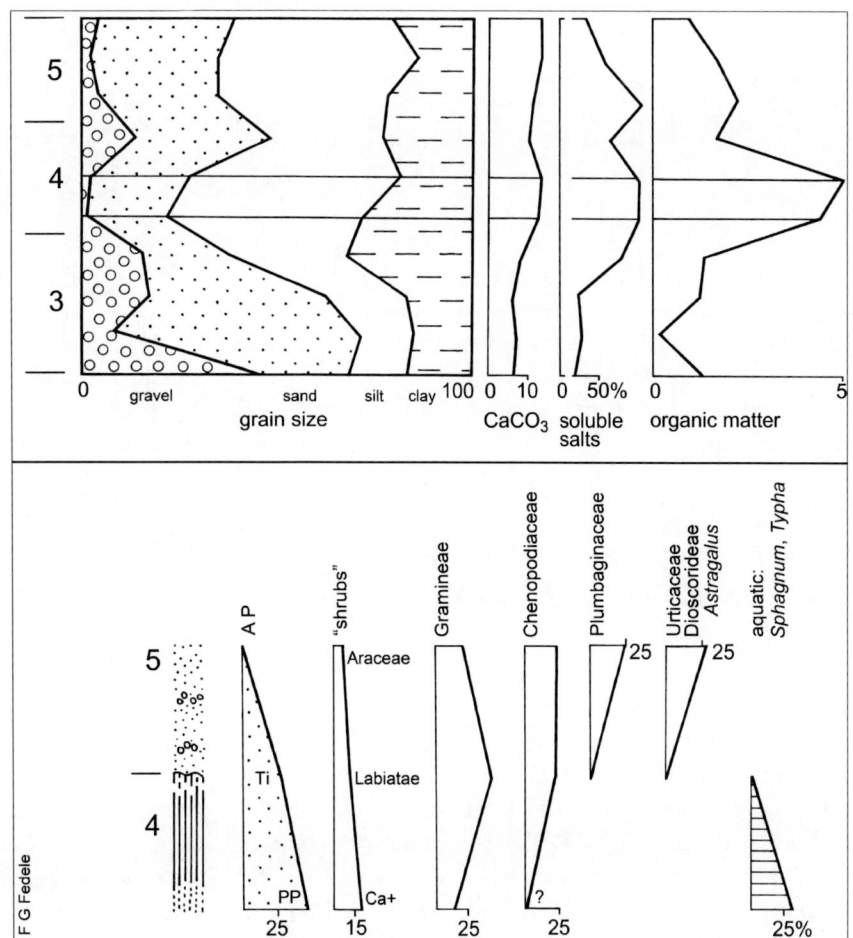

Fig. 5 Sediment and pollen analyses of a Holocene series from middle Wādī at-Tayyilah (after Fedele, 1990a, revised; data from Marcolongo et al., 1988). Strata as in Fig. 3

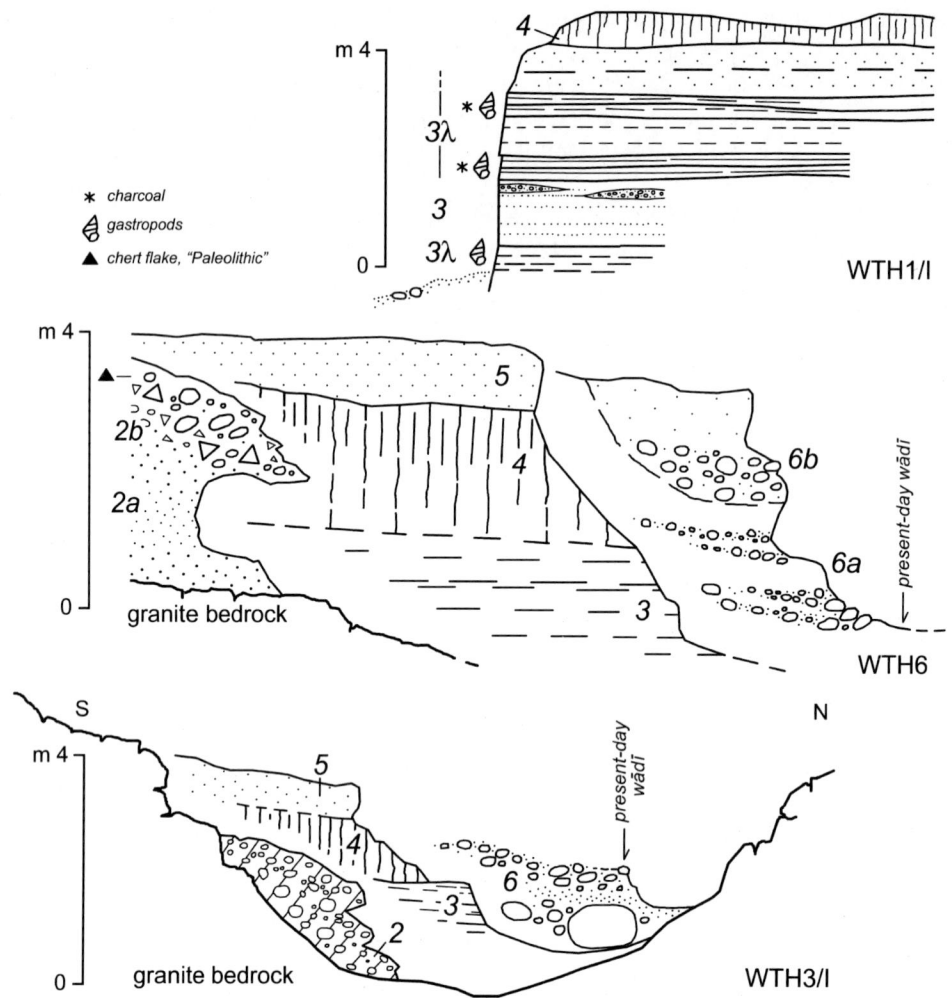

analogy from attributes of similar buried soils on the Plateau (e.g., French, 2003: 224–234). Widespread geomorphic stability contributed to this kind of landscape in the eastern highlands, before being rather abruptly ended by regional tectonic uplift and a concurrent new cycle of severe desiccation (Fedele, 1990a).

Site WTH3: Setting, Local Sequence, and Pre-Neolithic Components

Stratified pre-Neolithic evidence on the eastern Yemen Plateau was first obtained through soundings during the excavations of the Neolithic settlement at WTH3 between 1984 and 1986. These earlier components will be labelled "pre-Neolithic" for the sake of a noncommittal designation. As mentioned, no hint of earlier human activity could be perceived on the surface. Site WTH3 (44° 39′ 58″ E, 15° 10′ 00″ N) lies in a semi-desert landscape at an altitude of

2,025 m in the middle Wādī at-Tayyilah drainage, which is almost completely set within the Precambrian basement migmatite at the foothills of the limestone tableland of Jabal al-‘Urqūb. The bedrock and surrounding rocks at WTH3 have a distinct granite facies. Half-buried on the rock-strewn hillside and covering an estimated 70 by 90 m, the site coincides with a mildly sloping terrace in proximity of a watercourse, a standard Neolithic location in the region of Khawlān. The present-day wadi runs eastwards about a hundred meters north of the site and is flanked by a series of alluvial terraces; the third and topmost can possibly indicate the margin of the mid-Holocene riverbed (Fig. 4).

WTH3 has remained one of the very few Neolithic sites investigated on the Yemen Plateau (see Kallweit, 1996, for a further example). The site was excavated and recorded with geoarchaeological criteria and exacting procedures, unprecedented on the Plateau (e.g., Fedele, 1995). All sediments were dry screened with 4-mm mesh and expertly hand-picked for artifacts and ecofacts. Deposits from particular

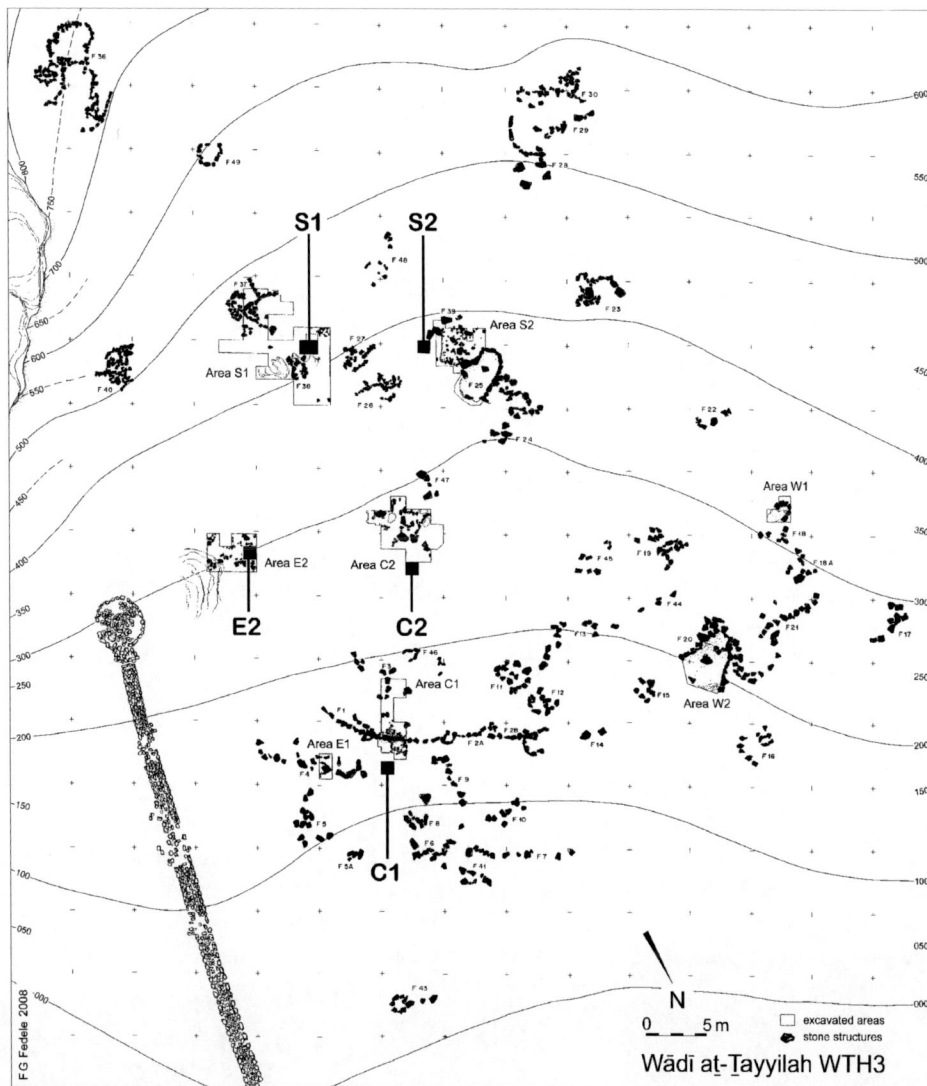

Fig. 7 Plan of Wādī at-Ṭayyilah site WTH3, showing above-ground features and position of the excavated pre-Neolithic levels (redrawn from V. Labianca and D. Picchi/Mairay, plan A16, 1986). The large structure to the east is a later tomb, unrelated to the Pre-Neolithic and Neolithic occupations

contexts were bagged for water sieving in Sanʿāʾ. Badly preserved faunal finds were block-lifted within their matrix for laboratory processing after consolidation in the field. As a consequence, WTH3 generated a controlled, very large collection of lithic finds and a valuable sample of archeofaunal material (Fedele, 2008, with references).

The most conspicuous feature of the site is its Neolithic occupation, apart from a later tomb (Fig. 7), and the excavation program was primarily aimed at revealing and understanding the Neolithic component. The attendant culture can be confidently labelled Neolithic from the well-documented presence of domestic animals (Fedele, 1988, 2008; a different opinion in Crassard and Khalidi, 2004). A total of about 120 m² were excavated, amounting – in spite of the effort – to perhaps 5% of the site. In order to sample spatial variation eight excavation areas were opened, and small soundings for the detection

of earlier deposits and cultural horizons – if any – were made in five of them: north to south, Areas C1, C2, E2, S2 and S1 (black squares in Fig. 7).

The existence of pre-Neolithic levels would not have been revealed without intentional sounding. Already during the initial testing in 1984 the site turned out to be stratified, up to 1-m-thick in Area C1, and to possess some evidence of earlier material than the Neolithic. It was subsequently found that the deposits reached a similar thickness in other parts of the site. The discovery of pre-Neolithic evidence opened up an entirely new subject within the program, which would have been developed in future field seasons. In fact, the earlier levels could only be explored on about six square meters in total.

The soundings below the main occupation had to be kept to a minimum considering not only research aspects, but time constraints, limited staff, and generally difficult field-work conditions. Under the circumstances, the exploration of the pre-Neolithic levels implied a vertical excavation strategy which could not be easily scheduled within the essentially horizontal strategy (open-area stripping or *déca-page*) demanded by work on the Neolithic, our research priority. In addition, it became clear from the outset that a correct evaluation of pre-Neolithic evidence embedded within sandy arkosic sediments, characterized by rather weak unit boundaries, would have required the most analytical and patient reading of lithostratigraphy, a task that we could only pursue on a small number of soundings. Given suitable conditions, however, I would urge researchers to take up the task of deep testing below Neolithic or Bronze Age occupations wherever possible, without stopping at presumed culturally sterile soil.

The detailed stratigraphic profiles from WTH3 can be correlated to the standard depositional sequence of the Thayyilah-NAB area (Fig. 3). Above the decayed migmatite/granite bedrock and related arkosic sand (A) there are 40–80 cm of colluvial and aeolian sediments, predominantly silty-sandy in texture, due to prolonged but discontinuous slope deposition. And here again this trend was punctuated by the formation of the Thayyilah Paleosol, which blankets the site and is indicated as stratum G (for gray) in the site's general profile (Fig. 8). The pre-Neolithic levels are associated with strata M and Y. Stratum M (for Italian *marrone*, brown) is possibly to be equated with poorly developed, slightly argillic brown-earth type soils of the earlier half of the Holocene

such as those studied in the Damār plains (French, 2003: 228–232). Stratum Y is a yellowish layer, sandy-silty in texture with variable amounts of grit, separated from overlying M by a generally weak upper boundary.

There appear to be more than one pre-Neolithic level. A particularly precise horizon is documented in all five test pits in the upper part of stratum M and will be provisionally called "Pre-Neolithic", with a capital P. In Areas E2 and S1 it is characterized by stone clusters and pits, associated with heavily weathered, leached burnt features that look like hearths (simple campfires?). Stone-filled hollows to be attributed to the same horizon, or the very base of the Neolithic, were recorded in Area C2 (de Maigret et al., 1988: 23).

A potential dating object was found within a pit in Area S1, a feature of mixed cultural-erosional origin that contained a pocket of dark ashy silts and piled stones, perhaps from a nearby hearth (Fig. 9). The finding is a partial figurine made of hardened, unfired clay (Fig. 11a), which may represent a female torso or two closely facing figures; it is at the moment the oldest piece of portable "art" in Yemen (Fedele, 1986, Fig. 28; Fedele, 2008, Fig. 8). The nearest parallels are probably to be found in the Pre-Pottery Neolithic B of the Levant (e.g., Jordan; Kuijt and Chesson, 2005, Figs. 8.2 and 8.4), and according to this hypothesis a date in the seventh millennium BC is tentatively proposed. Such an artifact might be of some relevance for Drechsler's (2007) model of the dispersal of the Neolithic into southern Arabia (see the last section of chapter).

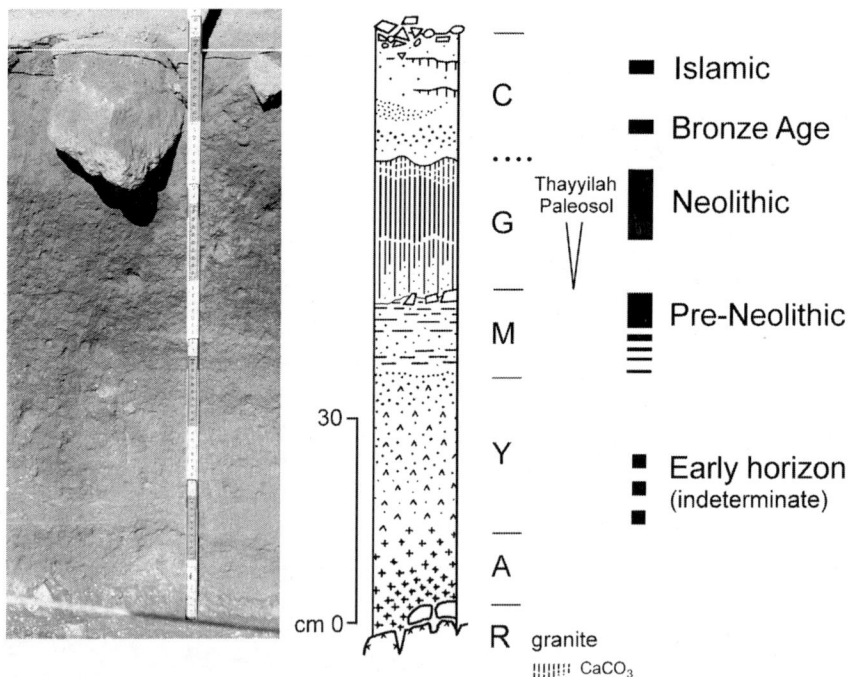

Fig. 8 General stratigraphy of site WTH3: litho-pedostratigraphic units and cultural stratigraphy. On the *left*: view of a representative section from the test pit in Area C1 (1984)

Archeofaunal samples, totalling about 140 pieces, come from deep contexts in Areas S1 and E2 (Table 1); a bone from E2 bears cut-marks. Preservation was mildly favored by rapid burial and slight charring, as in the locus of the clay figurine in Area S1, which gave bone remains from large bovids. An adult radius is metrically intermediate between wild and domestic cattle (Fig. 11b; Table 2): its estimated proximal width gives a logarithmic difference from a female European aurochs assumed as standard of c. −0.030, which means that although large the WTH3 *Bos* could be either wild or domestic (Fedele, 2008, with references; for a relevant diagram see Grigson, 1989, Fig. 5).

Fig. 9 Site WTH3: view of sounding in Area S1 (1985 excavations). The base of the test pit on the left shows the locus of the unfired clay figurine of Fig. 11; the Neolithic layer is exposed within the square on the *right*

Table 1 Archeofauna from the Pre-Neolithic horizon of site WTH3, Wādī at-Tayyilah: species composition and number of identified specimens (after Fedele, 2008)

	WTH3: Pre-Neolithic
Total number of specimens:	c. 140 (identified 6)
Wild or indeterminate status	
Bos sp., possibly wild	4
cf. *Gazella*, possibly gazelle	2
Only identified to size group	
Cattle-equid size group	c. 45
Caprine-gazelle size group	c. 20

Table 2 WTH3, Bos sp., adult radius no. 165.1. Coded measurements follow von den Driesch (1976)

	Measurements (mm)
Width of the proximal end, Bp	94 ± 1
Width of facies proximalis, BFp	~85
Depth of facies proximalis	38
Maximum depth of the proximal end	43

Since all the materials from the Pre-Neolithic appear to derive from large bovids and gazelle-sized animals, with domesticates not clearly present, I would suggest a wild fauna in which the aurochs may be dominant. Another wild species, buffalo, was reported from a mid-Holocene occupation at Sa'dah, next to a rock surface with depictions interpreted as the same species ("*Pelorovis antiquus*=*Bubalus arnee*"; Garcia et al., 1991; Garcia and Rachad, 1997). At WTH3, like at Sa'dah, we may be dealing with campsites where forager groups would bring butchered game, in the context of seasonal occupation by essentially mobile populations. It is unfortunate that such glimpses of mid-Holocene campsites as those provided by Sa'dah and WTH3 has remained until now isolated, and inevitably under-explored.

The test pit in Area C1 has produced a tiny amount of lithic finds from deeper levels than the Pre-Neolithic: provisionally these levels will be referred to as the WTH3 "Early horizon" (Fig. 8). Even allowing for short vertical migration in fine-grained sediments, the existence of a distinct earlier horizon well within stratum Y and at the A/Y contact was considered real, according to a reiterated, critical check of lithostratigraphy (Fig. 10). The collection only numbers four chert artifacts – three waste flakes and a blade – which on the basis of physical freshness suggest chipping in situ. Bone material was absent or not preserved. Unfortunately, no date can be offered for this deeper horizon.

The A/Y stratigraphic contact also gave a large, mildly worn denticulate scraper made of quartzite, wich is highly suggestive of redeposited Paleolithic material. This is a distinct possibility, since Area C1 likely lay very close to the riverbank at the time of the Pleistocene-Holocene transition, and elsewhere this kind of location corresponds to a frequent Paleolithic choice (for instance on the Yemen Tihāmah; Bulgarelli, 1985). Indeed in proximity, at findspot WTH3/III in October 1984, a fossilized fragment of long-bone diaphysis from a *Bos*-sized animal was found (Fig. 11c), recently eroded from cemented sands of this former riverbed of Wādī at-Tayyilah, Stratum 2 of the standard sequence. The bone fragment is relatively unworn, thus ruling out river transport, and shows splintering marks at one extremity which suggest human percussion rather than incidental breakage. Its correlation to the "early horizon" is only inferential, but not implausible.

The lithic artifacts from the Pre-Neolithic are nondiagnostic as well, unfortunately, and sample size is once again very small, a few dozen pieces (Fig. 12). Micro-waste from onsite working is common, with evidence of microblade technology, and there is a frequency of expediently utilized blanks as well as chert and granite macroliths. Obsidian was used, including a gray variety that is rare in the Neolithic, and whose source is unknown. Pending detailed re-analysis of the material (cf. Note 1) it is not possible to provide more

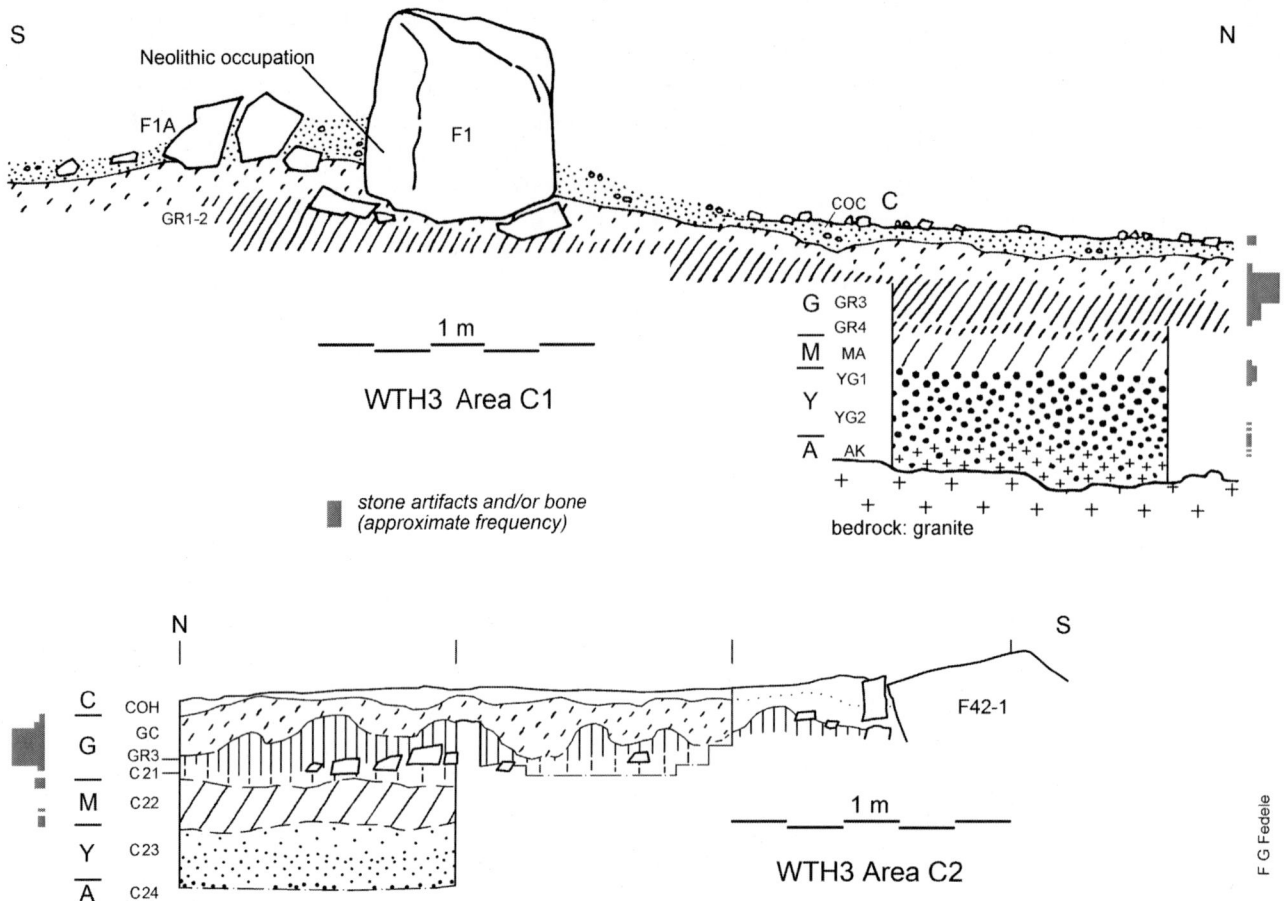

Fig. 10 Site WTH3: selected profiles from Areas C1 and C2 to show the pre-Neolithic cultural horizons. *Shaded histograms* on the side of profiles suggest the vertical distribution and frequency of cultural material

specific information, or evaluate the finds in the light of the current upsurge of innovative lithic technology studies. With suitable artifact samples, the analytic framework now constructed by Crassard (2007 and see also Crassard et al., 2006) should be borne in mind.

The impression on present evidence is that the Pre-Neolithic and Neolithic manifestations may be phases of a single continuum, an idea primarily supported by the apparent continuity of occupation at site WTH3. From the artifactual point of view, such a continuum might hint to similarities with the East African sequence rather than the Fertile Crescent, which would incline towards adopting an East African terminology (Fedele and Zaccara, 2005). The lack of pottery contributes to this impression. If so, the above lithic phases could be grouped under a "Late Stone Age" of the Yemen Plateau, a terminology already proposed for south-eastern Arabia by M. Uerpmann (1992). Further exploration of this issue is clearly necessary.

Other Pleistocene/Holocene Sites in the Thayyilah-NAB Area

Near WTH1, where the Early Holocene limnic series was observed (WTH1/I), the paleo-wadi had the character of a wide, meandering rivercourse with a tendency for overflowing. Our survey suggests the plausible existence of several associated sites, but actual reconnaissance was not possible. However, there is a hint of early human activity at WTH6, a thick terrace section on a small right tributary of the Tayyilah called Wādī Swayhāt (Fig. 6). Another exposure on this wadi had been sampled by Marcolongo and Palmieri (1990; de Maigret et al., 1989) during their initial reconstruction of recent geologic history. At WTH6, a chert flake was found in the exposed section within the breccia-conglomerate unit or Stratum 2. These gravels are especially rich in limestone and siliceous clasts from the wadi's headwaters, and prehistoric groups might have

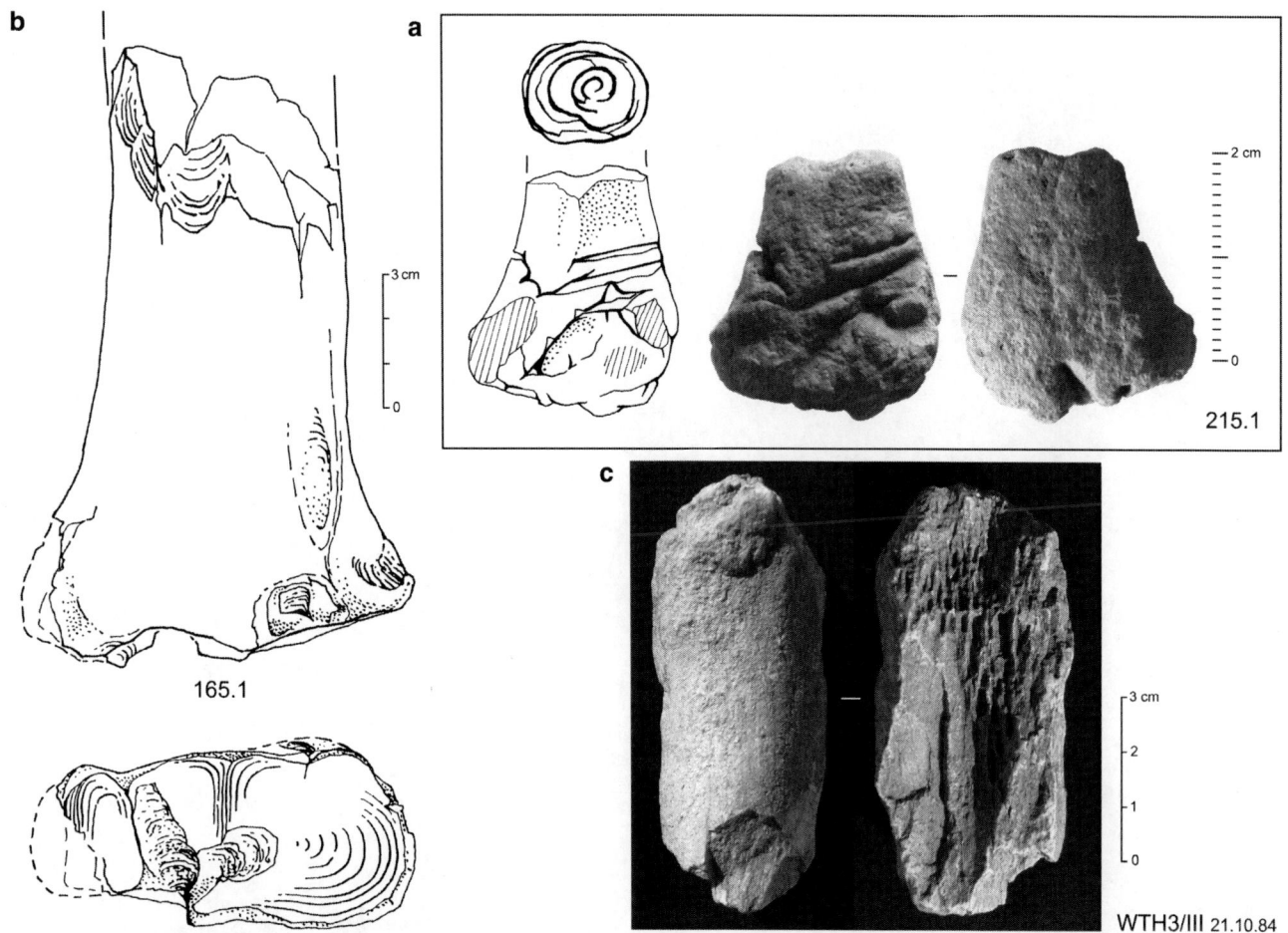

Fig. 11   Site WTH3, Pre-Neolithic: (**a**) part of figurine made of unfired clay; (**b**) *Bos* sp., proximal radius intermediate between wild and domestic; both from area S1. Site WTH3/III: (**c**) *Bos* sp., fragment of fossil bone from the cemented alluvial gravels (Stratum 2)

exploited chert cobbles as a resource. A Paleolithic site indeed exists on the limestone plateau overlooking the area from the west, Jabal al-ʻUrqūb. A flake alone is hardly diagnostic, but its origin from a discoid core as well as size correlate well with its Paleolithic appearance and stratigraphic date. To our knowledge this stratified find is still unique in the region.

Pre-Neolithic Evidence from the Wādī Khamar Basin (Jihānah)

The Area and Its Depositional–Environmental Sequences

Three sub-regions can be distinguished in the Khawlān at-Tiyāl, as historically defined by tribal territories. However, as indicated by de Maigret (1990: 3–4, 11), they also largely correspond to the three main geologic and lithologic zones (Figs. 1 and 2). East to west they are: Al-Aʻrūsh, in the Precambrian basement dominated by migmatite and granite; As-Suhmān, closely identified with the tableland made of Jurassic limestone, dolomite and calcarenite (Kohlan Sandstone and Amran Limestone formations, this latter in its shelf facies); and Jihānah, linked to Cretaceous-Paleocene sandstone (Tawilah and Medj-zir Sandstone formations) and Tertiary volcanics. Detailed geology can be seen in Kruck et al. (1996, Sheet 5 Sanʻaʼ, initially published 1991). It should be noted that the Jihānah designation is here taken in its geographic rather than administrative sense.

Wādī Khamar is a branching valley located in the sandstone country of the easternmost Jihānah district along the border with Suhmān, about 40 km east-southeast of Sanʻāʼ (Fig. 13). It corresponds to a small and rather secluded basin, about 11 km² in area, and is composed of a north-to-south valley discharging south. Two smaller, tributary catchments

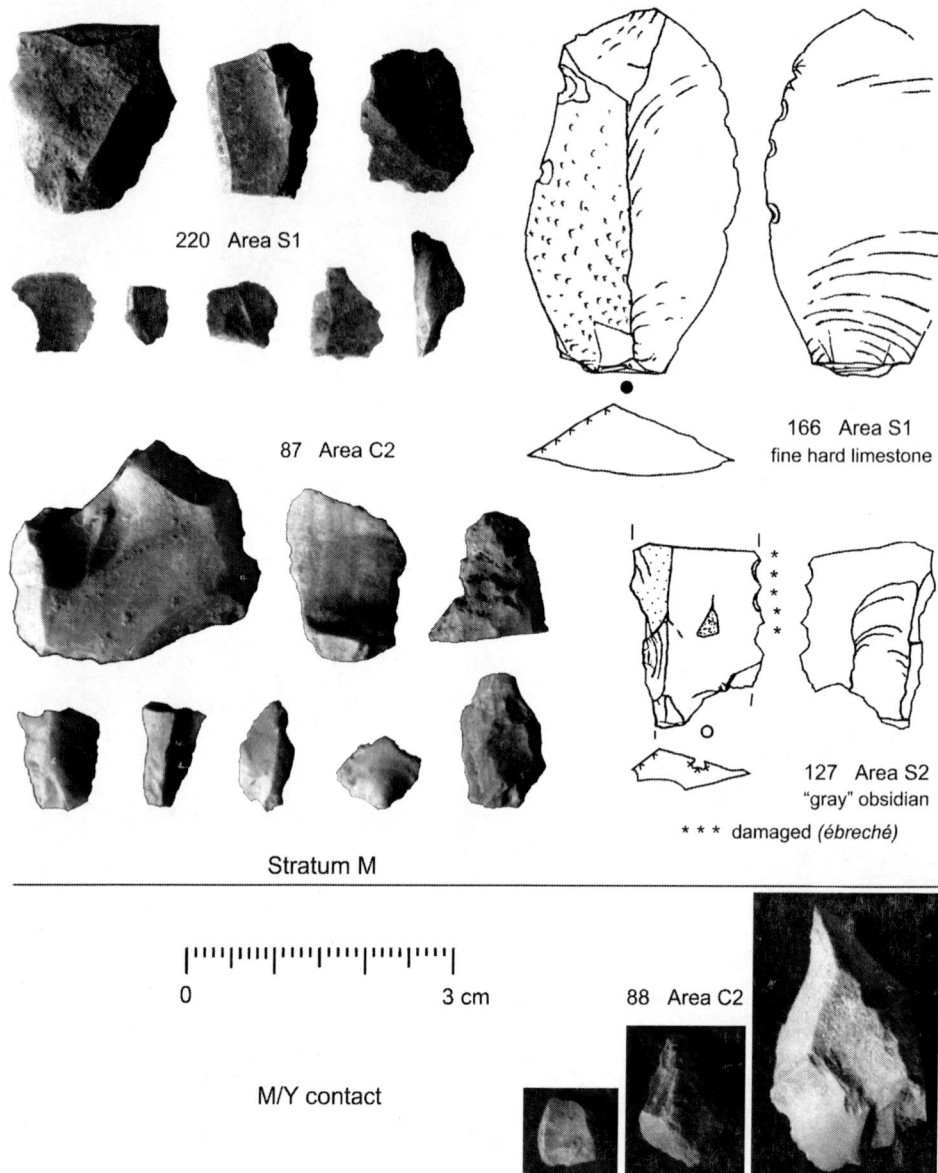

220   Area S1

87   Area C2

166   Area S1
fine hard limestone

127   Area S2
"gray" obsidian

* * * damaged (ébreché)

Stratum M

0   3 cm

M/Y contact

88   Area C2

Fig. 12 Site WTH3: selection of lithic artifacts from levels of the Pre-Neolithic horizon. All chert, unless indicated otherwise

are situated on the right side of the valley, Wādī al-Farʿ and Wādī al-Akhdād, slightly suspended above the main valley axis, at elevations of 2,160 m and 2,180–2,190 m, respectively. The lower sector of the valley is centred on the floodplain and hamlets of Bayt Abū Jaydāʿ, 2,110 m, and is separated from the upper sector, where the *qaryat* (village) of Khamar is located, 2,150 m, by a constriction of the valley and some rocky steps in the longitudinal profile.

A project was devised in 1987 to sample the archaeological occurrences within the sandstone belt of Khawlān (Kruck et al., 1996: 41–44), which had remained underexplored, and Wādī Khamar was selected for survey on the basis of air photo study. The attempt represented an extension of the initial research area of the Italian mission. Prospectively, a par-

ticular emphasis was put on the probable existence of caves and stratified cave deposits, a search for the earlier prehistory, and the presumed presence of early rock depictions ("rock art"). A single engraved rock had been previously recorded at the nearby sandstone-belt village of Al-Hisf, although without clues to its dating. All the theoretical expectations of the survey were indeed met with success, and several Neolithic and Bronze Age sites were recognized. However, it was not possible to set up camp at Khamar and spend more time in the field, and thus pursue our investigation beyond basic field reconnaissance. The survey took a net total of six days, one in November 1987 and five in February 1990; this latter phase coincided with an unusually rainy period, which provided insights about

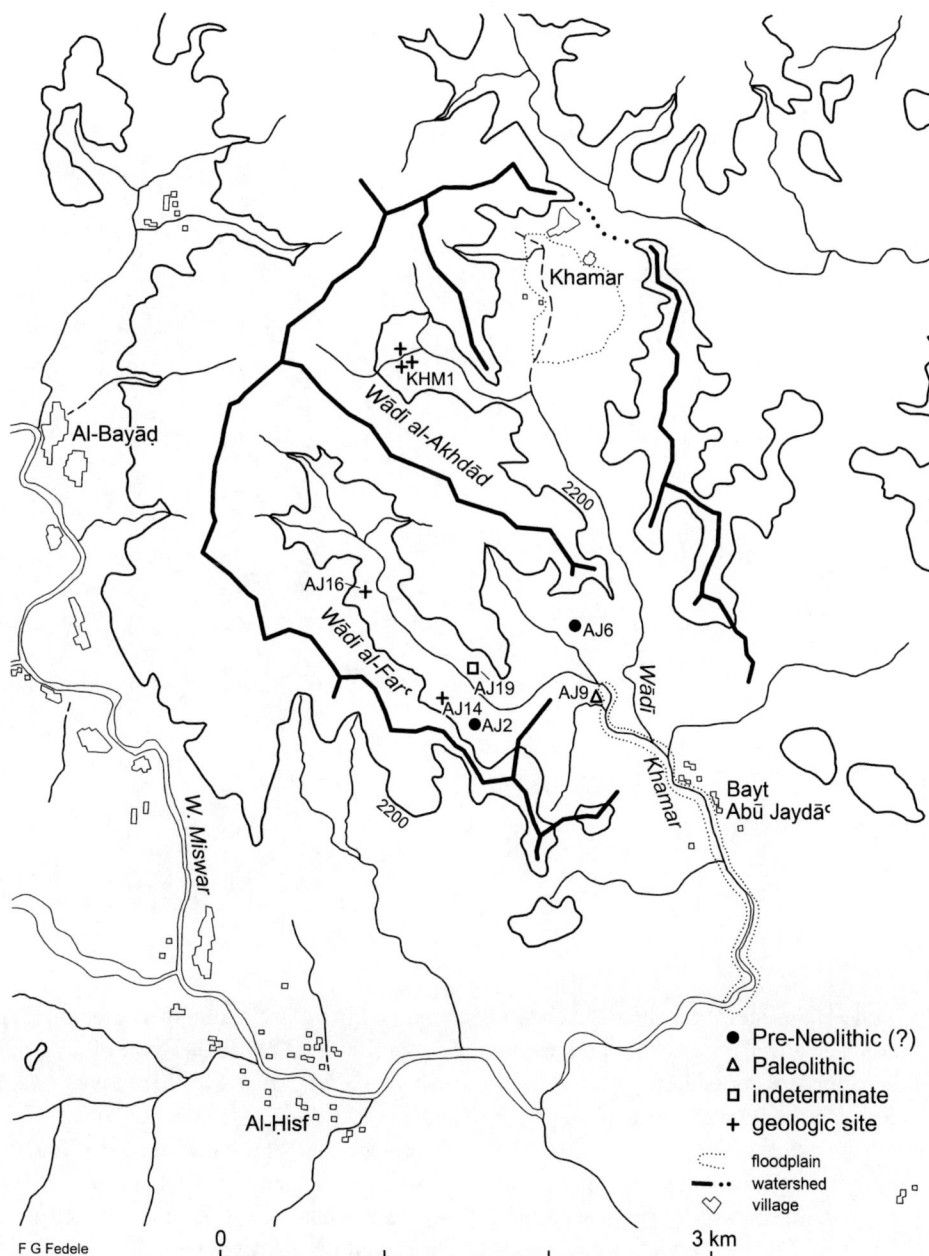

Fig. 13 Map of the Wādī Khamar basin. Prehistoric and geologic sites of interest are indicated

current processes of runoff and erosion. The survey's results have only been mentioned in a privately circulated report (Fedele, 1990b).

Among the acquisitions of interest here was the observation of numerous open-air, alluvial-colluvial stratified deposits in the Al-Far' and Al-Akhdād hanging basins, which almost perfectly replicate the typical stratigraphic succession of the Thayyilah-NAB area. Compared to that area, however, the Khamar sequences revealed an earlier unit below the alluvial conglomerate or gravels, thus expanding the environmental record back in time: this

unit is here designated Stratum 1. These sequences obviously span the Holocene and in some cases the Late Pleistocene periods: the main occurrences are plotted in Fig. 13. No less than a dozen individual outcrops were listed and partly recorded, the most notable in terms of extent, preservation and cultural potential being KHM1, AJ16, AJ14, and AJ2. This information confirms that the Thayyilah-NAB type-sequence has regional significance and reflects widespread environmental conditions, including controls at both regional and super-regional scales (e.g., climate).

Fig. 14 Holocene–Pleistocene depositional sequences of Wādī Khamar: different exposures at KHM1 (*left*, 1987 profile; *right*, 1990)

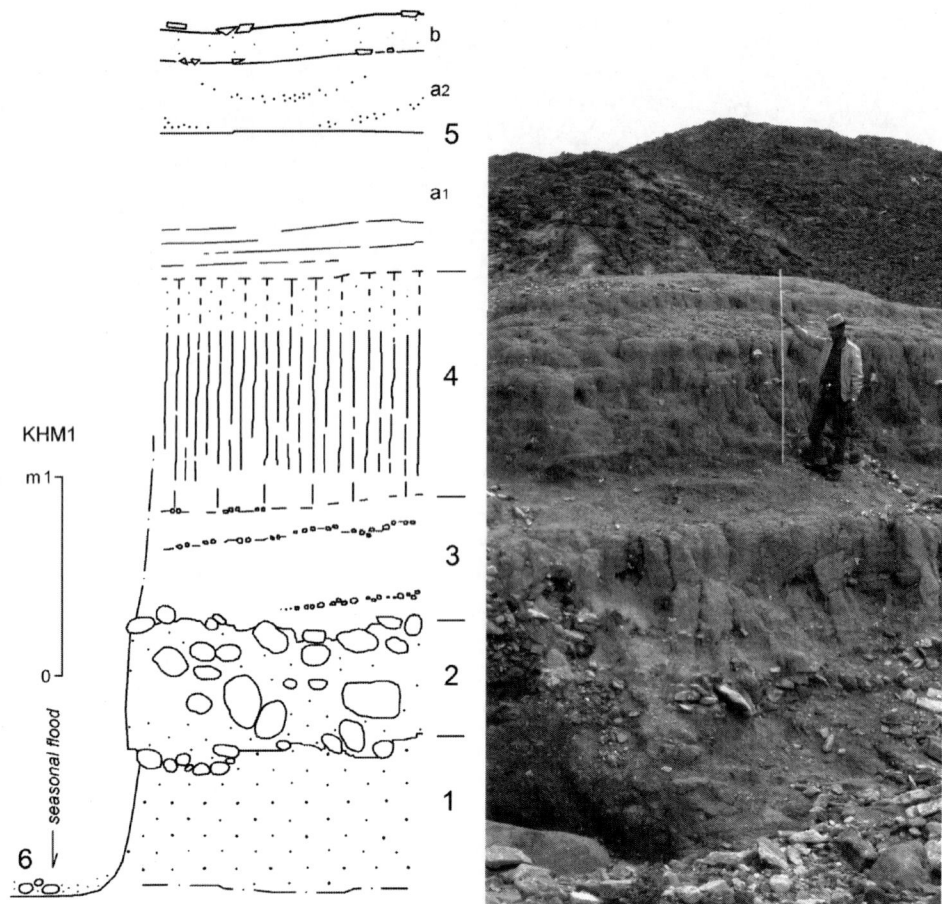

The Early and Middle Holocene sedimentation tends to blanket the valley bottom and lower slopes in both suspended basins, except where erosion was more active. The series is regularly and conspicuously dominated by a pedogenetic unit which I would equate with the Thayyilah Paleosol, the pedogenesis being superimposed on a band of often indurated, carbonate-rich, gray and dark gray silts (Stratum 4 in Figs. 14 and 15). These mid-Holocene units are normally sandwiched between a cover of yellow sands and silts, largely aeolian in origin (Stratum 5), associated with the topographic surface, and an underlying suite of pale silty-clayey units, gray to brown in color and varying locally in texture and structure (Stratum 3).

The above sequence is separated by the lower and earlier deposits by a marked discontinuity (2/3), usually the only significant lithostratigraphic break to be seen in the exposures. This discontinuity and the underlying deposits can only be inspected in wadi bottoms, where recent stream erosion has been cutting through the sediment cover, down to an earlier floodplain and/or its bedrock. One of the best examples of the whole series, and possibly the most comprehensive in the Khamar basin, can be observed at KHM1, a cluster of dissected terraces nested upvalley in Al-Akhdād (Fig. 14). Like elsewhere, the 2/3 discontinuity is clearly

erosional in nature and may also entail a significant hiatus. Stratum 2 is the usual layer of alluvial gravels, often cemented, indicating a former stream-bed. Below this layer and unique to KHM1, the top of a thick unit of yellow cemented sands can be observed ("Early sands", Stratum 1), which disappears beneath the active level of the present-day wadi. These sands are not necessarily alluvial in origin. They could be traced laterally for a short distance and turned out to interdigit with pale blue-greyish silts, lacustrine in all probability, a heteropic relationship; no detailed study was possible.

Together, Strata 1 and 2 can be interpreted as representing an interval within the Late Pleistocene. They might approximately date anywhere from between the later Pleniglacial and the end of the Pleistocene, before about 11 ka cal. BP; a date at the Pleistocene-Holocene limit for the heavy discharge associated with the gravels appears plausible. In principle, Strata 1 and 2 might be attributed to a period of alternating pluvial and arid regimes, with the Early sands marking a particularly arid phase within it such as – again in principle – the Late Glacial hyperaridity of about 20–12 ka cal. BP. Only investigations aimed at the specific conditions of Khawlān could bring precision, particularly because tectonic history may have locally modulated the action of climate,

Fig. 15 Holocene–Pleistocene depositional sequences of Wādī Khamar: Wādī al-Farʿ basin, sites AJ14 (*above*) and AJ2 (*below*, with 1990 test excavations)

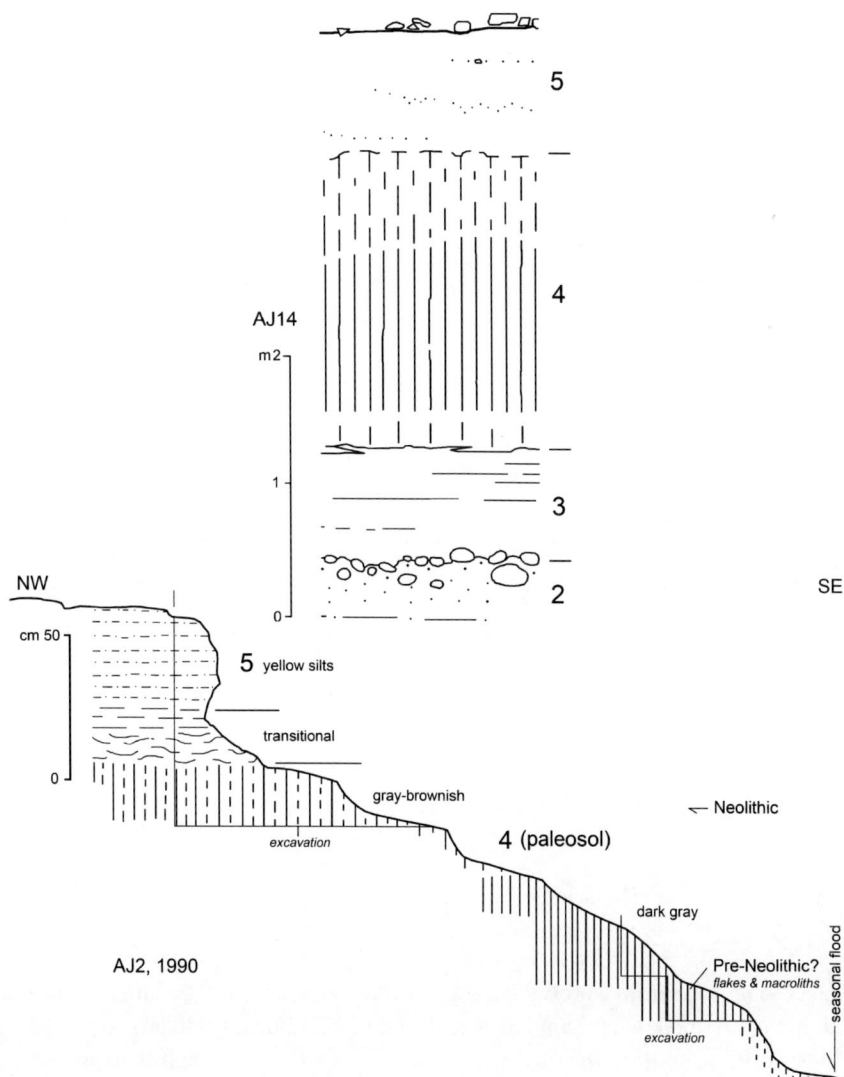

in depositional terms, as it did in the Holocene. Unfortunately, no artifacts were found in these lowermost units. Elements of lithic industry were only observed in the mid-Holocene sequence, where they were eroding from the "gray" paleosol of Stratum 4.

Early Holocene and Putative Late Pleistocene Sites

The sites of interest are all open-air. Two cave deposits of some significance were inferred from visible morphology and erosional windows in the sediment fills, which gave hints of deep sequences, but could not be tested. The open-air occurrences will be discussed according to their presumed chronological order starting with the most recent.

AJ6, AJ2. Site AJ6, near the flat or *hanaka* of Bayt Abū Jaydāʿ, is a small quarry and workshop site connected with

an outcrop of poor-quality chert. Chert appears to be rare or nonexistent in the Khamar basin and possibly in the sandstone belt in general, which explains why inferior but easily obtained material would be valued. Although obviously impossible to date on the basis of two brief inspections, the surface material is partly compatible with the Neolithic, this attribution being based on the Thayyilah and Qutrān "industries" already recognized on the eastern Plateau (Fedele, 1988; Fedele and Zaccara, 2005). However, some lithic clusters appear to be the product of a "Neolithic"-looking but less definite chipping tradition. This, and the type of site itself, make it possible that an Early Holocene component is also present. The possibility is strengthened by the stratified occurrence of some artifacts, including macroliths, at the base of the Thayyilah Paleosol dark-gray band at AJ2, one of the principal terrace exposures in Wādī al-Farʿ (1990 test excavation; Fig. 15).

AJ19. Also in Wādī al-Farʿ, this site is an exceptional spread of lithic artifacts located on an isolated rise (a terrace

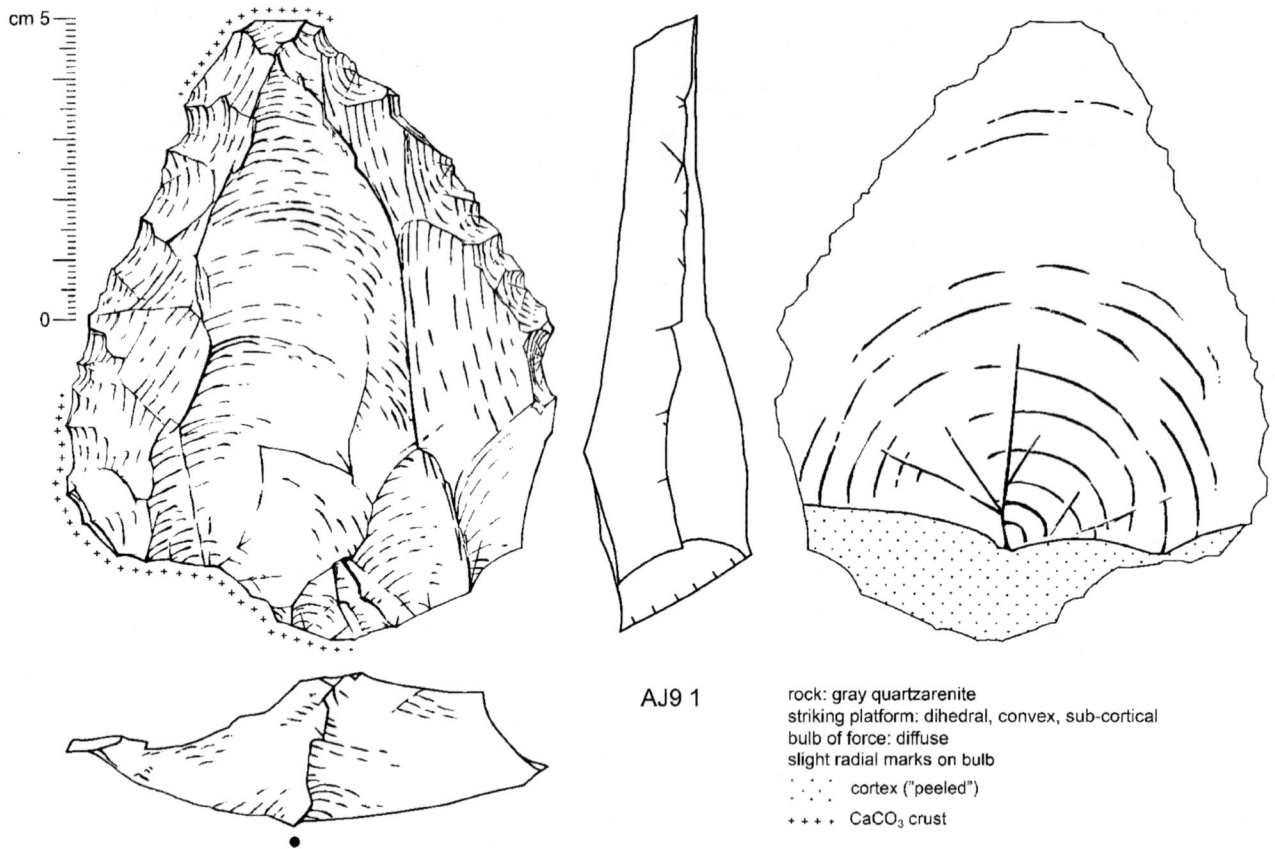

AJ9 1

rock: gray quartzarenite
striking platform: dihedral, convex, sub-cortical
bulb of force: diffuse
slight radial marks on bulb

⋰⋱ cortex ("peeled")

+ + + + CaCO₃ crust

Fig. 16 Wādī Khamar: Paleolithic tool from site AJ9

remnant?) upslope from a protruding rock. This lithic scatter is highly unusual relative to similar sites in Khawlān in terms of composition, artifact density, and to some extent size. Three varieties of red-brown contact-metamorphic rocks were imported and worked, but alongside workshop waste there is about 30% of utilized blanks and finished tools. It was not possible to resolve whether the site is a palimpsest – a mixing of various ages – or contains instead an essentially synchronous component. A distinct perception was gained during the survey that the site was different from what was already known in Khawlān. There is at least a component whose technology and typology do not match the Neolithic and Bronze Age lithic inventories of the region (for the Bronze Age see Di Mario, 1987, 1990). From a purely theoretical standpoint, a core and some large prismatic blades are formally compatible with an "Upper Paleolithic" affiliation. The condition and context of the site do not contradict this possibility.

AJ9. Rare artifacts, some of them apparently reflecting Paleolithic flaking technology, are eroding at AJ9 from yellowish, silty-sandy, non-organic sediments that look very different from the recent deposits. A polyhedral core, a denticulate, and an unusual flake tool were collected. This last artifact (Fig. 16) is a convergent scraper made on a large

flake of gray quartzarenite partly covered with a calcium carbonate crust; it has a convex, dihedral platform and a simple, slightly denticulate retouch. These surface finds come from eroded, gently rolling remnants of the highest terrace in the Abū Jaydā' area. The sediments from which they appear to derive resemble the Stratum 1 sands of the general sequence. Until further research this possible association must remain hypothetical.

A Note on the Paleolithic Evidence from Other Areas of Khawlān

To add perspective, the archaeological occurrences from other areas of Khawlān that have been attributed to the Paleolithic will be summarized. Lacking stratigraphy, the attribution was always based on patination and techno-morphological traits alone, with the indeterminacy and inherent limits of this procedure. All sites are open-air, and most correspond to rich lithic scatters. Several such sites were initially reported by de Maigret (1982, 1983) and Bulgarelli (1988; cf. de Maigret et al., 1984: 437–439), and additional observations and findings were made by the author during field operations that were principally concerned with Neolithic archaeology (unpublished records). The sites will be listed south to north (Figs. 1 and 2).

Sites of Al-A'mās, on the southern fringe of Khawlān (FGF, October 1984). On the tabular limestone plateau of Dulā' (or Zlā') al-A'mās, at least one site presents abundant *débitage* and retouched tools made of chert, coated with heavy patinas and aeolian lustre or "desert varnish" (site DA1). The particular setting is a surface mosaic of subangular to rounded clasts resembling a deflation pavement. Levallois characteristics are clear enough to indicate a probable Mousterian, in terms of the tool-making traditions of the Near East; this description, however, is not meant to imply cultural affiliation (see Petraglia and Alsharekh, 2003, for an overview of alternatives). Another similar site with a predominantly "Mousterian" composition, GSH3, is located on the flanks of the Jabal al-Watadah inselberg, facing the Jabal Sha'īr relief from the north, where long-stabilized semi-desert surfaces meet the basal slopes of the relief.

Sites of Suhmān. Several surface scatters including lithic artifacts of "Middle" Paleolithic appearance have been observed in 1982–1984 on the limestone tablelands comprising As-Suhmān (Fig. 2). Some have been interpreted as workshops by Bulgarelli (1988). The southernmost is HGN4 at Hammāt Ghawl an-Numayrī and was identified on the basis of a few tools made of nodular flint (Bulgarelli, 1984). Three similar sites, MAS2-MAS4, lie on the tableland of Al-Masannah, an area sloping towards the scenic "loop" of Wādī Habābid, or Habābiz, whose riverbed here is deeply entrenched in a canyon (plan in de Maigret, 1990, Fig. 4). Four more, HA1-HA4, lie to the southwest of the previous cluster on the tabular calcarenites of Humayd al-'Ayn, again bordering the wadi's canyon and loop (de Maigret, 1982; Bulgarelli, 1988). To the north, a site indicated by a few finds was recognized at Jabal al-Humaymah, GHU1 (Bulgarelli, 1984). Finally, a concentration exists on Jabal al-'Urqūb, GUR1, near the edge of the spectacular *mesa* overlooking the Thayyilah-NAB rift, where a tabular chert was exploited (FGF, October 1984 and subsequent visits). The scatters of lithic material suggest a mix of various technological complexes, including heavily patinated, "Middle" Paleolithic artifacts. In addition to manufacture from the local chert, exogenous flints are represented.

Dayq Qā' Jahrān. One Paleolithic locality of potential interest outside the region of Khawlān deserves to be mentioned, as it adds to the significant prehistoric evidence from the Qā' Jahrān plain, a flat *enclave* stretching north and south of Ma'bar in the central Plateau (e.g., Wilkinson, 1997; Wilkinson and Edens, 1999; French, 2003). Two findspots were located in 1983–1984 at the foothills of the Hayd Ahmad extinct volcano (QG1) and near Jabal Ghawl ar-Rā'i (QG2) a few kilometers south of Ma'bar (de Maigret et al., 1984: 439). A total sample of a hundred flake tools and

choppers were recovered, in addition to two ovoid handaxes, made of rhyolite and volcanic tuff, which together suggested an attribution to the "Late Acheulean" (Bulgarelli, 1988; a handaxe is pictured on p. 41).

Bulgarelli (1985: 360) comments: "Although all these discoveries prove interesting, lithic tools have been gathered on the surface, out of their original stratigraphical context. This situation makes their cultural and chronological attribution less easy and certain." It is also unfortunate that the above findings were neither documented nor formally published, and the apparent assemblages were never defined; the collections are housed at the National Museum in San'ā'. Similar remarks were made in the same period by Toplyn (1988: 84), who calls for extreme caution in attributing surface artifacts to "heretofore undefined Yemeni Acheulean and Mousterian lithic industries". In fact, such archaeological circumstances are not restricted to Yemen: the recovery of Paleolithic material from surface situations has been common practice throughout the Arabian peninsula up to this day.

Conclusions and Inferences

The above information is preliminary. However, as it has a bearing on the peopling of a region in the Yemen highlands, it represents a first step in investigation, and the evidence suggests viable strategies for possible future research. Early Holocene and terminal Pleistocene archaeology should be identified as a specific research goal and approached accordingly. What is needed to further our limited knowledge is additional material from controlled in situ proveniences. This would include artifacts or ecofactual elements from stratified occurrences, if not whole occupation layers; or surface finds for which some kind of evidence might suggest their primary context. Such obvious requirements have just only begun to be met in other parts of Yemen, away from the western mountains, for instance in the Hadhramawt (Crassard and Khalidi, 2004; Crassard et al., 2006). Admittedly, suitable working conditions and research arrangements to accomplish that task may not yet be available in parts of the highlands.

If anything, the information here presented shows how much there is still to record and understand in the highlands. Territorial surveying is badly needed, and Wādī Khamar provides an example. On a few days' inspection, a small territory can produce tangible sites and enough background data about paleoenvironmental conditions as to allow more penetrating archaeological discovery. Insights from circumstantial evidence should not be discounted, as there can be little doubt that such a remarkable sample of wadi terrace exposures as those in Wādī Khamar will bring out pre-Neolithic evidence when adequately explored. In the same perspective, the case of Neolithic site WTH3 should serve as a reminder of what can be revealed by careful testing below later prehistoric occupations, or even blind-testing in stratified Holocene deposits. In this chapter a brief mention was also made of the preceding, Late Pleistocene information, which in spite of its extreme paucity can be of promise for future research.

At the moment, two levels of inference can be envisaged from the primary evidence reported. At the regional level, i.e. Khawlān at-Tiyāl, a working hypothesis can be advanced that the Early Holocene peopling of the eastern highlands might have foreshadowed in importance that of mid-Holocene times. At least in the numerous Tayyilah-like or Khamar-like basins the environmental preconditions for population infiltration and sedentarisation were all in place. Such conditions reflected locally what is now becoming known about the Early Holocene climate of southern Arabia, as reconstructed in terms of changes in monsoon precipitation and the tropical rainfall belt in general (Fleitmann et al., 2007).

At the super-regional level, it is worth asking whether the limited data from the headwaters' basins of eastern Yemen can perhaps be generalized to the whole of the eastern Yemen Mountains, again as a working proposition. Elsewhere I speculated (Fedele, 1988, 2008; cf. Edens and Wilkinson, 1998: 63–71) that most of the eastern Plateau Neolithic might represent a regional tradition specific to the highlands and distinct from both desert (e.g., Di Mario et al., 1989) and coastal cultures (Durrani, 2005). This "upland Neolithic tradition" would have taken full shape with early cattle herders who co-adapted to the severe landscape and settled the highlands widely by use of pastoralism and particular toolkit inventories. An expanded model could now be suggested whereby the emergence of an upland tradition had roots in the Early Holocene – the "Pre-Neolithic" – if not slightly earlier.

In effect, the uplands may have served as a refuge area during the periods of more severe desiccation in the lowlands, and have thus contributed to a distinct population history already under Late Glacial conditions. That is an entirely hypothetical construct. However, if the idea is correct, the Yemen Mountains and perhaps Yemen at large ought to be viewed as the southern periphery of a deep-rooted cultural continuum specific to the western Arabian uplands. This continuum would have entailed adaptive invention and original developments, hence a degree of cultural autonomy from the rest of Southwest Asia. Furthermore, an upland-adapted tradition in Yemen might have more in common with the parallel and broadly coeval developments in the Ethiopian Highlands, or elsewhere round the Horn of Africa (e.g., Barnett, 1999, with references), than with the terminal Pleistocene and earlier Holocene of the northern Near East. How to accord such a scenario – a fundamental independence from the northern Near East – with the apparently indisputable fact that several domestic animals and other cul-

tural elements were of Near Eastern origin, is a problem for future research.

On the basis of simulation work, Drechsler (2007) has recently suggested that one of the two putative dispersal pathways of the Neolithic out of the Levant and into the Arabian peninsula was along the Red Sea coast. This wave or "branch" may have advanced rather rapidly, because of inferred higher human mobility in less favorable environmental conditions; south of about 20° latitude it apparently dispersed away from the coast and over the uplands, i.e. in Greater Yemen. A potential, distinct role for the western Arabian uplands in the spread of the Neolithic lifeways would thus emerge. The same study would support the view of an environmentally dependent process of dispersal, with rapid climatic change fostering the appearance of original Neolithic developments in southern Arabia after about 6 ka cal. BC. This conclusion suggests once again that the peopling of the Yemen Mountains during the Pleistocene-Holocene transition and immediately afterwards deserves close scrutiny, hence appropriate investigation.

The above problems evoke a wider context for our data, and at the same time, inevitably, point to the intrinsic limits of the evidence from the eastern Yemen Plateau. However, it should be obvious from the preceding account that such limits are contingent on the present state of research. There is potential for our work to be replicated and indeed expanded throughout the eastern highlands, or at least, predictably, in a broad south–north belt of local landscapes along the Precambrian and Jurassic-Cretaceous formations. The peculiar tectonic history that has led to a "fossilization" of the upper Wādī Danah drainage, and thus to a widespread preservation of Early Holocene geologic landscapes, certainly is not confined to central northern Yemen (see for instance Garcia et al., 1991, for an area further north). Future research is needed to clarify such possibilities, and eventually define areas of greater research potential. Given that our understanding of the Late Pleistocene and Early Holocene occupation of southern and western Arabia is still rudimentary, the proposition that the Yemen Mountains have the potential to contribute to this goal in their own distinctive way – as *mountains* – should be tested.

Acknowledgments Special thanks are due to Alessandro de Maigret for his intellectual and financial support within the Italian Archaeological Mission to Yemen. Permission for fieldwork was granted by the General Organisation of Antiquities, Manuscripts and Museums (Sanʿāʾ), then headed by Dr Yūsuf ʿAbdallāh. The excavations and survey in the WTH area were carried out with the assistance of Francesco Di Mario and the participation of Yemeni archaeologists from GOAMM and Italian students and technical staff. For the Wādī Khamar survey, in addition to colleagues from GOAMM and soldiers from the military post at Jihānah, valuable assistance was provided by Mario Mascellani and Jill Morris. The present chapter greatly benefited from the comments of three anonymous reviewers. I am grateful to Jeffrey A. Blakely, Ueli Brunner, Rémy Crassard, Philipp

Drechsler, Christopher Edens, Michel-Alain Garcia, Biagio Giaccio, Caroline Grigson, Marie-Louise Inizan, Joy McCorriston, Andrea Manzo, Bruno Marcolongo, and Tony J. Wilkinson (among others) for literature or comments, although the use here made of this information is my sole responsibility.

References

Barnett T. The emergence of food production in Ethiopia. Cambridge Monographs in African Archaeology, vol 45. Oxford: BAR; 1999.

Bulgarelli GM. Research on Pleistocene and Palaeolithic sites. In: de Maigret A, Bulgarelli GM, Costantini L, Cuneo P, Fedele FG, Francaviglia V, Marcolongo B, Palmieri A, Scerrato U, Tosi M, Ventrone G, editors. Archaeological activities in the Yemen Arab Republic, 1985. East and West (n.s.) 1985;35:360–63.

Bulgarelli GM. Evidence of Paleolithic industries in Northern Yemen. In: Daum W, editor. Yemen: 3000 years of art and civilisation in Arabia Felix. Innsbruck, and Umschau-Verlag, Frankfurt am Main: Pinguin-Verlag; 1988. p. 32–3.

Crassard R. Apport de la technologie lithique à la définition de la préhistoire du Hadramawt, dans le contexte du Yémen et de l'Arabie du Sud. 2 vols. Thèse de Doctorat, Université Paris I Panthéon-Sorbonne, Paris; 2007.

Crassard R, Khalidi L. De la pré-Histoire à la Préhistoire au Yémen. Des données anciennes aux nouvelles expériences méthodologiques. Chroniques Yéménites. 2004;12:1–18.

Crassard R, McCorriston J, Oches E, Bin 'Aqil A, Espagne J, Sinnah M. Manayzah, early to mid-Holocene occupations in Wādī Sanā (Hadramawt, Yemen). Proceedings of the Seminar for Arabian Studies. 2006;36:151–73.

de Bayle des Hermens R. Première mission de recherches préhistoriques en République Arabe du Yémen. L'Anthropologie 1976;80:5–37.

de Maigret A. Ricerche archeologiche italiane nella Repubblica Araba Yemenita. Notizia di una seconda ricognizione (1981). Oriens Antiquus. 1982;21:237–53.

de Maigret A. Activities of the Italian Archaeological Mission in the Yemen Arab Republic (1983 campaign). East West (n.s.). 1983;33:340–4.

de Maigret A. Arabia Felix: an exploration of the archaeological history of Yemen. London: Stacey International; 2002.

de Maigret A, Bulgarelli GM, Fedele FG, Marcolongo B, Scerrato U, Ventrone G. Archaeological activities in the Yemen Arab Republic, 1984. East West (n.s.). 1984;34:423–54.

de Maigret A, Fedele F, Di Mario F. Lo Yemen prima del regno di Saba. Le Scienze. 1988;40(234):12–23.

de Maigret A, Azzi C, Marcolongo B, Palmieri AM. Recent pedogenesis and neotectonics affecting archaeological sites in North Yemen. Paléorient. 1989;15:239–43.

Di Mario F. A new lithic inventory from Arabian peninsula: the North Yemen industry in Bronze Age [sic]. Oriens Antiquus. 1987;26(1–2):89–107.

Di Mario F. The Bronze Age lithic industry. In: de Maigret A, editor. The Bronze Age culture of Hawlān at-Tiyāl and Al-Hadā (Republic of Yemen): a first general report. Rome: Istituto Italiano per il Medio ed Estremo Oriente; 1990. p. 81–114.

Di Mario F, Costantini L, Fedele FG, Gravina F, Smriglio C. The western ar-Rub' al-Khālī "Neolithic": new data from the Ramlat Sab'atayn (Yemen Arab Republic). Annali, Istituto Universitario Orientale [Napoli] 1989;49:109–48.

Drechsler P. The Neolithic dispersal into Arabia. Proceedings of the Seminar for Arabian Studies. 2007;37:93–109.

Durrani N. The Tihamah coastal plain of South-West Arabia in its regional context c. 6000 BC–AD 600. Society for Arabian studies monographs, vol 4. Oxford: Archaeopress; 2005.

Edens C, Wilkinson TJ. Southwest Arabia during the Holocene: recent archaeological developments. Journal of World Prehistory. 1998;12:55–119.

Fedele FG. Neolithic and protohistoric cultures. Excavations and researches in the eastern Highlands. In: de Maigret A, Bulgarelli GM, Costantini L, Di Mario F, Fedele FG, Gnoli G, Marcolongo B, Palmieri A, Scerrato U, Tosi M, Ventrone G, editors. Archaeological activities in the Yemen Arab Republic, 1986. East West (n.s.) 1986;36:396–400.

Fedele FG. Research on the Neolithic and the Holocene of the Yemen Highlands, 1987. Unpublished circulated report for MAIRAY, Institute of Anthropology, University of Naples; 1987.

Fedele FG. North Yemen: the Neolithic. In: Daum W, editor. Yemen: 3000 years of art and civilisation in Arabia Felix. Innsbruck, and Umschau-Verlag, Frankfurt am Main: Pinguin-Verlag; 1988. p. 34–7.

Fedele FG. Man, land and climate: emerging interactions from the Holocene of the Yemen Highlands. In: Bottema S, Entjes-Nieborg G, van Zeist W, editors. Man's role in the shaping of the Eastern Mediterranean Landscape. Rotterdam and Brookfield: A.A. Balkema; 1990. p. 31–42.

Fedele FG. Prehistoric archaeology and ecology in the eastern Highlands of North Yemen. Report of activities conducted in February 1990, Unpublished circulated report for MAIRAY, Institute of Anthropology, University of Naples; 1990b.

Fedele FG. Archaeological stratification and the logic of excavation. In: Urbańczyk P, editor. Theory and practice of archaeological research. 2: Acquisition of field data at multi-strata sites. Warszawa: Institute of Archaeology and Ethnology, Polish Academy of Sciences; 1995. p. 81–106.

Fedele FG. Wādī at-Tayyilah 3, a Neolithic and Pre-Neolithic occupation on the eastern Yemen Plateau and its archaeofaunal information. Proceedings of the Seminar for Arabian Studies 38: forthcoming; 2008.

Fedele FG, Zaccara D. Wādī al-Tayyila 3: a mid-Holocene site on the Yemen Plateau and its lithic collection. In: Sholan AM, Antonini S, Arbach M, editors. Sabaean studies. Archaeological, epigraphical and historical studies in honour of Yūsuf M. 'Abdallāh, Alessandro de Maigret, Christian J. Robin on the Occasion of their sixtieth birthdays. San'ā': YICAR, Naples, University of San'ā' and Centre français d'archéologie et de sciences sociales; 2005. p. 213–45.

Fleitmann D, Burns SJ, Mangini A, Mudelsee M, Kramers J, Villa I, et al. Holocene ITCZ and Indian monsoon dynamics recorded in stalagmites from Oman and Yemen (Socotra). Quaternary Science Reviews. 2007;26:170–88.

French CAI. Geoarchaeology in action: studies in soil micromorphology and landscape evolution. London: Routledge; 2003.

Garcia M-A, Rachad M. L'Art des Origines au Yémen. Paris: Éditions du Seuil; 1997.

Garcia M, Rachad M, Hadjouis D, Inizan M-L, Fontugne M. Découvertes préhistoriques au Yémen, le contexte archéologique de l'art rupestre de la région de Saada. Comptes rendus de l'Académie des Sciences de Paris. 1991;313(série II):1201–6.

Grigson C. Size and sex: evidence for the domestication of cattle in the Near East. In: Milles A, Williams D, Gardner N, editors. The beginnings of agriculture. Symposia of the association for environmental archaeology. 8th ed. Oxford: BAR; 1989. p. 77–109.

Kallweit H. Neolithische und Bronzezeitliche Besiedlung im Wadi Dhahr, Republik Jemen. Eine Untersuchung auf der basis von Geländebegehungen und Sondagen. Inaugural-Dissertation der Philosophischen Fakultäten der Albert-Ludwigs-Universität zu Freiburg i. Br. (Published online 2001.); 1996.

Kruck W, Schäffer U, Thiele J. Explanatory Notes on the Geological Map of the Republic of Yemen – Western Part (Former Yemen Arab Republic), with 8 maps 1:250000. Geologisches Jahrbuch, B87. Bundesanstalt für Geowissenschaften und Rohstoffe und Geologische Landesämter in der Bundesrepublik Deutschland, Hannover; 1996.

Kuijt I, Chesson MS. Lumps of clay and pieces of stone: ambiguity, bodies, and identity as potrayed in Neolithic figurines. In: Pollock S, Bernbeck R, editors. Archaeologies of the Middle East. Critical perspectives. Oxford: Blackwell; 2005. p. 152–83.

Lézine A-M, Saliège J-F, Robert C, Wertz F, Inizan M-L. Holocene lakes from Ramlat as-Sabʻatayn (Yemen) illustrate the impact of monsoon activity in southern Arabia. Quaternary Research. 1998;50:290–9.

Marcolongo B, Palmieri AM. Palaeoenvironmental conditions in the areas of Wādī at-Tayyilah and Barāqiš: preliminary report. In: de Maigret A, Bulgarelli GM, Costantini L, Di Mario F, Fedele FG, Gnoli G, Marcolongo B, Palmieri A, Scerrato U, Tosi M, Ventrone G, editors. Archaeological activities in the Yemen Arab Republic, 1986. East West (n.s.) 1986;36:461–4.

Marcolongo B, Palmieri AM. Paleoenvironment history of western Al-Aʻrūsh. In: de Maigret A, editor. The Bronze Age culture of Hawlān at-Tiyāl and Al-Hadā (Republic of Yemen): a first general report. Rome: Istituto Italiano per il Medio ed Estremo Oriente; 1990. p. 137–43.

Marcolongo B, Palmieri AM, Lentini A. Environmental modification and settlement conditions in the Yalā area. In: de Maigret A, editor. The New Sabaean archaeological complex in the Wādī Yalā (Eastern Hawlān at-Tiyāl, Yemen Arab Republic): a preliminary report. Rome: Istituto Italiano per il Medio ed Estremo Oriente; 1988. p. 45–53.

McCorriston J. Early settlement in Hadramawt: preliminary report on prehistoric occupation at Shi'b Munayder. Arabian Archaeology and Epigraphy. 2000;11(2):129–53.

Overstreet WC, Grolier MJ. Summary of environmental background for the human occupation of the al-Jadidah basin in Wadi al-Jubah, Yemen Arab Republic. In: Grolier MJ, Brinkmann R, Blakely JA, editors. Environmental research in support of archaeological investigations in the Yemen Arab Republic, 1982–1987. Washington, DC: American Foundation for the Study of Man; 1996. p. 337–429.

Overstreet WC, Grolier MJ, Toplyn MR. The Wadi al-Jubah archaeological project. 4: geological and archaeological reconnaissance in the Yemen Arab Republic, 1985. Washington, DC: American Foundation for the Study of Man; 1988.

Parker A, Davies C, Wilkinson T. The early to mid-Holocene moist period in Arabia: some recent evidence from lacustrine sequences in eastern and south-western Arabia. Proceedings of the Seminar for Arabian Studies. 2006;36:243–55.

Petraglia MD, Alsharekh A. The Middle Palaeolithic of Arabia: implications for modern human origins, behaviour and dispersals. Antiquity. 2003;77(298):671–84.

Toplyn MR. Paleolithic artifacts from Wadi al-Jubah, Yemen Arab Republic. In: Overstreet WC, Grolier MJ, Toplyn MR, editors. The Wadi al-Jubah archaeological project. 4: Geological and archaeological reconnaissance in the Yemen Arab Republic, 1985. Washington, DC: American Foundation for the Study of Man; 1988. p. 77–84.

Uerpmann M. Structuring the Late Stone Age of southeastern Arabia. Arabian Archaeology and Epigraphy. 1992;3:65–109.

von den Driesch A. A guide to the measurement of animal bones from archaeological sites. Peabody Museum Bulletin, vol 1. Cambridge, MA: Peabody Museum of Archaeology and Ethnology, Harvard University; 1976.

Wilkinson TJ. Holocene environments of the high plateau, Yemen, recent geoarchaeological investigations. Geoarchaeology. 1997;12: 833–64.

Wilkinson TJ. Archaeological landscapes of the Near East. Tucson: The University of Arizona Press; 2003.

Wilkinson TJ. Soil erosion and valley fills in the Yemen highlands and southern Turkey: integrating settlement, geoarchaeology, and climate change. Geoarchaeology. 2005;20:169–92.

Wilkinson TJ, Edens C. Survey and excavation in the central highlands of Yemen: results of the Dhamār Survey Project, 1996 and 1998. Arabian Archaeology and Epigraphy. 1999;10:1–33.

Chapter 17
Southern Arabia's Early Pastoral Population History: Some Recent Evidence

Joy McCorriston and Louise Martin

Keywords Fauna • Grazing • Manayzah • Pastoralism • Shi'b Kheshiya • Southern Arabia

Introduction

Across the arid expanses of the Arabian peninsula and even at the margins of its limited upland farmlands in Northern Yemen and the Asir, pastoralism has proved an enduring and effective economic strategy through the later Holocene. Goats, camels, and cattle are the principal herd animals, with mixed strategies of goats and sheep, goats and camels, and to a lesser extent cattle and goats. Strategies have changed through time and across geographic and socio-political territories with the herding of particular animals such as cattle or camels conferring not only specific economic benefits and constraints but also playing significant roles in the establishing and differentiation of people's social identities and statuses. While it is not entirely clear when a fully pastoral commitment, that is, one that emphasized production of secondary animal products, appeared in Arabia, it is evident that there long remained groups with partial economic dependence on herd animals and still exploiting the rich interior game (e.g., gazelle, ibex) and coastal-estuarine resources (principally fish and shellfish). To the important questions of when and from where domesticated animals entered the Arabian peninsula therefore must be added the question of what constitutes a transition to true pastoralism in the ancient Arabian record. With new evidence from Southern Arabia, it is now possible to address these issues there.

J. McCorriston (✉)
Department of Anthropology, The Ohio State University,
4034 Smith Labs, 174 W 18th Ave, Columbus, OH, 43210, USA
e-mail: mccorriston.1@osu.edu

L. Martin
Institute of Archaeology, University College London, 31-34
Gordon Square, London, WC1H 0PY, UK
e-mail: louise.martin@ucl.ac.uk

Southern Arabia

Southern Arabia is a distinct and important province in geographic terms. It lies at the heart of the Indian Ocean, with possible connections to Africa, the Asian Subcontinent, and the Near East. From modern Oman's Dhofar region to the Ramlat Sabatayn depression, Southern Arabia is defined by limestone and shale mountains formed by the uplifting of the Arabian shield and deeply incised by Tertiary stream flow. To the north lies the Rub' al Khali, a natural barrier to human settlement and populations, as are the Wahiba Sands to the east. The mountainous regions of Dhofar, Mahra, and Hadramawt could have supported some of the densest Early Holocene populations and attracted desert foragers retreating from increasing aridification during the Middle Holocene (6-5 ka). Throughout the Later Pleistocene and Holocene, Southern Arabia has received precipitation from the Southwest Asian monsoon, which shifts the influence of tropical winds northward during summer insolation and pushes moisture-laden storms across the Indian Ocean and over the Arabian coastlines. The high mountainous country of Northern Yemen, with fertile soils and a relatively high precipitation determined by altitude, has long been home to settled agricultural peoples who tap summer rainfall and spring flow with indigenous irrigation systems. But the great indigenous civilizations of Arabia sprang up at the desert margins of Southern Arabia, making the prehistory of the peoples that lived in Southern Arabia itself one of the most compelling, if poorly understood, aspects of Arabian archaeology.

The Local Archaeological Record

Despite its importance, we know relatively little about Southern Arabian prehistory. Archaeological expeditions have been few and widely dispersed, and it was not until the 1980s that sustained prehistoric research began to fill vast chronological gaps with a few well-dated local sequences. Geographic coverage of Southern Arabia's prehistory is still very thin, and there remain large areas and long chronological stretches for which there exist no data. Tracking the introduction or expansions of human populations after Pleistocene aridification would require more data than is presently available. Given that domesticated plants and animals are found earlier outside of Arabia – the Levant, South Asia, probably Africa – it has seemed reasonable to assume that domesticates, whether plant or animal, were introductions to the Arabian peninsula. Fresh evidence from archaeological and paleoecological studies in highland Southern Arabia provides a good prehistoric sequence there and suggests that the introduction of domesticates and a local transition to full-scale pastoralism may be separated by at least a thousand years. But the data are without local parallels and must be viewed in the context of wider Arabian evidence for the introduction of domesticated animals and early pastoralism.

Other Data

The two major sources of information about the introduction of domesticated animals into Arabia are animal bones and rock art. The first of these is the more direct evidence, for where archaeological sequences have yielded good faunal assemblages and associated radiocarbon dates, it has been possible to document fully domesticated animals in prehistoric contexts (reviewed below). Rock art presents difficulties in dating and in interpretation but remains a much more widespread indicator of emergent pastoralism than the few Arabian faunal assemblages available to date. Complementary data from rock art and faunal remains in Arabia suggest that domesticated cattle and caprines were present from the seventh millennium BC.

Rock Art: Animals, Wusum, and Dating

There are major concentrations of rock art in the Arabian peninsula that indicate both hunting and herding activities in (Holocene) prehistory and throughout the historical period. Some of the first studies of Arabian rock art relied upon documentation from the Philby–Lippens–Ryckmans expedition of 1951 in southwest Saudi Arabia (Lippens, 1956; Anati, 1972, 1974). There exist impressive and multi-period panels of rock art in the northeastern Arabian peninsula (Khan, 1993) and in Northern Yemen (Garcia et al., 1991; Inizan and Rachad, 2007). Rock art also exists in Southern Arabia and is sometimes painted (al-Shahri, 1994; Bin 'Aqil, 2004; Keall, 2005; Crassard, 2006, 2008; Braemer et al., 2007), but there are few large or complex panels to compare with those from other Arabian sites. Across the peninsula, rock art depictions include humans, symbols, and animals, frequently

including cattle, camels, and ibex. The scenes represented and the choice of animals likely stem from ideological, ritual, and social inspiration, and therefore the images cannot be "read" in purely economic terms as documented emphasis on hunting or pastoralism. Humans are sometimes depicted with weapons and there are signs that have been interpreted as traps, nets, and *wusum* (tribal markings) (Khan, 2000). In view of the importance of caprine herding today and also in the past, it is surprising how few goats or sheep appear in rock art, unless one accepts that rock art was designed to evoke and commemorate non-economic practices.

Along with probable representations of ritual hunts, dance, territorial rights, raiding, wild fauna, and hunting (Khan, 1993: 142; 2000; Bin 'Aqil, 2004: 42; Rachad, 2007b), domesticated animals were nonetheless represented, particularly cattle (Garcia et al., 1991; Khan, 2000: 103; Rachad, 2007b: 84). Many cattle depictions seem to be of piebald and therefore domesticated animals (Zarins, 1992: 27). In Wadi Daum of northeast Arabian, where systematic work suggests a stylistic chronology (Khan, 1993), there may be further indication of domesticated cattle in depictions of different horn alignment, with forward-pointing horns on wild cattle and backward-pointing horns on domestic cattle (this difference could also depict bulls and cows as in early Africa (Grigson, 1991)).

Rock art chronology is everywhere challenging, and Arabia is no exception. Anati's first attempts at chronological organization of different image styles and techniques was tied to North African rock art, as understood in the 1960s. Since then, Anati's chronological inferences have been largely supplanted by other localized studies relying also upon superposition, patination, technique, and nearby archaeological remains (e.g., Garcia et al., 1991; Khan, 1993; Rachad, 2007a). Khan suggests that the earliest domesticated cattle in Wadi Daum depictions antedate a Middle Holocene aridification, and his premises are echoed in the rock art analyses of Northern Yemen (Garcia et al., 1991; Rachad, 2007a). In Yemen's Sa'ada highlands, rockshelters contain early (sixth millennium BC) rock art depictions of wild cattle, buffalo (*Syncerus antiquus/caffer*) and contemporary (but not associated) domesticated cattle and dogs (Rachad, 2007a: 80, 2007b: 91–92). The oldest stylistic motifs are presumed related to the archaeological remains of Middle Holocene hunters who appear to have also consumed domesticated and wild game (Inizan, 2007:63; Hadjouis, 2007).

In a subsequent stylistic phase (very loosely dated between 5,000–3,500 BC) in Wadi Daum, cattle motifs are accompanied by "*wusum*," or tribal markings (Khan, 1993: 109), suggesting ownership of the cattle and territorial rights to grazing lands. Style III – dominated by painting – in Yemen is associated with the Bronze Age (third to second millennium BC) and characterized by herds of cattle and signs. Wild bovids are no longer depicted (Rachad, 2007a:81–82), although faunal deposits show that they were still hunted (Inizan, 2007:

67). While Arabian rock art still leaves much in question, the roughly dated motifs provide a corroborative context for the early, albeit scant remains of domesticated animals, including those from Southern Arabia where little rock art and virtually no rock art studies are available.

Southern Arabia in the Chronology of Early Arabian Pastoralism

Nevertheless, new faunal evidence from archaeological excavations points to the significance of the region in the chronology of early pastoralism. With some of the earliest dates on domesticates, Southern Arabia stands out not only as the geographic homeland of complex societies but also as a region where domesticates were introduced early and where early pastoral societies developed. This chapter therefore examines three related issues in Southern Arabia:

1. Where did the earliest domesticates appear and when?
2. Where might these animals have come from? Were they introduced from the Levant, from Africa, or from South Asia?

3. What does the new faunal evidence imply about a transition to fully pastoral societies? Was pastoralism an introduction, possibly entailing a population movement from other areas where archaeological evidence suggests an earlier adoption of pastoral lifestyles? (See Uerpmann et al., 2009.) Or was pastoralism an Arabian transition, with adoption of domesticates into local economic strategies and a subsequent economic, social and political transformation of indigenous human populations hitherto dependent on foraging? Some archaeologists have preferred the former scenario (e.g., cf., Bellwood, 2005; Dreschler, 2007) and some the latter (e.g., Tosi, 1986; Zarins, 1992; Uerpmann et al., 2000; Cleuziou et al., 2002).

The RASA Project Background

Archaeological and paleoecological research by the RASA Project (Roots of Agriculture in Southern Arabia) between 1998 and 2008 has generated a new regional chronology and landscape history for the Southern Jol, the uplifted limestone plateau that lies between a narrow coastal plain and the deep Wadi Hadramawt of Southern Arabia (Fig. 1). Regional

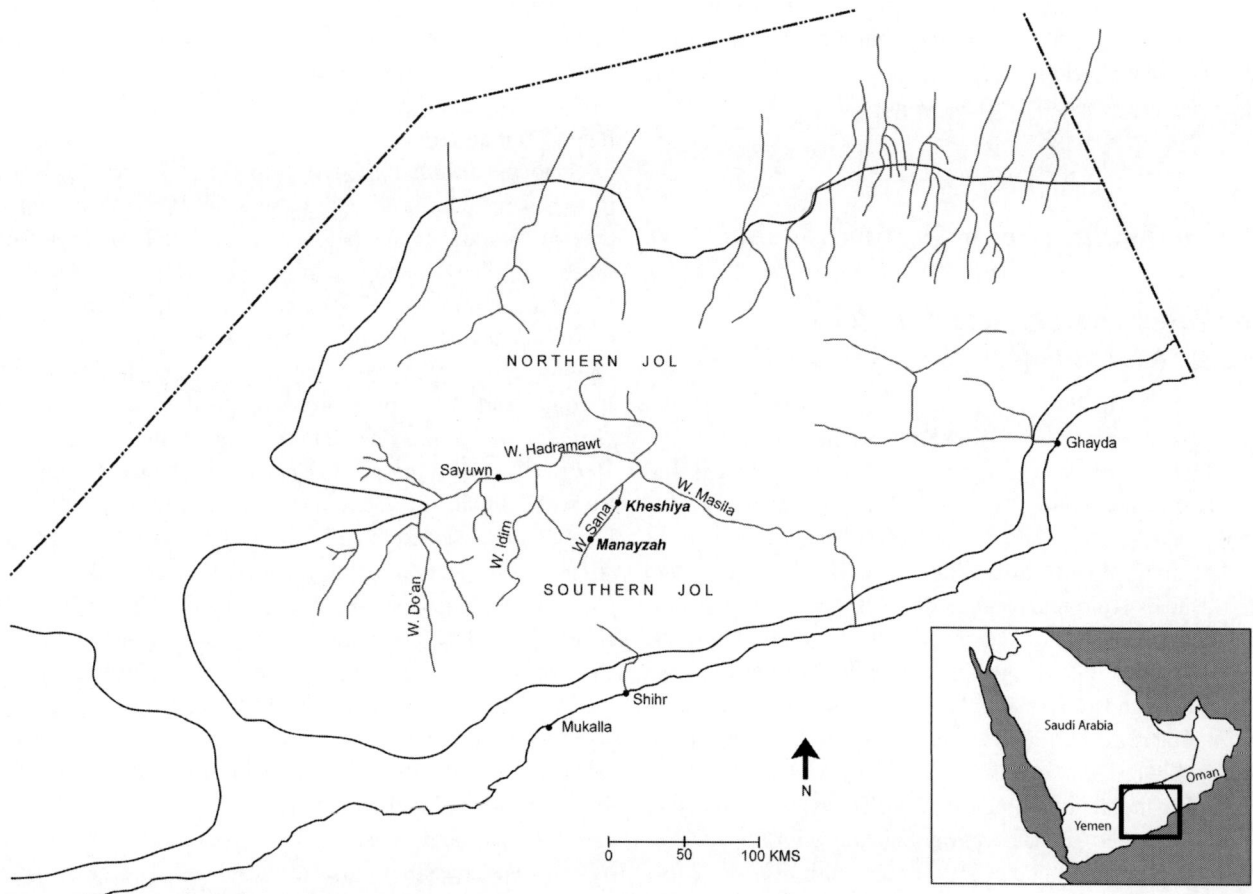

Fig. 1 Map showing Southern Arabia and RASA Project Area of the Wadi Sana drainage

survey and small-scale excavations took place with the aims of generating a basic cultural historical sequence appropriate for addressing broader problems of human adaptations to climate change and the transition to food production. In the highland drainage of Wadi Sana, archaeologists have established that there has never been significant permanent settlement and that the earliest agricultural technologies appeared after hunters and herders had camped in caves and along the marshy margins of annual flood zones for thousands of years. Paleoecological studies have been crucial in the reconstruction of flooding and down-cutting over thousands of years and in correlating a major shift in sedimentary regime (to incision) with the weakening of the Southwest Asian monsoon (Anderson et al., 2006). In turn, this local environmental history provides a crucial setting for understanding the archaeological record.

Archaeological remains in the Wadi Sana include numerous hearths, some buried in sediments and capable of yielding a radiocarbon date, others remnant as surface features or scatters of thermally altered rock. Other signs of human activity include buried burnt surfaces from anthropogenic firing of (formerly more dense) vegetation, occupations that include lithics, faunal bones, and features in rockshelters, minimal rock art and graffiti, water diversion structures for irrigation, tombs and other small-scale commemorative monuments, and rare artifact clusters as surface sites. Of these, several rockshelters, commemorative monuments, and a number of hearths have been excavated, yielding a good series of radiometric dates and significant associated material, including faunal remains.

Faunal Distributions in Southern Arabia

The Wider Context: Selected Arabian Faunal Assemblages

New faunal remains from Wadi Sana can best be understood in the wider context of early dated faunal assemblages and wild faunal distributions in Arabia. There are firm radiometric dates for caprine-and-cattle herders–foragers–fishers of the late sixth to early fifth millennium BC Arabian Gulf at Jebel Buhais (Uerpmann et al., 2000). The caprines are convincingly seen as introductions on zoogeographic grounds, and the cattle include both native wild *Bos primigenius* (Uerpmann and Uerpmann, 2008: 104) and domestic cattle originating in the Fertile Crescent (Uerpmann and Uerpmann, 2008: 112–113, 129 and note 9, see also Uerpmann et al., 2009). The highland Neolithic of Wadi Thayylah 3 (Khawlan, Northern Yemen) is dated by association with the moister conditions of the middle fifth millennium BC and also includes cattle and caprines. Two hundred kilometers north-

wards in the Sa'ada highlands of Northern Yemen lie the Sa'ada rockshelters in which archaeological soundings have yielded domesticated cattle bones dated to the late sixth millennium as well as wild *Bos* (Hadjouis, 2007: 51). These assemblages offer the strongest evidence for early pastoral groups in Eastern and Southwestern Arabia respectively, but they and the Wadi Sana finds must be understood in the context of native faunal distributions in Arabia.

Understanding the Early and Middle Holocene wild mammalian fauna in southern Arabia is problematic. The current documented wild ungulates in the region will be much diminished in both diversity and range by millennia of over-hunting, over-grazing, the loss of habitat to herded domesticates, and increasing aridification and environmental degradation over the past 5–6,000 years. Nevertheless, there is a need to create some prediction, if tentative, of a native fauna for the Early and Middle Holocene, in order to be able to identify human introductions, or domestications, when they occur. This can be achieved through a review of the current wild remnant fauna, a discussion of expectations based on neighbouring regions, combined with evidence from the few zooarchaeological datasets, although as can be seen here, there remain many questions about which wild taxa may have inhabited the area.

Concern here is not with the smaller mammals (rodents and lagomorphs) or the wide array of wild carnivores that certainly inhabited southern Arabia, since these were less likely to be mainstays of human subsistence. Instead the focus is on the larger mammals, which could have served as hunted human prey.

Amongst the smaller bovids (up to 50 kg), gazelles would have covered a wide array of habitats from deserts to steppe-deserts to moister steppes, although which species were present is less clear. Harrison and Bate (1991) show *G. gazella*, *G. dorcas*, *G. subgutturosa* and the Yemeni-specific *G. bilkis*, each inhabiting southern Arabia, with some overlapping distributions. It is often not possible to identify beyond genus level from archaeological remains, although *G. subgutturosa* has been identified at Umm an-Nar (Hoch, 1979: 616; Uerpmann, 1987) and probably *G. gazella* from Al-Buhais 18 (Uerpmann and Uerpmann, 2000: 43; Uerpmann and Uerpmann, 2008: 105). The Nubian ibex (*Capra nubiana*) is well attested in all the steep mountainous areas of Arabia, from Sinai to Hadramawt (Harrison and Bate, 1991: 182), preferring a habitat of rocky crags, while Uerpmann and Uerpmann have recently suggested that the cooler-adapted wild goat, *Capra aegagrus*, ranged as far south as northern Oman in the Neolithic (Uerpmann and Uerpmann, 2008: 105–111) although it would have been at the limits of its climatic tolerance.

In similar size range to wild goats, the Arabian Thar (*Hemitragus jayakari*) must be considered. This small goat-like animal is currently only found in the mountainous areas of Oman where it receives protection (Harrison and Bates, 1991: 180) but if not under threat from domestic caprine competitors or hunters, it may have seen a wider distribution in the past, in other steep mountain slopes and bare cliffs (Uerpmann, 1987: 111–3). Identification in archaeological assemblages is rare due to the difficulty of separating the Thar from other goats, although Uerpmann (1987: 113) finds remains from Neolithic Ras al-Hamra in Oman.

Of the larger bovids (60–200 kg), the desert-adapted Arabian oryx (*Oryx leucoryx*) finds its homeland in the Arabian peninsula, and is likely to have been widespread in the past (Harrison and Bate, 1991: 189). It was found at Al-Buhais 18 in Sharjah (Uerpmann and Uerpmann, 2000; Uerpmann and Uerpmann, 2008: 104–5) and at third millennium BC Umm an-Nar (Hoch, 1979: 616–19). There is no direct evidence that the addax (*Addax nasomaculatus*) ever inhabited the deserts and semi-deserts of Arabia, as it did those of Egypt and Sudan, although since habitats would have been similar, the possibility cannot be ruled out (Uerpmann, 1987: 83). Similarly, the grassland-dwelling hartebeest (*Alcelaphus buselaphus*) cannot be completely excluded, although no finds have been made in Arabia, and its need for open water seems to limit it to riverine valleys and oases. Whether the kudu (*Tragelaphus imberbis*), which is native of north-east Africa, ever inhabited Arabia is debatable. While a couple of specimens are known from the twentieth century, doubt exists as to whether they were indigenous or introduced (Harrison and Bates, 1991: 192). They are known from dense thicketed habitats.

Of the very large sized bovids (>200 kg), the African buffalo (*Syncerus caffer*), which ranges from arid regions to wetlands and tolerates both dense cover and open woodlands, may have been part of the native fauna of southern Arabia, although this has not been confirmed with direct evidence. The case for the presence of wild aurochs (*Bos primigenius*) in southern Arabia in the Early Holocene has been strengthened by recent finds. The aurochs can exist in a wide range of environments, from forest and woodlands to open steppe, being limited mainly by the need for standing water every few days (Russell, 1988: 55–59; Van Vuure, 2002). While McClure (1984: 179–82 and Plate 32) found large bovid remains from Upper Pleistocene lakes in western Arabia, it has been unclear whether these belonged to *Bos* or other of the larger taxa mentioned above such as oryx (see also Grigson, 1996). Uerpmann and Uerpmann (2008: 104–105), however, report two erupting cattle pre-molars which fall in the large size range *Bos primigenius* from a fifth millennium BC relict population in Eastern Arabia. Also, Inizan (2007: 67) and Hadjouis (2007: 51) describe 24 bones as belonging to *Bos primigenius* from the northern Highlands of Yemen, suggesting a wider distribution of the species.

In terms of domestic cattle, Bokonyi identified a cattle horncore from the seventh millennium cal BC shell midden site of Ash-Shumah in the Tihama (Cattani and Bökönyi, 2002), and claims it derives from domestic stock on the basis of its thin wall – an insecure criterion since horncores increase in size with an animal's age. The Ash-Shumah assemblage deposited in a shell mound is 92% wild equids, and thus appears very unlike pastoralism.

Excavations at Jebel Makhruq in the Sa'ada region of North Yemen found the bones of domesticated cattle in Early Holocene cave occupations dating to the late sixth millennium BC (Hadjouis, 2007: 51). The larger samples of cattle found from Neolithic levels Wādī at-Tayyilah 3 show that there were not only convincingly smaller domesticates present by the mid fifth millennium BC cal on the eastern Yemen Plateau (Fedele, 2008), but also some larger *Bos* remains from "Pre-Neolithic" levels, which may be from wild aurochsen (Fedele, 2008). This "Pre-Neolithic" phase is more difficult to date without radiometric means, and the excavators can only suggest a date through one stylistic comparison to seventh millennium BC Pre-Pottery Neolithic B Levantine figurines (Fedele, 2008: 159, 163). The excavators find the domestic status of the cattle in this earlier phase to be ambiguous, but what seems clear is that domestic cattle are known from southwest Arabia from the late sixth millennium BC, and there is the possibility of wild aurochsen in the area too.

The Arabian peninsula may have been part of the range of real wild asses (*Equus africanus*). Uerpmann (1987: 30) states that the wild range must have included eastern, southern, and western parts of Arabian peninsula, and ass-sized equids are identified in assemblages from across the region (e.g., Uerpmann and Uerpmann 2000: 41 identify *E. africanus* teeth from Al-Buhais 18; Fedele, 2008). Wild asses would have avoided the sand deserts, but can be assumed to have inhabited rocky parts of Central Arabia. They were found at Ash-Shumah (see Cattani and Bökönyi, 2002 – who claim it as *Equus hemionus*), Late Neolithic Ras al-Hamra LN in Oman, and Bronze Age Hili 8 in the United Arab Emirates.

In sum, gazelles and wild asses can be assumed to have lived in relatively high densities in the more arid areas of southern Arabia, with the ibex common in more craggy locales. There is far less certainty about which of the medium and larger bovids would have been found over much of the peninsula, and whether those taxa with a current African or more northerly distribution extended into Arabia at times of increased moisture and lusher vegetation. The case for indigenous wild cattle, for example, has been made, although it is unclear as to how widespread populations may have been. One mammalian taxon which has never been associated with the wild fauna of Arabia, which is of key interest in discussion below, is sheep (*Ovis* sp). The wild sheep (*Ovis orientalis*) has its native distribution in the well-watered foothills and grassy plains of the Fertile Crescent (Garrard et al., 1996) and is known outside this area only as an introduced domesticate.

Wild Fauna Distributions Inferred from Vegetative Expectations

Indirect evidence for potential wild animal habitats comes from botanical evidence and vegetation ecology. Studies of modern vegetation and ancient charcoal fragments from hearths, occupation layers, and burnt surfaces show that the floristic components of vegetative cover have changed little, with the same woody species growing today as in the past (Table 1). Yet the density of vegetation and particularly of grasses, sedges, rushes, and early growth was much greater in the Early Holocene than it is today. In this environment humans and animals (wild and domesticated) could find permanent standing water in the dry winter season.

Wadi Sana's Vegetative History

Wood charcoal fragments from archaeological contexts prior to 5,000 years ago show that the same genera selected by Early Holocene inhabitants for cooking or heating fires are present today in Wadi Sana. These data provide an important qualitative understanding of the Early Holocene vegetation, for they are plants adapted then as now to monsoon rainfall during the warmest season in July and suggest that the margins of annually flooded marshy areas must have sustained woody species that were not submerged each year. Significantly, they complement regional paleoclimate proxies that indicate monsoon precipitation in the warm summer months (Lezine et al., 1998; Fleitmann et al., 2003, 2007; Davies CP 2006).

Another significant aspect of the paleo-vegetation is evident in the archaeologically documented burnt surfaces. These indicate anthropogenic firing of a more dense cover

Table 1 Woody vegetation in the Wadi Sana

Archaeobotanical specimens	Modern Woody Taxa in Wadi Sana
Cadaba sp.	*Cadaba heterotricha*
Cadaba/Maerua type	*Maerua crassifolia*
Anogeissis sp.	*Anogeissis benthii*
Acacia sp., *A. odorata*	*Acacia hamulosa, A. mellifera, A. odorata*
Zizyphus sp.	*Zizyphus leucoderma*
Cyphostemma sp.	*Cyphostemma crinitum*
Ficus sp.	*Ficus salicifolia*
Delonix sp.	*Delonix elata*
	Moringa peregrine
Tamarix sp.	*Tamarix* sp.
	Indigofera spp.
	Lycium shawii
Calotropis sp.	*Calotropis procera*
	Commiphora gileadensis, C. kataf
	Boswelia sacra

than is extant today (McCorriston, 2008). One burnt surface has yielded a date of 5,880 ± 55 BP (OS16689) and all others can be stratigraphically dated to the Early Holocene (prior to 5,000 uncal BP). Burnt surfaces resulted from deliberate human set fires to enhance grasses and other nutritious graze that would feed herd animals or attract game. That such conflagrations could alter the magnetic signature of underlying sediments (McCorriston et al., 2002: 66) indicates stands of woods and grassland much denser than today's extremely sparse cover. This denser vegetation would have provided an important resource for herd animals and humans and especially would be compatible with the grazing requirements of *Bos*, which also requires a ready access to water.

Geomorphological Evidence

Dense vegetation and deep alluvial sediments imply that there was much more water available in the Early Holocene of Wadi Sana. Indeed, the preliminary results of geomorphological studies (Anderson et al., 2006) detect the presence of a paleochannel with permanent standing water in the middle Wadi Sana between 6,000 and 5,500 uncal BP (early fifth millennium BC). Frequent overbank flooding during the rainy season deposited several meters of silty sediments across the Wadi Sana and supported the dense vegetation mats and gallery forest in times of relative land surface stability. The organically enriched and slightly darkened bands that mark paleosols in the natural modern cut-banks of the Wadi Sana attest to vegetation growth in the Early and Middle Holocene. Much of this environment was marshland, annually shrinking during the winter dry season and re-filling during the seasonal replenishment of summer months.

Marsh Analog Vegetation in Wadi Harou

Although there is nothing left of this environment in Wadi Sana, a continuous seep over ten years in the Wadi Harou (water trickling from a large tank maintained for local water tankers) has ponded in a small catchment and created a local marsh like ancient Wadi Sana's (Fig. 2). Here the water is permanent, and there are tamarisks and other perennial species in a gallery forest. The surface has a thick cover of *Juncus*, *Cyperus*, and perennial grasses. Because grazing animals have been excluded, the cover provides a particularly rich and useful analog for what Wadi Sana may have looked like only 40 km to the east, especially in the areas where permanent springs fed marshy basins.

Fig. 2 Modern marsh vegetation in Wadi Harou

The Evidence from Manayzah and Shi'b Kheshiya

In this rich Early and Middle Holocene environment herd animals were pursued and tended. Two sites in the Wadi Sana sequence have supplied faunal assemblages of very different character and composition. There is no single archaeological site in Wadi Sana that provides a full Holocene stratigraphic sequence, but extensive geomorphological research in combination with more than 80 radiocarbon and optically-stimulated luminescence dates has provided a good regional context for most archaeological deposits.

Manayzah

In the lower (southern) part of Wadi Sana where the channel is relatively constrained by ancient deep incision into hard limestone, a deep and relatively narrow gorge has still today a few high seasonal springs and seeps along its sides (Fig. 3).

In a narrow rockshelter and its broad terrace next to one such spring are more than 2.5 m of occupation debris that may begin in the Pleistocene and have upper layers well-dated to the Early Holocene (8,000–6,900 BP uncal, 7,300–5,700 cal BC) by radiometric assay on wood charcoals in hearths or on occupation surfaces (Table 2). A 4 × 4 m horizontal exposure

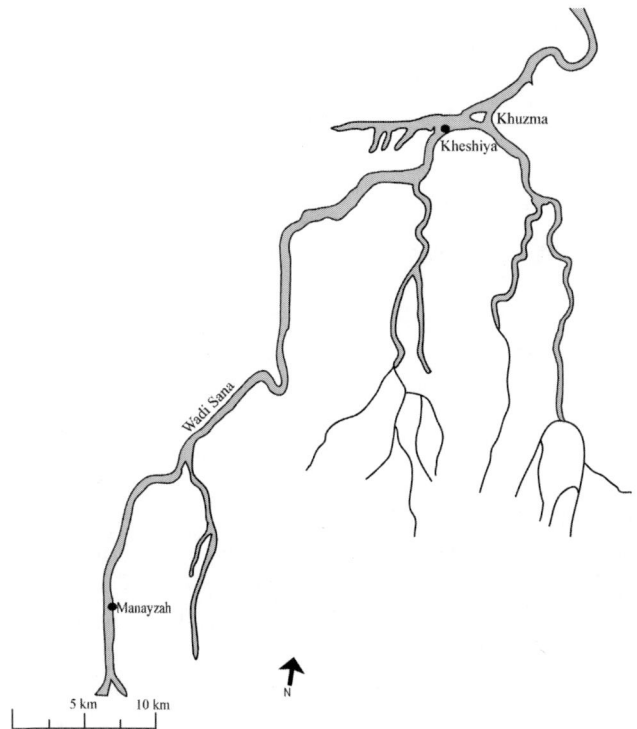

Fig. 3 Map of lower and middle Wadi Sana drainage with archaeological sites indicated

revealed a series of occupation levels with abundant charcoal and knapping debris and contiguous with features like stone-lined hearths, pits, and postholes (Crassard et al., 2006). These darker surfaces inter-digitate with sandy orange-yellow mostly sterile lenses probably from intermittent overbank flooding linked to the wettest years when wadi flow reached particular highs. The moist seep from a crevice in the rockshelter itself supplied a thin calcite crust over all the site, protecting fragile occupation sediments from the Late Holocene flooding episodes that have cut and eroded Early Holocene sediments elsewhere in Wadi Sana. The site deserves further excavation and offers promise of intra-site patterning and variability. For example, there are multiple fire-pit hearths constructed on the same surface, suggesting contemporary social and economic sub-groups among the occupants. The faunal assemblages are likely the remains of food preparation and consumption at the site and accumulated along with substantial (slightly charred) dung and charcoal in occupation layers. The lowest levels,

Table 2 Radiocarbon dates for Manayzah rockshelter (calibrations with Oxcal 4)

Context	Material	Lab number	Intercept (uncal)	2 s range (uncal)	Calibrated BC
I14, C009-10	Charcoal	AA66684	6,981	7,032–6,930	5,984–5,976
L9, A010-15	Charcoal	AA66683	6,987	7,044–6,930	5,986–5,746
K9 N½, Hearth 1	Charcoal	AA59570	6,902	6,943–6,861	5,886–5,716
K9-017	Charcoal	AA66685	7,133	7,184–7,082	6,085–5,896
K9-020	Charcoal	AA66686	8,072	8,151–7,993	7,306–6,702

reached in a 1 × 1 m probe (K9), yielded few bones because of the restricted volume of excavated sediments, but the resulting assemblage contains some significant markers of domesticated herd animals, dated by stratigraphic association with charcoal samples.

Manayzah Fauna

The faunal remains derive from clear occupation layers, built-up presumably from frequent re-visiting of the same location. Some of the material is highly burnt, some appears water-worn, and all highly fragmented. Certain levels show little disturbance to have occurred since deposition since bones were still in articulation upon excavation.

Of the over 1,600 fragments, most were undiagnostic to genus, but the small sample of 69 that could be identified consisted of roughly equal numbers of gazelle and caprines, with fewer cattle (Martin et al., 2009) (Table 3). Of the caprines, four bones could be further identified as belonging to goat and one to sheep (following criteria in Boessneck, 1969). While bones identified as goat could potentially belong to the hunted ibex, and as discussed above, there is perhaps a possibility that wild cattle existed in the area, the sheep bone would certainly indicate an imported domesticate, and this find casts a different light on the assemblage. Neolithic animal imports into the southern Levant and Africa tended to see sheep and goats treated as a 'package' (Horwitz et al., 2000; Martin, 2000), and thus if Manayzah was even in part a herding camp, it would be likely that sheep and goats were tended together. It is tempting to also see the cattle bones as herded domesticates, and they are very small compared to known wild cattle (Martin et al., 2009).

The earliest domesticated fauna, from level K-16, slightly postdate a radiocarbon date on charcoal from underlying K17 at 6,085–5,896 cal BC (Martin et al., 2009), dating the first domesticates to the final decades of the seventh millennium BC and beginning of sixth millennium BC. With wild fauna still present, the Manayzah assemblage seems to show a combination of both hunting (of gazelle) and herding (caprines and probably cattle) activities.

Table 3 The number of identified mammalian specimens (NISP) from Manayzah

Taxa	NISP
Cattle (*Bos* sp.)	10
Gazelle/Caprine (*Gazella* sp./*Ovis* sp./*Capra* sp.)	21
Gazelle (*Gazella* sp.)	18
Caprine (*Ovis* sp./*Capra* sp.)	15
Goat (*Capra* sp.)	4
Sheep (*Ovis* sp.)	1
Total NISP	69
Undiagnostics	c.1600

Manayzah Tools for Hunting

Further inference may be suggested from the assemblages of stone tools recovered from excavations at Manayzah. Techno-typological analysis has identified the manufacture of several discrete styles of projectile points now clearly dated in stratigraphic sequence and belonging to the so-called "Rub' al Khali" tradition, a vague terminology that can now be supplanted with local types and dating (Crassard, 2008). It is striking that much of the manufacture and skill in tool production at Manayzah was directed toward projectile points, presumably for hunting. In the case of fluted points, great expertise was invested with the non-functional fluting (Crassard et al., 2006), suggesting prowess, aesthetic, and perhaps social focus on hunting, regardless of the meat yield suggested by the faunal assemblage. The fine, highly-skilled knapping disappeared from the archaeological record by the Middle Holocene, suggesting either that hunting technologies had changed (perhaps to netting and trapping) or that hunting itself no longer engaged the labor, skills, and perhaps importance of previous times.

Albeit a small number of bones, the Manayzah faunal assemblage with both likely imported domesticates and wild prey is consistent with a diverse economy using both hunting and consumption of herded animals. With Manayzah as the earliest dated Holocene occupation in the region, it is suggestive of a pioneer strategy with goats, sheep, and cattle supplementing wild game in the winter months beside permanent water.

Shi'b Kheshiya

Shi'b Kheshiya lies buried within alluvial sediment alongside a Middle Wadi Sana old paleo-channel that retained water throughout the dry season for at least 500 years between 5,970 and 5,400 BP uncal (Anderson et al., 2006) (Fig. 4). The site does not bear the usual hallmarks of occupation, for in addition to many hearths constructed and used during this period, the only structures appear to have a significant ritual character. There were at least two phases (the earlier not excavated) of stone structures with standing stones erected outside. The later of these structures was a tear-drop shaped ring of uprights set into the sides of a shallow pit. Although there was no door and no evidence for how it was roofed, the small structure of only a meter width was briefly occupied, then deliberately filled with flat-lying slabs to create a monumental platform less than a meter in height. The entire structure and adjacent campsite hearths were sealed by alluvial sediments that ceased aggrading around 4,400 uncal BP (McCorriston et al., n.d.; McCorriston, 2006).

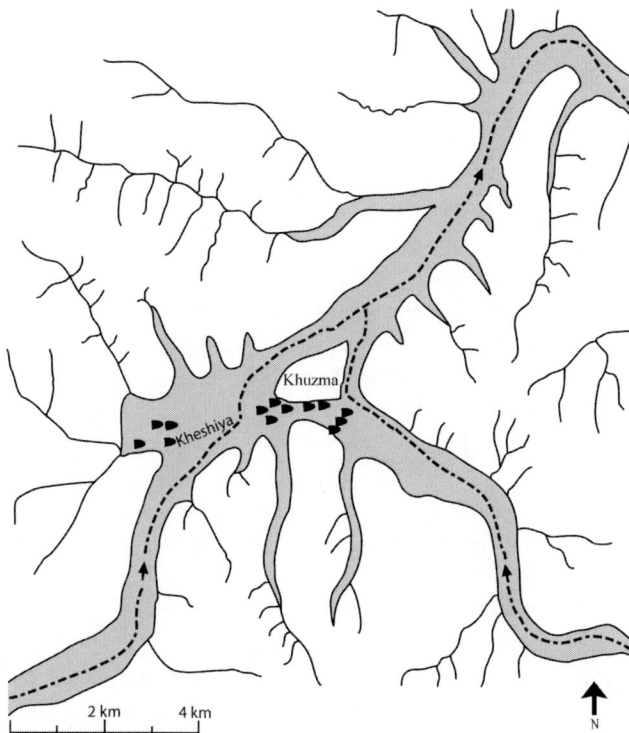

Fig. 4 Map of middle Wadi Sana with locations of tear-drop or D-shaped structures indicated

Fig. 5 Overview of Shi'b Kheshiya ritual site

A remarkable faunal assemblage accompanies this structure. An oval of about 40 cattle skulls was set up by pushing the noses into moist sediments beside the paleochannel (Fig. 5).

It is evident that there were a number of active hearths at the time because a layer of sediment enriched with ash, charcoal, thermally altered rock, and occasional chipped stone filled the skull ring and provided a middle fifth millennium BC terminal date for its construction (5,514 ± 48 uncal BP, 4,457–4,263 cal BC). The rare nature of the assemblage, its deliberate selection as construction materials for a bone ring,

and its association with a commemorative monument and *bayt-al*, or standing stone, make it clear that this assemblage is ritual rather than economic in nature. It nevertheless offers some important information about South Arabian pastoral population history.

Kheshiya Fauna

The first notable point about the Kheshiya faunal assemblage is that excavations produced only a single cattle mandible fragment alongside the c.40 cattle skulls, and no other skeletal elements were recovered. Thus only a very particular ritual deposit is represented, without the evidence for where or how the rest of the carcasses were consumed or deposited.

Thirty-five of the forty skulls were in a sufficiently complete condition to allow recording of morphometric and dental ageing data. Analysis focused on establishing whether the cattle skulls came from the European taurine cattle or Asian zebus, and whether they were wild or domestic. Following criteria established by Grigson (1974, 1976, 1980) for cattle skulls and horncores, morphology is seen to be most consistent with *Bos taurus*, and hence the animals are linked with the Levant or Africa, rather than Asia; the small size of the skulls compared to the range of European *Bos primigenius* strongly suggests they were domesticates.

Dental ageing (via tooth eruption and wear patterns following Grant, 1982) and skull fusion patterns showed all skulls to belong to animals culled as prime adults, suggesting that either animal size or age-status was important for inclusion in the slaughter. Both the stratigraphy and close examination of the cattle skulls for taphonomic signatures of weathering and exposure suggest that all the skulls were interred synchronously; the clustered ages-at-death also point towards a deliberate mass slaughter, rather than the bringing together of skulls from separate individual culls. Horncores were left exposed and had mostly eroded away, but would clearly have been the main visible part of the skull ring for a long period after its construction.

Kheshiya Territories

The sacrificed cattle and accompanying ritual construction of monuments in Wadi Sana suggest that tribal territories already existed in the Middle Holocene. The arguments for this social inference receive greater elaboration elsewhere, but here it may be simply stated that the sacrifice of 40 animals provided a large quantity of meat for which there exists at present no evidence for smoking and preservation and strong suggestion in multiple hearths for grilling and immediate consumption (McCorriston et al., 2005. Available evidence for sacrifice and monument construction seems to point to a collective event that drew large numbers of people not normally resident in the Middle Wadi Sana with its restricted winter grazing. The feasters, normally disbursed, clearly herded domesticated cattle and possibly also mixed herds with caprines. Whatever the economic strategy for herding, the numbers of animals required to cull 40 prime bulls and sustain support of humans consuming them would far exceed the grazing limits of wintertime Wadi Sana with its restricted soils constrained by the relatively narrow wadi channel. The stone monument and cattle ring commemorated the convergence of a social group or several social groups whose practice of ritual sacrifice emphasized their community ties (e.g., Bell, 1992; Jones, 2003). It is a practice very consistent with the behavior of tribes-people whose social relationships may be defined by lineage or territory (Evans Pritchard, 1940; Tapper, 1990; McCorriston and Bin 'Aqil, 2009) and often both. One of the significant aspects of community gatherings for mobile peoples is the opportunity to reestablish social ties and affirm communal rights to land and other resources.

Grazing Limits and Human Behavioral Ecology

Much of the local evidence for territories comes from rituals at stone structures, which in Wadi Sana can be dated as appearing in the fifth millennium BC. Anthropogenic burning provides another contemporary indication of people-packing to suggest that by this time the grazing resources of Middle Wadi Sana were fully exploited and that land management and resource intensification strategies were in place (McCorriston et al., 2005: 150–151). Other wider Arabian data such as *wusum* in rock art also suggest Middle Holocene territories inhabited by discrete social groups that we call tribes.

Human behavioral ecology (HBE) has usefully served to construct testable models of (economically rational) human choices within environmental and resource parameters (e.g., Smith, 1991; Hawkes et al., 1982, 1995; Kaplan and Hill, 1992; Winterhalder and Kennett, 2006). HBE models predict that human territorial behavior emerges as population densities rise (Rosenberg, 1990, 1998; Winterhalder and Smith, 2000) and when people defer short-term gains (e.g., hunted meat) to conserve resources for longer-term benefits (e.g., herding animals) (Alvard and Kuznar, 2001). In Wadi Sana, a transition to greater dependence on herded animals from the pioneer hunting-herding strategy at Manayzah would entail an increased need to ensure adequate graze and herd protection.

From Where Were Domesticates Introduced?

Although it is not possible to ascertain whether the herders at Shi'b Kheshiya were indeed the cultural and biological descendents of their Manayzah forebears, there do seem to be developmental differences in the strategies these human groups practiced. The cattle at Shi'b Kheshiya could have been from the descendent herds from Manayzah animals, but the evidence is inconclusive. There is a possibility that new herding people arrived in Wadi Sana bringing new animal stock, but this hypothesis too has insufficient evidence to test. As with the first introductions around 6,000 BC at Manayzah, the question is from where did such animals come? The point is important because the Manayzah domesticates are the oldest yet documented in Arabia (Martin et al., 2009), and the Kheshiya cattle show an ideological focus consistent with (albeit not conclusively pointing to) specialized cattle pastoralism.

With the presence of wild *Bos* populations in Arabia comes the question of indigenous domestication, but none of the genetic evidence in modern *Bos* populations would seem to suggest this as a likelihood (Hanotte et al., 2002), nor is there sufficient archaeofaunal material to trace indigenous domestication through rigorous metric analysis of reductions in cattle size. Fully domesticated taurine cattle could have arrived in Arabia from the Levant (Uerpmann et al., 2009) or possibly from Africa, and it is important to recognize that the Manayzah and Shi'b Kheshiya data may point to such introductions with differing human populations and population densities and very different adoption strategies at different times.

Whereas domesticated cattle could have arrived in Arabia from the Levant, African archaeology provides an important broader context for the emergence of pastoralism in Southern Arabia. In both Africa and Southern Arabia, cattle-herding and ultimately full pastoralism developed in the absence of agriculture, and in the Arabian case, in the absence of any evident plant collection or cultivation. There is extensive evidence to suggest a long history of cattle pastoralism in Northeast Africa where the crucible of cattle pastoralism provided not only important economic resources but also generated ideological and social frameworks throughout prehistory and history (Di Lernia, 2006; Wengrow, 2006). Archaeofaunal remains of cattle from the region of Nabta Playa do not clearly indicate domesticated species until the middle sixth millennium BC (Grigson, 2000; Wengrow, 2003: 200; Gifford-Gonzalez, 2005), but there may have been mutualistic associations between humans and wild cattle as early as the ninth millennium BC that resulted in independent domestication in Africa (Gautier, 1984; Wendorf et al., 1987; Close and Wendorf, 1992; Close and Wendorf, 2001: 70, Gautier, 2001: 631–632). If independent domestication in Africa did occur, it is as likely that cattle were needed for ritual purposes (sacrifices and burial rites) as for food or stability of food supply (Marshall and Hildebrand, 2002: 111, 113; Russell and Martin, 2005: 55–56, see also Cauvin, 2000; Bar-Yosef and Bar-Yosef Mayer, 2002: 349). Archaeologists emphasize better evidence for fully domesticated cattle in Africa after 6,500 BC (Gautier, 2001; Gifford-Gonzalez, 2005: 196), somewhat before the first Arabian (Manayzah) domesticated cattle (Martin et al., 2009). From this time onwards, cattle played an important role in ritual practices linked to the social and cosmic world of pastoralism. Goats did not. They were apparently introduced to Africa from the Levant thereafter (Vermeersch et al., 1994; Hassan 2000, 2002; Shirai, 2005), although they did spread through the Horn of Africa as a compliment to cattle pastoralism (Marshall and Hildebrand, 2002; Gifford-Gonzalez, 2005).

Nabta Playa in the southern Egyptian Sahara has the earliest African dates for domesticated cattle used in rituals, although there are elsewhere Late Paleolithic burials that incorporated the horns of wild cattle (Wendorf, 1968). At Nabta Playa, an entire domesticated cow was deliberately interred in the middle sixth millennium BC in Tumulus E-94-1N of the Late Neolithic Nabta Playa Ceremonial Complex (somewhat later than Arabia's first domesticated cattle but a full millennium before the cattle skull ring at Kheshiya). Other nearby tumuli also contained cattle bones along with jackel, gazelle or caprine, sheep, and in one case isolated from any fauna, human remains. Nabta Playa's tumuli date to the Late Neolithic phase, contemporary with other Saharan deliberate interments of cattle (Applegate et al., 2001). By the middle fifth millennium BC pastoralists with cattle and goats used the rich Nile Valley floor for fishing, seasonal pasture, and harvesting wild tubers and grasses (e.g., sorghum) (Jesse, 2004: 39–40). Sites of the Khartoum Neolithic included cattle buchrania and probably cattle hides in the burials of high status individuals (Reinhold, 1994: 95, 2000: 64, 72–76; Caneva, 1988: 22–27; Gautier, 1998: 59–61). Thereafter, cattle, humans, and burial continued to be interwoven in important symbolic frameworks that can be traced throughout Eastern Sahara and Nile prehistory (e.g., Chaix, 1988; Di Lernia, 2006; Wengrow, 2006). Such Nilotic and Eastern Saharan cultures shared only a symbolic focus on cattle with their Arabian contemporaries, which neither included cattle in burials nor adopted ceramics and plant-based subsistence.

Do the South Arabian (Wadi Sana) Sites Indicate True Pastoralism?

Definitions of pastoralism vary between those who hold the very appearance of domestic herd animals in an area to signify the arrival of pastoralism (e.g., Harris, 1996), and those who reserve the term for more specialist forms of animal keeping, which are separate from agriculture and tend to be more specialized, involving primarily a single species moving over long distances, hence associated with highly mobile

populations (Khazanov, 1994). From current evidence, it would seem that Manayzah represents a frequently re-visited camp where caprine (and maybe also cattle) herders also engaged in hunting local game. Subsistence at Manayzah was apparently not highly specialized, but appears to have been more of an opportunistic mixed strategy. There is little evidence on population mobility or sedentism except from the site location itself: Manayzah's dangerous setting in a steep-walled and narrow canyon makes it inappropriate for summer occupation when flash-flooding could occur. The faunal assemblage presently available from the site is too small to provide definitive indications about whether cattle truly lag behind an earlier introduction of sheep and goat.

Kheshiya is a specialist site for ritual and may not be representative of the full range of animal-based economic activities taking place in the southern highlands of Yemen at that time. Yet, if such large-scale culling was sustainable, some form of specialist herding of cattle is likely to have underlain it. Cattle have particular ecological needs (greater water and forage requirements) distinct from caprine herds, and the cattle skull ring site may hint that cattle herding was conducted separately from the herding of other animals. The cattle ring at Kheshiya is consistent with specialized pastoralism, although there is no indication whether secondary products were part of that strategy. The site must also be considered in the context of other middle fifth millennium BC cattle herding in Arabia, with mixed caprine-cattle strategy and use of coastal resources at Jebel Buhais and mixed cattle-caprine assemblages also at Wadi Ath-Thayyilah. With the present evidence, it seems possible that specialized cattle herding could have developed locally from an original mixed herding and hunting strategy at Manayzah, but the data are inconclusive on this point.

Conclusions

While there remain many questions and full presentation of the Wadi Sana data sets elsewhere upon which to elaborate debate, several conclusions are apparent. First of all, domesticated animals, including sheep and goat, were available to hunters and very likely herded in Southern Arabia's highlands by the late seventh millennium BC. Cattle were certainly present by the early sixth millennium BC. The Manayzah radiocarbon dates are earlier than other Arabian assemblages with domesticated animals, raising the question of why and how the first domesticated animals appear in Southern Arabia. The answer may be as simple as a paucity of research at early stratified sites elsewhere. With fuller publication and more work at Manayzah, it may be possible to use an expanded data set from earlier levels to test in Southern Arabia the hypothesis that Levantine PPNB herders migrated in through the Arabian peninsular interior and eastern coast (Dreschler, 2007; Uerpmann et al., 2009). At present it seems that the Manayzah dates and lithics do not point to an introduction of pastoral economies or people closely related either in time or in material culture to the Levantine PPNB (e.g., Crassard, 2008).

A second important point is that the Wadi Sana sequence provides a glimpse, albeit imperfect, into the local development of herding strategies. The earliest herd animals were probably not introduced by fully committed pastoralists but as a pioneering strategy among hunters (e.g., Zeder, 1994; Garcea, 2004, 2006: 205–207). One might suggest that the introduction of herd animals into Southern Arabia did not entail significant human population incursions as pastoralists from other areas and most particularly not from the Levant where hunting remained an economic mainstay in the desert fringes until the fifth millennium BC (Tchernov and Bar-Yosef, 1982; Herskovitz et al., 1994; Bar-Yosef and Bar-Yosef Mayer, 2002; Martin, 1999; cf., Uerpmann et al., 2009).

Indeed, once specialized pastoralism did emerge – and we tentatively suggest here that by the middle fifth millennium BC there were specialized cattle herders sacrificing bulls at Shi'b Kheshiya – cattle pastoralism appears to have more in common with contemporary African herding systems and the cattle cults of the Sahara than with the Levantine agro-husbandry and caprine herding practices around settled villages to the north. Of course there are vast cultural differences between African cattle pastoral peoples and the herders at Shi'b Kheshiya. It seems clear that such introductions would have been of animals and possibly of economic strategies, not human pastoralists moving en masse from Africa. Indeed, the gap between late seventh to early sixth millennium BC Manayzah and middle fifth millennium BC Kheshiya could mean that specialized cattle pastoralism developed locally from the mix of domesticates pioneered at Manayzah and that further archaeological exploration may recover new evidence to test such a hypothesis.

Acknowledgments The authors thank first and foremost the RASA team for their field and laboratory labors. We are especially indebted to Rick Oches, Abdalaziz Bin 'Aqil, Remy Crassard, Mike Harrower, Catherine Heyne, and Lisa Usman, upon whose hard work this chapter greatly depends. We thank the General Organization of Antiquities and Museums in Yemen under the direction of Youssef Abdulla and of Abdallah BaWazir and the American Institute for Yemeni Studies with Resident Director Chris Edens. Research was sponsored principally by the US National Science Foundation (Grant numbers BSC 0211497, BSC 0624268) although the views expressed here are those of the authors. The RASA team has also gratefully accepted very generous logistical support from Canadian Nexen Petroleum Yemen and Yemen's military forces and patient hospitality from the Bayt Al-Aly bedouin in Wadi Sana. Thanks go to Jen Everhart for illustrations and editorial assistance. We also acknowledge the Arizona AMS facility for radiocarbon dates and our home institutions The Ohio State University and University College London for research support.

References

Al-Shahri AA. Kayf ibtidiyna wa kayf irtiqiyna bil-hadhara al insana min shaba al-jazira al-'arabiya (Zufar): Kitabatiha wa naquwshiha al-qadiyma. 1st ed. Salalah: Private Printing; 1994.

Alvard MS, Kuznar L. Deferred harvests: the transition from hunting to animal husbandry. American Anthropologist. 2001;103:295–311.

Anati E. Rock-Art in Central Arabia 3. Corpus of the Rock Engravings I and II. Publications de l'Institut Orientaliste de Louvain, 4. Expédition Philby-Ryckmans-Lippens en Arabie. Université Catholique de Louvain, Louvain-la-Neuve; 1972.

Anati E. Rock-art in central Arabia 4. Corpus of the rock engravings III and IV. Publications de l'Institut Orientaliste de Louvain, 6. Expédition Philby–Ryckmans–Lippens en Arabie. Louvain-la-Neuve: Université Catholique de Louvain; 1974.

Anderson S, Oches EA, Sander K, McCorriston J, Harrower M. Fluvial sediments record Middle Holocene climate change in southern Yemen. Poster presented at the Geological Society of America 2006 Annual Meetings; 2006.

Applegate A, Gautier A, Duncan S. The North Tumuli of the Nabta Late Neolithic ceremonial complex. In: Wendorf F, Schild R et al., editors. Holocene settlement of the Egyptian Sahara, vol 1. The archaeology of Nabta Playa. New York: Kluwer/Plenum; 2001. p. 468–88.

Bar-Yosef O, Bar-Yosef Mayer DE. Early Neolithic tribes in the Levant. In: Parkinson WA, editor. The archaeology of tribal societies. Ann Arbor, MI: International Monographsin Prehistory; 2002. p. 340–71.

Bell C. Ritual theory, ritual practice. New York: Oxford University Press; 1992.

Bellwood P. First farmers: the origins of agricultural societies. Malden: Blackwell; 2005.

Bin 'Aqil AJ. Qaniysu al Wa'lun fi Hadramawt. Sana'a: 'Asma al Thaqafi al Arabiya; 2004.

Boessneck J. Osteological differences between sheep (Ovis aries Linne) and goat (Capra hircus Linne). In: Brothwell D, Higgs E, Clark G, editors. Science in archaeology. 2nd ed. London: Thames and Hudson; 1969. p. 331–58.

Braemer F, Bodu P, Crassard R, Manqûsh M. Chapitre IX Jarf al-Ibil et Jarf al-Nabîrah, deux sites rupestres de la région d'al-Dâli'. In: Inizan M-L, Rachad M, editors. Art Rupestre et Peoplements Préhistoriques au Yémen. Sana'a: CEFAS Editions; 2007. p. 95–100, 118–27.

Caneva I. El Geili: The history of a Middle Nile Environment, 7000 BC–AD 1500. British Archaeological Report International Series, 424. Oxford: British Archaeological Reports; 1988.

Cattani M, Bökönyi S. Ash-Shumah. An Early Holocene settlement of desert hunters and mangrove foragers in the Yemen Tihama. In: Cleuziou S, Tosi M, Zarins J, editors. Essays on the late prehistory of the Arabian peninsula. Serie Orientale Roma 93. Roma: Istituto Italiano per L'Africa e L'Oriente; 2002. p. 31–54.

Cauvin J. The birth of the gods and the origins of agriculture [trans. Trevor Watkins]. Cambridge: Cambridge University Press; 2000.

Chaix L. Le monde animal à Kerma (Soudan). Sahara. 1988;1:77–84.

Cleuziou S, Tosi M. Hommes, climates, et environments de la péninsularearabique à l'Holocène. Paléorient. 1997;23:121–36.

Cleuziou S, Tosi M, Zarins J. Introduction. In: Cleuziou S, Tosi M, Zarins J, editors. Essays on the late prehistory of the Arabian peninsula. Serie Orientale Roma 93. Roma: IstitutoItaliano per L'Africa e L'Oriente; 2002. p. 9–27.

Close AE, Wendorf F. The beginnings of food production in the eastern Sahara. In: Gebauer AB, Price DT, editors. Transitions to agriculture in prehistory. Madison: Prehistory Press; 1992. p. 63–72.

Crassard R. Preliminary report on wâdî Ibn 'Alî 1 site, near Shibâm, Hadramawt. In: Chroniques Yéménites en langue arabe, 3. Sana'a: CEFAS; 2006. p. 3–10.

Crassard R. Apport de la technologie lithique à la définition de la préhistoire duHadramawt, dans le contexte du Yémen et de l'Arabie du Sud. Unpublished Ph.D. dissertation, Université de Paris; 2007.

Crassard R, McCorriston J, Oches E, Bin'Aqil A, Espagne J, Sinnah M. Manayzah, early to mid-Holocene occupations in Wadi Sana (Hadramawt, Yemen). Proceedings of the Seminar for Arabian Studies. 2006;36:151–73.

Davies CP. Holocene paleoclimates of southern Arabia from lacustrine deposits of the Dhamar highlands, Yemen. Quaternary Research. 2006;66:454–64.

Di Lernia S. Building monuments, creating identity: cattle cult as social response to rapid environmental changes in the Holocene Sahara. Quaternary International 2006;151:50–62.

Di Mario F. The western ar-Rub' al-Khali "Neolithic:" new data from the Ramlat Sab'atayn. Annali Istituto Universario Orientale Napoli 1989;49:109–48.

Dreschler P. The Neolithic dispersal into Arabia. Proceedings of the Seminar for Arabian Studies. 2007;37:93–109.

Evans Pritchard EE. The Nuer. Oxford: Oxford University Press; 1940.

Fedele F. Man, land and climate: emerging interactions from the Holocene of the Yemen Highlands. In: Bottema E-N, van Zeist W, editors. Man's role in the shaping of the Eastern Mediterranean Landscape. Rotterdam, the Netherlands: A.A. Balkema, 1990. p. 31–42.

Fedele FG. Wadi al-Tayyila 3: a Mid-Holocene site on the Yemen plateau and its Lithic Collection. In: Sholan AM, Antonini S, Arbach M, editors. Sabaean studies. Archaeological, epigraphical and historical studies in honour of Yusuf M. Abdallah, Alessandro de Maigret and Christian J. Robin on the occasion of their sixtieth birthdays. Naples-Sana'a: Universita degli Studi di Napoli "L'Orientale."; 2005. p. 214–45.

Fedele FG. Wadi al-Tayyilah 3, a Neolithic and Pre-Neolithic occupation on the eastern Yemen Plateau, and its archaeofaunal information. Proceedings of the Seminar for Arabian Studies. 2008;38:153–72.

Fleitmann D, Burns SJ, Mudelsee M, Neff U, Kramers J, Mangini A, et al. Holocene forcing of the Indian monsoon recorded in a stalagmite from southern Oman. Science. 2003;300:1737–9.

Fleitmann D, Burns SJ, Mangini A, Mudelsee M, Kramers J, Villa I, et al. Holocene ITCZ and Indian monsoon dynamics recorded in stalagmites from Oman and Yemen (Socotra). Quaternary Science Reviews. 2007;26:170–88.

Garcea EAA. An alternative way towards food production: the perspective from theLibyan Sahara. Journal of World Prehistory. 2004;18:107–53.

Garcea EAA. Semi-permanent foragers in semi-arid environments of North Africa. World Archaeology. 2006;38:197–219.

Garcia MA, Rachad M. L'Art des Origins au Yémen. Paris: Editions du Seuil; 1997.

Garcia MA, Rachad M, Hadjouis D, Inizan M-L, Fontugnes M. Découvertes préhistoriques au Yémen, le contexte archéologique de l'art rupestre de la region de Saada. Comptes Rendus de l'Académie des Sciences de Paris 313, série II; 1991. p. 1201–6.

Garrard A, Colledge S, Martin L. The emergence of crop cultivation and caprine herding in the "Marginal Zone" of the southern Levant. In: Harris D, editor. The origins and spread of agriculture and pastoralism in Eurasia. London: University College of London Press; 1996. p. 204–26.

Gautier A. Notes on the animal bone assemblage from the Early Neolithic at Geili. In: Caneva I, editor. El Geili. The history of a middle Nile environment, 7000 BC–AD 1500. British archaeological reports: International Series 424. Oxford: British Archaeological Reports; 1988. p. 57–64.

Gautier A. The early to late Neolithic archeofaunas from Nabta and Bir Kiseiba. In: Wendorf F, Schild R, Associates, editors. Holocene settlement of the Egyptian Sahara, vol 1. The archaeology of Nabta Playa. New York: Kluwer/Plenum; 2001. p. 609–35.

Gell A. The anthropology of time. Oxford: Berg; 1992.

Gifford-Gonzalez D. Pastoralism and its consequences. In: Stahl AB, editor. African archaeology. Oxford: Basil Blackwell; 2005. p. 187–224.

Grant A. The use of tooth wear as a guide to the age of domestic animals. In: Wilson B, Grigson C, Payne S, editors. Ageing and sexing of animal bones from archaeological sites, BAR British Series 109. Oxford: British Archaeological Reports; 1982. p. 91–108.

Grigson C. The craniology and relationships of four species of Bos. I. Basic craniology: Bos taurus L. absolute size. Journal of Archaeological Science. 1974;1:353–79.

Grigson C. The craniology and relationships of four species of Bos. III. Basic craniology: Bos taurus L. Sagittal profiles and other non-measurable characters. Journal of Archaeological Science. 1976;3:115–36.

Grigson C. The craniology and relationhsips of four species of Bos. V. Bos indicus L. Journal of Archaeological Science. 1980;7:3–32.

Grigson C. Early cattle around the Indian Ocean. In: Reade JE, editor. The Indian Ocean in antiquity. London: Kegan Paul International and British Museum; 1996. p. 41–74.

Hadjouis D. Chapitre V La faune des grands mammifères. In: Inizan M-L, Rachad M, editors. 2007 Art Rupestre et Peoplements Préhistoriques au Yémen. Sana'a: CEFAS Editions; 2007. p. 51–60.

Hanotte O, Bradley DG, Ochieng JW, Verjee Y, Hill EW, Rege JEO. African pastoralism: genetic imprints of origins and migrations. Science. 2002;296:336–9.

Harris D. The origins and spread of agriculture and pastoralism in Eurasia. London: University College of London Press; 1996.

Harrison DL, Bate PJJ. The mammals of Arabia. Sevenoaks, Kent: Harrison Zoological Museum; 1991.

Hassan F. Holocene environmental change and the origins and spread of food production in the Middle East. Adumatu. 2000;1:7–28.

Hassan F. Holocene environmental change and the transition to agriculture in southwest Asia and northeast Africa. In: Yasuda Y, editor. The origins of pottery and agriculture. New Delhi: Roli Books; 2002. p. 55–68.

Hawkes K, Hill K, O'Connell JF. Why hunters gather: optimal foraging and the Aché of Eastern Paraguay. American Ethnologist. 1982;9(2):379–98.

Hawkes K, O'Connell JF, Blurton Jones NG. Hadza children's foraging: juvenile dependency, social arrangements, and mobility among hunter-gatherers. Current Anthropology. 1995;36(4):688–700.

Herskovitz I, Bar-Yosef O, Arensburg B. The pre-pottery Neolithic populations of south Sinai and their relations to other circum-Mediterranean groups: ananthropological study. Paléorient. 1994;20(2): 59–84.

Hoch E. Reflections on prehistoric life at Umm an-Nar (Trucial Oman) based on faunal remains from the third millennium BC. In: Taddei M, editor. South Asian archaeology 1977 – Instituto Universitario Orientale Seminario di Studi Asiatici, Series Minor VI, vol. 1. Naples: Instituto Universitario Orientale; 1979. p. 589–638.

Horwitz L, Tchernov E, Ducos P, Becker C, von den Driesch A, Martin L, et al. Animal domestication in the southern Levant. Paléorient. 2000;25(2):63–80.

Inizan M-L. Chapitre VI des occupations préhistoriques à Sa'ada. In: Inizan M-L, Rachad M, editors. 2007 art rupestre et peoplements préhistoriques au Yémen. Sana'a: CEFAS Editions; 2007. p. 61–72.

Inizan M-L, Rachad M, editors. Art Rupestre et Peoplements Préhistoriques au Yémen. Sana'a: CEFAS Editions; 2007.

Jesse F. The Neolithic. In: Welsby DA, Anderson JR, editors. Sudan: ancient treasures. London: The British Museum; 2004. p. 35–45.

Jones A. Technologies of remembrance: memory, materiality, and identity in Early Bronze Age Scotland. In: Williams H, editor. Archaeologies of remembrance: death and memory in past societies. New York: Kluwer/Plenum; 2003. p. 65–88.

Jumaily MM. Mammals of the Republic of Yemen. Fauna of Arabia. 1998;17:477–500.

Kaplan H, Hill K. The evolutionary ecology of food acquisition. In: Smith EA, Winterhalder B, editors. Evolutionary ecology and human behavior. New York: Aldine; 1982. p. 167–203.

Keall EJ. Rock-shelter paintings in the Tihâma foothills. In: Sholan AM, Antonini S, Arbach M, editors. Sabaean studies. Archaeological, epigraphical and historical studies in honour of Yusuf M. Abdallah, Alessandro de Maigret and Christian J. Robin on the occasion of their sixtieth birthdays. Naples-Sana'a: Universita degli Studi di Napoli "L'Orientale"; 2005.

Khan M. Prehistoric rock art of northern Saudi Arabia. Riyadh: Ministry of Education Department of Antiquities and Museums; 1993.

Khan M. Wusum: the tribal symbols of Saudi Arabia. Riyadh: Ministry of Education; 2000.

Khazanov A. Nomads and the outside world. 2nd ed. Madison, WI: The University of Wisconsin Press; 1994.

Lezine A-M, Saliège J-F, Robert C, Wertz F, Inizan M-L. Holocene lakes from Ramlat as-Sab'atayn (Yemen) illustrate the impact of monsoon activity in southern Arabia. Quaternary Research. 1998;50:290–9.

Lippens P. Expédition en Arabie Centrale. Paris: Adrien-Maisonneuve; 1956.

Marshall F, Hildebrand E. Cattle before crops: the beginnings of food production in Africa. Journal of World Prehistory. 2002;16(2):99–143.

Martin L. Mammal remains from the eastern Jordanian Neolithic, and the nature of caprine herding in the steppe. Paléorient. 2000;25(2): 87–104.

Martin LA, McCorriston J, Crassard R. Early Arabian pastoralism at Manayzah. In: Wadi Sana, Hadramawt. Proceedings of the Seminar for Arabian Studies 39 [forthcoming]; 2009.

McClure HA. Late Quaternary Palaeoenvironments of the Rub' al Khali. Unpublished Ph.D. dissertation, University College London, 1984.

McCorriston J. Breaking the rainfall barrier and the tropical spread of Near Eastern agriculture into southern Arabia. In: Kennett DJ, Winterhalder B, editors. Behavioral ecology and the transition to agriculture. Berkeley, CA: University of California Press; 2006. p. 217–36.

McCorriston J. Anthropogenic burning strategies in highland southern Arabia: pastoral territories, Middle Holocene climate change, and resource intensification. Paper presented at the 6th Deutsches Archäologes Kongress, Mannheim; 18 May 2008.

McCorriston J, Oches E, Walter DE, Cole K. Holocene paleoecology and prehistory in Highland Southern Arabia. Paleorient. 2002;28(1): 61–88.

McCorriston J, Harrower M, Oches E, Bin 'Aqil A. Foraging economies and population in the Middle Holocene highland of southern Yemen. Proceedings of the Seminar for Arabian Studies. 2005a;35:143–54.

McCorriston J, with contributions by Heyne C, Harrower M, Patel N, Steimer-Herbet T, Al Amary I, Barakany AK, Sinnah M, Ladah R, Oches E, Crassard R, Martin L. Roots of Agriculture (RASA) Project 2005: a season of excavation and survey in Wadi Sana, Hadramawt. Bulletin of the American Institute for Yemeni Studies 2006;47:23–8.

Rachad M. Chapitre VIII Thèmes de l'art rupestre. In: Inizan M-L, Rachad M, editors. 2007 Art rupestre et peuplements préhistoriques au Yémen. Sana'a: CEFAS Editions; 2007a. p. 83–94.

Rachad M. Chapitre VII Chronologie et styles de l'art rupestre. In: Inizan M-L, Rachad M, editors. Art rupestre et peoplements préhistoriques au Yémen. Sana'a: CEFAS Editions; 2007b. p. 73–82.

Reinhold J. Le cimitière néolithique KDK 1 de Kadruka (Nubie soudanoise). Premiers resultants et essai de correlation avec les rites du Soudan central. In: Bonnet C, editor. Etudes nubiennes. Conférence de Genève: Actes du VIIe Congrès International d'études nubiennes 3–8 Septembre 1990. Volume II communications. Genève: Charles Bonnet; 1994. p. 93–100.

Reinhold J. Archéologie au Soudan. Paris: Editions Errance; 2000.

Rosenberg M. The mother of invention: evolutionary theory, territoriality, and the origins of agriculture. American Anthropologist. 1990;92:399–415.

Rosenberg M. Cheating at musical chairs: territoriality and sedentism in an evolutionary context. Current Anthropology. 1998;39:653–64.

Russell KW. After Eden. The behavioral ecology of early food production in the Near East and North Africa. British Archaeological Reports, International Series 391. Oxford: British Archaeological Reports; 1988.

Russell N, Martin L. Çatalhöyük mammal remains. In: Hodder I, editor. Inhabiting Çatalhöyük: reports from the 1995–99 seasons. Cambridge: McDonald Institute Monographs/British Institute of Archaeology at Ankara; 2005. p. 33–98.

Shirai N. Walking with herdsmen: in search of the material evidence for the diffusion of agriculture from the Levant to Egypt. Neolithics. 2005;1(5):12–7.

Smith EA. Inujjuamiut foraging strategies: evolutionary ecology of an arctic hunting economy. New York: Aldine; 1991.

Tapper R. Anthropologists, historians, and tribespeople on tribe and state formation in the Middle East. In: Khoury PS, Kostiner J, editors. Tribes and state formation in the Middle East. Berkeley, CA: University of California Press; 1990. p. 48–73.

Tchernov E, Bar-Yosef O. Animal exploitation in the Pre-Pottery Neolithic B period at Wadi Tbeik, Southern Sinai. Paléorient. 1982;8:17–37.

Tosi M. The emerging picture of prehistoric Arabia. Annual Reviews in Anthropology. 1986;15:461–90.

Uerpmann H-P. The ancient distribution of ungulate mammals in the Middle East. Wiesbaden: Dr. Ludwig Reichert Verlag; 1987.

Uerpmann M. Structuring the Late Stone Age of southern Arabia. Arabian Archaeology and Epigraphy. 1992;3:65–109.

Uerpmann M, Uerpmann H-P. Faunal remains of Al-Buhais 18: an Aceramic Neolithic site in the Emirate of Sharjah (SE Arabia) – excavations 1995–1998. In: Mashkour M, Choyke AM, Buitenhuis H, Poplin F, editors. Archaeozoology of the Near East IVB. Publicatie 32. Gronginen: Center for Archaeological Research and Consultancy; 2000. p. 40–9.

Uerpmann H-P, Uerpmann M. Neolithic faunal remains from al-Buhais 19 (Sharjah, UAE). In: Uerpmann H-P, Uerpmann M, Jasim SA, editors. The natural environment of Jebel al-Buhais: past and present. The archaeology of Jebel Al-Buhais Sharjah United Arab Emirates, vol. 2. Tuebingen: Kerns Verlag; 2008. p. 97–132.

Uerpmann H-P, Uerpmann M, Jasim S. Stone age nomadism in SE Arabia – palaeo-economic considerations on the Neolithic site of Al-Buhais 18 in the Emirate of Sharjah, U.A.E. Proceedings of the Seminar for Arabian Studies. 2000;30:229–34.

Uerpmann H-P, Potts DT, Uerpmann M. Holocene (re-) occupation of Eastern Arabia. In: Petraglia MD, Rose JI, editors. The evolution of human populations in Arabia: paleoenvironments, prehistory and genetics. The Netherlands: Springer; 2009. p. 205–14.

Van Vuure CT. History, morphology and ecology of the Aurochs (Bos taurus primigenius). Lutra. 2002;44:3–18.

Vermeersch PM, van Peer P, Moyersons J, van Neer W. Sodmein Cave, Red Sea mountains (Egypt). Sahara. 1994;6:31–40.

Wendorf F. Late Paleolithic sites in Egyptian Nubia. In: The prehistory of Nubia. Dallas: Fort Burgwin Research Center and Southern Methodist University Press;1968. p. 791–953.

Wendorf F, Close AE, Schild R. Early domestic cattle in the Eastern Sahara. In: Coetzee JA, editor. Palaeoecology of Africa. Rotterdam: A.A. Balkema; 1987. p. 441–8.

Wengrow D. On desert origins for the ancient Egyptians. Antiquity. 2003;77:597–601.

Wengrow D. The archaeology of early Egypt. Social transformations in north-east Africa, 10000–2650 BC. Cambridge: Cambridge University Press; 2006.

Winterhalder B, Kennett DJ. Behavioral ecology and the transition from hunting and gathering to agriculture. In: Kennett DJ, Winterhalder B, editors. Behavioral ecology and the transition to agriculture. Berkeley, CA: University of California Press; 2006. p. 1–21.

Winterhalder B, Smith EA. Analyzing adaptive strategies: human behavioral ecology at twenty-five. Evolutionary Anthropology. 2000;9(2):51–72.

Martin LA (n.d.) Neolithic Cattle at Shi'b Kheshiya, Hadramawt, Yemen. [Manuscript]

McCorriston J, Bin 'Aqil A (n.d.) Convergences in the ethnography, semantics, and archaeology of prehistoric small scale monument types in Hadramawt (southern Arabia). [Manuscript]

McCorriston J, Martin LA, Oches EA, Harrower M (n.d.) Neolithic cattle sacrifice at the place of Khuzma. Hadramawt Province, Yemen. [Manuscript]

Zarins JA. Pastoral nomadism in Arabia: ethnoarchaeology and the archaeological record – a case study. In: Bar-Yosef O, Khazanov A, editors. Pastoralism in the Levant: archaeological materials in anthropological perspectives. Madison, WI: Prehistory; 1992. p. 219–40.

Zeder M. After the revolution: post Neolithic subsistence in northern Mesopotamia. American Anthropologist. 1994;96:97–126.

Chapter 18
Archaeological, Linguistic and Historical Sources on Ancient Seafaring: A Multidisciplinary Approach to the Study of Early Maritime Contact and Exchange in the Arabian Peninsula

Nicole Boivin, Roger Blench, and Dorian Q. Fuller

Keywords Domesticates • Exchange • Linguistics • Maritime • Seafaring • Trade

Introduction

The Arabian subcontinent sits at a critical juncture in the Old World, surrounded to the west, north and east respectively by the African landmass, the Levant (with the European world beyond it), and the Asian continent. While its ancient and historical development has certainly been shaped by this positioning relative to the great continents, however, Arabia is equally defined by its near circumspection by the sea, which wraps itself around some 80% of its perimeter, and has served as both barrier and bridge to the surrounding regions since the emergence of modern humans out of Africa at ca. 80–60 ka (Petraglia and Alsharekh, 2003; Petraglia et al., 2007; Bailey, 2009). An increasing weight of evidence suggests that the three main bodies of water that surround Arabia – the Red Sea, the Persian Gulf and the Arabian Sea – not only offered a rich resource base for thousands of years of human occupation in the subcontinent, but also witnessed some of the world's earliest seafaring and maritime exchange activities. Evidence for maritime contact over long distances is for this arena also amongst the oldest in the world. At the same time, the sea has also sometimes served to distance

N. Boivin (✉)
Research Laboratory for Archaeology and the History of Art
University of Oxford, Dyson Perrins Building,
South Parks Road, Oxford, UK, OX1 3QY
e-mail: nicole.boivin@rlaha.ox.ac.uk

R. Blench
Kay Williamson Educational Foundation, 8 Guest Road,
Cambridge, CB1 2AL, UK
e-mail: R.Blench@odi.org.uk

D.Q. Fuller
Institute of Archaeology, University College London,
31–34 Gordon Square, London, WC1H 0PY, UK
e-mail: d.fuller@ucl.ac.uk

Arabia from her neighbors, helping to shape a distinctive trajectory within the subcontinent.

The Arabian peninsula's unique maritime position and relationship with the sea obviously deserve close investigation. And in the past few decades, research into Arabia's prehistoric and subsequent maritime activities and seafaring contact with other regions has certainly intensified. While it is true that evidence for many relevant areas remains patchy – for example, for much of the littoral region of the Red Sea, for the Horn of Africa, for significant parts of the southern Arabian peninsular coast, and for the littoral regions of Iran and Pakistan – it is also true that many inroads have been made, particularly on the Arabian side of the Gulf and the Gulf of Oman, in recent years. Increasing evidence from inland regions of the subcontinent also adds to the emerging picture. What still remains to be done, however, is to examine the diverse early maritime activities evident in different parts of the subcontinent and its neighboring regions, and to link them together into a syncretic framework – to examine Arabia and its developments as part of the wider pre- and protohistory of the Arabian Sea. The regional specialization that has to some degree prevented such a development is of course understandable in light of the detailed and intensive research still required in many of these understudied regions. But the paucity of studies on the subcontinent's overall maritime development is likely also a reflection of a general terrestrial bias that often precludes the kind of maritime-oriented analysis that led Braudel so successfully to link diverse regions through a study of the people of the Mediterranean Sea (Braudel, 1995).

While many challenges face any attempt to create a pre- and protohistory of the northern Arabian Sea at this point in time, we nonetheless feel that such an enterprise is now essential. It is required if scholars are to endeavor to appreciate better the important and diverse roles of maritime activities in human societies (Erlandson, 2001; Cooney, 2003; Bailey, 2004). It is also necessary if researchers are going to properly resolve the question of how some very long distance biological, material, and linguistic translocations in the prehistoric Indian Ocean actually transpired (e.g., Meyer et al., 1991; Blench, 1996, 2003; Mbida et al., 2000; Fuller,

2003a; Magnavita, 2006; Zeder, 2006; Adelaar, 2009; Cleuziou and Tosi, 1989, 2007; Sauer, 1952). Arabia has undoubtedly long been a key maritime player in the Old World and in the Arabian Sea in particular. This role is clear from Pre-classical times, when the "Incense Road" connected Egypt and the Levant to western Arabia (Yemen), and the Sabaean Lane placed Arabia at the heart of a network of Indian Ocean trade routes (Groom, 2002). Such trade links were well-established and travelled by the time that Greco-Roman sailors began to add their own vessels to the Indian Ocean network in the last centuries BC, as recorded in the first century AD *Periplus Maris Erythraei* (Miller, 1968; Casson, 1989; Ray, 1998; Cappers, 2006). Many maritime routes of trade and contact had significant antiquity by the time they were recorded in Classical sources, and Arabia's position as a central node within them makes detailed consideration of its prehistoric, proto-historic and Classical period maritime activities, contacts, exchange links and trade networks essential.

In this chapter we will look at the emergence and intensification of Arabia's maritime orientation, and summarize a chronology for maritime development across the peninsula as a whole. In doing so, we will develop a number of key themes, having to do with trade, the spread of domesticates, and also contact between ethnolinguistic communities in the Arabian region. We will consider the development of Arabian maritime activity through a number of phases, starting in the Early to Mid-Holocene period with evidence for the first intensively coastally focused communities and the beginnings of maritime trade and seafaring. We will then address the emergence of the first state-level societies in the region, and the role played by Arabian communities in the expansion and intensification of maritime trade during the early Bronze Age, and subsequently the shifts in contact and trade patterns that take place in the middle then late Bronze Age and early Iron Age. We will also address the dispersal of domesticates into the Arabian subcontinent, focusing in particular on those that were likely introduced wholly or partly by sea routes, and their implications for our understanding of maritime activities and patterns of contact. Finally, we also turn to the evidence of language, both in Classical texts and in the present-day distribution of languages in and around the subcontinent. We thus take a multidisciplinary approach, that draws together archaeology with the findings of the archaeological and environmental sciences, Classical studies and historical linguistics.

One of the other key themes that will constitute a focus here is the role of small-scale societies in both the emergence of maritime contact and exchange, and the later more systematic and regular Bronze Age trade that developed in the Red Sea, the Gulf and the Arabian Sea. There has been, as Mark Horton has observed, not only a tendency to focus on textual evidence for trade in the Indian Ocean, but also a marked bias

towards looking at the trade activities of the larger, state-level societies (Horton, 1997). This is perhaps understandable in light of the generally broader variety of evidence ancient states are able to bring to the table – not only historical records, but also greater concentrations of goods, better preservation, depictions of maritime activities in art and iconography, and both longer-term and larger-scale excavation all contribute to a greater wealth of evidence. As we will aim to show, however, there is increasing evidence for both local processes and indigenous communities in early maritime contact and exchange – including contact over long distances.

As researchers whose regional foci have been on the regions adjacent to Arabia – Africa and the Indian subcontinent[1] – our interest in Arabia was initially driven by the need to know more about what separated these two areas, within the context of a wider study into prehistoric patterns of trade and contact in the Indian Ocean. It has rapidly become obvious to each of us that Arabia has a fascinating record that is interesting not just for what it can tell us about wider Indian Ocean connections and exchange patterns, but also for its own sake. We wish to acknowledge the myriad of researchers whose studies we have drawn upon here, and in particular the important syntheses of Cleuziou and Tosi (1989 and 2007) and Potts (1990). We also emphasize that what we have outlined here should be considered a preliminary sketch whose details – and in some cases even broad outlines – will need subsequent further working out. A general comparative chronology for the Arabian peninsula and adjacent regions, which may offer useful reference to the reader is provided in Fig. 1[2]. This provides a summary of the various cultural phases by region for a significant part of the Old World, and indicates the wide geographical and chronological scale at which we have considered Arabia's ancient maritime past in this chapter.

Physical Geography and Paleoecology

Geography is always a key element in the internal evolution and external relations of past societies. As a peninsula with water on three sides, that is separated by the vast Arabian

[1] Details of this work can be found in published articles by Boivin, Fuller and Blench (e.g., Blench, 1996, 1997, 1999, 2003, 2006, 2007b; Fuller, 1998, 1999, 2002, 2003a, 2004, 2005, 2006, 2007b; Blench and MacDonald, 2000; Boivin, 2000, 2004a, 2004b, 2007; Boivin et al., 2007, 2008; Fuller et al., 2001, 2007).
[2] A generally accepted chronological framework remains to be achieved (note the absence of a chart in reviews by Potts, 1990, 1997), and thorough review of the radiometric evidence for the Arabian peninsula is beyond the scope of the present chapter. As noted by Cleuziou (2002) there are chronological discrepancies that derive from matching radiocarbon evidence with historical chronologies, and for the latter there are both short and long chronologies that must be contended with.

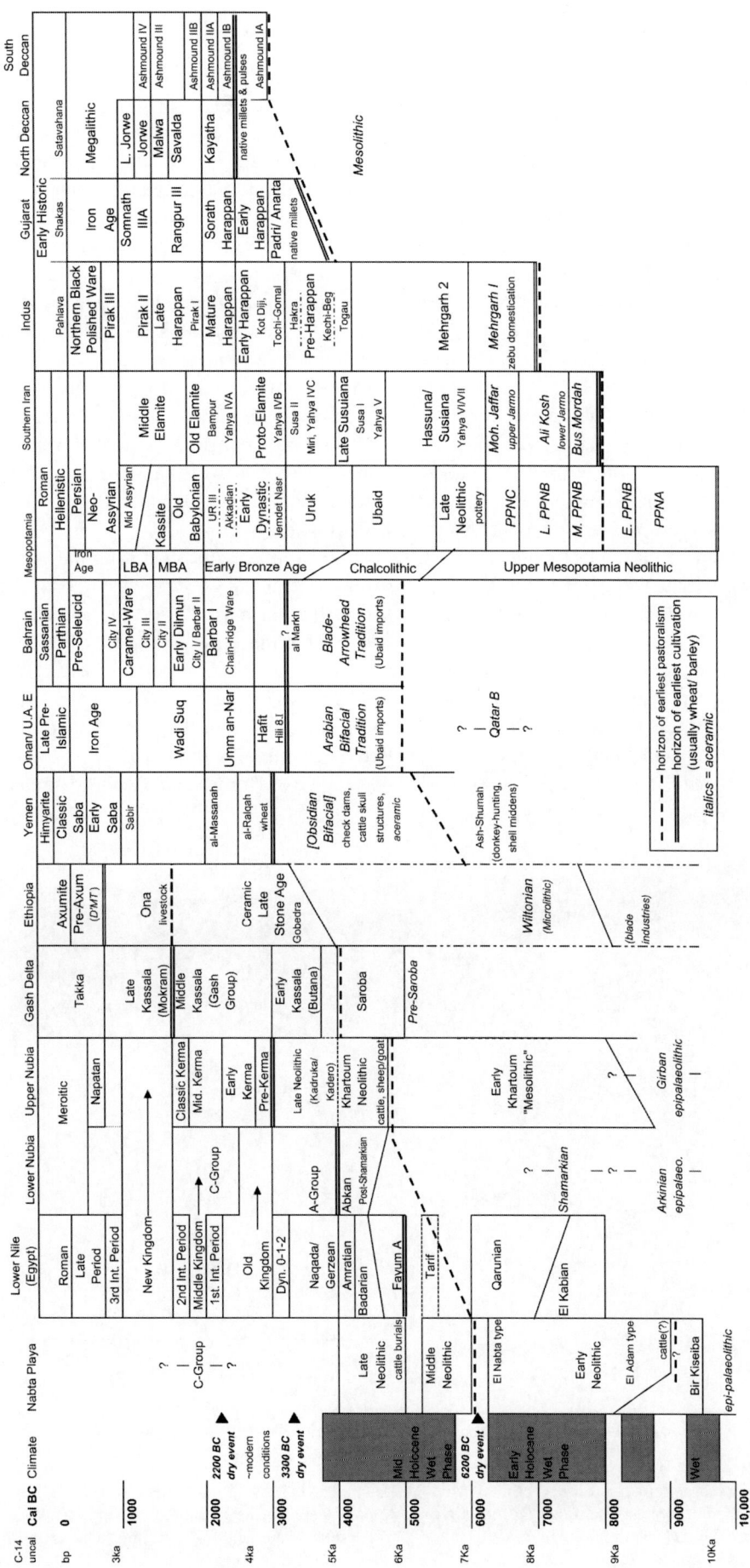

Fig. 1 A general comparative chronology for the Arabian peninsula and adjacent regions. Inferred horizons for the beginnings of pastoralism and plant cultivation are indicated. Divisions between phases and correlations are approximate only; the precise chronology in many regions is open for debate

desert from the populous lands of the 'Fertile Crescent' to the north, including Mesopotamia and the Levant, it is perhaps unsurprising that Arabia should have developed a maritime focus. Another key geographical factor that also impacted the emergence of cultural and maritime patterns in the peninsula is the distribution, both spatially and seasonally, of fresh water, informed by monsoon rainfall patterns, topography and the distribution of rivers and springs.

The wind patterns and currents that are in large part driven by the Indian Ocean monsoon cycle are critical to the issue of maritime contact and trade in the Arabian Sea. In general terms, the Indian Ocean monsoon phenomenon is the result of the differential warming of air over land and sea (Webster and Yang, 1992; Schott and McCreary, 2001; Mitchell, 2005). In the northern summer, from June to September, land warms faster than the ocean, causing Eurasian continental air masses to rise. This creates a low pressure zone, that results in a steady wind blowing toward the land, bringing the moist near-surface air over the oceans with it. The Earth's axial rotation deflects this air such that it blows from the southwest. In the winter, the situation reverses, and the wind blows from the northeast (retreating monsoon). The result is that sailors aware of this consistent pattern could use the monsoon winds to propel their ships from the Red Sea eastwards in the Indian Ocean in the summer, and then back again in the winter. Pliny, in his *Natural History*, described how sailors

exploited this pattern of winds in his description of the spice trade (Miller, 1968).

In general, surface water currents reflect those of wind direction, and the broad patterns are highlighted in Fig. 2. Intimately connected to the monsoon is the Somali current, which carries water north and east along the Somali coast in summer, and reverses this current in the winter. As the Arabian peninsula became part of ever wider interaction zones, the main east–west Indian Ocean currents that reversed direction every 6 months became increasingly relevant by enabling return voyages between Africa/Yemen and India or Southeast Asia beyond. Monsoon currents also cause regions of coastal ocean upwelling (Fig. 2), making certain coastal areas biotically rich, and providing for rich fishing. These are focused near the tip of the Horn of Africa, and along the eastern Yemeni and Omani coasts (Schott and McCreary, 2001: Fig. 8).

In the Red Sea, the wind pattern divides the sea into two main zones, making it very easy to sail out of the Red Sea southwards for most of the year, and correspondingly difficult to sail northwards up it (Facey, 2004). Journeys north of the line roughly between modern-day Jiddah and Aydhab would have been both dangerous and tedious, and it is this fact that likely led to the gradual southward creep of many Red Sea ports over time. Thus the Romans developed the ports of Myos Hormos and Berenice, both quite a way down

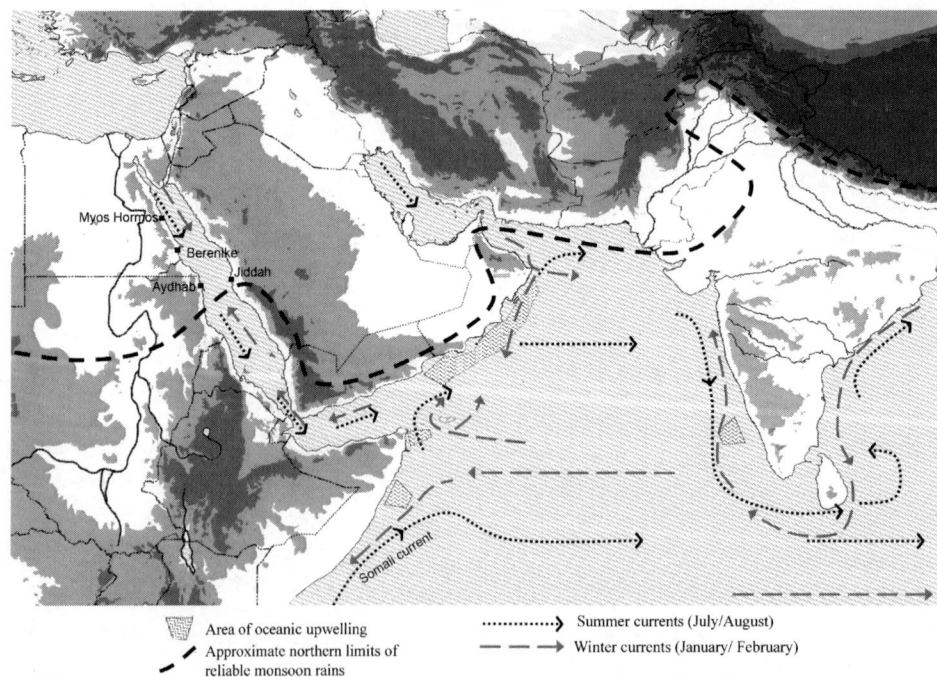

Fig. 2 Arabian peninsula, wind patterns, and broad climatic division of monsoonal region. *Arrows* indicate the major current directions in the summer (*black dotted lines*) and winter (*grey long dashed lines*) (based on Schott and McCreary, 2001; Facey, 2004; Mitchell, 2005). Major summer ocean upwelling regions indicated. Approximate northern limited of significant monsoon rainfall, in which some summer dry cropping is possible, indicated by *thick black dashed line*

the coast from Suez and served by well-maintained routes from the Nile Valley as a means to get around these challenging winds (Facey, 2004: 11).

Rivers are also important from the point of view of how they structure the distribution of neighboring foci of civilization. As observed by Facey (2004), major rivers with associated civilizations (i.e., the Indus and Mesopotamia) flowed into the Gulf, giving these civilizations a more immediate and direct orientation toward the Arabian Sea. By contrast, Egypt and the highland civilization of Ethiopia were separated from the sea by hills and desert, as the Blue Nile and Lower Nile flowed northwards to the Mediterranean. River systems have also been important for enabling the spread of crops. Thus, as we discuss later, boats bringing crops from Africa eastwards would have found riverine communities of farmers ready to try new seeds in South Asia, whereas farming on the northeast African coast was focused on rivers far inland. This contrast may in part help to explain why there seem to have been many more crops that spread early on from Africa and became established in South Asia, rather than vice-versa (Blench, 2003).

Human occupation and maritime activity on the Arabian peninsula have also been shaped by climatic and ecological change. In particular, monsoon intensity has changed in the past, altering summer insolation over Eurasia, linked to orbital precession (Kutzbach, 1981; Ruddiman, 2006). Useful datasets from which to infer climatic changes come from lakes and paleolakes in Arabia and the Qunf Cave stalagmite in southwest Oman (Lézine et al., 1998, 2007; Fleitman et al., 2003; Parker et al., 2004, 2006a, 2006b). These in turn can be correlated with the general patterns recorded in East African lakes (Gasse, 2000), the Eastern Sahara (Hassan, 1997), lakes in the Thar Desert in northwestern India, and Arabian sea sediments that relate to the Indus river discharge (for a recent review of these and other South Asian datasets, see Madella and Fuller, 2006). The correlations between a selection of these sources are shown in Fig. 3, and the location of the sites are plotted in Fig. 4. In broad terms, we see that after the return to glacial-like conditions during the Younger Dryas, during which time deserts were drier, the Early and Middle Holocene period was characterised by higher water/rainfall levels from ca. 9000 BC to 2500 BC, although this was punctuated by numerous dry episodes. The impact of higher rainfall would have been most dramatic in the desert and semi-desert regions, like those in the Sahara and most of the Arabian peninsula. In the Eastern Sahara, for example, increases in rainfall of 150–200 mm, linked to northward shifts in latitudinal vegetation belts of as much as 600 km, are inferred (Neumann, 1989; Hassan, 1997). This increase shifted the transition from savannah to desert (the boundary of monsoon climate, as indicated in Fig. 2) from the central Sudan to southwestern Egypt, allowing colonization of the southern Sahara by groups of hunter-

gatherer-fishers of the early ceramic horizon. Similar changes occurred in Arabia, with evidence for human settlement associated with paleolakes of inner Arabia (McClure, 1976; Masry, 1997; Lézine et al., 1998), and rich settlement evidence associated with the wadi systems of interior Yemen, such as Wadi Sana (McCorriston and Martin, 2009). The richer vegetation of the semi-desert and savannah interior of Early to Mid-Holocene Arabia supported a more diverse and extensive fauna that could in turn support hunters, as well as good habitat for cattle and caprine herding, which began sometime after 6000 BC (see below).

Nevertheless, the Early and Mid Holocene were punctuated by dry episodes when the desert would have expanded. The first focused on ca. 6800 BC, and is particularly marked in the Al-Hawa data from Yemen (see Fig. 3). This arid event appears merged with the later dry event of 6200–6000 BC in the East Africa datasets, but it is clear from the Thar and Arabian evidence that there was a recovery of rainfall in between. The major dry episode of 6200–6000 BC now appears to have been a more or less global climatic event, reflected also in the Greenland ice-cores, as well as East African and South Asian datasets (Alley et al., 1997; Gasse, 2000; Alley and Ágústdóttir, 2005; Madella and Fuller, 2006; Kobashi et al., 2007). In the Al-Hawa data, for example, lower lake levels are reached closer to 5900 BC, a time when Hassan (1997) infers a peak in aridity for the Egyptian desert. There seems little basis to conclude, as Potts (2008) does, that Arabia was significantly more attractive for human settlement than the Levant or Mesopotamia during this arid interval.

During the subsequent mid-Holocene there were additional dry episodes, in particular at 4300 BC, perhaps 3300/3200 BC and the late third millennium (the 2200 BC event). Of particular relevance to eastern Arabia is a more localized dry phase from 3800 BC. Evidence from northern Oman, United Arab Emirates and the An-Nafud region beginning around this time suggests a particularly marked period of aridity and decline in settlement evidence, which has been called the "Dark Millennium" (Uerpmann, 2003). This period was first postulated on the basis of the poor evidence for human occupation, except for a few seasonal coastal sites, during this post-Ubaid period (Potts, 1993; Uerpmann, 2003). The recent paleoenvironmental reconstruction from the Awafi paleolake in United Arab Emirates indicates two peaks in aridity, at ca. 3900 BC and 3200 BC (Parker et al., 2006). While these downward trends are evident in the Qunf speleothem (see Fig. 3), it is also clear that this period is not recorded as arid further afield in East Africa or South Asia, nor probably in southwest Arabia. The impact of vegetation in eastern Arabia is suggested by first a sharp decline in woody vegetation followed by its near disappearance, as inferred from Awafi phytoliths ratios (Parker et al. 2004). The absence of paleolake stands in An-Nafud at this

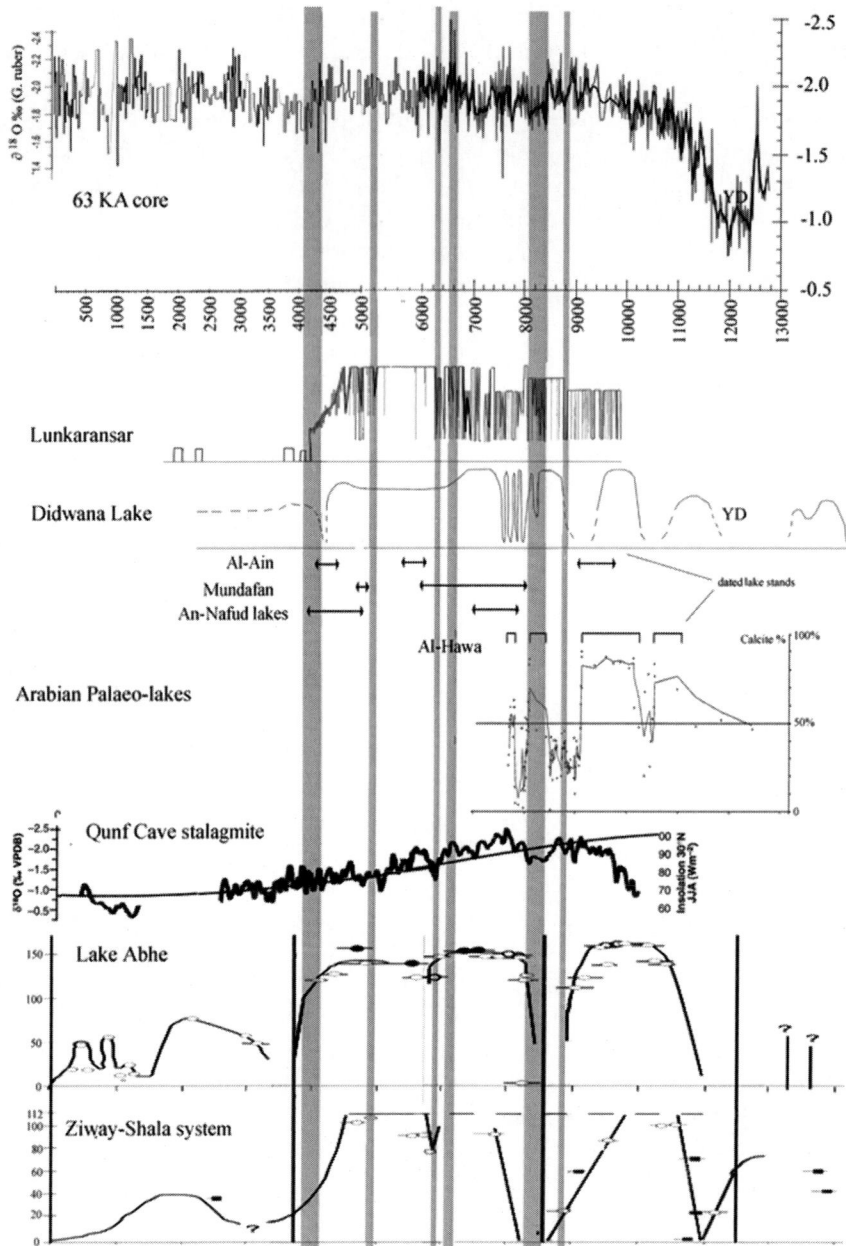

Fig. 3 Correlation of paleoclimatic proxies for the Arabian peninsula, northwestern South Asia and East Africa. From *top* to *bottom*: O-18 isotopic variation from Pakistan continental margin, core 63 KA (after Staubwasser et al., 2002, 2003); lake level data from Lunkaransar (after Enzel et al., 1999); lake levels from Didwana lake (after Wasson et al., 1984); dated high lake stands from selected Arabian paleolakes (after Lézine et al., 1998); lake level proxy calcite data from Al-Hawa paleo-lake, Yemen (after Lézine et al., 2007); O-18 isotopic record from the Qunf Cave stalagmite (after Fleitman et al., 2003); lake level data from Abhe and Ziway Shala in Ethiopia (after Gasse, 2000)

time suggests aridity was particularly marked in northern and eastern Arabia.

Agriculture and thus much Holocene occupation in the Arabian peninsula has been shaped by environmental and hydrological conditions. For most of the peninsula, rainfall has been insufficient to support agriculture directly. However, subsurface water reservoirs (aquifers), which are slowly topped up by rains, provide water at natural seepages, which form oases, and can be tapped by wells (Edens, 1993; Blau, 1999).

Much traditional oasis agriculture is thus based on tapping these below ground sources, and advances in the methods for doing so have been important to the development of agriculture in the subcontinent. Of particular importance was the development of *falaj* (or *qanat*) irrigation systems by the early Iron Age (see discussion below). In some mountain areas, however, sufficient water can be derived from run-off of the limited rains, derived from the summer monsoon. Thus two regimes, of aquifers and run-off, define

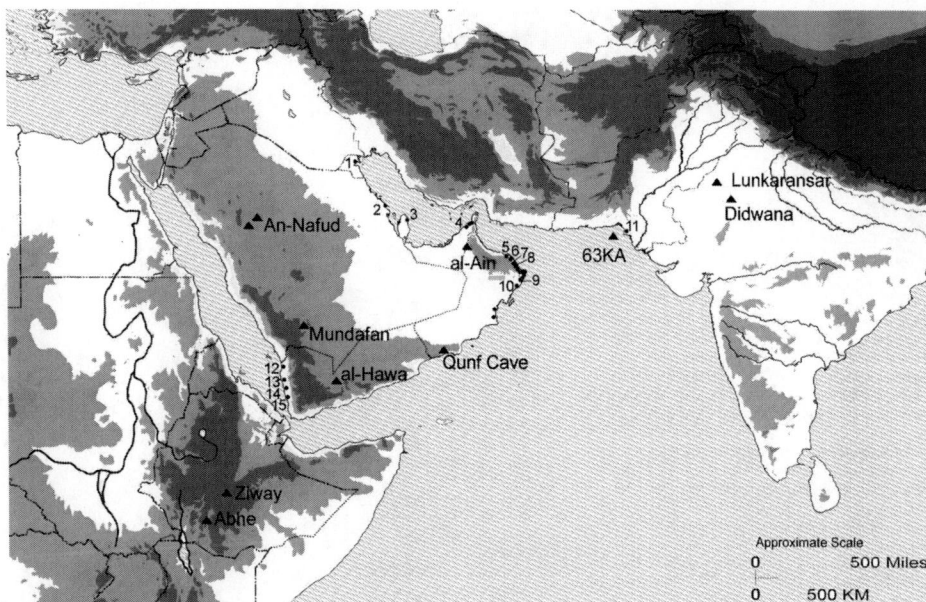

Fig. 4 Map showing the distribution of major paleoclimatic datasets discussed in the text (indicated by *triangles*; see also Fig. 3, above), and the general distribution of Mid-Holocene shell midden sites (indicated by *circles*). Sites with chronometric evidence (see Fig. 5) are numbered: 1. H3; Kuwait; 2. Dosariyah, Saudi Arabia; 3. Khor D & Khor FB, Qatar; 4. ar-Ramlah 6 (RA 6), UAE; 5. Ras al-Hamrah sites and Saruq, Oman; 6. Wadi Wuttaya (WW), Oman; 7. Bandar Khayran, Oman; 8. Daghmar, Oman; 9. Suwayh (SWY-11), Oman; 10. SAQ-1, Oman; 11. Daun-1, Pakistan; 12. Jizan area shell middens; 13. Wadi Sardud; 14. Hodeidah area shell middens; 15. Ash-Shumah, Yemen

the potential centers of agriculture in Arabia. In the Gulf, the two Bronze Age civilizations can be related to development of each of these (Edens, 1993), with *Dilmun* (Bahrain and adjacent) focused on aquifers and oases, while *Magan* (Oman and adjacent) drew upon run-off. The proximity of these centers of potential agriculture to the Gulf may help to account for the early extensive development of maritime trade along the eastern peninsular littoral. Meanwhile, in interior Yemen, zones of both these types were important in the economy of the classic Sabaean civilization (Robin, 2002; Wilkinson, 2002). However, in contrast to the situation the Gulf side, in Yemen these agricultural zones were on the inside of the mountains, oriented towards the desert.

In addition to seeing important climatic alterations, the Early and Mid-Holocene was also a period of major coastline change, with sea levels rising at the end of the Late Pleistocene glacial melt. In the relatively shallow Gulf, sea-level rise had dramatic consequences (Lambeck, 1996), with sea-levels reaching modern levels at 5400 BC, and their highest point at 5000 BC, creating what we now recognize as the Gulf (see Fig. 5). Rising sea levels had less impact on the Red Sea, which is based on a much deeper rift (part of the African rift valley tectonic fault complex). From the point of view of human populations in Arabia, the rising sea-levels, which subsequently fell slightly over the mid-Holocene, together with the aridification of the inland deserts, meant that populations would have become increasingly restricted to a narrow coastal zone near the modern coastline. Sea level rise can also be expected to have had a taphonomic effect on

sites (Bailey, 2004). It may explain why evidence is basically lacking for coastal occupation in the Pleistocene and Early Holocene prior to the dry event of ca. 6200–6000 BC. Other taphonomic factors active in the Arabian Sea and its subsidiary water bodies include coastal sedimentation, river shift, Late Holocene sea level fall, erosion, and tectonic activity (Shroder, 1993; Chandramohan et al., 2001; Mathur et al., 2004; Sanil Kumar et al., 2006; Gaur and Sundaresh, 2007; Shajan et al., 2008).

The First Ichthyophagi and the Emergence of Seafaring

The earliest evidence for maritime activity in the Arabian peninsula occurs in the form of shell middens, which appear roughly simultaneously at sites around the peninsular littoral in the seventh millennium BC (Fig. 5), and indicate the presence of 'Ichthyophagi' (primitive 'Fish-Eaters', as described in Classical sources like the *Periplus of the Erythraean Sea* [see also Biagi et al., 1984; Horton, 1997; Beech, 2004]). Shell midden and coastal sites like Suwayh (SWY-11) and Wati Wuttaya (WW), an inland site with shells, both in the Gulf of Oman, have dates that calibrate back to 5900 BC (Biagi, 1994, 2006; Biagi and Nisbet, 2006). Sites like Ras al-Hamra (RH-7) and Dosariyah, in Saudi Arabia, begin in the mid sixth millennium BC, suggesting a later emergence for maritime exploitation to the north. These eastern littoral

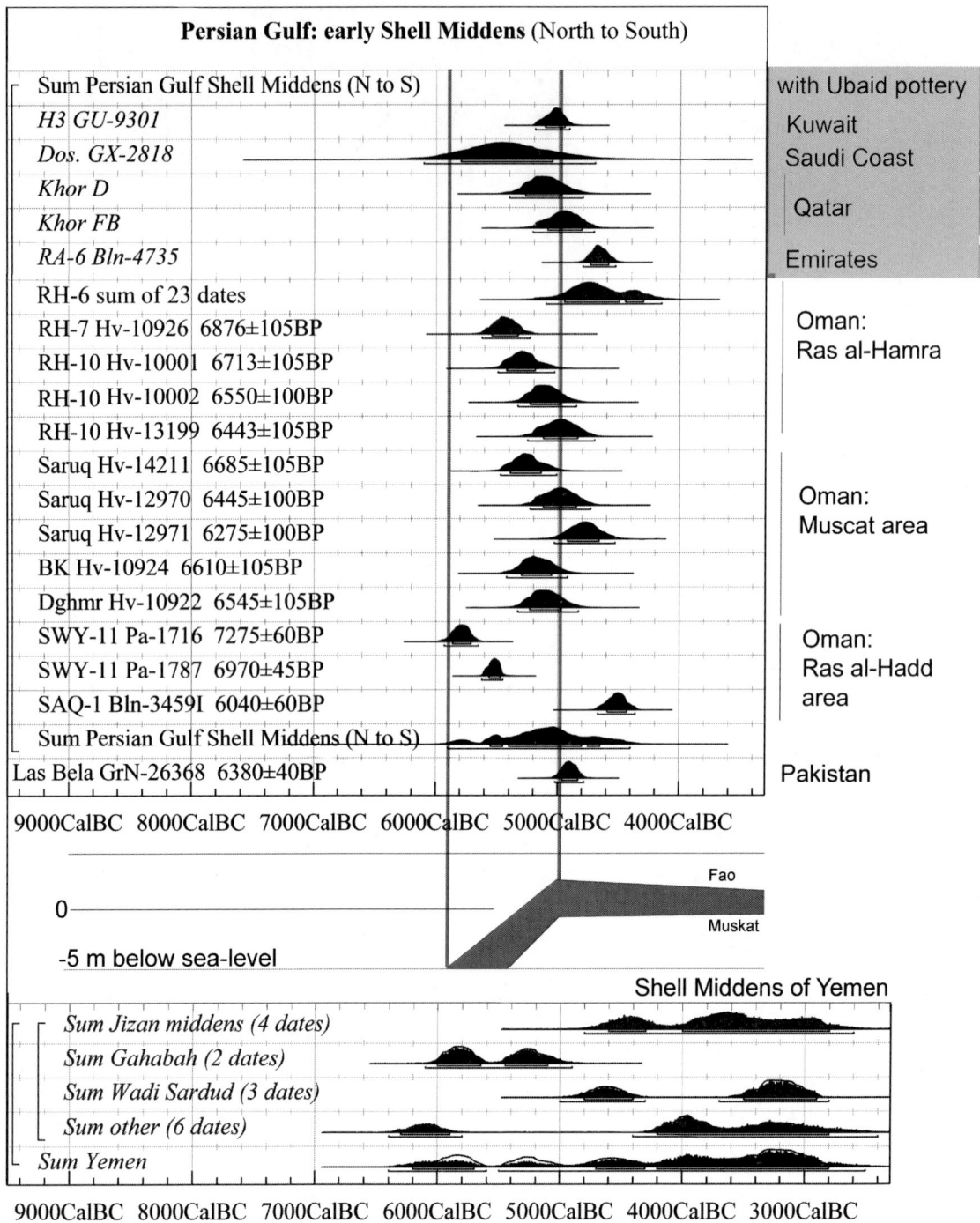

Fig. 5 The probability distribution of calendrical ages of representative early Arabian shell middens (>6000bp) compared with Persian Gulf sea level rise (after Lambeck 1996, converted to a calibrated time scale: the gray curve indicates inferred variation in sea level rise between Fao and Muscat). As almost all dates were on shells, fishbones (and some charcoal may derive from mangrove), marine reservoir corrections were used, with ΔR derived from the Queens University Belfast database (http://intcal.qub. ac.uk/marine/). For Persian Gulf dates, ΔR = 230 ± 65, was derived from 3 datasets (map # 256, 581, 584), while for Red Sea dates, ΔR = 188 ± 73, was derived from 7 datasets (map # 253, 582–3, 585–7). Calibrations were performed with OxCal 3.10 (Bronk Ramsey 2005). Dates for RH5 from Biagi and Nisbet 1992; other Persian Gulf dates from Biagi 1994; 2006. Data on Yemen shell middens from Edens and Wilkinson (1998) or Durrani (2005). For locations of sites, see Figure 4

sites have a significant food-producing component, with evidence for both sheep/goat and cattle bones consistently found from the earliest strata (Biagi et al., 1984; Biagi, 2006; Potts, 2008).

In the Red Sea, sites bearing Arabian Bifacial Tradition technology dating back to the late seventh millennium BC have been found on the Tihama plain, and are also frequently associated with shell middens (Tosi, 1985, 1986a; Edens and

Wilkinson, 1998; Phillips, 1998; Cattani and Bökönyi, 2002; Durrani, 2005; Khalidi, 2007; Munro and Wilkinson, 2007). Few have been excavated, but it has been suggested that the decrease in bifacial elements in the lithic toolkit may mark a distinctive coastal adaptation (Uerpmann, 1992). Sites are frequently 5–10 km inland, and associated with exploitation of mangrove environments. As on the eastern Arabian coast, the sense is of a variety of diverse economic strategies, focused on shellfish gathering and also fishing, but also incorporating hunting activities (at Ash Shumah, hunted wild donkeys make up around 90% of the faunal assemblage [Cattani and Bökönyi, 2002]). As on the eastern Arabian littoral, domesticates are present from an early date (the sixth and perhaps also seventh millennia), and indicate a mixed and by no means strictly hunting and gathering lifestyle. Shell midden dates continue up to the later fourth millennium BC and beyond, indicating a fairly stable economic system based on a mixture of hunting, herding and shell fish collection.

The earliest evidence for seafaring activity in the peninsula also appears roughly simultaneously in the Gulf and Red Sea, some thousand years later, in the sixth millennium

BC. Such evidence also attests the first movements of material objects across the sea, probably as a result of local and regional exchange activity. Evidence for maritime exchange is better for the Gulf than the Red Sea, and may indicate more active exchange networks in this geographically more favorable arena (although patterns of archaeological focus are also certainly relevant). The evidence in the Gulf is in the form of Ubaid pottery, from Mesopotamia, which is introduced in the late sixth millennium BC onto Neolithic sites of the Arabian Bifacial Tradition (Oates et al., 1977; Potts, 1990; Roaf and Galbraith, 1994). Ubaid pottery has now been found at over 60 Arabian Neolithic sites (Carter, 2006), usually but not always located on the coast (as well as a number of off-shore islands – for example, Dalma and Bahrain), from Ras al-Sabiyah in the north to the approach of the Straits of Hormuz in the south (Phillips 2002; Fig. 6). While a number of earlier interpretations of the Ubaid pottery – which archaeometric analyses demonstrate was manufactured in Mesopotamia (Oates et al., 1977; Roaf and Galbraith, 1994) – read it as an indication of Mesopotamian contact with Gulf inhabitants, or even the remnants of Mesopotamian maritime expeditions (e.g., Oates et al., 1977; Potts, 1990;

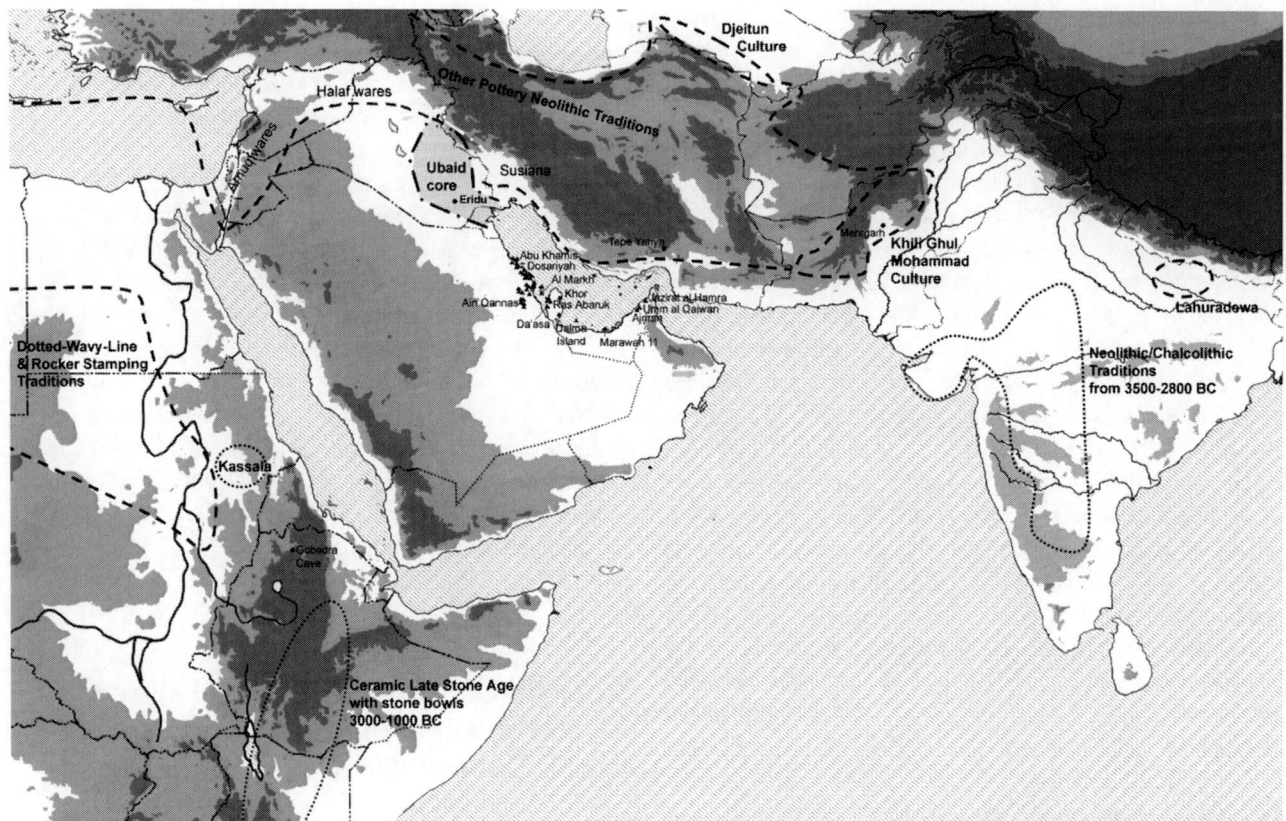

Fig. 6 Finds of Ubaid ceramics in the Gulf (after Crawford, 1998; Carter, 2006), in relation to the core range of Ubaid pottery in Mesopotamia and other early ceramic traditions. The *dashed line* indicates the extent of early ceramic traditions of ca. 6000 BC prior to the development of Ubaid. *Dotted areas* indicate important regional developments where ceramics were later, beginning between 3500 and 2500 BC (East African stone bowl traditions: Barnett, 1999; South Asian traditions: Fuller, 2006; Sahara-Sudan traditions: Jesse, 2003; Kasalla: Sadr, 1991)

Lawler, 2002), increasing evidence suggests a potentially more active role for Arabian Neolithic peoples in moving the ceramics (Roaf and Galbraith, 1994; Vogt, 1994; Kallweit, 2002; Cleuziou, 2003; Carter, 2006). Robert Carter has emphasized that the Ubaid pottery is an intrusive element on sites whose material culture is otherwise overwhelmingly Neolithic and Arabian (Carter, 2006; see also Roaf and Galbraith), and suggestive of mixed hunting-gathering, fishing and pastoral activities (Beech, 2002, 2003, 2004; Kallweit, 2002; Beech and al-Husaini M, 2005). Burial patterns at the site of UAQ-2 on the UAE shoreline, where a cemetery with Ubaid ceramics appears to be that of a local population (Phillips, 2002), emphasize the indigenous flavor of Ubaid-related sites in the Gulf (see also Vogt, 1994). Carter has drawn upon available evidence to suggest the operation of local exchange networks in which Ubaid ceramics featured as prestige goods, possibly exchanged in ceremonial contexts that played an important role in the negotiation of power and status within and between groups (Carter, 2006).

Also circulated and exchanged in the Gulf's Neolithic maritime exchange economy were items like bitumen beads, stone and stone artifacts (especially flint and obsidian), and probably also pearls, shell and mother of pearl jewellery and beads, ochre and a wide range of perishable goods (e.g., hides, fish [both fresh and dried], and livestock, including cattle) (Flavin and Shepherd, 1994; Beech, 2002, 2004; Phillips, 2002; Beech and al-Husaini M, 2005; Connan et al., 2005; Carter, 2006). Some sites, like H3 in Kuwait, also show evidence for some degree of craft specialization (Beech and al-Husaini M, 2005), and increasing degrees of sedentism are suggested by various lines of evidence, including more substantial structures, seen most notably at the island site of Marawah in the UAE (Anonymous, 2004; Beech et al., 2005) and at H3 (Carter and Crawford, 2003). Maritime movement was apparently by reed-built boat. Excavations at H3 have unearthed evidence of what may well be the world's earliest boat remains (Lawler, 2002), consisting of over 50 pieces of bituminous amalgam, mostly with reed-impressions and/or barnacle encrustations (Carter, 2006). These accompany a ceramic model of a reed-bundle boat and, especially notable, a painted disc depicting a sailing boat, demonstrating employment of the sail by the Ubaid 3 period.

In the Red Sea meanwhile, initial evidence for seafaring activity comes primarily from source studies of obsidian artifacts (Francaviglia, 1989; Zarins, 1990, 1996; Khalidi, 2009). Obsidian first appears on Tihama sites beginning in the sixth millennium BC, and indicates direct or indirect contact with source traps in the central or southern highlands of Yemen, and/or the Horn of Africa beginning at this time. Preliminary source studies suggest that much of the Tihama obsidian may have come from the Eritrean/Ethiopian highlands (Zarins, 1990, 1996). The impression of a maritime origin is strength-

ened by recent coastal survey indicating that obsidian densities are highest at sites right on the coastline and decrease at sites along the river deltas leading to the coastal interior (Durrani, 2005; Khalidi, 2007, 2009).

There are probably parallels between trade in exotic Mesopotamian ceramics in the Gulf and the putative trade in exotic obsidians in southwestern Arabia. As Carter argued for the Ubaid ceramics, these materials likely featured as prestige goods whose acquisition and redistribution conferred status in the context of gradually emerging hierarchies. Both intra-group (gender, age) and inter-group (lineage, kin-group) differences may have been increasingly articulated. In Oman, such processes perhaps climaxed in the highly visible Hafit-type cairn burials of the late 4th millennium BC (Cleuziou and Tosi, 1997). Fourth millennium BC communities in Oman used boats to fish large deep water species like tuna and jacks (Beech, 2004; Biagi and Nisbet, 2006), and appear to have led a more sedentary than mobile existence – they were probably seasonally sedentary (Charpentier, 1996, 2002; Uerpmann, 2003; Biagi and Nisbet, 2006). In the Red Sea, increasing complexity and intensification in exchange in prestige goods is most visible with the Egyptians, who began to participate in obsidian trade in the Predynastic period (5000–3100 BC), when silver, lapis lazuli, turquoise, galena, malachite, svenite, specular iron and 'resins', as well, undoubtedly, as perishable items, were also traded, possibly via Red Sea routes (Zarins, 1989, 1996). Maritime trade appears to date back to the Naqada I period (ca. 4000–3500 BC), and to have become well established by the Naqada II period (3500–3200 BC). Obsidian objects are initially small – simple blades and flakes, or beads, for example – and unlikely to have been the focus of trade. The Egyptians travelled to the Red Sea via the Wadi Hammamat, a desert corridor where depictions of ships have been identified. The boats – perhaps papyrus or reed, but according to Ward probably already wooden sewn types (Ward, 2006) – were dismantled and dragged overland to the coast. Based on the evidence already outlined for obsidian trade networks in the southern Red Sea, it is likely that the Egyptians were simply tapping into an existing exchange network (Zarins, 1996) and that trade over long distances was still indirect (Burstein, 2002; Kitchen, 2002).

Expansion and Intensification of Maritime Contact and Exchange in the Early Bronze Age (ca. 3500–2000 BC)

In the mid-fourth millennium BC, the emergence of the first major state-level civilizations of the Old World in a number of regions bordering the Arabian peninsula impacted upon the development of maritime activity in the region (see Fig. 7).

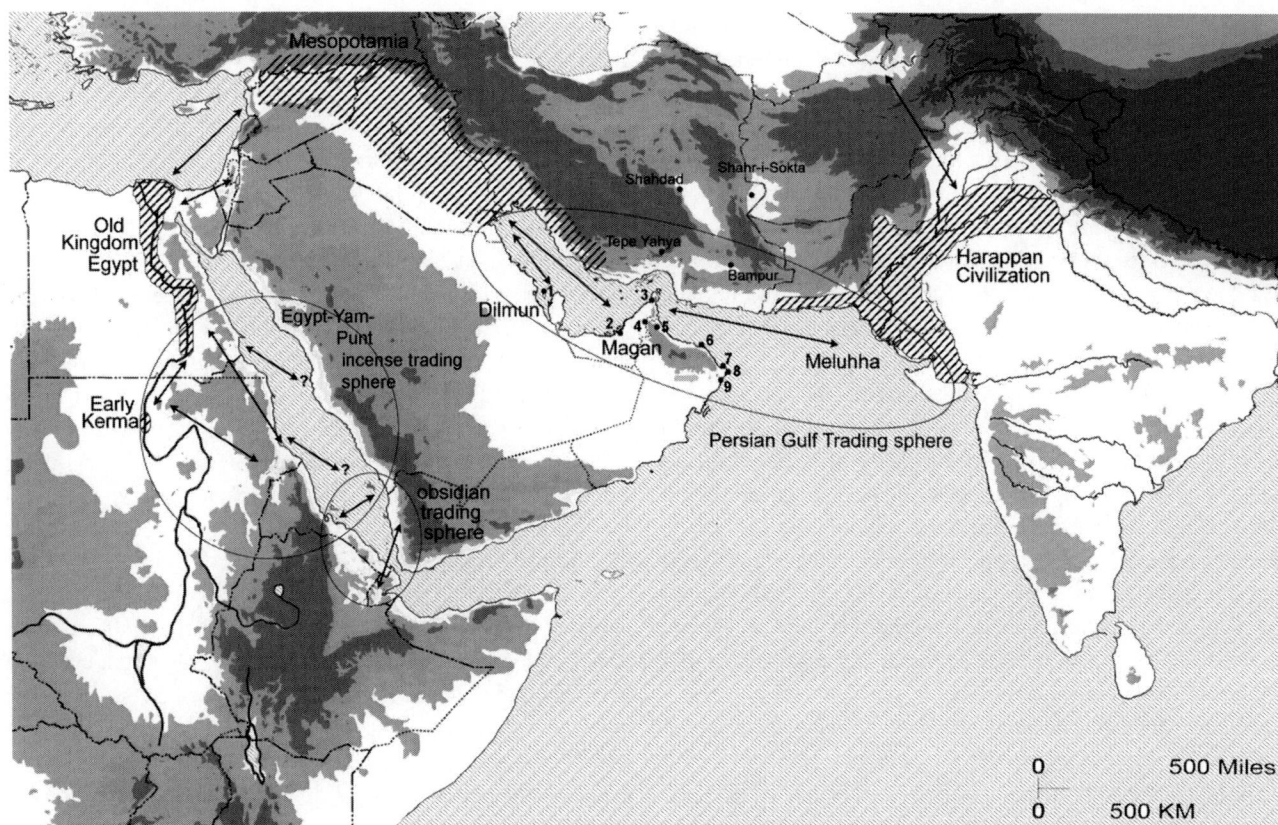

Fig. 7 Third millennium trading spheres map with selected sites indicated: 1. Barbar; 2. Umm-an-Nar; 3. Tall Abraq; 4. Hili; 5. Wadi Suq; 6. Ras al-Hamra; 7. Ras al-Hadd; 8. Ras al-Jinz; 9. as-Suwayh

In Arabia, we also track the emergence of more intensive agricultural production and new modes of social organization at this time. Linked to this are signs of both increasingly intensive, and increasingly far-reaching maritime trade activities. While the urbanized states are clearly major players in this trade, there are also, as we shall see, intriguing indications that coastal communities and local merchants had an important role to play. In addition, date palm-focused oasis settlements, together with donkeys for transport, likely supported increased movement through the interior, and to and from the coast, beginning at this time.

The Red Sea at the outset of the Bronze Age is dominated by the Egyptian record, which now provides more direct evidence on maritime activities in the region, in the form of iconographic and textual records, as well as actual preserved boat remains. The middle of the fourth millennium BC probably saw the Egyptians shift from reed or papyrus to wooden boats for Red Sea transport, as well as the introduction of a form of the sail by the Late Predynastic period, ca. 3100 BC (Stieglitz, 1984; Fabre, 2005: 89; Ward, 2006). Egyptian rulers continued to promote long-distance trade for prestige and political purposes (Zarins, 1996) and large watercraft appear to have had a key role to play in

social competition. Spectacular wooden boat burials are found in Egyptian funerary contexts from the First Dynasty (ca. 3000 BC), and their prestige value likely derived from the resources, technical skill and craft specialists necessary to build them, and their important role in acquiring exotic goods and controlling regional exchange networks (Arnold, 1995; Ward, 2006). Improvements in wood sources, building techniques and sail rigging all appear to have contributed to the construction of larger boats increasingly well suited to the open sea in the third millennium BC (Faulkner, 1941; Fabre, 2005: 89–92). It is probably no coincidence that it is in this period, during the reign of Sahure, that the first sea voyage to *Punt* is recorded (Faulkner, 1941; Kitchen, 1993; Harvey, 2003).

The Egyptians referred to *Punt* as a 'mining region' and imported a variety of products from it, including incense (frankincense or myrrh), electrum, staves (wood, perhaps ebony), pygmies, and probably slaves, as well as exotic animals and leopard skins (Lucas, 1930, 1937; Dixon, 1969; Kitchen, 1993; Phillips, 1997; Meeks, 2003). The location of *Punt* has long been a source of debate, and is directly relevant to discerning patterns of maritime trade and exchange in the Red Sea region. Most scholars are agreed that *Punt* lay southwards and was reached by the Red Sea, and that it was the main source of incense. The current orthodoxy of Egyptological opinion is generally that *Punt* was situated in eastern Sudan and northern Eritrea (Kitchen, 1993, 2002; Mitchell, 2005: 78). In the second millennium BC mortuary temple of Queen Hatshepsut, who boasted about the expedition she sent to *Punt*, the presence in relief frescoes of stilted huts, as well as animals like the giraffe and the rhinoceros, has been taken to support an African location. An alternate theory situates *Punt* along the eastern side of the Red Sea, in Arabia and Yemen (Meeks, 2003).

While many have seen textual records of contact with *Punt*, and archaeological finds of exotica, as an indication that Egypt itself was an active maritime player in the Red Sea, it is also possible that Egypt's role has been overemphasized relative to that of the smaller scale communities in the Red Sea region. Both Burstein and Kitchen have argued convincingly that Egypt in fact undertook relatively limited forays into the Red Sea, and that local trade networks were responsible for much of the movement of goods seen in the archaeological record (Burstein, 2002; Kitchen, 2002). Depictions like that of two 18th Dynasty Theban officials, which show Egyptians meeting laden animal skin boats with triangular sails from the Land of *Punt* somewhere in a desert, suggest that peoples living along the southern Red Sea were regularly involved in moving commodities (Bradbury, 1996; Meeks, 2003: 61–63; Mitchell, 2005: 79). While significantly-sized watercraft are clearly present and important in Egypt from the very start of the Dynastic period, we have seen that their value was partly symbolic (Fabre, 2005; Ward, 2006), suggesting both their control by elites and the relative expense and rarity of expeditions like that undertaken by Hatshepsut. This scenario of indirect trade would seem to be supported by archaeological evidence for more intensive exchange across the Red Sea, between small-scale societies, beginning in the third millennium BC (Khalidi, 2007, 2009). Such regional exchange by local maritime-oriented communities may have increased in scale not only to meet the demands of increasingly powerful elites of Egypt, but also the needs of the increasingly hierarchical societies emerging in the hilly interior of Yemen, where terrace systems and possibly megaliths began to be constructed by this period (Edens and Wilkinson, 1998).

Seafaring and maritime exchange also intensify in the Gulf during the early Bronze Age, and move out into the wider Arabian Sea (Oppenheim, 1954; Lamberg-Karlovsky, 1972; Edens, 1993; Possehl, 1996; Ratnagar, 2001, 2004; Ray, 2003), although here too the activities of smaller-scale societies must also be taken into account. Third millennium BC Mesopotamian textual records clearly identify sea trade with and between the regions of *Dilmun, Magan* (*Makkan*) and *Meluhha*, linked through descriptions and archaeological finds to the real regions of present-day

Bahrain (and/or variously Falaika and the eastern Arabian littoral), the Oman peninsula, and the Indus Valley and Gujarat, respectively. These extended sea routes were complemented by riverine and overland routes that connected the coastal sites and ports to a rich array of inland sites both close to and far from the coast, and contributed to the formation of what Possehl has referred to as a Middle Asian Interaction Sphere (Possehl, 1996, 2002, 2007).

Overall, as Edens has outlined, trade in the eastern Arabian peninsular region in the first half of the third millennium BC was relatively small in bulk and centered on a variety of luxuries (Edens, 1992). For example, small quantities of Mesopotamian pottery and other small finds made their way to Arabian grave and settlement sites (During Caspers, 1971; Edens, 1992; Potts, 1993; Vogt, 1996), and Harappan small finds, shell from India, as well as perhaps cardamom from the Nilgiris reached Mesopotamia by this time (Ratnagar, 2004; Keay, 2006). During this period, copper nonetheless emerged already as a key traded item in the Gulf. Arabian oasis settlements like Hili engaged in the production and exchange of copper (Cleuziou, 1996). Coastal Arabian communities at this time – as at Ras al-Jins, Ras al-Hadd, and Ras Shiyah, for example – also began to import and use copper, as well, apparently, as sesame oil and minor quantities of exotic Mesopotamian pottery (Potts, 1993; Cleuziou, 1996). They also began to catch more deep-water species of fish (Beech, 2004), possibly indicating greater maritime proficiency. Cleuziou hypothesizes that the fish that they continued to process in various ways now began to see production for larger scale export (Cleuziou, 1996). Such communities also produced a range of raw materials (like shell and fertilizer, the latter vital to palm-grove cultivation) and goods (jewellery, beads, fish hooks, khol containers, etc.) that were traded both locally and further afield (Charpentier, 1996, 2002). Interaction with southeastern Iran is hinted at by ceramic and copper production parallels that indicate technological borrowings (Cleuziou, 1996; Cleuziou and Méry, 2002). South Asian materials are also attested on Oman sites, reinforcing the impression that maritime trade with the still emergent Indus Valley civilization began at an earlier date than we currently have good evidence for (During Caspers, 1979; Parpola, 1977; Cleuziou, 1992; Ratnagar, 2004). Nonetheless, during this period, trade between Oman and Mesopotamia appears to have been mediated via Dilmun (Edens, 1992; Cleuziou and Méry, 2002), which now moved towards urbanism and statehood (Cleuziou, 1996).

The second half of the third millennium saw many important changes to trade and its socio-political context, following the emergence, by 2750 BC, of the Umm an-Nar cultural entity in Oman (Tosi, 1986b; Cleuziou, 1996; Cleuziou and Méry, 2002) and the Mature Harappan in the Indus valley (Possehl, 2002). After a gradual intensification of trade (Edens, 1992; Vogt, 1996), with *Dilmun* continuing to act as an intermediary, we see evidence for the emergence of direct contacts between the main trade participants (Edens, 1992). This is subsequently and famously reflected in Sargon of Akkad's (2334–2270 BC) boast that he has moored in his harbor ships from or destined for *Meluhha*, *Makkan* and *Dilmun* (Oppenheim, 1954). A lesser known late Sargonic tablet (datable to ca. 2200 BC) also mentions a man with an Akkadian name entitled "the holder of a *Meluhha* ship", while an Akkadian cylinder seal bears the inscription "Su-ilisu, *Meluhha* interpreter" (Parpola, 1977). Indus seals begin to appear in the Mesopotamian archaeological record at this time (Parpola, 1977; Possehl, 2002). Ratnagar notes the paucity of material evidence for any Mesopotamian presence in Oman during this period, however, despite the military incursions by Akkadian kings (Ratnagar, 2004). Mesopotamian pottery is evidently no longer desired by Oman communities for use in burial contexts after around 2600 BC, and the subsequent use of Mesopotamian jars at coastal settlements only continues until the beginning of the Akkadian period (Cleuziou and Méry, 2002).

Coastal societies in Oman were nonetheless heavily dependent on trade in the second half of the third millennium BC (Cleuziou, 1996; Cleuziou and Tosi, 1994, 1997). During the Umm an-Nar period, Oman seems to have been very strongly linked to southeastern Iran and to the Indus Valley (Edens, 1993; Potts, 1994; Vogt, 1996; Cleuziou and Méry, 2002). Items of supposed Harappan provenience or inspiration, meanwhile, are found from all over the Oman peninsula, and include a wide range of items, suggesting the import into *Magan* of both basic commodities and luxury items (Vogt, 1996). Along with carnelian (some etched), combs, shell and shell objects, metal and metal objects, seals, and weights of more or less clear Harappan provenience (Edens, 1993; Possehl, 1996; Ratnagar, 2004; Vogt, 1996), there is also a rich testimony of ceramic sherds, in particular of the widely distributed Indus black-slipped ware (Cleuziou, 1992; Potts, 1993, 1994; Vogt, 1996; Cleuziou and Méry, 2002). The potsherds of this black-slipped ware belong to a highly standardized category of large-volumed vessels that appear to be storage jars. The black-slipped jars are more common on the coast than the interior, and particularly on the coast of Oman rather than the Gulf (Cleuziou and Méry, 2002). Their mineralogy supports an Indus origin (Cleuziou and Méry, 2002).

Trade in the final centuries of the third millennium BC saw an important shift from the predominantly luxury-oriented system that probably extended back several millennia, to a mixed trade that began to include necessities (Edens, 1992). Trade also became less direct at this time, with significantly less evidence for first-hand interaction between Mesopotamia and India. Ur III documents from the site of Ur record the activities of seafaring merchants who took textiles, wool, leather, sesame oil and barley to *Magan*, which

seemed to develop as a primary trading center (Oppenheim, 1954). Mesopotamian ceramics and other artifacts are found in Oman from this time period, though primarily on the coast (Vogt, 1996). The focus of Mesopotamian merchants seems to have been on obtaining copper, which came to be used for increasingly utilitarian purposes during the Akkadian period and after (Edens, 1992). Available evidence suggests that Oman's interaction and trade with the Harappan civilization increased in the last few centuries of the third millennium BC (Vogt, 1996). At this time, the evidence for direct contact with the Harappan culture is better than for the subsequent period, though it is focused on coastal sites (Edens, 1993). Charpentier has argued that coastal communities played an intermediary role between inland and Indus civilization populations, acquiring and supplying goods to both parties (Charpentier, 1996).

Disruption, Transformation and Intensification of Trading Spheres in the Middle and Late Bronze Age (2000–ca. 1200 BC)

The Early to Middle Bronze Age transition represents a period of political instability and upheaval across much of the area under consideration here. This is seen in Egypt (Baines and Malek, 1980; Connah, 2005), southwest Asia (Matthews, 2005), and the Indus (Possehl, 2002). Some so-called peripheral zones like Arabia also exhibit changes, highlighting their close relations with 'cores'. In Oman, for example, there is an abrupt shift from the Umm an-Nar to the Wadi-Suq cultural phase (Cleuziou, 1996; Potts, 1997). The broader regional trend towards instability corresponds to a period of climatic shifts towards drier and more volatile conditions in the region, starting with the 2200 BC event (Weiss et al., 1993; Staubwasser et al., 2002; Staubwasser and Weiss, 2007). The extent to which this 2200 BC event was a prime-mover of cultural change probably varied significantly from region to region, however, and it is clear that events were neither perfectly synchronous nor uniformly destabilizing across the area, with some regions seeing growth and trends towards increased stability. In Bahrain, for example, stability continued, with *Dilmun* emerging as a state by the end of the third millennium BC, and experiencing the "culmination of trends in population growth and urbanization" (Edens, 1992: 94).

Evidence for continuity of maritime activities is variable around the Arabian peninsular littoral; there are indications that trade and exchange patterns continued in some areas, but altered significantly or reduced drastically in intensity over time in others. The evidence continues to be stronger during this period for the Gulf side of the peninsula than the Red Sea side. There we find continued signs of regular, albeit altered, trade at the beginning of this period. In the Wadi Suq period, the Harappan evidence for the eastern Arabian wider region shifts from Oman to Bahrain, for example. Harappan or Harappan-style material culture falls off rapidly in Omani assemblages at this time, and is replaced by imports mediated through Barbar or Kassite Bahrain (Vogt, 1996). Thus *Dilmun* appears to have supplanted *Magan* as a trade entrepot (though see Potts, 1993), and to have monopolized Harappan trade with Mesopotamia, a fact corroborated by Mesopotamian textual sources (Oppenheim, 1954). In both Mesopotamia and Bahrain, trade shifted to the hands of private entrepreneurs (Oppenheim, 1954; Matthews, 2005), and the risks it involved likely made it the prerogative of a very small, elite, component of society. The Harappan relationship with *Dilmun* seems, not surprisingly, to have been different than its relationship to neighboring *Magan*. It led not to the import of large quantities of Harappan goods, but rather to the incorporation of Harappan administrative and ideological frameworks. Thus when sealing procedures were implemented, it was the stamp seal of the Indus Valley, rather than the cylinder seal of Mesopotamia that was adopted (Eidem and Højlund, 1993; Vogt, 1996; though the influence may also be from the round stamp seals of Iran). The Indus weight system was also used, and later became known to the Mesopotamians as the "standard of *Dilmun*" (Vogt, 1996).

Evidence for Harappan trade continues into the Late Harappan period, as evidenced by both archaeological finds and textual sources like the Mari letters (Carter, 2001; Warburton, 2007). As discussed below, and indicated by ceramic parallels (Potts, 1994; Carter, 2001) trade was during this period with the Late Harappan communities of Gujarat rather than the now disintegrated society of the Indus valley proper. But after the first quarter of the second millennium BC, trade in the Gulf region diminished greatly in volume and probably geographic scope, even if signs of contact remained for some time (Potts, 1994). *Dilmun* lost contact with the mining centers of *Magan* (Oppenheim, 1954), and copper seems to have entered Mesopotamia from the north (Edens, 1992; Warburton, 2007). *Dilmun* similarly lost contact with the regions that supplied it with stone and timber, and essentially reverted to being an island famous for its dates and sweet water (Oppenheim, 1954). Interruptions of archaeological sequences for at least several centuries suggest regional social disintegration in the Gulf (Edens, 1992). The relationship between the end of the early *Dilmun* civilization and the final disappearance of the Harappan civilization remains to be clarified (Carter, 2001)

In Western Arabia, there is also evidence for changes in patterns of maritime contact and trade, coinciding in particular with lapses in Egypt's power. The first such lapse occurred in the First Intermediate Period, when Egypt entered a period of instability and regionalism. Subsequently, however, Egypt resumed its Red Sea trade during the Middle Kingdom period, sending fresh expeditions to Punt via Hammamat and the Port of Mersa Guweisis, to bypass the now powerful kingdom of Kush on the Nile (Kitchen, 2002). Excavations at Mersa Guweisis have yielded remains of expedition ships, as well as a few exotic ceramics from the Tihama and remains of African ebony (Bard et al., 2007). After a second lapse in power during the Second Intermediate Period, leading for a time to a fully independent Kush, a reunited and resurgent Egypt projected her rule south during the New Kingdom Period, and evidence for trade again increases (Kitchen, 2002). The first visual glimpses of trade with Punt are also seen through Queen Hatshepsut's record of her expedition (Phillips, 1997), discussed earlier. The expedition was explicitly undertaken to cut out middlemen in the southern Nubian trade, via the Nile (Kitchen, 2005). After this, maritime trade in the Red Sea became so habitual that kings ceased to boast about it, and it is generally only referred to in passing (Kitchen, 2005). Around 1200 BC, the first pepper appears in the Egyptian record, positively identified from the dried fruits in the nostrils of the mummy of Ramses II (Plu, 1985). This is the first indication of possible contact between Egypt and India, though by what route remains unclear.

Further south in the Red Sea, contact also continues, and intensifies, perhaps partly in response to disruptions in the north. The middle and perhaps early part of the second millennium BC is linked by a number of researchers to the emergence of shared ceramic affinities across the southern part of the Red Sea. Variably referred to as the Afro-Tihama culture (Kitchen, 2002), Afro-Arabian cultural complex (Fattovich, 1997), or Tihama cultural complex (Fattovich, 1999), this sphere of interaction perhaps represents an intensification of an earlier engagement traced through shared lithic sources and techniques by scholars like Zarins, Khalidi and Crassard. Sites from Sihi to Subr (Sabir) along the west and southern coasts of Arabia (de Moulins et al., 2003; Durrani, 2005: 62–67), for example, exhibit pottery that is seen to have parallels with older C-group and Kerma cultures of the Middle Nile (Phillips, 1998; Kitchen, 2002). The Sabir culture itself, which began in the early second millennium BC, was clearly a sea-oriented coastal culture (Ray, 2003: 84). The recently discovered Bronze Age megalithic site of al-Midamman in Yemen, which seems to span the late third to early first millennium BC, has also been argued to have parallels not only with the Sabir culture, but also with material on the African side (Giumlia-Mair et al., 2002; Keall, 2004). However, caution is warranted as most material culture, and particularly ceramic comparisons have been made at a very general rather than typologically specific level, and further research is needed (Durrani, 2005: 107–112). Nevertheless, in general, we can infer close contacts with Africa, which were to intensify in the first millennium BC and are presumably connected with the ethnolinguistic relationships described below. These trans-Red Sea exchanges are regarded, albeit controversially, as one of the key catalysts in the emergence of complex societies in Eritrea and Ethiopia (Phillipson, 1998: 41–49; cf. Durrani, 2005: 114–125; Curtis, 2008).

Transport Innovations and the Emergence of Pan-Arabian and Arabian Sea Trade in the Iron Age

The Bronze/Iron Age transition in the Arabian peninsula saw a number of important changes to trade patterns. These took place within the context of important socio-cultural, technological and economic changes in the peninsula and surrounding region. In the western part of the peninsula, at the same time that Egyptian power and Red Sea navigation simultaneously fell into decline (Fattovich, 2005), a number of prosperous trading kingdoms arose. These included the 'incense kingdoms' of Sabaea, Qataban, Hadhramaut and Ma'in in Arabia, the Ethiopian state of Axum, and to the northeast of the Red Sea, the kingdom of Nabataea, with Petra as its capital (see Scarre, 1988). These kingdoms owed their emergence in part to a transport revolution brought about by the domestication and spread of the dromedary camel. While dromedaries were presumably wild in Arabia, and are known to have been hunted during the Bronze Age (Uerpmann and Uerpmann, 2002), it is only in the Late Bronze Age and the start of the Iron Age that they become attested in adjacent regions and can be argued to be domesticated. Camels not only greatly enhanced overland trade within Arabia and to adjoining land areas, especially in incense, but may also have helped to promote further competitive development of maritime trade. It has certainly been argued that contact across the southern part of the Red Sea, between Africa and southern Arabia, further intensified at this time, and some have even seen the Eritrean pre-Aksumite kingdom of Da'amat as a Sabaean colonization (Kitchen, 2002; Fattovich, 2005). While linguistic, epigraphic and monumental evidence have all been called on to support such claims, they remain controversial (e.g., Schmidt and Curtis, 2001).

In the eastern Arabian peninsula, the early Iron Age saw the revitalization of trade after a period of relative isolation (Magee and Carter, 1999). While a significant level of regionalism suggests that regional interaction was still limited, it is nonetheless clear that relatively intensive exchange was being undertaken, involving trade with the Elamites, Iran and perhaps Central Asia (Magee and Carter, 1999). From ca. 1000 BC, there is an explosion in settlements in the record, and fortification appears (Potts, 2001; Magee, 2004), changes which may probably be linked to the emergence of *falaj* (*qanat*) irrigation and the impact of the camel (Magee, 2004). A pendant from Tell Abraq of this period has been argued to carry the earliest depiction of a lateen sail, otherwise not depicted in the region until the Sasanian period and absent in the Mediterranean until 900 AD (Potts, 2001); the image is very stylised, however, and confirmation must await further evidence.

Another transport revolution during this period likely involved the first regular use of the monsoon winds for long-distance sea transport between India and Arabia. It became possible for Indian goods to reach Egypt and the eastern Mediterranean basin entirely by sea, as well as by the millennia old river and caravan routes running through Mesopotamia and Syria (Burstein, 2002). This period accordingly witnessed the beginning of the Asian spice trade. Black pepper, from its limited source area in south India, was especially prominent in this trade, as suggested by Roman era records both written and archaeological (Miller, 1968; Cappers, 2006). Also important at this time was the emergence of local textile production in the Arabian peninsula, which might also be expected to have contributed to intensifying trade. On the one hand, local production would have highlighted regional differences in quality and design, creating new complex demands for different textiles within local systems of social signification. In addition, it can be suggested to have promoted further diversification in trade towards other high value commodities such as spices. The extent to which different trade strategies in this period were developed by different communities around the Arabia peninsula, and how this laid the foundation for the later development of trade in spices and textiles, deserves further study.

The Dispersal of Domesticates into Arabia: Implications for Maritime Contact and Exchange

The arrival of domesticated plants and animals in the Arabian subcontinent naturally greatly impacted on cultural and economic developments in the region. Not only did domesticated species provide the basis for new types of societies in Arabia, they also catalysed new patterns of contact, trade and exchange. Domesticates moved into Arabia by both maritime and land routes, and accordingly hold clues to maritime routes of contact and trade. The earliest arrival of domesticates into Arabia appears to have been primarily by land, however. Livestock seem, on present evidence, to have spread initially in the absence of plant-based agriculture in Arabia (Edens and Wilkinson, 1998; Uerpmann et al., 2000; see also McCorriston and Martin, 2009; Uerpmann et al., 2009), much as was the case in Saharan and East Sudanic Africa (Marshall and Hildebrand, 2002; Garcea, 2004), as well as parts of savannah India (Fuller, 2006: 58). Secure finds of domesticated cattle generally coincide with the arrival of sheep and goat, and occur at roughly the same time in eastern and western Arabia, after ca. 6000 BC, making it likely that domesticated cattle were ultimately introduced from the Near East rather than Africa. The sheep/goat/cattle triad appeared at roughly the same time in Egypt and western Arabia, suggesting parallel processes of dispersal, moving from the Levant through the Sinai region. This suggests a hunter-forager-herder economy in Arabia, as in the Sahara, but with possible precursors in the Pre-Pottery Neolithic C period of desert margins of eastern Jordan (Martin, 2000; Wengrow, 2006: 25).

The earliest field crops in Arabia were the cereals wheat and barley, which also originated in the Near East (Zohary and Hopf, 2000), although they may not have arrived in Arabia until the fourth millennium BC. The earliest hard archaeobotanical evidence for these cereals dates from ca. 3000 BC, both in Yemen, at al-Raqlah, Hayt al-Suad and Jubabat al-Juruf (see Costantini 1990; de Moulins et al., 2003; Ekstrom and Edens, 2003), and in the United Arab Emirates, as at Hili (see Cleuziou and Costantini, 1980; Tengberg, 2003). These cereals were accompanied by the Near Eastern pulses, pea and lentil. While these continued to be the dominant cereals, and important pulses, through pre-history and historical times, there is an intriguing difference between western and eastern Arabia that suggests agricultural interaction spheres focused on the Gulf and the Red Sea, respectively. Sites on the eastern side of the peninsula more often have evidence for free-threshing wheat, likely bread wheat (*Triticum aestivum*), which was also the most common wheat in the Indus region beyond. By contrast, sites in Yemen sites have more often produced the glume wheat emmer (*Triticum diococcum*), which was also the wheat that dominated ancient Egypt (Murray, 2000), Nubia (Fuller, 2004) and the limited archaeobotanical record for Ethiopia (Boardman, 2000) and Eritrea (D'Andrea et al., 2008). Maritime activities may have been responsible for introducing these Near Eastern crops (and others such as chickpea, grasspea and flax) as well as sheep and goat into the Ethiopian highlands, where they have been the basis of a plow based agricultural system throughout history.[3] The association of this agriculture with speakers of Ethiosemitic languages may be indicative of a prehistoric agricultural system-language co-dispersal from Arabia across the Red Sea (see below).

Maritime activities are meanwhile very clearly implicated in the transfers of domesticates between East Africa and South Asia, although the potential role of Arabia in these

remains unclear. The question of how precisely African crops, such as sorghum, pearl millet, finger millet, hyacinth bean and cowpea, reached India by around 2000 BC is one of the major outstanding questions of archaeobotanical research, and has been a recurrent focus of discussion and debate (e.g., Possehl, 1986, 1996; Weber, 1998; Mehra, 1999; Fuller, 2003a; Misra and Kajale, 2003; Tengberg, 2003; Potts, 1990). Early reports of domesticated sorghum from Arabia are botanically problematic (Rowley-Conwy et al., 1997; Fuller, 2002: 281–282; Tengberg, 2003), and this species should probably not be regarded as an important contributor to prehistoric subsistence in Oman or Yemen. As summarized in Table 1, none of the other African crops have yet been found in Arabia for this time period (de Moulins et al., 2003; Tengberg, 2003). The African crops are, however, unambiguously in Gujarat and Baluchistan in the second millennium BC, suggesting that Gujarat maritime contacts were by this period no longer only with Oman and Dilmun but also extended westwards around Arabia towards Yemen and Africa. At present count, some 33 archaeological sites in South Asia dating from the Middle Bronze Age (ca. 2000 BC) through the Iron Age (to ca. 300 BC) have evidence for crops of African origin for which botanical identity is acceptable (data augmented from Fuller, 2003a; with Chanchala 2002; Saraswat and Pokharia, 2003; Saraswat, 2004, 2005; Cooke et al., 2005). It is likely that the lack of farming communities on the coastal rim of Arabia (in contrast to coastal Gujarat) had a major role to play in the failure of the African crops to transfer to Arabia in the Bronze Age.

Moving in counterflow from Asia to Africa, in this case via Arabia, were the Asian common millet *Panicum milaceum* and zebu cattle. *Panicum miliaceum* was an early crop in China, from ca. 6000 BC, but it is first found in southern Asia and Arabia in the third or early second millennium BC. In northwestern South Asia in appears ca. 1900 BC, with evidence for other crops and artifacts that suggest diffusion from East Asia (cf. Fuller, 2006: 36). Somewhat earlier third millennium BC dates for *Panicum* occur on the Arabian peninsula (Costantini, 1990; Ekstrom and Edens, 2003) and at Tepe Yahya in southern Iran (Costantini and Biasini, 1985), suggesting that the start of this line of contact was across the Gulf already in the later third millennium BC and thence to Yemen. The continuation of this line of diffusion to Africa is indicated by evidence for *Panicum miliaceum* in Nubia at Ukma from the Kerma period (Van Zeist, 1987). The other domesticate which moved between the Indian subcontinent and Africa, probably via Arabian maritime links, was the South Asia-derived zebu cattle (*Bos indicus*). Zebu genetic data show a pattern of inter-regional introgression in which eastern and southern Africa, together with the Arabian peninsula near Africa, show a genetic cline, especially in Y-chromosome data, that indicates much higher zebu bull input than is the case for Mesopotamia and more northerly

---

[3] Nevertheless, there appears to have been at least one local parallel domestication, in the case of the Ethiopian Pea (*Pisum abyssinicum*), which was likely native to Ethiopia, or perhaps Yemen (Butler, 2003; Kosterin and Bogdanova, 2008).

Table 1 Selected crops of African, South Asian and East Asia origin cultivated in Arabia. Local Arabic names from Mason (1946) or Varisco (1994)

Crop, with common names in English and Arabic	Region of origin and earliest evidence there	Cultivation in Arabia, historical evidence
Sorghum bicolor, Sorghum, *dhura*; in Socotra: *makedhīra*, for the grain: *habb, ta`am*	Eastern Sudanic savanna zone, by third millennium BC(?) (Fuller, 2003a)	?Hili, Oman; ?Yemen finds, all botanically dubious; wild sorghum at Sabir, ca. 900 BC; Medieval staple in Yemen with numerous varieties (Varisco, 1994)
Pennisetum glaucum, Pearl millet, *dukhn* (but see also, *Panicum miliaceum*)	West Africa Sahel, by mid third millennium BC (Fuller, 2007a; Finucane et al., 2008)	*Dukhn* cultivated in Medieval Yemen (Varisco, 1994: 167)
Eleusine coracana, Finger millet, *keneb* sometimes *dukhn*	Ethiopia, by late second millennium BC(?) (Fuller, 2003a)	
Eragrostis tef, Tef, *tahaf*	Ethiopia and Eritrea, by later first millennium BC (Boardman, 2000; D'Andrea et al., 2008)	Hajar Bin Humeid, first millennium BC; Cultivated in present day in Arabia
Panicum miliaceum, Broomcorn millet, *dukhn* (sometimes), *bakūr, siyal*	China by ca. 6000 BC (Crawford, 2006)	Yemen by late third millennium BC
Setaria italica, Foxtail millet, *msebeli* or *keneb* (but see also *Eleusine coracana*)	China by ca. 6000 BC (Crawford, 2006)	Cultivated in present day in Arabia
Vigna unguiculata, Cowpea, *lūbiyā', dijr/dujr*	West Africa, Ghana by 1700 BC (D'Andrea et al., 2007); has spread to India also at this time (Fuller, 2003a)	Medieval Yemen (Varisco, 1994: 190)
Lablab purpureus, Hyacinth bean, *hurtimān, kishd*	East Africa, by early second millennium BC; in India by 1700 BC; south India by 1600 BC (Fuller, 2003a; Fuller et al., 2007)	Medieval Yemen (Varisco, 1994: 189)
Vigna radiata, Mungbean, *qusheri*	India: northwest and south, by late third millennium BC (Fuller and Harvey, 2006)	Cultivated in present day in Arabia
Vigna mungo, Urd bean, black gram, *māsh, dizur awad*	India: Gujarat/northern peninsula by 2500 BC (Fuller and Harvey, 2006); Eastern and Southern India, by 1400 BC (Fuller and Harvey, 2006)	Medieval Yemen (Varisco, 1994: 181)
Cajanus cajan, Pigeon pea, *qishta,* at Aden: *turai*		Cultivated in present day in Arabia
Sesamum indicum, Sesame, *simsim, juljul/jiljil*	Pakistan, by Harappan times (2500 BC) (Fuller, 2003b)	First millennium BC Yemen (Sabir, Hajar Bin Humeid); cultivated in Yemeni mountains in Medieval times (Varisco, 1994: 195)
Gossypium arboreum, Tree cotton, *qutun, `otb*	Pakistan, by 5000 BC (Moulherat et al., 2002; Fuller, 2008)	On Bahrain (*Tylos*) according to Theophrastus, ca. 350 BC
Musa sapientum, Bananas	New Guinea/ Indonesia, by 4000 BC (De Langhe and De Maret, 1999; Kennedy, 2008); Indus Valley by 2000 BC (Fuller and Madella, 2001).	Cultivated in Dhofar and foothills in Medieval times (Varisco, 1994: 190)
Areca catechu, Betel-nut, areca-nut, *faufal*	Mainland/ Island Southeast Asia, by 2000 BC(?)	Cultivated in Yemen and Batinah of Oman (Mason, 1946: 594); *Piper betle* is also found cultivated occasionally
Cocos nucifera, Coconut, *jauz hindi* ("Indian walnut"), *jauz narjīl*	Island Southeast Asia	Cultivated in coastal gardens of Aden area (Mason, 1946: 594)

areas (Zeder, 2006). While translocated crops were presumably not themselves the commodities of trade, but moved in boats as food for long voyages, with leftovers used for planting, zebu cattle, on the other hand, moved as bulls (see MacHugh et al., 1997; Loftus and Cunningham, 2000), were presumably rare commodities of high value. Archaeozoological evidence for *Bos indicus* has been reported from Tell Abraq by the Wadi Suq period and possibly in the Umm

an-Nar phase (Uerpmann, 2001). A recent review for Africa suggests no major influx of zebu, but rather occasional occurrences in Africa, based mainly on depictions rather than osteological evidence, and probably indicating rare imports. These occur in Egypt beginning between 2000 and 1500 BC, in Niger in the second millennium BC and in the Chad Basin in the first millennium BC (Magnavita, 2006). Recent archaeological evidence from the poorly studied Horn suggests that zebu was present there by at least the first millennium BC (Schmidt and Curtis, 2001).

Beyond those crops for which there is archaeobotanical evidence, are many others that moved between East Africa, Arabia and South Asia, highlighting the recurrent role of maritime contacts in the spread of crops (see, e.g., Engels and Hawkes, 1991; Blench, 2003). A selection of the various crops that have been introduced into Arabia is provided in Table 1. Meanwhile, two other animal domesticates deserve mention due to their importance in the overland trade that complemented maritime systems of commerce and social exchange, namely donkeys and camels. The donkey is evidenced in wild form at Early Holocene sites in Yemen and Oman (Edens and Wilkinson, 1998: 67; Uerpmann et al., 2000; Cattani and Bökönyi, 2002; see also McCorriston and Martin, 2009), but does not appear to have been locally domesticated. Based on modern genetic data, donkeys appear to have been domesticated twice, from the two disjunct wild populations, the Nubian and Somali subspecies (Vilà et al., 2006). Historical linguistics also suggests more than one origin (Blench, 2000). The earliest archaeozoological evidence for donkeys that were probably domesticated comes from the Late Neolithic and Predynastic of Egypt, from sites such as Maadi (ca. 4500 BC) and Hieronkopolis (ca. 3500 BC). Figurines indicate that donkeys were by this time being used as pack animals and were presumably important in trade between urban Mesopotamia and the emerging Egyptian state (Wengrow, 2006). Donkey trade can be inferred to have moved southwards as well towards sources of incense in Yemen and/or Ethiopia (Wengrow, 2006). Another species of interest is the camel, which is presumably native to the Arabian peninsula. Its domestication, perhaps in the Late Bronze Age (late second millennium BC), and spread, by the early Iron Age (Zeuner, 1963; Köhler-Rollefson, 1996), would have had a major impact on trade by making cross-desert travel easier, thereby increasing connectedness across the region, including in seasons when ocean currents and winds were unfavorable.

Overall then, the pattern of domesticate dispersals into and around Arabia suggests that maritime processes had an important role to play in moving domesticated plants and animals. From the Middle Bronze Age, there is clear evidence for the movement of species between lands as far as South Asia and Africa, and this gradually expanding network of contacts and trade routes helped to make the subsistence and farming systems of Arabia an ever more diverse mosaic. The transport of crop plants and animals probably involved significant transport by boat around the coasts of Arabia. While maritime Gujarat was almost certainly involved in some of these translocations, it is also very likely that Arabian seafarers had an important role to play. As we have seen, sea-capable and trade-oriented societies of various types had clearly emerged on the Arabian littoral by the time that these trans-

locations began to take place. When the crop translocations between Africa and South Asia occur, for example, there is clear evidence that Oman had already adopted the use of plank-built wooden boats (Cleuziou and Tosi, 1997). While their role is often overlooked, it is very likely that small-scale coastal communities also played a dynamic role in the intensive trade systems that developed in the Bronze Age (Charpentier, 1996; Cleuziou, 2003).

Greco-Roman Period Trade, and the Classical Records

The last centuries BC and first centuries AD saw the arrival onto the Arabian scene of a number of new powers, including the Greeks, Romans and Sassanians. The period witnessed radical transformations in maritime activity and trade, with the emergence of new greatly expanded trading spheres and regular, long-distance maritime travel in the wider Indian Ocean. Increased use of shipping along the Red Sea tipped the balance of power and prosperity in southern Arabia in favor of those states with control of the major ports, such as Qana, Muza and Eden, and the East African kingdom of Axum accordingly thrived (see Scarre, 1988). Classical sources from this time begin to offer new insights into maritime experience and the voyages undertaken and places seen by maritime explorers and traders. We outline a selection of these with the aim of highlighting the utility of further integration of these sources with archaeological and other findings pertaining to maritime activity and trade by and with Arabians, and in the wider region.

One of the earliest Classical records is a story in Herodotos (ca. 500 BC), Book IV, 44, of the voyages of Scylax of Caryanda, who was sent by the Persian emperor Darius to find the mouth of the Indus. Records of this journey are also preserved by Hecataeus of Miletus[4] (ca. 550–ca. 476 BC), who mentions an encounter with the land of Maka (Oman) and the Farasan islands (possibly Socotra and the Kurya-Murya islands). This is the first historical record of coasting around the Arabian peninsula, although the Gulf seems to have remained the most popular route between India and the Mediterranean for another few centuries (Keay, 2006). After this, we have Theophrastus correctly identifying Saba, Hadramaut and Qataban among the incense producing regions of Arabia (Keay, 2006). Thereafter, more intimate knowledge of the Arabian peninsula by the Classical world had to await Ptolemaic initiative. The Ptolemies resumed the

[4] Hecataeus is said to have compiled a *Ges Periodos* ('World Survey'), but his work only survives in some 374 fragments, quoted in the Ethnika of Stephanus of Byzantium (sixth century AD).

Pharaonic hunt for exotica, and revived the commercial enterprise of rulers like Hatshepsut, opening trading stations down the Red Sea (Keay, 2006: 48). The Ptolemaic merchant fleet explored the Arabian and Ethiopian coastlines, and vividly, if not always accurately, described the tribes they encountered, including the Ichthyophagi (Keay, 2006: 49). They referred to western Arabia as 'Eudaimon Arabia', the Greek form of 'Arabia Felix' or 'Fortunate Arabia' – described by Agatharchides, writing in the second century BC, as bearing "most of the products considered valuable by us" (Keay, 2006: 51). Interestingly, Agatharchides also describes 'white cattle' and 'walled cities' on what is probably the island of Socotra, where ships from the port that Alexander built on the Indus River (Patala, near the Pakistani city of Hyderabad) are encountered, as well as others from Persia and Arabia (Keay, 2006: 52). These vessels may already have carried the lateen sail (Keay, 2006: 55), as also discussed above.

Another intriguing, but difficult to interpret reference occurs in Diodorus Siculus (first century BC). He recounts the story of Yambulos, a Nabataean spice-merchant who travelled through Arabia in search of spices and was captured by pirates, probably in what is now northern Somalia. He was sent by them in a small boat and after 4 months reached a huge island of 'happy and wise' people where there were giant but harmless snakes (Kobishchanow, 1965). Stechow argued that this was an early record of a visit to Madagascar (Stechow, 1944) and certainly evidence from translocated murids is beginning to suggest Greco-Roman contact at about this period (Blench, 2007a). Nonetheless, the story contains clear mythic elements and is too imprecise to more than hint at early Mediterranean presence in the Arabian Sea. As the Sabaean kingdom in southern Arabia collapsed, however, to be replaced by the Himyarite kingdom in the late second century BC, and Arab power in the southern Red Sea weakened, direct contact between Ptolemaic Egypt and India became possible. Discovery by Classical sailors of the monsoon winds in the first century BC subsequently led to major changes in the scale, organization and conduct of Indian Ocean trade, including an increase in the scale and value of trade to a level clearly beyond that of a limited trade in 'luxuries', and the emergence of more common direct contact between India and the Mediterranean through the Red Sea (Burstein, 2002). From this period onwards, Classical references multiply as the Greco-Roman quest for spices and aromatics expanded, and Beeston (1979) summarizes the Classical evidence for the routes to South Arabia, which encompassed both coastal and inland routes. In 24 BC the Romans sent a (failed) naval expedition to try and capture the source areas of frankincense and myrrh. Strabo and Ptolemy reflect the expansion of geographical knowledge as trade to Arabia and India increased. Faller presents a detailed examination of the sources for routes and knowledge of Taprobane

(Sri Lanka) based on movement down the Red Sea and the Persian Gulf (Faller, 2000). Pliny is able to recount details of the spice-trade in the Horn of Africa, and Theophrastus the properties of its medical plants. The *Periplus*, a first century seaman's guide to the East African coast and other areas of the Indian Ocean, records ports in Arabia, India, and Africa as far south as modern-day Tanzania.

The Red Sea continued to be the Classical world's most important entry into the spice route for several centuries, especially as hostilities between Rome and the Parthian and then Sassanid rulers of Persia made the Gulf route unsafe (Keay, 2006: 15). In the fourth century BC, however, the situation was reversed, as Mediterranean power transferred from Rome to Constantinople, shifting the spice route north to the Gulf (Keay, 2006). Two later travellers and historians from the sixth century AD provide important records of the Indian Ocean trade as it pertained to Arabia at that time. Procopius, the prolific Byzantine historian, describes in the *De bello persico* (after 550 AD) the trade between the Ethiopian kingdom of Axum and South Arabia based around the port of Adulis. He notes the use of ships without nails and recounts a complex story of how the Ethiopian kingdom tried to outflank Persian control of the silk trade with India. Cosmas Indicopleustes [whose name means 'India sea-voyager'] published his *Christian topography* in 550 AD (Winstedt, 1909), recounting a voyage to the Malabar coast in the 520s. Cosmas similarly describes the trade between Adulis and India, mounted by Ethiopians in their own vessels. These and earlier Classical sources hold an important key to understanding the place of Arabia in wider Indian Ocean maritime trade developments.

Ethnolinguistic Geography and Historical Linguistics

The linguistic geography of the Arabian peninsula and adjacent area also provides some intriguing clues to early settlement patterns and population movements in the region, as well as, possibly, to maritime migrations. Today, the languages of the Arabian peninsula are wholly drawn from the Semitic branch of the Afroasiatic phylum (see Fig. 8). Essentially, the pattern is that Arabic dominates most of the land area, but all along the southern coast in the Hadramaut and Oman as well as on the island of Socotra, a set of archaic and rather diverse Semitic languages, are spoken, the so-called 'South Semitic' branch (Fig. 9; Johnstone, 1977, 1981, 1987; Simeone-Semelle, 1991). Most linguists consider that these languages would formerly have been much more widespread in the peninsula prior to the expansion of Islam and consequently Arabic in the seventh century. Epigraphic materials survive in the so-called 'Sabaean' languages which are

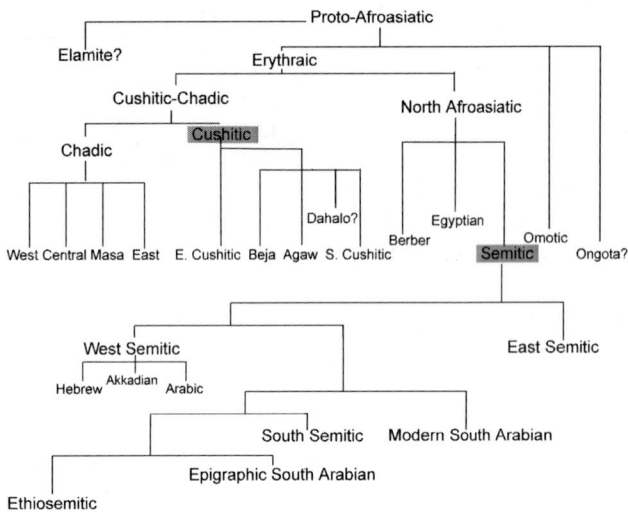

Fig. 8 Afroasiatic classification (modified from Blench, 2006). *Highlighted branches* are those that occur on either side of the Red Sea

generally considered ancestral to modern South Semitic (Beeston, 1981; Korotayev, 1995). These include Sabaean, Minaean and Qatabanian inscriptions and are generally dated to between the eighth century BC and the sixth century AD (Versteegh, 2000). However, the assumption is that these languages were spoken much earlier still as their closest relatives outside Africa are the 'West Semitic' languages, which include both Arabic and Hebrew, but also all the epigraphic languages of the Near East, such as Akkadian.

The ultimate homeland of Afroasiatic is Africa and most probably Ethiopia, where its most diverse branches, Omotic and Cushitic, are spoken. Despite its aura of antiquity, Semitic is a relatively late branching from Afroasiatic, as testified by the relative closeness of all Semitic languages. As a consequence, the dominance of Semitic in the Arabian peninsula is comparatively recent. It must be the case that other quite different languages were spoken prior to Semiticisation several thousand years ago. There is no evidence as to the

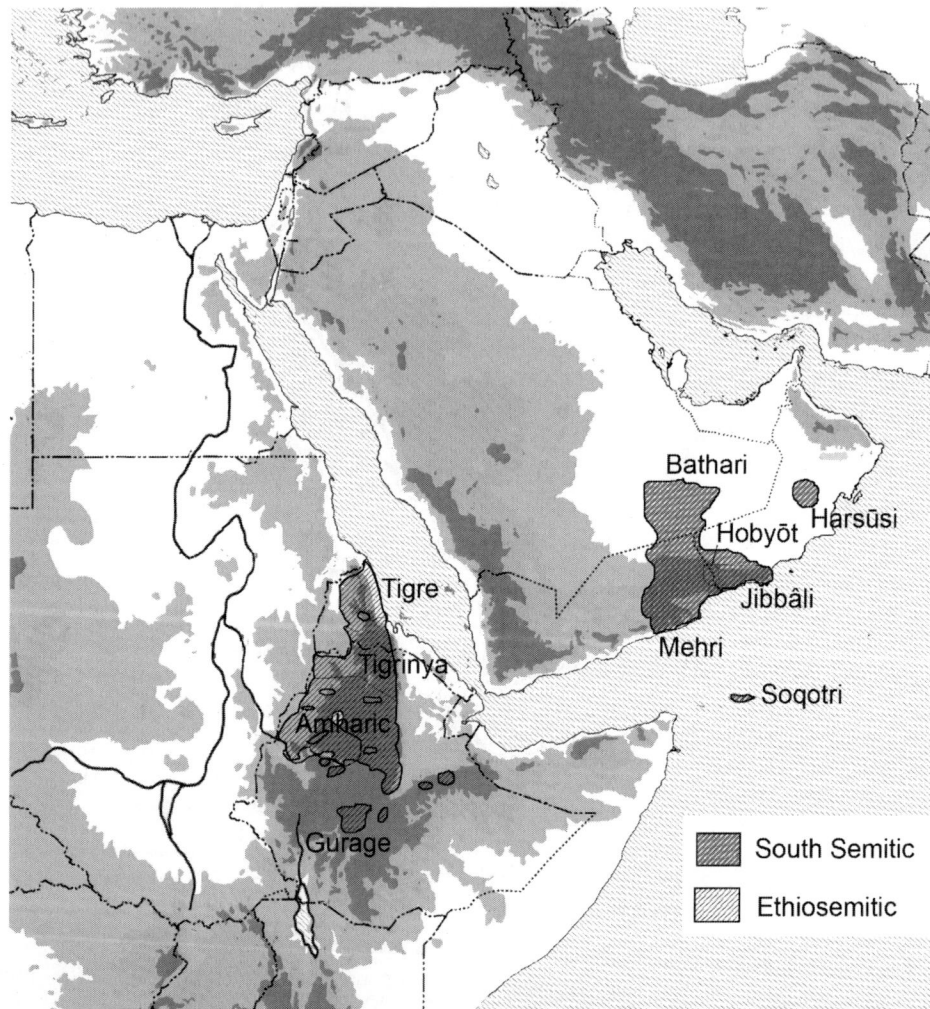

Fig. 9 Distribution of South Semitic and Ethiosemitic languages in Arabia and Africa

nature of these languages or their affiliation; although such a major cultural transformation must have left traces in regional archaeology, no proposals have been made as to the 'signal' of the Semitic expansion.

Across the Red Sea, in Ethiopia and the Horn of Africa, the pattern of languages is quite different. With limited exceptions in the west and north of Ethiopia, all the languages spoken are also Afroasiatic. However, the dominant branches are Cushitic and Omotic, much more internally diverse subgroups of considerable antiquity. Figure 8 shows the internal classification of the Afroasiatic phylum and the highlighted fill indicates the branches that occur on either side of the Red Sea. The pattern of Semitic in Ethiopia represents something of a puzzle. The highlands and northeast are dominated by an extensive group of languages usually called 'Ethiosemitic' and including well-known languages such as Tigrinya and Amharic. Comparisons between South Semitic and Ethiosemitic suggest that the Ethiopian languages are a branch of the epigraphic languages of South Arabia, and that it is therefore likely that the ancestors of the Amhara migrated back across the Red Sea within the last few millennia. Bender argues that the South Arabian languages share a number of innovations with Ethiosemitic (Bender, 1970). There are also significant bodies of oral tradition; the story of King Solomon and the Queen of Sheba [from Yemen] is virtually an Ethiopian national myth and artifacts in Axum have South Arabian inscriptions.

This hypothesis is also suggested by linguistic geography; Ethiosemitic forms a coherent territorial bloc imposed upon and acting to fragment the *in situ* Cushitic and Omotic languages in the highlands from the northeast. This migration was potentially driven by the development of an innovative type of agriculture: the seasonal cultivation of cereals based on the plow. Ethiopia has a characteristic plow, an ard which fractures and disturbs the soil, which was perhaps introduced following the migrations of Ethiosemitic speakers across from Arabia. McCann points out that rock-drawings in Eritrea point to the use of the plow as early as 500 BC and that it shows similarities to implements in South Arabia (McCann, 1995: 40). Historical evidence points to a north-south spread of Semitic in Ethiopia. The Amharic term for plow, *maräša*, has been borrowed into all the main languages of Ethiopia. Barnett canvasses the idea of introductions of the plow from Arabia or Egypt 3,000–4,000 BP (Barnett, 1999: 24), but the linguistic evidence suggests a more recent date.

Almost all classifications of Ethiosemitic languages treat them as a single branch. However, in the south, there is a distinctive subgroup, the Gurage cluster, which is significantly more diverse than all the other Ethiosemitic languages (Leslau, 1979). It used to be thought that Ethiosemitic was a single subgroup, but more recently its internal diversity has led scholars to question this. It may be that the origin of the Gurage languages is different, either they are a core Semitic group that stayed behind after the break-up of North Afroasiatic or they represent an earlier and different migration from Arabia. Features that the Gurage languages have in common with the Amharic group would thus be the result of long interaction rather than direct genetic affiliation.

The implications of this overall linguistic geographical pattern are as follows: the Semitic languages are likely to have expanded southwards into the Arabian peninsula from the Near East. This expansion is likely to have had both a maritime, coastal component and an overland component, perhaps based on livestock. As we have indicated, the early Sabaeans developed an elaborate literate culture and were in intensive contact with the Ethiopian coast through interactions across the Red Sea. At some point, it appears that there was a significant population migration from Arabia, presumably in the region of modern-day Eritrea, which transformed the economy of highland Ethiopia. Cultural contacts across the Red Sea seem to have stimulated the development of an indigenous maritime trading culture reaching as far as India, whose members acted independently as brokers in the aromatics trade.

Concluding Remarks

We have attempted, in this chapter, to take a very broad approach to maritime prehistory in the Arabian subcontinent, drawing together the findings of diverse scholars, regions, time periods and disciplines. Such a broad brush approach obviously comes with caveats, and omissions, oversights and errors are not unlikely in the preceding discussion. Nonetheless, this approach has also been useful in providing a general synthesis and overview of developments pertaining to maritime subsistence, seafaring and trade in and around the Arabian subcontinent. It has highlighted in particular important similarities and differences between the western and eastern littorals of the peninsula, and their maritime trajectories. Our broad summary also, by tracing developments and transformations across a wider area than is often addressed, enables firmer placement of the subcontinent's trajectory within the wider Indian Ocean framework. Such steps are necessary for resolving the still unclear question, alluded to in the introduction, and in parts of this chapter, of how the earliest translocations of crops, animals and material culture in the Indian Ocean were effected.

The data presented here offer further support to arguments that effective maritime proficiency and regular, even long-distance maritime trade in the Indian Ocean significantly predated the Greco-Roman sea adventures and commercial activities so evident in the wealth of Classical period texts. The Arabian data not only highlight the emergence of extremely early seafaring and maritime trade activities in the

Red Sea, the Gulf and the Arabian Sea, but also emphasize the problematic nature of the kind of core-periphery models that reliance on textual sources can often encourage. Small-scale fishing, trading and coastal-dwelling communities were clearly not only responsible for the early emergence of seafaring and maritime trade in the region, but also had a significant role to play in maritime activities even after the arrival on the scene of the large Bronze Age states. Thus, it begins to look increasingly likely that the sometimes impressive, and even spectacular translocations across the Arabian Sea and the wider Indian Ocean that we find evidenced in the archaeological, linguistic and genetic records are at least partly attributable to the activities of relatively small-scale societies. What should not be overlooked, however, is the large-sized ambitions that likely attended such activity; social competition, personal achievement, elite political manoeuvring and prestige good economies have undoubtedly played as much a role in deep water seafaring and maritime trade for small-scale societies as large ones throughout human history.

The maritime pre- and protohistory of the Arabian subcontinent has been greatly clarified by intensive archaeological survey and excavation in the region over the past few decades. While many questions have been answered, however, many more remain and other new ones have emerged. The challenge of addressing these certainly lies partially in continued archaeological endeavors in the peninsula and surrounding regions. However, it is also increasingly clear that new disciplines, like molecular genetics, and emerging syntheses, like that between archaeology, genetics and historical linguistics, offer new ways of addressing the questions that archaeologists and others want to answer. Our chapter has attempted to make some preliminary headway with such an interdisciplinary approach, but due to limitations of time and space has elected not to consider the genetics literature, except in passing, in this particular discussion. Animal, plant and human genetic data nonetheless have their own insights to offer to the developments, patterns and questions we have addressed here. The challenges of more multidisciplinary approaches are many, but the effectiveness of such multi-stranded methodologies is being increasingly demonstrated, and we see the future of Arabian archaeology in this direction. It is our hope that the synthesis offered here, and the emerging multidisciplinary paradigm it suggests, will help to provide a base from which such exciting new studies may be undertaken in the years ahead.

Acknowledgments We have benefited from discussions with Remy Crassard, Lamya Khalidi, Louise Martin, Michael Petraglia, Greg Possehl and Dave Wengrow. We are also grateful to Paolo Biagi, Mark Beech and Adrian Parker for supplying useful information. We would like to acknowledge the assistance of the European Research Council, which is funding the authors' research into maritime prehistory in the Indian Ocean.

References

Adelaar A. Towards an integrated theory about the Indonesian migrations to Madagascar. In: Peregrine PN, Peiros I, Feldman M, editors. Ancient human migrations: an interdisciplinary approach. Salt Lake City, UT: University of Utah Press (in press).

Alley RB, Ágústdóttir AM. The 8k event: cause and consequences of a major Holocene abrupt climate change. Quaternary Science Reviews. 2005;24:1123–49.

Alley RB, Mayewski PA, Sowers T, Stuiver M, Taylor KC, Clark PU. Holocene climatic instability: a prominent, widespread event 8200 yr ago. Geology. 1997;25:483–6.

Anonymous. Houses dating to 5000 BC discovered in U.A.E. Current World Archaeology. 2004;3.

Arnold JE. Transportation innovation and social complexity among maritime hunter-gatherer communities. American Anthropologist. 1995;97:733–47.

Bailey G. World prehistory from the margins: the role of coastlines in human evolution. Journal of Interdisciplinary Studies in History and Archaeology. 2004;1:39–50.

Bailey G. The Red Sea, coastal landscapes, and hominin dispersals. In: Petraglia MD, Rose JI, editors. The evolution of human populations in Arabia: paleoenvironments, prehistory and genetics. The Netherlands: Springer; 2009. p. 15–37.

Baines J, Malek J. Atlas of ancient Egypt. New York: Facts on File; 1980.

Bard KA, Fattovich R, Ward C. Sea port to Punt: new evidence from Marsā Gawāsīs, Red Sea (Egypt). In: Starkey J, Starkey P, Wilkinson TJ, editors. Natural resources and cultural connections of the Red Sea. BAR International Series 1661. Oxford: Archaeopress; 2007. p. 143–8.

Barnett T. The emergence of food production in Ethiopia. BAR International Series 763. Oxford: Archaeopress; 1999.

Beech MJ. Fishing in the 'Ubaid: a review of fish-bone assemblages from early prehistoric coastal settlements in the Arabian Gulf. Journal of Oman Studies. 2002;12:25–40.

Beech MJ. The development of fishing in the U.A.E.: a zooarchaeological perspective. In: Potts DT, Al Naboodah H, Hellyer P, editors. Proceedings of the First International Conference on the Archaeology of the U.A.E. London: Trident Press; 2003.

Beech MJ. In the land of the Ichthyophagi: modelling fish exploitation in the Arabian Gulf and Gulf of Oman from the 5th Millennium BC to the late Islamic period, BAR International Series 1217. Oxford: Archaeopress; 2004.

Beech MJ, al-Husaini M. Preliminary report on the vertebrate fauna from site H3, Sabiyah: an Arabian Neolithic/'Ubaid site in Kuwait. In: Archaeozoology of the Near East VI: proceedings of the sixth international symposium on the Archaeozoology of Southwestern Asia and adjacent areas. Groningen: ARC-Publicaties 123; 2005. p. 124–38.

Beech MJ, Cuttler R, Moscrop D, Kallweit H, Martin J. New evidence for the Neolithic settlement of Marawah Island, Abu Dhabi, United Arab Emirates. Proceedings of the Seminar for Arabian Studies. 2005;35:37–56.

Beeston AFL. Some observations on Greek and Latin data relating to South Arabia. Bulletin of the School of Oriental and African Studies. 1979;42:7–12.

Beeston AFL. Languages of Pre-Islamic Arabia. Numéro Spécial Double: Études de Linguistique Arabe. Arabica. 1981;28:178–86.

Bender ML. A preliminary investigation of South Arabia. In: Proceedings of the third international conference of Ethiopian studies. Addis Ababa: Haile Selassie I University; 1970. p. 26–37.

Biagi P. A radiocarbon chronology for the aceramic shell-middens of coastal Oman. Arabian Archaeology and Epigraphy. 1994;5:17–31.

Biagi P. The shell-middens of the Arabian Sea and Persian Gulf: maritime connections in the seventh millennium BP? Adumatu. 2006;14:7–16.

Biagi P, Nisbet R. Environmental history and plant exploitation at the aceramic sites of RH5 and RH6 near the mangrove swamp of Qurm (Muscat – Oman). Bulletin de la Societé Botanique Fancaise. 1992;139:571–8.

Biagi P, Nisbet R. The prehistoric fisher-gatherers of the western Arabian Sea: a case of seasonal sedentarization? World Archaeology. 2006;38:220–38.

Biagi P, Torke W, Tosi M, Uerpmann H-P. Qurum: a case study of coastal archaeology in northern Oman. World Archaeology. 1984;16:43–61.

Blau S. Of water and oil: exploitation of natural resources and social change in eastern Arabia. In: Gosden C, Hather J, editors. The prehistory of food: appetites for change. London: Routledge; 1999. p. 83–98.

Blench R. The ethnographic evidence for long-distance contacts between Oceania and East Africa. In: Reade J, editor. The Indian Ocean in antiquity. London: Kegan Paul; 1996. p. 417–38.

Blench R. Language studies in Africa. In: Vogel JO, editor. Encyclopaedia of Precolonial Africa. Walnut Creek: Altamira; 1997. p. 90–100.

Blench R. The languages of Africa: macrophyla proposals and implications for archaeological interpretation. In: Blench R, Spriggs M, editors. Archaeology and language, IV. London: Routledge; 1999. p. 29–47.

Blench R. A history of donkeys, wild asses and mules in Africa. In: Blench RM, MacDonald KC, editors. The origins and development of African livestock: archaeology, genetics, linguistics and ethnography. London: UCL Press; 2000. p. 339–54.

Blench R. The movement of cultivated plants between Africa and India in prehistory. In: Neumann K, Butler A, Kahlheber S, editors. Food, fuel and fields: progress in African archaeobotany. Koln: Heinrich Barth Institut; 2003. p. 273–92.

Blench R. Archaeology, language and the African past. Lanham: Alta Mira Press; 2006.

Blench R. New palaeozoogeographical evidence for the settlement of Madagascar. Azania. 2007a;XLII:69–82.

Blench R. Using linguistics to reconstruct African subsistence systems: comparing crop names to trees and livestock. In: Denham T, Iriarte J, Vrydaghs L, editors. Rethinking agriculture: archaeological and ethnographic perspectives. Walnut Creek: Left Coast Press; 2007. p. 408–38.

Blench R, MacDonald KC, editors. The origins and development of African livestock: archaeology, genetics, linguistics and prehistory. London: UCL Press; 2000.

Boardman S. Archaeobotany. In: Phillipson DW, editor. Archaeology at Aksum, Ethiopia, 1993–7, vol II. London: The Society of Antiquaries; 2000. p. 363–369, 412–414.

Boivin N. Life rhythms and floor sequences: excavating time in rural Rajasthan and Neolithic Çatalhöyük. World Archaeology. 2000;31:367–88.

Boivin N. Landscape and cosmology in the south Indian Neolithic: new perspectives on the Deccan ashmounds. Cambridge Archaeological Journal. 2004b;14:235–57.

Boivin N. Rock art and rock music: petroglyphs of the south Indian Neolithic. Antiquity. 2004a;78:38–53.

Boivin N. Anthropological, historical, archaeological and genetic perspectives on the origins of caste in South Asia. In: Petraglia MD, Allchin B, editors. The evolution and history of human populations in South Asia. Dordrecht: Springer; 2007. p. 341–61.

Boivin N, Brumm A, Lewis H, Robinson D, Korisettar R. Sensual, material, and technological understanding: exploring prehistoric soundscapes in south India. Journal of the Royal Anthropological Institute (N.S.). 2007;13:267–94.

Boivin N, Fuller DQ, Korisettar R, Petraglia MD. First farmers in south India: the role of internal processes and external influences in the emergence of the earliest settled societies. Pragdhara. 17 (in press)

Bradbury L. kpn-boats, Punt trade, and a Lost Emporium. Journal of the American Research Center in Egypt. 1996;33:37–60.

Braudel F. The mediterranean and the mediterranean World in the Age of Philip II, vol I. Berkeley, CA: University of California Press; 1995.

Bronk Ramsey C.: OxCal 3.10. http://www.rlaha.ox.ac.uk/orau/calibration.html (2005).

Burstein SM. Kush, Axum and the ancient Indian Ocean trade. In: Bács T, editor. A tribute to excellence: studies offered in honor of Ernö Gaál. Budapest: Ulrich Luft, and Lásló Török Studia Aegyptiaca XVII; 2002. p. 127–137.

Butler A. The Ethiopian pea: seeking the evidence for a separate domestication. In: Neumann K, Butler A, Kahlheber S, editors. Food, fuel and fields: progress in African archaeobotany. Frankfurt: Heinrich-Barth-Institut; 2003. p. 37–48.

Cappers R. Roman foodprints at Berenike. Los Angeles, CA: Cotsen Institute of Archaeology; 2006.

Carter R. Saar and its external relations: new evidence for interaction between Bahrain and Gujarat during the early second millennium BC. Arabian Archaeology and Epigraphy. 2001;12:183–201.

Carter R. Boat remains and maritime trade in the Persian Gulf during the sixth and fifth millennia BC. Antiquity. 2006;80:52–63.

Carter R, Crawford HEW. The Kuwait–British archaeological expedition to as-Sabiyah: report on the fourth season's work. Iraq LXV, 2003; p. 77–90.

Cattani M, Bökönyi S. Ash-Shumah: an Early Holocene settlement of desert hunters and mangrove foragers in the Yemeni Tihamah. In: Cleuziou S, Tosi M, Zarins J, editors. Essays on the Late Prehistory of the Arabian peninsula, Serie Orientale Roma XCIII. Rome: Istituto Italiano per l'Africa e l'Oriente; 2002. p. 31–53.

Chanchala S. Botanical remains. In: Tewari DP, editor. Excavations at Chard. Lucknow: Jarun Prakashan; 2002. p. 166–94.

Chandramohan P, Jena BK, Sanil Kumar V. Littoral drift sources and sinks along the Indian coast. Current Science. 2001;81:292–7.

Charpentier V. Archaeology of the Erythraean Sea: craft specialization and resource optimizations as part of the coastal economy on eastern coastlands of Oman during the 4th and 3rd millennia BC. In: Afanas'ev GE, Cleuziou S, Lukacs JR, Tosi M, editors. The prehistory of Asia and Oceania, Colloquia 16. Forli: UISPP; 1996. p. 181–92.

Charpentier V. Archéologie de la côte des Icthyophages: coquilles, squales et cétacés du site IV-IIIe millénaire de Ra's al-Jinz. In: Cleuziou S, Tosi M, Zarins J, editors. Essays of the Late Prehistory of the Arabian peninsula. Rome: Serie Orientale Roma XCIII; 2002. p. 73–99.

Cleuziou S. The Oman peninsula and the Indus civilization: a reassessment. Man and Environment. 1992;17:94–103.

Cleuziou S. The emergence of oases and towns in eastern and southern Arabia. In: Afanas'ev GE, Cleuziou S, Lukacs JR, Tosi M, editors. The prehistory of Asia and Oceania, Colloquia 16. Forli: UISPP; 1996. p. 159–65.

Cleuziou S. Early Bronze Age trade in the Gulf and the Arabian Sea: the society behind the boats. In: Potts DT, Al Naboodah H, Hellyer P, editors. Archaeology of the United Arab Emirates: Proceedings of the first international conference of the U.A.E. London: Trident Press; 2003.

Cleuziou S, Costantini L. Premiers elements sur l'agriculture protohistorique de l'Arabie orientale. Paleorient 1980;6.

Cleuziou S, Méry S. In-between the great powers: the Bronze Age Oman peninsula. In: Cleuziou S, Tosi M, Zarins J, editors. Essays on the Late Prehistory of the Arabian peninsula. Rome: Istituto Italiano per l'Africa e l'Oriente; 2002. p. 273–316.

Cleuziou S, Tosi M. Black boats of Magan: some thoughts on Bronze Age water transport in Oman and beyond from the impressed bitumen slabs of Ra's al-Junayz. In: Parpola A, Koskikallio P, editors. South Asian Archaeology 1993. Suomalainen Tiedeakatemia, Helsinki; 1994.

Cleuziou S, Tosi M. Evidence for the use of aromatics in the Early Bronze Age of Oman: Period III at RJ-2 (2300–2200 BC). In: Avanzini A, editor. Profumi d'Arabia. Saggi di Storia Antica 11. Roma; 1997. p. 57–81.

Connah G. Holocene Africa. In: Scarre C, editor. The human past: world prehistory and the development of human societies. London: Thames & Hudson; 2005.

Connan J, Carter R, Crawford H, Tobey M, Charrié-Duhaut A, Jarvie D, et al. A comparative geochemical study of bituminous boat remains from H3, As-Sabiyah (Kuwait), and RJ-2, Ra's al-Jinz (Oman). Arabian Archaeology and Epigraphy. 2005;16:21–66.

Cooke M, Fuller DQ, Rajan K. Early historic agriculture in southern Tamil Nadu: archaeobotanical research at Mangudi, Kodumanal and Perur. In: Franke-Vogt U, Weisshaar J, editors. South Asian archaeology 2003: Proceedings of the European Association for South Asian Archaeology Conference. Aachen: Linden Soft; 2005. p. 329–34.

Cooney G. Introduction: seeing land from the sea. World Archaeology. 2003;35:323–8.

Costantini L. Harappan agriculture in Pakistan: the evidence of Naursharo. In: Taddei M, editor. South Asian archaeology 1987. Rome: Istituto Italiano per il Medio ed Estremo Oriente; 1990. p. 321–32.

Costantini L, Biasini LC. Agriculture in Baluchistan between the 7th and 3rd millenium B.C. Newsletter of Baluchistan Studies. 1985;2:16–37.

Crawford H. Dilmun and its neighbours. Cambridge: Cambridge University Press; 1998.

Curtis MC. New perspectives for examining change and complexity in the Northern Horn of Africa during the 1st Millennium BC. In: Schmidt PR, Curtis MC, Teka Z, editors. The archaeology of ancient Eritrea. Asmara: The Red Sea Press; 2008. p. 329–48.

D'Andrea AC, Kahlheber S, Logan AL, Watson DJ. Early domesticated cowpea (Vigna ungiuculata) from Central Ghana. Antiquity. 2007;81:686–98.

D'Andrea AC, Schmidt PR, Curtis MC. Paleoethnobotanical analysis and agricultural economy in early first millennium BCE sites around Asmara. In: Schmidt PR, Curtis MC, Teka Z, editors. The archaeology of ancient Eritrea. Asmara: The Red Sea Press; 2008. p. 207–16.

De Langhe E, De Maret P. Tracking the banana: its significance in early agriculture. In: Gosden C, Hather J, editors. The prehistory of food: appetites for change. London: Routledge; 1999. p. 377–96.

de Moulins D, Phillips CS, Durrani N. The archaeobotanical record of Yemen and the question of Afro-Asian contacts. In: Neumann K, Butler A, Kahlheber S, editors. Food, fuel and fields: progress in African archaeobotany. Africa Praehistorica. 2003;15:213–228.

Dixon DM. The transplantation of Punt incense trees in Egypt. Journal of Egyptian Archaeology. 1969;55:55–65.

During Caspers E. New archaeological evidence for maritime trade in the Persian Gulf during the Late Protoliterate Period. East and West. 1971;21:22–55.

During Caspers E. Sumer, coastal Arabia and the Indus Valley in Protoliterate and Early Dynastic eras: supporting evidence for a cultural linkage. Journal of the Economic and Social History of the Orient. 1979;22:121–35.

Durrani N. The Tihamah coastal plain of South-West Arabia in its regional context c. 6000 BC–AD 600. BAR International Series 1456. Oxford: Archaeopress; 2005.

Edens C. Dynamics of trade in the ancient Mesopotamian world system. American Anthropologist. 1992;94:118–39.

Edens C. Indus-Arabian interaction during the Bronze Age: a review of evidence. In: Possehl GL, editor. Harappan civilization: a recent perspective. 2 revisedth ed. New Delhi: Oxford & IBH; 1993.

Edens C, Wilkinson T. Southwest Arabia during the Holocene: recent archaeological developments. Journal of World Prehistory. 1998;12:55–119.

Eidem J, Højlund F. Trade or diplomacy? Assyria and Dilmun in the eighteenth century. World Archaeology. 1993;24:441–7.

Ekstrom H, Edens CM. Prehistoric agriculture in highland Yemen: new results from Dhamar. Bulletin of the American Institute of Yemeni Studies. 2003;45:23–35.

Engels JMM, Hawkes JG. The Ethiopian gene centre and its genetic diversity. In: Engels JMM, Hawkes JG, Worede M, editors. Plant genetic resources of Ethiopia. Cambridge: Cambridge University Press; 1991. p. 23–41.

Enzel Y, Ely L, Mishra S, Ramesh R, Amit R, Lazar B, et al. High resolution Holocene environmental changes in the Thar Desert, northwestern India. Science. 1999;284:125–7.

Erlandson JM. The archaeology of aquatic adaptations: paradigms for a new millennium. Journal of Archaeological Research. 2001;9: 287–350.

Fabre D. Seafaring in Ancient Egypt. London: Periplus; 2005.

Facey W. The Red Sea: the wind regime and location of ports. In: Lunde P, Porter A, editors. Trade and travel in the Red Sea region: Proceedings of the Red Sea Project I. BAR International Series 1269. Oxford: Archaeopress; 2004. p. 6–17.

Faller S. Taprobane im Wandel der Zeit. Da Śrî Lankâ-bild in Griecheischen und Lateinischen quellen zwischen Alexanderzug und Spätantike. Stuttgart: Franz Steiner; 2000.

Fattovich R. The Near East and Eastern Africa. In: Vogel JO, editor. Encyclopedia of precolonial Africa: archaeology, history, languages, cultures and environments. Walnut Creek: Altamira Press; 1997. p. 484–9.

Fattovich R. The development of urbanism in the northern Horn of Africa in ancient and medieval times. In: Sinclair P, editor. The development of Urbanism from a global perspective. Uppsala: Uppsala Universitet; 1999.

Fattovich R. The archaeology of the Horn of Africa. In: Raunig W, Wenig S, editors. Afrikas Horn: Akten der Ersten Internationalen Littman-Konferenz 2. Wiesbaden: Harrasowitz Verlag; 2005. p. 3–29.

Faulkner RO. Egyptian seagoing ships. The Journal of Egyptian Archaeology. 1941;26:3–9.

Finucane B, Manning K, Toure M. Late Stone Age subsistence in the Tilemsi Valley, Mali: stable isotope analysis of human and animal remains from the site of Karkarichinkat Nord (KN05) and Karkarichinkat Sud (KS05). Journal of Anthropological Archaeology. 2008;27:82–92.

Flavin K, Shepherd E. Fishing in the Gulf: preliminary investigations at an Ubaid site, Dalma (UAE). Proceedings of the Seminar for Arabian Studies. 1994;24:115–34.

Fleitman D, Burns SJ, Mudelsee M, Neff U, Kramers J, Mangini A, Matter A. Holocene forcing of the Indian monsoon recorded in a stalagmite from Southern Oman. Science. 2003;300(5626):1737–1739.

Francaviglia VM. Obsidian sources in ancient Yemen. In: De Maigret A, editor. The Bronze Age culture of Khawlan at-Tiyal and al-Hada (Yemen Arab Republic). Rome: IsMEO; 1989. p. 129–34.

Fuller DQ. Palaeoecology of the Wadi Muqaddam: a preliminary report on the significance of the plant and animal remains. Sudan and Nubia. 1998;2:52–60.

Fuller DQ. A parochial perspective on the end of Meroe: changes in cemetery and settlement at Armina West. In: Welsby DA, editor. Recent research on the Kingdom of Kush. London: British Museum Press; 1999. p. 203–17.

Fuller DQ. Fifty years of archaeobotanical studies in India: laying a solid foundation. In: Settar S, Korisettar R, editors. Indian archaeology in retrospect, volume III: archaeology and interactive disciplines. Delhi: Manohar; 2002. p. 247–363.

Fuller DQ. African crops in prehistoric South Asia: a critical review. In: Neumann K, Butler A, Kahlheber S, editors. Food, fuel and fields: progress in African archaeobotany, Africa Praehistorica 15. Köln: Heinrich-Barth Institut; 2003a. p. 239–71.

Fuller DQ. An agricultural perspective on Dravidian historical linguistics: archaeological crop packages, livestock and Dravidian crop vocabulary. In: Bellwood P, Renfrew C, editors. Examining the farming/language dispersal hypothesis. Cambridge: McDonald Institute for Archaeological Research; 2003b. p. 191–213.

Fuller DQ. Early Kushite agriculture: archaeobotanical evidence from Kawa. Sudan and Nubia. 2004;8:70–4.

Fuller DQ. Ceramics, seeds and culinary evolution: models of micro-diffusion in prehistoric Indian agriculture. Antiquity. 2005;79:761–77.

Fuller DQ. Agricultural origins and frontiers in South Asia: a working synthesis. Journal of World Prehistory. 2006;20:1–86.

Fuller DQ. Contrasting patterns in crop domestication and domestication rates: recent archaeobotanical insights from the Old World. Annals of Botany. 2007a;100:903–24.

Fuller DQ. Non-human genetics, agricultural origins and historical linguistics in South Asia. In: Petraglia MD, Allchin B, editors. The evolution and history of human populations in South Asia. Dordrecht: Springer; 2007b. p. 393–443.

Fuller DQ. The spread of textile production and textile crops in India beyond the Harappan zone: an aspect of the emergence of craft specialization and systematic trade. In: Osada T, Uesugi A, editors. Linguistics, archaeology and the human past. Indus Project. Kyoto: Research Institute for Humanity and Nature; 2008. p. 1–26.

Fuller DQ, Harvey E. The archaeobotany of Indian pulses: identification, processing and evidence for cultivation. Environmental Archaeology. 2006;11:241–68.

Fuller DQ, Madella M. Issues in Harappan archaeobotany: retrospect and prospect. In: Settar S, Korisettar R, editors. Indian archaeology in retrospect, vol II. Protohistory. New Delhi: Manohar; 2001. p. 317–90.

Fuller DQ, Korisettar R, Venkatasubbaiah PC. Southern Neolithic cultivation systems: a reconstruction based on archaeobotanical evidence. South Asian Studies. 2001b;17:171–87.

Fuller DQ, Boivin N, Korisettar R. Dating the Neolithic of South India: new radiometric evidence for key economic, social and ritual transformations. Antiquity. 2007;81:755–78.

Garcea EAA. An alternative way towards food production: the perspective from the Libyan Sahara. Journal of World Prehistory. 2004;18:107–54.

Gasse F. Hydrological changes in the African tropics since the Last Glacial Maximum. Quaternary Science Reviews. 2000;19:189–211.

Gaur AS, Sundaresh. Evidence of shoreline shift on the northern Saurashtra coast: study based on the submerged temple complex at Pindara. Current Sciences. 2007;92:733–5.

Giumlia-Mair A, Keall EJ, Shugar AN, Stock S. Investigation of a copper-based hoard from the Megalithic site of al-Midamman, Yemen: an interdisciplinary approach. Journal of Archaeological Science. 2002;29:195–209.

Groom N. Trade, incense and perfume. In: Simpson J, editor. Queen of Sheba: treasures from ancient Yemen. London: The British Museum Press; 2002. p. 88–94.

Harvey SP. Interpreting Punt: geographic, cultural and artistic landscapes. In: O'Connor D, Quirke S, editors. Mysterious lands, encounters with ancient Egypt. London: UCL Press; 2003. p. 81–92.

Hassan FA. Holocenes palaeoclimates of Africa. African Archaeological Review. 1997;14:213–30.

Horton MC. Mare Nostrum – a new archaeology in the Indian Ocean? Antiquity. 1997;71:753–5.

Jesse F. Early ceramics in the Sahara and the Nile Valley. In: Krzyzankiak K, Kroeper Kobusiewicz M, editors. Cultural markers in the later prehistory of Northeastern Africa and recent research. Poznan: Polish Academy of Sciences, Poznan Archaeological Museum; 2003. p. 35–50.

Johnstone TM. Harsusi Lexicon and English-Harsusi Index. Oxford: Oxford University Press; 1977.

Johnstone TM. Jibbāli Lexicon. Oxford: Oxford University Press; 1981.

Johnstone TM. Mehri Lexicon and English-Mehri Word-List. London: SOAS, University of London; 1987.

Kallweit H. Remarks on the Late Stone Age in the U.A.E. In: Potts DT, al-Naboodah H, Hellyer P, editors. Archaeology of the United Arab Emirates. Proceedings of the First International Conference on the Archaeology of the U.A.E. London; 2002. p. 56–63.

Keall EJ. Possible connections in antiquity between the Red Sea coast of Yemen and the Horn of Africa. In: Lunde P, Porter A, editors. Trade and travel in the Red Sea region: Proceedings of the Red Sea Project I. BAR International Series 1269. Oxford; 2004. p. 43–55.

Keay J. The spice route: a history. Berkeley, CA: University of California Press; 2006.

Kennedy J. Pacific bananas: complex origins, multiple dispersals? Asian Perspectives. 2008;47:75–94.

Khalidi L. The formation of a southern Red Sea landscape in the late prehistoric period: tracing cross-Red Sea culture-contact, interaction and maritime communities along the Tihama coastal plain, Yemen, in the third to first millennium BC. In: Starkey J, Starkey P, Wilkinson T, editors. Natural resources and cultural connections of the Red Sea. Oxford: BAR International Series 1661; 2007.

Khalidi L. Holocene obsidian exchange in the Red Sea region. In: Petraglia MD, Rose JI, editors. The evolution of human populations in Arabia: paleoenvironments, prehistory and genetics. The Netherlands: Springer 2009. p. 279–91.

Kitchen KA. The Land of Punt. In: Shaw T, Sinclair P, Andah B, Okpoko A, editors. The archaeology of Africa: food, metals and towns. London: Routledge; 1993. p. 587–608.

Kitchen KA. Egypt, Middle Nile, Red Sea and Arabia. In: Cleuziou S, Tosi M, Zarins J, editors. Essays on the Late Prehistory of the Arabian peninsula. Rome: Istituto Italiano per l'Africa e l'Oriente; 2002. p. 383–401.

Kitchen KA. Ancient peoples of the Red Sea in Pre-Classical antiquity. In: Starkey J, editor. People of the Red Sea: Proceedings of the Red Sea Project II. Oxford: BAR International Series 1395; 2005. p. 7–14.

Kobashi T, Severinghaus JP, Brook EJ, Barnola J-M, Grachev AM. Precise timing and characterization of abrupt climate change 8200 years ago from air trapped in polar ice. Quaternary Science Reviews. 2007;26:1212–22.

Kobishchanow YM. On the problem of the sea voyages of ancient Africans in the Indian Ocean. The Journal of African History. 1965;6:137–41.

Köhler-Rollefson I. The one-humped camel in Asia: origin, utilization and mechanisms of dispersal. In: Harris D, editor. The origins and spread of agriculture and pastoralism in Eurasia. London: UCL Press; 1996.

Korotayev A Ancient Yemen: some general trends of the evolution of Sabaic language and Sabaean culture. Journal of Semitic Studies, Suppl 5. New York: Oxford University Press;1995

Kosterin OE, Bogdanova VS. Relationship of wild and cultivated forms of Pisum L. as inferred from an analysis of three markers of the plastid, mitochondrial and nuclear genomes. Genetic Resources and Crop Evolution. 2008;55:735–55.

Kutzbach JE. Monsoon climate of the Early Holocene: climate experiment with Earth's orbital parameters for 9000 years ago. Science. 1981;214:59–61.

Lambeck K. Shoreline reconstructions for the Persian Gulf since the Last Glacial Maximum. Earth and Planetary Science Letters. 1996;142:43–57.

Lamberg-Karlovsky CC. Trade mechanisms in Indus-Mesopotamian interrelations. Journal of the American Oriental Society. 1972;92:222–9.

Lawler A. Report of oldest boat hints at early trade routes. Science. 2002;296:1791–2.

Leslau W. Etymological dictionary of gurage. Wiesbaden: Harrassowitz; 1979.

Lézine A-M, Saliège J-F, Robert C, Wertz F, Inizian M-L. Holocene lakes from Ramlat as-Sab'atayn (Yemen) illustrate the impact of monsoon activity in southern Arabia. Quaternary Research. 1998;50:290–9.

Lézine A-M, Tiercelin J-J, Robert C, Saliège J-F, Cleuziou S, Inizan M-L, et al. Centennial to millennial-scale variability of the Indian monsoon during the Early Holocene from a sediment, pollen and isotope record from the desert of Yemen. Palaeogeography, Palaeoclimatology, Palaeoecology. 2007;243:235–49.

Lightfoot DR. The origin and diffusion of qanats in Arabia: new evidence from the northern and southern peninsula. The Geographical Journal. 2000;166:215–26.

Lucas A. Cosmetics, perfume and incense in ancient Egypt. Journal of Egyptian Archaeology. 1930;16:41–53.

Lucas A. Notes on myrrh and stacte. Journal of Egyptian Archaeology. 1937;23:27–33.

Madella M, Fuller DQ. Paleoecology and the Harappan Civilisation of South Asia: a reconsideration. Quaternary Science Reviews. 2006;25:1283–301.

Magee P. The impact of southeast Arabian intra-regional trade on settlement location and organization during the Iron Age II period. Arabian Archaeology and Epigraphy. 2004;15:24–42.

Magee P, Carter R. Agglomeration and regionalism: southeastern Arabia between 1400 and 1100 BC. Arabian Archaeology and Epigraphy. 1999;10:161–79.

Magnavita C. Ancient humped cattle in Africa: a view from the Chad Basin. African Archaeological Review. 2006;23:55–84.

Marshall F, Hildebrand E. Cattle before crops: the beginnings of food production in Africa. Journal of World Prehistory. 2002;16:99–143.

Martin L. Mammalian remains from the eastern Jordanian Neolithic, and the nature of caprine herding in the steppe. Paleorient. 2000;25:87–104.

Mason K, editor. Western Arabia and the Red Sea. London: Naval Intelligence Division; 1946.

Masry AH. Prehistory in Northeastern Arabia: the problem of interregional interaction. London: Kegan Paul International; 1997.

Mathur UB, Pandey DK, Bahadur T. Falling Late Holocene sea-level along the Indian coast. Current Science. 2004;87:439–40.

Matthews R. The rise of civilization in Southwest Asia. In: Scarre C, editor. The human past: world prehistory and the development of human societies. London: Thames & Hudson; 2005. p. 432–71.

Mbida CM, Van Neer W, Doutrelepont H, Vrydaghs L. Evidence for banana cultivation and animal husbandry during the first millennium BC in the forest of southern Cameroon. Journal of Archaeological Science. 2000;27:151–62.

McCann J. People of the Plow: an agricultural history of Ethiopia, 1800–1990. Wisconsin: University of Wisconsin Press; 1995.

McClure HA. Radiocarbon chronology of the Late Quaternary lakes in the Arabian Desert. Nature. 1976;263(5580):755–756.

McCorriston J, Martin L. Southern Arabia's early pastoral population history: some recent evidence. In: Petraglia MD, Rose JI, editors. The evolution of human populations in Arabia: paleoenvironments, prehistory and genetics. The Netherlands: Springer 2009, p. 237–50.

Meeks D. Locating Punt. In: O'Connor D, Quirke S, editors. Mysterious lands, encounters with Ancient Egypt. London: UCL Press; 2003. p. 53–80.

Mehra KL. Subsistence changes in India and Pakistan: the Neolithic and Chalcolithic from the point of view of plant use today. In: Gosden C, Hather J, editors. The prehistory of food: appetites for change. London: Routledge; 1999. p. 139–46.

Meyer C, Todd JM, Beck CW. From Zanzibar to Zagros: a copal pendant from Eshnunna. Journal of Near Eastern Studies 1991;50(4):296–7.

Miller JI. The spice trade of the Roman Empire. Oxford: Oxford University Press; 1968.

Misra VN, Kajale MD, editors. Introduction of African crops into South Asia. Pune: Indian Society for Prehistoric and Quaternary Studies; 2003.

Mitchell P. African connections: archaeological perspectives on Africa and the wider world. Walnut Creek: Altamira; 2005.

Moulherat C, Tengberg M, Haquet J-F, Mille B. First evidence of cotton at Neolithic Mehrgarh, Pakistan: Analysis of mineralized fibres from a copper bead. Journal of Archaeological Science. 2002;29: 1393–401.

Munro RN, Wilkinson T. Natural resources and cultural connections of the Red Sea. Oxford: BAR International Series 1661; 2007.

Murray MA. Fruits, vegetables, pulses and condiments. In: Nicholson P, Shaw I, editors. Ancient Egyptian materials and technology. Cambridge: Cambridge University Press; 2000. p. 609–55.

Neumann K. Holocene vegetation of the eastern Sahara: charcoal from prehistoric sites. African Archaeological Review. 1989;7:97–116.

Oates J, Davidson TE, Kamilli D, McKerrell H. Seafaring merchants of Ur? Antiquity. 1977;51:221–35.

Oppenheim AL. Seafaring merchants of Ur. Journal of the American Oriental Society. 1954;74:6–17.

Parker AG, Goudie AS, Stokes S, White K, Hodson MJ, Manning M, et al. A record of Holocene climate change from lake geochemical analyses in southeastern Arabia. Quaternary Research. 2006;66: 465–76.

Parpola A. The Meluhha village: evidence of acculturation of Harappan traders in late third millennium Mesopotamia? Journal of the Economic and Social History of the Orient. 1977;20:129–65.

Petraglia MD, Alsharekh A. The Middle Palaeolithic of Arabia: implications for modern human origins, behaviour, and dispersals. Antiquity. 2003;77:671–84.

Petraglia MD, Korisettar R, Boivin N, Clarkson C, Ditchfield P, Jones S, et al. Middle Palaeolithic assemblages from the Indian subcontinent before and after the Toba super-eruption. Science. 2007;317: 114–6.

Phillips J. Punt and Aksum: Egypt and the Horn of Africa. Journal of African History. 1997:38.

Phillips CS. The Tihamah c. 5000 to 500 BC. Proceedings of the Seminar for Arabian Studies, 1998. p. 28

Phillips CS. Prehistoric middens and a cemetary from the Southern Arabian Gulf. In: Cleuziou S, Tosi M, Zarins J, editors. Essays on the Late Prehistory of the Arabian peninsula. Rome: Istituto Italiano per l'Africa e l'Oriente; 2002. p. 169–86.

Phillipson DW. Ancient Ethiopia. London: The British Museum Press; 1998.

Plu A. Bois et graines. In: Balout L, Roubet C, editors. La Momie de Ramsès II. Paris: Contribution Scientifique à l'Égyptologie. Éditions Recherches sur les Civilisations; 1985. p. 166–74.

Possehl GL. African millets in South Asian prehistory. In: Jacobson J, editor. Studies in the archaeology of India and Pakistan. New Delhi: Oxford & IBH and the American Institute of Indian Studies; 1986. p. 237–56.

Possehl GL. Meluhha. In: Reade JE, editor. The Indian Ocean in Antiquity. London: Kegan Paul International; 1996. p. 133–208.

Possehl GL. The Indus civilization: a contemporary perspective. Walnut Creek: Altamira; 2002.

Possehl GL. The middle Asian interaction sphere. Expedition. 2007;49: 40–2.

Potts DT. The Arabian Gulf in antiquity. Volume 1: from prehistory to the Fall of the Achaemenid Empire. Oxford: Clarendon; 1990.

Potts DT. Rethinking some aspects of trade in the Arabian Gulf. World Archaeology. 1993;24:423–39.

Potts DT. South and Central Asian elements at Tell Abraq (Emirate of Umm al-Qaiwain, United Arab Emirates), c. 2200 BC–AD 400. In: Parpola A, Koshikallio P, editors. South Asian archaeology 1993: Proceedings of the Twelfth International Conference of the European Association of South Asian Archaeologists. Helsinki: Suomalainen Tiedeakatemia; 1994. p. 615–28.

Potts DT. Before the Emirates: an archaeological and historical account of developments in the region ca. 5000 BC to 676 AD. In: Al Abed I, Hellyer P, editors. United Arab Emirates: a new perspective. London: Trident Press; 1997. p. 28–69.

Potts DT. Arabian peninsula. In: Pearsall D, editor. Encyclopedia of archaeology. New York: Elsevier; 2008. p. 827–34.

Prabha Ray H. The Archaeology of seafaring in ancient South Asia. Cambridge: Cambridge University Press; 2003.

Ratnagar S. The Bronze Age: unique instance of a pre-industrial world system? Current Anthropology. 2001;42:351–79.

Ratnagar S. Trading encounters: from the Euphrates to the Indus in the Bronze Age. revised 2nd ed. New Delhi: Oxford University Press; 2040.

Roaf M, Galbraith J. Pottery and p-values: 'seafaring merchants of Ur?' re-examined. Antiquity. 1994;68.

Robin C. Saba and the Sabaeans. In: Simpson SJ, editor. Queen of Sheba: treasures from ancient Yemen. London: British Museum Press; 2002. p. 51–8.

Rowley-Conwy P, Deakin W, Shaw CH. Ancient DNA from archaeological sorghum (*Sorghum bicolor*) from Qasr Ibrim, Nubia: implications for domestication and evolution and a review of the archaeological evidence. Sahara. 1997;9:23–34.

Ruddiman WF. What is the timing of orbital-scale monsoon changes? Quaternary Science Reviews;2006.

Sanil Kumar V, Pathak KC, Pednekar P, Raju NSN, Gowhaman R. Coastal processes along the Indian coastline. Current Science. 2006;91:530–6.

Saraswat KS. Plant economy of early farming communities. In: Singh BP, editor. Early farming communities of the Kaimur (excavations at Senuwar). Jaipur: Publication Scheme; 2004. p. 416–35.

Saraswat KS. Agricultural background of the early farming communities in the Middle Ganga Plain. Pragdhara. 2005;15: 145–77.

Saraswat KS, Pokharia AK. Palaeoethnobotanical investigations at Early Harappan Kunal. Pragdhara. 2003;13:105–40.

Scarre C, editor. Past worlds: atlas of archaeology. London: Times Books; 1988.

Schmidt PR, Curtis MC. Urban precursors in the horn: early 1st-millennium BC communities in Eritrea. Antiquity. 2001;75:849–59.

Schott FA, McCreary JPJ. The monsoon circulation of the Indian Ocean. Progress in Oceanography. 2001;51:1–123.

Shajan KP, Cherian PJ, Tomber R, Selvakumar V. The external connections of Early Historic Pattanam, India: the ceramic evidence. Antiquity 82. http://antiquity.ac.uk/Projgall/tomber/index.html (2008).

Shroder JFJ, editor. Himalaya to the sea: geology, geomorphology and the Quaternary. London: Routledge; 1993.

Simeone-Semelle M-C. Récents développements des recherches sur les langues sudarabiques modernes. In: Mukarovsky HG, editor. Proceedings of the Fifth International Hamito-Semitic Congress. Vienna: Universitat Wien; 1991. p. 321–37.

Staubwasser M, Weiss H. Corrigendum to introduction: Holocene climate and cultural evolution in late prehistory–early historic West Asia [Quaternary Research 66 (2006) 372–387]. Quaternary Research. 2007;68:175.

Staubwasser M, Sirocko F, Grootes PM, Erlenkeuser H. South Asian monsoon climate change and radiocarbon in the Arabian Sea during the early and middle Holocene. Paleoceanography. 2002;17: 1–12.

Staubwasser M, Sirocko F, Grootes PM, Segl M. Climate change at the 4.2 ka BP termination of the Indus Valley Civilization and Holocene South Asian monsoon variability. Geophysical Research Letters. 2003;30:1425.

Stechow E. Kannte das Altertum die Insel Madagaskar? Petermanns Mitteilungen. 1944;90:84–5.

Stieglitz RR. Long-distance seafaring in the ancient Near East. Biblical Archaeologist. 1984;47:134–42.

Tengberg M. Archaeobotany in the Oman peninsula and the role of Eastern Arabia in the spread of African crops. In: Neumann K, Butler A, Kahlheber S, editors. Food, fuel and fields: progress in African archaeobotany. Koln: Heinrich-Barth Institut; 2003. p. 229–38.

Tosi M. Archaeological activities in the Yemen Arab Republic, 1985: Tihamah coastal archaeological survey. East and West. 1985;35:363–9.

Tosi M. Archaeological activities in the Yemen Arab Republic, 1986: Neolithic and protohistoric cultures, survey and excavations in the coastal plain (Tihama). East and West. 1986a;36:400–15.

Tosi M. The emerging picture of prehistoric Arabia. Annual Review of Anthropology. 1986b;15:461–90.

Uerpmann M. Structuring the Late Stone Age of southeastern Arabia. Arabian Archaeology and Epigraphy. 1992;3:65–109.

Uerpmann M. Remarks on the animal economy of Tell Abraq (Emirates of Sharjah and Umm al-Qaywayn, UAE). Proceedings of the Seminar for Arabian Studies. 2001;31:227–34.

Uerpmann M. The dark millennium – remarks on the Final Stone Age in the Emirates and Oman. In: Potts DT, Naboodah H, Hellyer P, editors. Archaeology of the United Arab Emirates: Proceedings of the First International Conference on the Archaeology of the U.A.E. London: Trident; 2003. p. 74–81.

Uerpmann H-P, Uerpmann M. Camel in south-east Arabia. Journal of Oman Studies. 2002;12:235–60.

Uerpmann M, Uerpmann H-P, Jasim S. Stone Age nomadism in Southeast Arabia – palaeo-economic considerations on the Neolithic site of Al-Buhais 18 in the Emirate of Sharjah, U.A.E. Proceedings of the Seminar for Arabian Studies. 2000;30:229–34.

Uerpmann H-P, Potts DT, Uerpmann M. Holocene (re-) occupation of Eastern Arabia. In: Petraglia MD, Rose JI, editors. The evolution of human populations in Arabia: paleoenvironments, prehistory and genetics. The Netherlands: Springer; 2009. p. 205–14.

Van Zeist WA. The plant remains. In: Vila A, editor. Le Cimetière Kermaique d'Ukma Ouest. Paris: CNRS; 1987. p. 247–55.

Varisco DM. Medieval agriculture and Islamic science. the almanac of a Temeni Sultan. Seattle: University of Washington Press; 1994.

Versteegh CHM. The Arabic Language. Edinburgh: Edinburgh University Press; 2000.

Vilà C, Leonard JA, Beja-Pereira A. Genetic documentation of horse and donkey domestication. In: Zeder MA, Bradley DG, Emshwiller E, Smith BD, editors. Documenting domestication: new genetic and archaeological paradigms. Berkeley, CA: University of California Press; 2006. p. 342–53.

Vogt B. In search for coastal sites in pre-historic Makkan: mid-Holocene shell-eaters in the coastal desert of Ras al-Khaimah, U.A.E. In: Kenoyer JM, editor. From Sumer to Meluhha: contributions to the archaeology of South and West Asia in memory of George F. Dales, Jr. Madison: Wisconsin Archaeological Reports; 1994.

Vogt B. Bronze age maritime trade in the Indian Ocean: Harappan traits on the Oman peninsula. In: Reade J, editor. The Indian Ocean in Antiquity. London: Kegan Paul; 1996.

Warburton DA. What happened in the Near East ca. 2000 BC? In: Seland EH, editor. The Indian Ocean in the ancient period: definite places, translocal Exchange. Oxford: BAR International Series 1593; 2007.

Ward C. Boat-building and its social context in early Egypt: interpretations from the First Dynasty boat-grave cemetary at Abydos. Antiquity. 2006;80:118–29.

Wasson RJ, Smith GI, Agrawal DP. Late Quaternary sediments, minerals and inferred geochemical history of Didwana Lake, Thar Desert, India. Palaeogeography, Palaeoclimatology, Palaeoecology. 1984;46:345–72.

Weber SA. Out of Africa: the initial impact of millets in South Asia. Current Anthropology. 1998;39:267–74.

Webster PJ, Yang S. Monsoon and ENSO: selectively interactive systems. Quarterly Journal of the Royal Meteorological Society. 1992;118:877–926.

Weiss H, Courty MA, Wetterstrom W, Guichard F, Senior L, Meadow R, et al. The genesis and collapse of third millennium north Mesopotamian civilization. Science. 1993;261:995–1004.

Wengrow D. The archaeology of early Egypt: social transformations in North-East Africa, 10000 to 2650 BC. Cambridge: Cambridge University Press; 2006.

Wilkinson TJ. Agriculture and the countryside. In: Simpson SJ, editor. Queen of Sheba: treasures from ancient Yemen. London: The British Museum Press; 2002. p. 102–8.

Winstedt EO. The Christian topography of Cosmos Indicopleustes. Cambridge: Cambridge University Press; 1909.

Zarins J. Ancient Egypt and the Red Sea trade: the case for obsidian in the Predynastic and Archaic periods. In: Leonard A, Williams B, editors. Essays in ancient civilization presented to Helene J. Kantor, studies in oriental civilization. Chicago: Oriental Institute; 1989. p. 339–68.

Zarins J. Obsidian and the Red Sea trade. In: Taddei M, editor. South Asian archaeology 1987. Rome: Istituto Universitario Orientale; 1990. p. 509–41.

Zarins J. Obsidian in Predynastic/Archaic Egyptian Red Sea trade. In: Reade J, editor. The Indian Ocean in antiquity. London: Kegan Paul; 1996.

Zeder M. Central questions in the domestication of plants and animals. Evolutionary Anthropology. 2006;15:105–17.

Zeuner FE. A history of domesticated animals. London: Hutchinson; 1963.

Zohary D, Hopf M. Domestication of plants in the Old World. 3rd ed. Oxford: Oxford University Press; 2000.

Chapter 19
Holocene Obsidian Exchange in the Red Sea Region

Lamya Khalidi

Keywords Geometric Microliths • Microlithic • Obsidian • Red Sea • Tihamah • Yemen

Introduction

Arabia holds a particularly interesting geographic position for our knowledge of population dispersals and exchanges. It occupies the southern end of a peninsula with access to two of the most heavily exploited maritime channels in antiquity as well as to desert routes linking it to the Near East. In addition it is close to the African continent and thus to the birthplace of our human ancestors. Due to its central geographic position, Arabia must be considered as a migration route during different periods of early human prehistory. In the same light, Arabia also cannot be ignored as a major crossroads of inter-continental Holocene human interaction and movement.

Unfortunately, knowledge of the Paleolithic record of Arabia is still in its infancy, as the Pleistocene human fossil record is non-existent and stratified sites are scant as a result of poor preservation, taphonomic processes, and relatively poor research coverage (Amirkhanov, 1997; Petraglia, 2003; Crassard, 2007). These issues are by no means limited to Pleistocene remains, but also affect those of the Early and Middle Holocene which have, in southwest Arabia in particular, suffered a similar fate (Crassard and Khalidi, 2005). As in many regions of the world with a recent history of archaeological investigation, the field of Arabian archaeology is fueled by what is most visible: ancient monuments, the complex societies who built and inscribed them, and the classical texts which evoke them. It is for this reason that one must often tread backward in time to reconstruct the choices that were made by prehistoric people but affecting those very

L. Khalidi (✉)
Centre d'Études Préhistoire, Antiquité, Moyen Age,
UMR 6130-CNRS Université de Nice, Sophia Antipolis,
250, rue Albert Einstein, Sophia-Antipolis,
F.-06560, Valbonne, France
e-mail: lamya.khalidi@gmail.com

pathways that made ancient history memorable. It is those prehistoric pathways and, particularly, the conduits provided by the Red Sea that will be considered in this chapter from the perspective of obsidian exploitation in the region.

The southern Red Sea region, which includes the Arabian peninsula and the Horn of Africa, is dotted with obsidian source areas that provided prehistoric people with a sought-after resource for tool production and luxury objects. The majority of northern Anatolian and Trans-Caucasian obsidian sources, like those of the Mediterranean, have been analyzed, and the production of the material and its diffusion effectively traced to a number of Near Eastern and Central Asian prehistoric sites. However, the next major source area to the south which happens to be that of the southern Red Sea zone (Renfrew, 1998: 5), remains virtually unexamined.

This chapter reviews the state of obsidian research in the southern Red Sea zone. In addition, it addresses new data recovered along the Arabian Red Sea coastal plain (Tihamah) and highlands and its implications for Holocene human interaction.

Geological Obsidian in the Southern Red Sea Zone

Obsidian is a natural volcanic glass of rhyolite composition (defined by a silica content of around 70% by weight). Although it is generally translucent and black, it can have different hues (for example: green, gray, brown and red) and opacity as a result of the presence of different metallic oxides, vesicles (former gas bubbles), and devitrification (a re-structuring of the glass that occurs over time). Due to its numerous qualities – physical and aesthetic – it is well known both as a luxury item and tool material throughout the prehistoric and historic world. Obsidian is the sharpest available raw material (producing edges as sharp as 3 nm) and is homogeneous in nature, making it easily workable and desirable for tool production (Piel-Desruisseaux, 2004: 47)

Moreover, the geochemical homogeneity of each obsidian source allows for obsidian to be traceable back to its source at outcrop. This requires geochemical characterization of

M.D. Petraglia and J.I. Rose (eds.), *The Evolution of Human Populations in Arabia*, Vertebrate Paleobiology and Paleoanthropology,
DOI 10.1007/978-90-481-2719-1_19, © Springer Science+Business Media B.V. 2009

sources in order to establish the origins of obsidian artifacts. While this process appears quite simple, and has proven an effective method for tracing exchange routes and reconstructing complete *chaînes opératoires*, geochemical fingerprinting of obsidian is not always conclusive or straightforward.

Obsidian is produced during both effusive and explosive volcanic eruptions. The basic starting material for building volcanoes is a molten rock (magma) called basalt, which once set is typically dark and fine-grained, containing small crystals of various silicate minerals. Typically, it contains around 45% or so silica by weight, substantially less than obsidian. Most of the Earth's volcanic rocks are basaltic but sometimes basalt magmas sit in the Earth's crust for thousands of years or longer before eruption. During these long periods, the magma cools and crystallizes, and dense iron, magnesium-rich, and silica-poor minerals separate out. This leaves behind an ever-cooling residue that becomes richer in silica as well as volatile compounds such as water and carbon dioxide. The result is a magma that becomes increasingly viscous. On eruption, it may either freely lose its now gaseous volatiles and flow sluggishly out of the vent (forming a lava *coulée* or dome), or, if the gas cannot escape in time, explode due to rapidly expanding bubbles that violently fragment the magma (Francis and Oppenheimer, 2004).

Both kinds of eruptions can end up generating glassy obsidian outcrops but generally it is the domes and coulées that prevail as sources of desirable lithic materials. Individual obsidian sources typically display significant trace element variability if sampled at different points of the dome or *coulée* (Cauvin et al., 1991:8–10). An understanding of the micro-variability of these flows is crucial to the success of the results afforded by trace-element analyses of obsidian sources. Acquiring the abovementioned signatures requires systematic sampling strategies of the different flows that exist for each source.

The identification of distinctive chemical profiles depends greatly on both a volcanological assessment of the eruption history of each obsidian-rich volcano, followed by the sampling methodology used to recover geological obsidians. Random sampling of a source can provide only a partial reading of the trace element composition variability of that source. Consequently, if a source flow is not sampled comprehensively, it is possible to mistakenly conclude that certain archaeological obsidians are not derived from it. By properly isolating and separately analyzing the numerous flows belonging to the two major highland Yemeni obsidian sources, Francaviglia's (1990b: 131) pioneering work characterizing Arabian obsidian sources circumvented the problems arising from such geochemical variability within individual sources.

Moreover, different origins of obsidian may have different physical and chemical properties. Obsidian in the southern Red Sea zone can be found in short, thick flows originating from isolated volcanic features (punctiform), in pyroclastic deposits, or in extensive sub-horizontal sheets that belong to the Oligocene (~30 million years old) Trap series of volcanic rocks (Francaviglia, 1990b: 129–130; Peate et al., 2005). Many of the isolated obsidian sources were erupted much more recently, during the Quaternary (Kabesh et al., 1980; Sanlaville, 2000: 135; Wilkinson, 2003: 16).

In Yemen alone, the Trap series covers 40,000 km^2 (Lipparini, 1954; Francaviglia, 1990b: 130). While the identification and sampling of Trap obsidians could add a rather large area of inquiry to the recovery of geological obsidians in Arabia, it is generally the case that these obsidians are devitrified due to their age (~30 million years). Most Trap obsidians in Yemen ceased to be suitable for knapping long before occupation of the area. This fact does not entirely disqualify certain Trap obsidians from having been exploited in antiquity as different flows and obsidian found in pyroclastic deposits can have differential devitrification.

Francaviglia (1990a: 46, 1990b: 132) sampled an obsidian level from the Trap series in the Yemeni highlands that had a slightly different mineralogical composition (non comenditic) from the two main analyzed obsidian sources in Yemen (peralkaline comenditic). Because of its poor quality, Francaviglia concluded that such obsidian was probably not heavily exploited for tool production in prehistory.

While the southern Red Sea zone is rich in obsidian, it is important to isolate those volcanic features that contain obsidian flows that were accessible and of a quality suitable for tool production in the prehistoric period. It is very likely, for example, that the majority of Trap obsidians in Arabia were of poor quality (Francaviglia, 1990a: 46) and purposefully disregarded by prehistoric populations in favor of better quality Quaternary sources. However, it is also important to account for and analyze all potential obsidian sources, including those of inferior quality such as the Trap series obsidian. It is also likely that prehistoric populations exploited nearby sources for their accessibility (economical strategies of introduction) rather than travel distances in search of quality material.

Arabian Tihamah Obsidian: Arabian or African Origin?

Many obsidian sources exist on either side of the southern Red Sea, making it a region that would have been an axis of exchange in prehistory (Fig. 1). Only a fraction of these has been sampled and geochemically characterized. One of the main archaeological study areas that has provided new obsidian data is the Tihamah coastal plain.

Fig. 1 Map of obsidian source areas and documented obsidian geometric microliths on both shores (western Yemen, Eritrea and Djibouti) of the southern Red Sea region

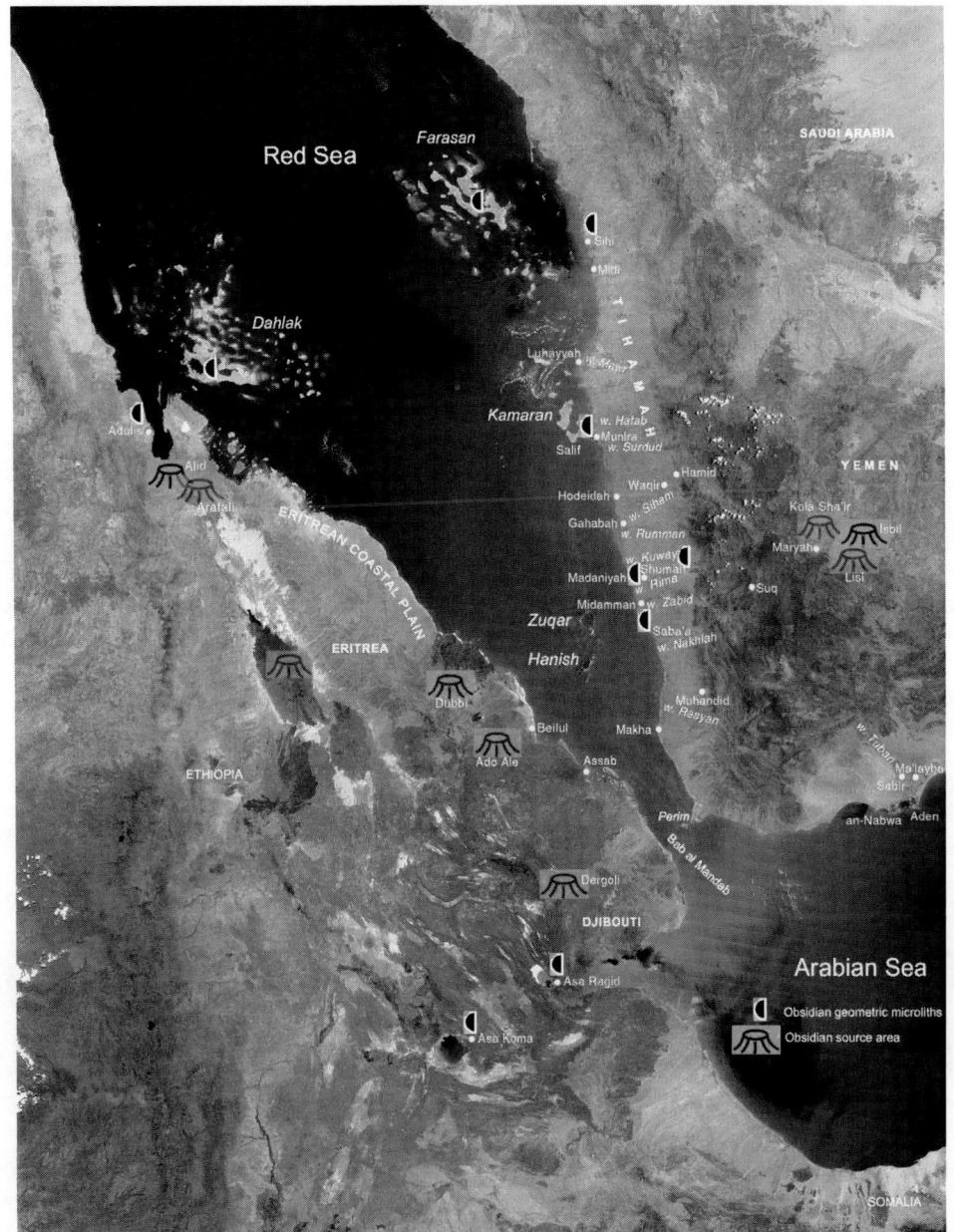

Spanning an area 30–60 km wide by 415 km along the Arabian coast of the Red Sea, this region is devoid of obsidian sources. To its east lies the escarpment of the Yemeni plateau which steeply ascends to the vast highland plains where two major obsidian sources exist at over 2,000 m above sea level and at a distance, as the crow flies, of roughly 170 km. On the other side of the Red Sea, at least five obsidian sources are interspersed along the Eritrean coastal plain (the closest only about 110 km by sea to the opposing coast and another 38 km overland to the source) and a number of additional sources dot the Ethiopian highlands beyond. Lying on the Red Sea littoral and straddled on each side by accessible obsidian source areas, the Tihamah – an area that produced a relatively large quantity of archaeological obsidian – is a pertinent region for discussions of prehistoric interaction across continental and geographic divides.

Obsidian in the Tihamah was predominantly used as a resource for tool manufacture, and appeared in the area during the Early Holocene. Obsidian tools first occur in the archaeological record of the Tihamah around the sixth millennium BC (Tosi, 1986: 407; Zarins, 1989, 1990; Cattani and Bokonyi, 2002: 44; Khalidi, 2006a: 140, 2007).

The early presence of obsidian in the region implies contact with the central or southern highlands or with the Horn of Africa coast where known obsidian sources exist. Its continued and increased use into the late prehistoric and historic periods illustrates an established long-term system of material procurement and exchange.

Although it appears to be a simple matter of matching lithic samples to their source outcrops, the dearth of geological source material has hampered tracing prehistoric obsidian procurement in southwest Arabia. A brief discussion of the present state of our knowledge on obsidian in the region follows.

Previous Obsidian Research

Cann and Renfrew's (1964) article, followed by the work of Renfrew and Dixon on obsidian distribution throughout western Asia, were the first attempts to relate the known sources throughout the Near East and the Arabian peninsula to the sites where obsidian has been found (Cann and Renfrew, 1964; Renfrew et al., 1968; Renfrew and Dixon, 1976). Only a few archaeological samples were analyzed from western Arabia and the Red Sea. All of these were among the peralkaline obsidians classified into groups 4d (defined by low Barium and high Zirconium) and 6 (lj) (defined by low Zirconium and high levels of Barium) (Renfrew and Dixon, 1976). Renfrew and Dixon state that all of these specimens must have come from sources near the Red Sea coasts. However, no sources in the Tihamah are known, and those known sources on the Eritrean side have yet to be sampled or sent for geochemical analysis.

Since the seminal work of Renfrew, Dixon and Cann in the 1960s and 1970s, much progress has been made in the field of obsidian source characterization in the Mediterranean and Near East, so much so that wider obsidian trade patterns throughout the Mediterranean world, Mesopotamia, Anatolia and the Trans-Caucasus have been confidently identified (Fornaseri et al., 1975–1977; Francaviglia, 1984, 1990a, 1994; Cauvin et al., 1986; Cauvin et al., 1991; Bader et al., 1994; Chataigner, 1994; Gratuze, 1994; Cauvin et al., 1998; Balkan-Atli et al., 1999; Cauvin, 2002).

In 1989, Zarins (1989) re-assessed previous work on obsidian trade and origin in the Red Sea region. At this time, only two source areas had been sampled in the Arabian peninsula. These included samples from three obsidian-rich volcanoes (Jebel Isbil and Jebel Lisi in Yemen and Jebel Abyad in Saudi Arabia), that were analyzed by Francaviglia (Francaviglia, 1989, 1990a, b, 1996). The situation in Arabia remains problematic to this day. As Zarins (1989: 58) clearly demonstrates:

"Francaviglia rightly suggests that to assess source area characterizations for obsidian several different approaches are necessary. First both standard chemical and trace element analyses are important and both should be published since differentiating obsidian groups solely on the basis of trace elements remains complex and elusive. In addition, as many samples as possible should be run to establish patterns and ranges of variation (Francaviglia, 1984). In evaluating our Red Sea sources, the lack of standardization is especially hard felt. Of our 32 reported sites, we have standard chemical sample reports only from ten. Sites with more than one sample analyzed number only five. Trace element analysis is available from only four sites and the number of sites which have both standard chemical and trace element analyses is confined to the two major sites from Arabia (Jebel Abyad and Dhamar-Reda)."

Many potential obsidian source localities have been mentioned by Zarins (more than 20 in Yemen, 21 in Eritrea and Djibouti, and 18 in Ethiopia). While certain Ethiopian sources have been analyzed (Francaviglia, 1990a: 47), such as that of Balchit lying south of Addis Ababa (Muir and Hivernel 1976), Fantale volcano and the Cañon de l'Aouache (Teilhard de Chardin P and Lamare, 1930a, b), most of those in Arabia and Eritrea appear neither to have been confirmed on the ground nor sampled (Zarins, 1989: 346, Fig. 44).

Analysis by Francaviglia has established that many of the Yemeni highland and interior archaeological obsidian specimens dating to the Neolithic and Bronze Age do not match the three analyzed Arabian sources (Francaviglia, 1990a: 48, b: 133–134). While Francaviglia demonstrates that a few samples from eastern highland plains sites (including Wadi Yana'im, Gabal Qutran and Wadi an-Nagid al-Abyad [De Maigret 1990]) match the neighboring Jebel Lisi volcano, and even fewer may pertain to an obsidian horizon from the Trap series, the majority originate from unknown sources in distant regions (Francaviglia, 1990b: 134). He advances that although there are possibilities that the obsidian may originate from a source near Aden (El-Hinnawi, 1964), or from un-analyzed elusive sources in Saudi Arabia, it most probably originates from the Eritrean/Ethiopian source areas known for their abundant Quaternary record of volcanism (Francaviglia, 1990a: 46, 1990b: 133–134; Wiart and Oppenheimer, 2005).

This emphasizes the need to confirm the presence of other potential Yemeni source areas, such as those in the southern regions of Aden and Ta'iz and the region north of Sana'a. Nonetheless, geochemical analysis of 136 obsidian specimens, both primary and knapped, recovered in Yemen, the Saudi Tihamah, the Yemeni Tihamah, Dhofar, Eritrea and Ethiopia has demonstrated that most of the knapped specimens do not have a known Arabian origin (Francaviglia, 1990a, b, 1996). This verifies that it is essential to also look elsewhere and especially in the Horn of Africa where large numbers of sources exist in proximity to Arabia and certainly to the Tihamah.

Arabian and African Obsidian Source Areas

Arabia

The two obsidian sources that have been characterized in Yemen are the sites of Jebel Isbil and Jebel Lisi, both located in the Dhamar-Rada' plains in the central highlands south of Sana'a (Francaviglia, 1990a, b). The third source to have been characterized is the site of Jebel Abyad in Saudi Arabia (Baker et al., 1973). The analyzed samples have shown that the obsidian from these sources fits into the peralkaline group characteristic of southwestern Arabia and the Horn of Africa (Francaviglia, 1990a, b) (see Fig. 1). A third western highland source was documented by the Dhamar Survey project in 1996 (Wilkinson et al., 1997: 122) and recently sampled by Khalidi and Lewis in 2006 and in more detail by Khalidi and Oppenheimer in 2008. The obsidian source of Jirab al-Souf (DSF06-030A-B), which makes up part of the Kowlat Sha'ir volcano, is located on the edge of the highland western escarpment and awaits analysis (Lewis and Khalidi, 2008).

Africa

> After about 800 stades (from Adulis) comes another, very deep, bay near whose mouth, on the right, a great amount of sand has accumulated; under this, deeply buried, obsidian is found, a natural local creation in that spot alone.
>
> – Unknown Egyptian-Greek merchant (A.D. 40–70) (1980: 53)

The obsidian sources that are of additional importance to the Red Sea littoral remain un-sampled. While three Ethiopian highland sources have been sampled and characterized, Francaviglia (1990a: 46–47) maintains that the only analyses that are useable for the three aforementioned characterized sources, are those published by Muir and Hivernel (1976). Despite this fact, there is a good deal of information as to the location of most East African obsidian outcrops from local sources, historical sources and archaeological and geological reports and articles. However, it must be kept in mind that the presence of an obsidian outcrop does not imply that the obsidian is of a workable quality or that the source was exploited. The identification of an exploited source of obsidian should combine a range of chemical analyses with survey of the surrounding area in order to determine whether there were, in fact, activity areas where the obsidian was quarried, prepared, and in some cases knapped in the vicinity, such as

those studies carried out in Anatolia and Armenia (Chataigner, 1994; Cauvin et al., 1998; Balkan-Atli et al., 1999; Barge and Chataigner, 2003; Chataigner and Barge 2005).

Three obsidian source areas in Eritrea are relevant to the discussion. The first is that of Arafali or Hawakil source 19 (Zarins, 1989). An obsidian specimen from the area of Arafali (see Fig. 1) was analyzed by Renfrew and assigned to group 6 (lj) (Renfrew and Dixon, 1976). It is interesting to note that two specimens from the Dahlak Islands (Eritrea), one from the Farasan Islands, and several from Hureidha in the Hadramawt in eastern Yemen, match the Arafali specimen (Zarins, 1990).

The second is the Alid volcano (source 20), which is located in the vicinity of Adulis (see Fig. 1). This site lies along the Gulf of Zula and reached its fame as a major Red Sea port in the Aksumite period. However, it is quite probable that the site was occupied as early as the second to third millennia BC and that it was used as a port contemporaneously (Fattovich, 1985: 459). Zarins points out that it is likely the site was used as a port from which obsidian was exported (Zarins, 1989: 360). The two sources discussed above are located in the Buri peninsula source area and are across from the northern Yemeni Tihamah and the Saudi Arabian Tihamah.

The third source area is located in the southern Red Sea region of Eritrea and lies directly across from the central Tihamah. Two source localities (22 and 23) in this area are mentioned by Zarins and correspond to the Dubbi and Ado Ale sources (see Fig. 1) (Zarins, 1989: 351). Although the author did not confirm the obsidian sources on a reconnaissance mission to the region, the entire volcanic zone was dense with obsidian scatters. The prehistoric site of Beilul (identified by the author on a reconnaissance mission to Eritrea in 2003) is located in this general area. The site is characterized by a number of circular stone-built tomb structures, rock art and obsidian debitage demonstrating that there was an association between Eritrean prehistoric sites and the adjacent obsidian sources.

Until the southern Red Sea region's Quaternary rhyolitic outcrops are systematically surveyed and obsidian sources suitably sampled and analyzed, the state of obsidian research and attempts at locating the origins of obsidian lithics are limited and prone to criticism. However, this fact does not undermine methods that have been used to isolate patterns and to group types of obsidian together. Given the characterized source areas in the Yemeni highlands and Saudi Arabia, it is possible to recognize whether samples of obsidian collected from sites in the region originated from these localities or not. In addition, the interpretation of different elemental plots and the analysis of the spatial and density distribution of obsidian and obsidian tool kits in specific areas can also provide clues as to the origin of archaeological obsidians.

Obsidian from the Central Tihamah Survey Area: Analysis

In 2003 a joint University of Cambridge – Yemeni survey project directed by the author, was carried out in the central Tihamah region of Yemen. The 2003 season was funded by a Fulbright IIE Grant and an American Institute for Yemeni Studies Fellowship. A second 2003 and a 2004 season were carried out in collaboration with the Yemeni Ministry of Environment, the Yemeni General Organization for Antiquities and Museums (GOAM), the Ministry of Public Works and Highways and its contractors: the Consulting Engineering Center (Sajdi & Partners) and the Regional Reef Group.

This project's aim was to apply systematic survey strategies to a bounded area (~90 km²) composed of a variety of landscapes in order to statistically assess settlement pattern, distribution and settlement variability in the context of land-use patterns and geomorphological processes affecting the Tihamah region. The 2003 study area was bound by the wadi Zabid to the south and the wadi Kuway' to the north. Its eastern boundary lay mid-way across the coastal plain in the area of the major north–south Hodeidah to Ta'iz highway and its western boundary was defined by the modern Red Sea littoral. Using handheld Geographical Positioning Systems and following a UTM grid system, north–south and east–west transects were walked every 500 m to begin and later at larger intervals in areas deemed poor for site preservation or for settlement location.

The Tihamah Survey (Khalidi, 2005, 2006b, 2007, 2008) documented a large number of sites (159) dating to the Early to Middle Holocene, the majority of which had a notable amount of obsidian in the form of debitage and tools. The relative dating of surface sites was aided by material culture parallels, geomorphological (e.g., presence of paleosols) and climatological (e.g., presence of molluscan species) indicators which corresponded to carbon dates recovered by Tosi (1986), Zarins and Al-Badr (1986) and Keall (2000) on sites previously excavated in the immediate region. Those sites pre-dating the sixth millennium BC were devoid of obsidian. From the sixth to the first millennium BC obsidian is ubiquitous on sites in the region, attaining preferred status as a tool material by the third millennium BC. Coinciding with the increase in obsidian is the production of obsidian geometric microliths (Fig. 2) and pièces esquillées (Fig. 3) on Tihamah late prehistoric sites (from the third to end of the first millennium BC). This particular lithic assemblage can be traced to contemporaneous sites across the Red Sea on the coastal plain of the Horn of Africa.

The analysis of samples of obsidian from the Tihamah survey, demonstrates an African origin for the materials from all periods documented, and further implies contact and interaction between these two areas. The introduction of African obsidian on the shores of the Tihamah goes back to sometime in the sixth millennium BC and its presence is confirmed until the second to first millennia BC. The changes that took place in exchange trajectories between African and Arabian shores throughout this long period of Red Sea interaction, remains to be studied. Such a study requires detailed sampling of the great number of obsidian sources located directly across the Red Sea in Eritrea, Ethiopia and Djibouti. However, data from the Tihamah confirms an increase of the material sometime in the third millennium BC alongside the appropriation of an African microlithic tradition.

The late prehistoric lithic assemblage of the Tihamah has demonstrated that obsidian was the primary tool material, and that the 'tool-kit' was dominated by obsidian geometric microliths and pièces esquillées and devoid of bifacial shaping. A technological study carried out by Crassard on the lithic assemblage from a contemporaneous excavated site, al-Midamman (Ciuk and Keall, 1996; Keall, 1998, 1999, 2000, 2004, 2005; Giumlia-Mair et al., 2000, Rahimi, 2001), agrees with the conclusions drawn for the surveyed sites. Sites exhibit small exhausted non-cortical bipolar microcores made on obsidian. While evidence of a micro-burin blow technique is absent, the production of predetermined bladelets and elongated flakes was made on anvil. These products provided the blanks for backed geometric microliths (Crassard, in press).

The small size of the obsidian nodules recovered on sites of the central Tihamah is noteworthy as it expresses a limited access to or availability of the material, as Crassard argues (Crassard, in press) or else a desire for small nodules which seems less likely. Either way, the presence of large amounts of small debitage and small tools exhibits an economical use of the material available. The paucity of cortical obsidian presumes a reduction scheme whereby nodules were decortified before arriving on Arabian shores. Unfortunately, surface collections are generally inadequate for in depth studies of core reduction schemes. While the material recovered appears to point to an exchange in small decortified nodules, it remains difficult to ascertain in what exact form, the nodules or cores were traded or brought over.

In addition to providing evidence of a widespread spatial distribution of this bipolar flaking technology, the Tihamah survey mapped patterns of obsidian density for sites of the late prehistoric period. The densities of obsidian are highest near the Tihamah littoral and remain consistently moderate to high along the riverbanks. On the other hand, densities are low along the inter-fluvial steppe and towards the interior of the coastal plain, especially beyond the main bifurcation of the deltas. Activity areas can be isolated along the wadi branches, and the general lithic assemblages from many sites suggest that obsidian was being worked at these sites.

Fig. 2 Geometric microliths from sites documented by the Central Tihamah Coastal Survey (drawings by L. Khalidi)

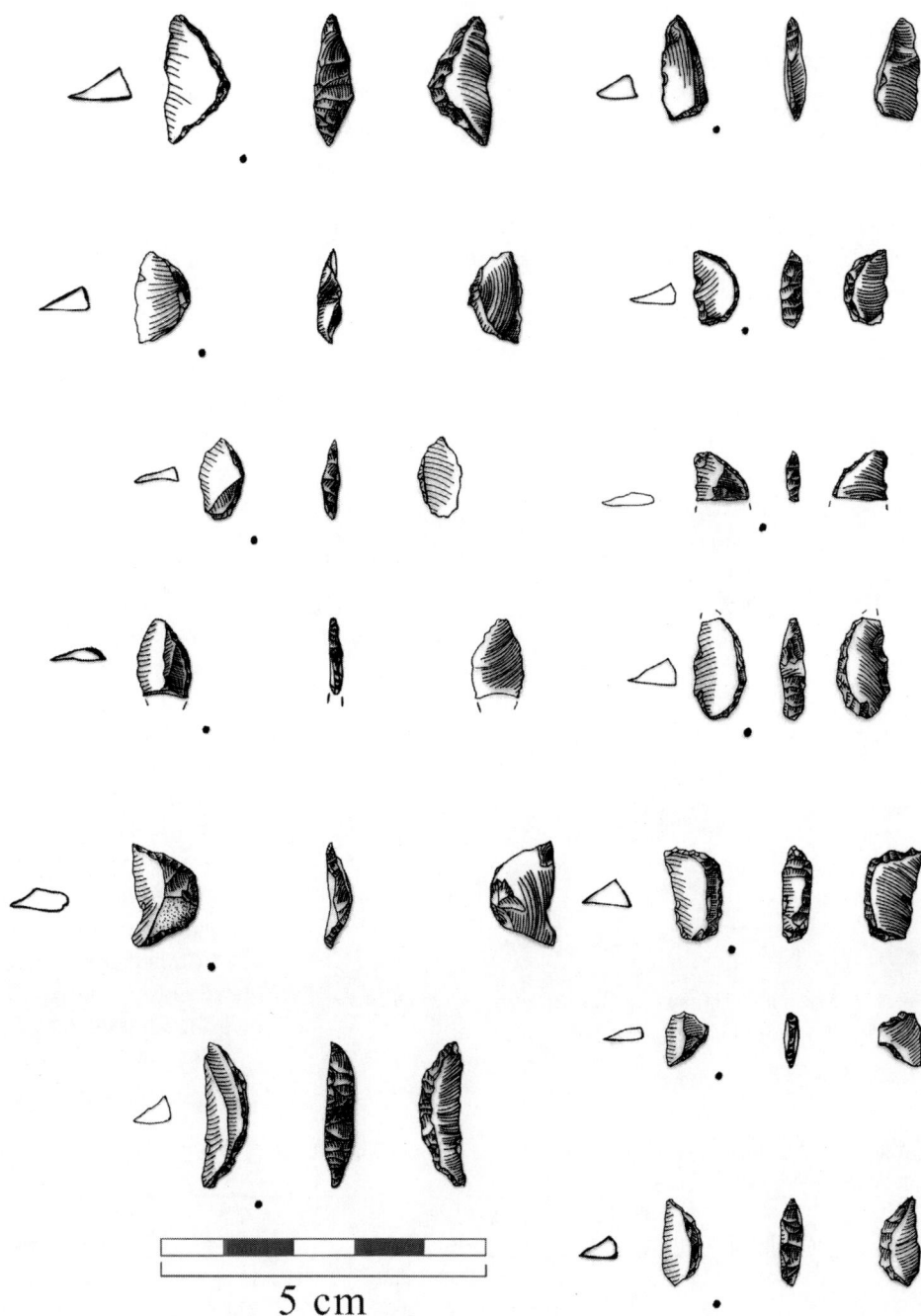

Furthermore, the main study area surveyed (Wadi Zabid to Wadi Kuway') has shown a general trend of higher densities (>80 samples) of obsidian on evenly dispersed sites along the coastline, and slightly more moderate densities (>50 samples) on evenly dispersed sites along the river banks (Khalidi, 2007: 38, map 2:2). Such densities suggest the obsidian was arriving by sea and traveling short distances upstream along the fluvial systems, eventually ending up at the large sites located midway between the littoral and the foothills. Samples of obsidian from fourteen sites (Table 1)

were analyzed and compared with potential sources by Francaviglia, previous Head of Research at the Institute of Applied Technology, CNR – Istituto per le Tecnologie Applicate ai Beni Culturali in Rome.

In Table 1, the sites from which the analyzed specimens were collected are listed, along with their projected dates. The sampled sites are evenly distributed along the coast and the two major *widyan* surveyed. Most of these sites constitute activity areas, an interpretation that was not solely based on the density or amount of obsidian recovered, but also on evidence for on-site tool production. All of the sites designated activity areas are characterized by a high percentage of obsidian debitage including cores and waste. Only a representative sample of obsidian was collected from each site.

A range of periods of occupation and site types are represented. The majority of the sites are late prehistoric in date, congruent with the fact that obsidian became more available and widely used during this period.

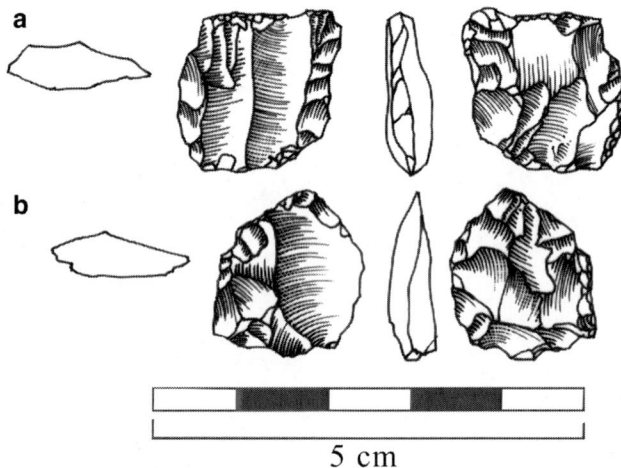

Fig. 3 *Pièces esquillées* from sites documented by the Central Tihamah Coastal Survey (drawings by J. Espagne)

Obsidian Analysis and Interpretation

The central Tihamah survey obsidian samples (see Table 1) analyzed by Francaviglia are summarized here. Several analytic approaches were used for the sourcing analyses including traditional X-ray fluorescence (XRF) spectroscopy, radioactive XRF, and tube-XRF, in order to retrieve major, minor and significant trace elements. The raw data were subjected to discriminant analysis and conventional bi-elemental plots. All of the samples showed homogeneity (e.g., bottle green in color). All of the Tihamah samples are of peralkaline composition and most lie in the field of comendites while four samples are pantellerites.

According to Francaviglia, the results of the plots show that, of the samples chosen from these sites, the majority definitely do not originate from known sources in Yemen (i.e., Jebel Lisi and Jebel Isbil) (see Fig. 1). In the case of Renfrew's Nb/Zr plot, some ten samples fall in the range of archaeological obsidians from Wadi Surdud, in the Tihamah, while 14 samples are very similar to obsidian samples from other sites in Tihamah and Eritrea (Francaviglia 1996: 70).

The samples originating from the Yemeni obsidian sources Isbil and Lisi as well as archaeological samples from other sites in Yemen, namely Sa'dah and Shabwah, have no direct correlation with the majority of Tihamah obsidians. On the other hand, when plotted alongside samples from Yemen, Eritrea (including the Alid source), and Ethiopia, two of the Tihamah samples are totally isolated as is the Eritrean Alid sample. The remaining 30 samples form isolated clusters unrelated to those of the Yemeni sources. However, some of these 30 samples can be interpreted as

Table 1 Analyzed obsidian samples from sites in the central Tihamah survey area. Type C and D tools refer to geometric microliths and *pièces esquillées*, respectively

Site #	# Analyzed	Total obsidian collected	Projected date BC	River system	Site morphology	Production area	Type tool
AJ4X4	10	24	Fifth–third millennium	Wadi Rima'	Shell midden	Yes	
AJ4Z	9	97	Third–first millennium	Wadi Rima'	Shell midden	Yes	C/D
AJ5Z	7	28	Third–first millennium	Wadi Kuway'	Artifact scatter	Yes	C/D
AJ6	5	56	Sixth–second millennium	Wadi Kuway'	Shell midden Artifact scatter	Yes	
AJ9B	2	29	Third–first millennium	Wadi Rima'	Shell midden	Yes	
AJ9B4	2	5	Seventh–fifth millennium	Wadi Rima'	Shell midden		
AJ11B	1	24	Seventh & third–first millennium	Wadi Kuway'	Shell midden Artifact scatter	Yes	C/D
AJ12	3	29	Fourth–third millennium		Shell midden Artifact scatter	Yes	
BF1F	3	12	Third–first millennium	Wadi Rima'	Shell midden	Yes	C/D
BF2A	1	6	Fourth–third millennium	Wadi Rima'	Shell midden		D
BF7C	2	20	Seventh & Third–first millennium		Shell midden	Yes	C
BF11	3	66	Third–first millennium	Wadi Rima'	Artifact scatter	Yes	
TH8	5	33	Third–first millennium	Wadi Rima'	Artifact scatter		D
TH16	8	169	Third–first millennium		Shell midden	Yes	C

having a clearer correlation to certain Ethiopian and Eritrean samples (Francaviglia, pers. comm., 2004).

The fact that a large proportion of the archaeological samples are similar to samples from Eritrea and Ethiopia is significant, since a large obsidian source area exists directly across the Red Sea (see Dubbi and Ado Ale in Fig. 1). The two isolated Tihamah samples are not related to any analyzed archaeological or source samples from either direction. Considering that more archaeological samples from the Arabian side have been analyzed and compared to the central Tihamah survey material, it is likely that these isolated obsidians pertain to un-characterized source areas in the southern Red Sea region, and most likely in the Beilul area. Although an African origin is not certain for all of the obsidian specimens analyzed, a discussion of microlithic technologies that are directly related to an African tradition establishes an undeniable contact between the two coasts of the Red Sea.

Obsidian Microlithic Technologies

Obsidian Geometric Microliths

In most parts of the world geometric microliths are synonymous with a Paleolithic and Mesolithic stone tool technology. The occurrence of these tools in such late contexts in Yemen is perplexing when viewed out of their geographic context. These tools occurred in tandem with *pièces esquillées* throughout the Tihamah coastal plain on contemporaneous sites of the third to first millennium BC. Furthermore, both technologies were made from obsidian only. The late occurrence of obsidian geometric microliths and *pièces esquillées* is contemporaneous with a similar phenomenon across the Red Sea.

The African Example

The occurrence of geometric microliths made on a variety of materials is well-known from a plethora of sites throughout sub-Saharan and North Africa. In the south and the interior these traditions are well-known from the Paleolithic period as a successor to the Levallois technique, while in parts of northern and eastern Africa they have a longer life-span and continue slightly later into the Epipaleolithic and Mesolithic (Phillipson, 1993: 60–61). In the Horn of Africa sequences, however, a later occurrence of microlithic technologies is evident in coastal areas, while the interior and highland instances are Neolithic in date or else contemporary with the rest of Africa. The Horn of Africa provides the youngest presence of geometric microlithic industries on the African continent.

As Phillipson (1993: 85–86) demonstrates, East Africa and especially the Horn of Africa have a continuous use of this tool type from as early as the Paleolithic into Neolithic contexts and found alongside bifacial industries. These Early Holocene industries can be found in North Africa, the Sudan, the Nile valley, and greater East Africa in concurrence with lake, river and coastal margin communities, some demonstrating the first ceramics and alternating hunter–forager and early domestication strategies. While such modes of subsistence are clearly related to the lush environments that characterized the Early Holocene, the explanation of the persistence of microlithic tool kits is less clear. It is possible that it corresponds to the exploitation of a variety of microenvironments and a resulting specialized hunting and incipient plant domestication, for which lighter hunting equipment and economical use of the materials and the tools was more advantageous. Because of their function as composite tools, microlithic blades could easily be repaired or replaced without replacing the entire implement, a factor that supports an opportunism and expediency as relates to the materials and to tool making (Phillipson, 1993: 99–100).

Late Prehistoric Period Geometric Microliths in the Horn of Africa

At the large site of Erkowit in Sudan, soundings and survey have yielded a microlithic tradition dated to the third millennium BC based on an associated ceramic assemblage falling within the 'Atbai Ceramic Tradition' (Callow and Wahida, 1981: 276; Zarins, 1989: 359; Fattovich, 1997).

The sites of Kokan and Ntanei in the Agordat area of Eritrea have yielded an abundance of obsidian debitage including flakes and cores, as well as a microlithic technology consisting of retouched lunates (Arkell, 1954). It is suggested that the microlithic technology may belong to the mid-second millennium BC 'Jebel Mokram Group' (Zarins, 1989: 359). At the port site of Adulis in Eritrea (see Fig. 1), deep soundings excavated at the turn of the century yielded a large amount of obsidian flakes and microliths (Paribeni, 1907). Several dates have been suggested for the Adulis assemblage. While Fattovich assigns a date in the first millennium BC, Zarins suggests a second millennium BC date based on ceramic parallels with Sihi, Sabir, and SLF-1 in southwest Arabia (Paribeni, 1907; Fattovich, 1985; Zarins and Al-Badr, 1986; Zarins, 1989: 359). A second millennium BC date is highly plausible given the occurrence of the tool type across the Red Sea in this period.

At the site of Asa Koma in Djibouti, 23% of the tool assemblage (consisting of approximately 16,000 obsidian tools) (Joussaume, 1995: 33–36) consisted of obsidian geometric microliths including both arch-shaped and circle segment samples. In addition a large number of *pièces esquillées* are represented in association with *bâtonnets*, a waste produced from the bipolar flaking technique (Joussaume, 1995: 35–6). Based on C-14 samples, the dates obtained for this site fall in the second millennium BC (Joussaume, 1995: 32). All of the characteristics of the lithic tradition of this site match those of the Tihamah. Although the site differs in many ways (stone accessibility, different ceramic tradition) the contemporaneity suggests a parallel mode of existence that is clearly linked by the Red Sea.

Similarly geometric microliths are attested from the sites of Asa Ragid on the coast of Djibouti, and Goda Ondji and Rachid Hussein in the Ethiopian Harar, all dating to the early to mid-Holocene transition (Joussaume, 1995: 30, 52–53). A handful of other sites, stretching from the Sudan in the north to Kenya in the south and dated to the third to second millennium BC have provided evidence of such a toolkit (Poisblaud, 1999).

A microlithic tradition was documented at Er Rih island in Sudan (Callow and Wahida, 1981: 36), at the site of 'Aqiq (Zarins, 1989: 359; Fattovich, 1997: 276) and the Red Sea Farasan islands (Zarins, 1989: 359). A tradition of geometric microliths made from obsidian with associated obsidian *pièces esquillées* was documented on the Dahlak Islands in Eritrea. Blanc attributed this occurrence to Wiltonian industries of the Mesolithic period, as at the time geometric microliths were thought to be more ancient (Blanc, 1952: 355–357). Blanc (1952: 357) found the occurrence of the industries on the Red Sea Island during the Mesolithic period perplexing given that paleographic conditions would not have allowed for occupation on this island, given the rise in sea levels during this period.

Given our current state of knowledge, it can now be demonstrated that such a tool kit was used in later periods. This has been confirmed by the parallel use of geometric microliths and (where they have been identified) *pièces esquillées* made on obsidian at all of the sites mentioned above, as well as those across the Red Sea along the Yemeni Tihamah. These are dated to between the third and the late first millennium BC and are a good indication that the Dahlak tool kit documented by Blanc in the 1950s and originally dated to the Mesolithic, is also late prehistoric in date.

Obsidian Circulation in the Yemeni Highlands and Tihamah

A combination of evidence (i.e., the exchange of obsidian as primary material; its decreased distribution away from the Arabian littoral; the association between the lithic technologies and tool types utilized simultaneously along the African and Arabian shores of the southern Red Sea) confirms long-distance interaction and exchange between African and Arabian populations. Recent data substantiates Francaviglia's arguments that distance and 'geographic boundaries' were not necessarily barriers influencing obsidian procurement and exchange in prehistoric Arabia. This appears to be the case in regions such as the Yemeni highlands where a majority of the obsidian was not always procured from the nearest source, but often came from very distant sources (Francaviglia, 1990a, b).

Yet, highland studies have shown the presence of obsidian production areas near highland sources. They have also demonstrated that local obsidian distribution networks were present, such as the production site of al 'Irr (DS217) near the Jebel al-Lisi source (Wilkinson and Edens, 1999: 6), or the more recently surveyed Maryah region with numerous localized production areas located near and along the obsidian source of Jirab al-Sawf (DSF06-030A-B)(Lewis and Khalidi, 2008). Though source analyses require further study, recent data in the Maryah area appears to parallel Wilkinson and Edens' view that the highland sources "… strongly suggest a relatively limited circulation of material from each source, with moderately steep drop-off with distance from the sources." (Wilkinson et al., 1997: 122).

In this case what were the factors influencing obsidian procurement and exchange in the prehistoric period? Were there small localized obsidian networks embedded within larger systems of obsidian exchange, or are we evidencing micro-regional variability in what relates to both inter-regional interaction and cultural preference?

It is still too early to answer such questions. Nonetheless, the data that are available have important implications for intra- and inter-regional contact and exchange. The majority of the highland archaeological obsidians analyzed by Francaviglia would have had to originate across the Red Sea in Eritrea (nearest source at 327 km as the crow flies), or in Saudi Arabia (nearest source 347 km distant). One can posit that long distances were traveled either for the procurement of better quality obsidian or as a result of pre-existing long-distance pathways of interaction and exchange. The non-confirmed sources in Yemen (Aden, Ta'iz and Saada), like the Trap obsidians, are likely to have been of a poorer quality. Consequently, obsidian quality may have played a large role in highland obsidian procurement and motivated exploitation from more distant, yet more homogeneous, sources. The combination of this evidence and the presence of cross-Red Sea contact linked to obsidian exchange, provide a strong argument for people having transported African obsidians as far as the Yemeni highlands.

It is possible that two modes of obsidian circulation existed in the prehistoric Yemeni highlands. The first is that of localized and limited circulation from certain highland sources. The second is that of circulation over long distances which would have had to be undertaken across very mountainous terrain (a distance of 115–215 km to non-confirmed source areas north of Sana'a and near Aden in Yemen, 327 km to the nearest Eritrean sources, or 347 km to the nearest Saudi sources).

Highland obsidian that appears to come from non-highland Yemen sources would have had to be procured at relatively long distances through extreme mountainous landscapes, and make the distances needed for a Red Sea crossing pale in comparison. For the obsidian in the Tihamah study area to have arrived from the nearest Eritrean source, that of Ado Ale or Dubbi (Wiart and Oppenheimer, 2000), it would have traveled 38 km overland, from the sources on the Eritrean side, then 110 km across the sea, with the crossing bridged by the island chain of Hanish and Zuqar lying halfway in between. A maximum distance of 148 km across the Red Sea and across flat coastal terrain to the Eritrean sources, as opposed to an ascent of 180 km along rugged mountain terrain to the nearest highland source, is evidently the nearest and more efficient choice for those inhabiting the Tihamah in the prehistoric periods.

That obsidian geometric microliths, in tandem with a bipolar flaking technology, are contemporary on both coasts of the Red Sea further validates that it was traversed in the late prehistoric period and supports the potential for maritime activity as early as the sixth millennium BC when the first non-Yemeni archaeological obsidians appear on the Tihamah coast.

A complex system of material procurement and human interaction is at play in the Arabian Holocene. The application of models of obsidian circulation remains difficult without better geological data from sources throughout the wider southern Red Sea obsidian zone. The factors that influenced prehistoric populations to exploit certain obsidian sources and not others and to move through diverse territories, sometimes at very large distances, is possibly representative of larger patterns of socio-economic behavior and cultural appropriation. It is likely that the patterns we are beginning to identify were shaped by more ancient human orientations or pathways as well as shaping later ones.

The studies undertaken in other nearby regions have demonstrated the advantage of obsidian research in reconstructing the history and movement of prehistoric populations. The success of such studies emphasizes the need for more systematic research in areas such as the southern Red Sea where the potential is great, but the data scarce.

Acknowledgments The Central Tihamah Coastal Survey was financed by a Fulbright IIE grant. It is thanks to the Yemeni General Organization for Antiquities and Museums (GOAM), Dr. Krista Lewis, Ahmed al Mosabi and Essam Hamana as well as the logistical support offered by the American Institute for Yemeni Studies, that this fieldwork was made possible. The results presented in this paper would not have been possible without the analysis carried out by Dr. Francaviglia of the CNR-Rome, and the intellectual support and advice of M.-L. Inizan, E. Keall, T. Wilkinson and R. Crassard. Finally, I would like to thank the Centre Français d'Archéologie et de Sciences Sociales de Sana'a (CEFAS), the Centre d'Etudes Préhistoire, Antiquité et Moyen Age (CEPAM-UNSA), the Fyssen Foundation, Dr. C. Oppenheimer, Dr. B. Gratuze, Dr. F. Marshall, and Dr. K. Lewis for their contributions and investment in the future of this project.

References

Amirkhanov H. L'Acheuléen de l'Arabie du Sud (en russe). Sovietskaya Arkeologiya 1997;4.

Arkell AJ. Four occupation sites at Agordat. Kush. 1954;2:33–62.

Bader NO, Merpert NJ, Munchaev RM. Les importations d'obsidienne sur les sites des IXᵉ-VIIᵉ millénaires B.P. du Djebel Sinjar, Iraq. Paléorient 1994;20(2):6–8.

Baker PE, Brosset R, Gass IG, Neary CR. Jebel al Abyad: a recent alkalic volcanic complex in western Saudi Arabia. Lithos. 1973;6:291–314.

Balkan-Atli N, Binder D, Cauvin MC, Biçaki E, Der Aprahamian G, Kuzucuoglu C. Obsidian: sources, workshops and trade in Central Anatolia. In: Özdogan M, Basgelen N, editors. Neolithic in Turkey: the cradle of civilization. Istanbul: Arkeoloji Ve Sanat Yayinlari; 1999. p. 133–43.

Barge O, Chataigner C. The procurement of obsidian: factors influencing the choice of deposits. Journal of Non-Crystalline Solids. 2003;323:172–9.

Blanc, A-C. L'Industrie sur obsidienne des iles Dahlac (Mer Rouge). Congres Panafricain de Préhistoire Algers; 1952. p. 355–57.

Callow P, Wahida G. Fieldwork in Northern and Eastern Sudan, 1976–1980. Nyame Akuma: Bulletin of the Society of Africanist Archaeologists. 1981;18:34–6.

Cann JR, Renfrew C. The characterization of obsidian and its application to the Mediterranean region. Proceedings of the Prehistoric Society. 1964;30:111–31.

Cattani M, Bokonyi S. Ash-Shumah: an Early Holocene settlement of desert hunters and mangrove foragers in the Yemeni Tihamah. In: Cleuziou S, Tosi M, Zarins J, editors. Essays on the late prehistory of the Arabian peninsula. Rome: Serie Orientale Roma, ISIAO; 2002. p. 3–52.

Cauvin MC. L'obsidienne et sa diffusion dans le Proche-Orient Néolithique. In: Guilaine J, editor. Matériaux, productions, circulations du Néolithique à l'Age du Bronze. Séminaire du Collège de France. Paris: Errance; 2002. p. 13–7.

Cauvin MC, Balkan N, Besnus Y, Saroglu F. Origine de l'obsidienne de Cafer Höyük (Turquie): premiers résultats. Paléorient. 1986;12(2): 89–97.

Cauvin MC, Besnus Y, Tripier J, Montigny R. Nouvelles analyses d'obsidiennes du Proche-Orient: modèle de géochimie des magmas utilisé pour la recherche archéologique. Paléorient. 1991;17(2):5–20.

Cauvin MC, Gourgaud A, Gratuze B, Arnaud N, Poupeau G, Poidevin J-L, et al., editors. L'obsidienne au Proche et Moyen Orient: Du volcan à l'outil. Oxford: Archaeopress; 1998.

Chataigner C. Les propriétés géochimiques des obsidiennes et la distinction des sources de Bingöl et du Nemrut Dag. Paléorient. 1994;20(2):9–17.

Chataigner C, Barge O. L'analyse de la circulation de l'obsidienne dans le nord du Proche-Orient préhistorique. In: Berger J-F, Bertoncello F, Braemer F, Davtian G, Gazenbeek M, editors. Temps et espaces de l'homme en société: Analyses et modèles spatiaux en archéologie. XXVe rencontres internationales d'archéologie et d'histoire d'Antibes. Antibes: Éditions APDCA; 2005. p. 405–9.

Ciuk C, Keall EJ. Zabid Project Pottery Manual 1995: Pre-Islamic and Islamic Ceramics from the Zabid area, North Yemen. Oxford: Tempvs Reparatvm; 1996.

Crassard R. Apport de la Technologie Lithique à la Définition de la Préhistoire du Hadramawt, dans le Contexte du Yémen et de L'Arabie du Sud, Vol. 1 and 2. Ph.D. dissertation, UFR 03 – Art & Archéologie – Anthropologie, Ethnologie, Préhistoire Université Paris 1 – Panthéon – Sorbonne, Paris; 2007.

Crassard R. Obsidian lithic industries from al-Midamman (Tihama coast, Yemen). In: Keall EJ, editor. Pots, rocks and megaliths. BAR International Series, Archaeopress, Oxford (in press)

Crassard R, Khalidi L. De la pré-Histoire à la Préhistoire au Yémen: des données anciennes aux nouvelles expériences méthodologiques. Chroniques Yéménites. 2005;12:1–18.

El-Hinnawi E. Petrochemical characters of African volcanic rocks. Part I: Ethiopia and the Red Sea Region (including Yaman & Aden). Neues Jahrbuch für Mineralogie. Monatshefte. 1964;3:65–81.

Fattovich R. Elementi per la Preistoria del Sudan Orientale e dell'Etiopia Settentrionale. In: Liverani M, Palmieri AM, Peroni R, editors. Studi di Paletnologia in Onore di Salvatore M. Puglisi. Rome: Università di Roma, La Sapienza – Dipartimento di Scienze istoriche; 1985. p. 451–63.

Fattovich R. The contacts between Southern Arabia and the Horn of Africa in Late Prehistoric and Early Historical Times: a view from Africa. In: Avanzini A, editor. Profumi D'Arabia, Saggi di Storia Antica. Roma: "L'ERMA" di BRETSCHNEIDER; 1997. p. 273–86.

Fornaseri M, Malpieri L, Palmieri AM, Taddeucci A. Analyses of obsidians from the Late Chalcolithic levels of Aslantepe (Malatya). Paléorient 1975–1977;3:231–46.

Francaviglia VM. Characterization of Mediterranean obsidian sources by classical petrochemical methods. Preistoria Alpina. 1984;20:311–32.

Francaviglia VM. Les gisements d'obsidienne hyperalcaline dans l'Ancien Monde: étude comparative. Revue d'Archéométrie. 1990a;14:43–64.

Francaviglia VM. Obsidian sources in ancient Yemen. In: De Maigret A, editor. The Bronze Age culture of Khawlan at-Tiyal and al-Hada (Yemen Arab Republic). Roma: IsMEO; 1990b. p. 129–36.

Francaviglia VM. L'origine des outils en obsidienne de Tell Magzalia, Tell Sotto, Yarim Tepe et Kül Tepe, Iraq. Paléorient. 1994;20(2):18–31.

Francaviglia VM. Il existait déjà au Néolithique un commerce d'obsidienne à travers la mer Rouge. In: L'Archéométrie dans les pays Européens de langue latine et l'implication de l'archéométrie dans les grands travaux de sauvetage archéologique. Actes du colloque d'Archéométrie 1995, Supplément à la Revue Archéométrie, Pole Editorial Archéologique de l'Ouest (P.E.A.O.), Perigueux; 1996. p. 65–70.

Francis P, Oppenheimer C. Volcanoes. 2nd ed. Oxford: Oxford University Press; 2004.

Giumlia-Mair A, Keall EJ, Stock S, Shugar A. Copper-based implements of a newly identified culture in Yemen. Journal of Cultural Heritage. 2000;1:37–43.

Gourgaud A. Géologie de l'obsidienne. In: Cauvin MC, Gourgaud A, Gratuze B, Arnaud N, Poupeau G, Poidevin J-L, Chataigner C, editors. L'obsidienne au Proche et Moyen Orient: du volcan à l'outil. BAR International Series 738 and Maison de l'Orient Méditerranéen. Archaeopress, Oxford; 1998. p. 15–29.

Gratuze B. Contribution à l'analyse d'outils provenant du site archéologique de Magzalia. Paléorient. 1994;20(2):32–4.

Joussaume R. Les premières sociétés de production. In: Joussaume R, editor. Tiya- l'Éthiopie Des Mégalithes: Du biface à l'art rupestre dans la Corne de l'Afrique. Poitiers: P. Oudin; 1995. p. 15–63.

Kabesh M, Refaat AM, Abdallah Z. Geochemistry of Quaternary volcanic rocks, Dhamar-Rada' Field, Yemen Arab Republic. Neues Jahrbuch für Mineralogie, Monatshefte. 1980;138:292–311.

Keall EJ. Encountering megaliths on the Tihamah coastal plain of Yemen. Proceedings for the Seminar for Arabian Studies. 1998; 28:139–47.

Keall EJ. Archäologie in der Tihamah: Die Forschungen der Kanadischen Archäologischen Mission des Royal Ontario Museum, Toronto. Zabid und Umgebung. Jemen-Report: Mitteilungen der Deutsch-Jemenitischen Gesellschaft e.V. 1999;30(1):27–32.

Keall EJ. Changing settlement along the Red Sea coast of Yemen in the Bronze Age. In: Matthiae P, Enea A, Peyronel L, Pinnock F, editors. Proceedings of the First International Congress on the Archaeology of the Ancient Near East, Rome; 2000. p. 719–29.

Keall EJ. Possible connections in antiquity between the Red Sea coast of Yemen and the Horn of Africa. In: Lunde P, Porter A, editors. Trade and travel in the Red Sea region. Society for Arabian studies monographs No. 2 and BAR International Series 1269. Oxford: Archaeopress; 2004. p. 43–55.

Keall EJ. Placing al-Midamman in time. The work of the Canadian Archaeological Mission on the Tihama coast, from the Neolithic to the Bronze Age. In: Archäologische Berichte aus dem Yemen, Deutsches Archäologisches Institut, Verlag Phillip Von Zabern, San'a', Mainz am Rhein; 2005. p. 87–99.

Khalidi L. The prehistoric and early historic settlement patterns on the Tihamah coastal plain (Yemen): preliminary findings of the Tihamah coastal survey 2003. Proceedings of the Seminar for Arabian Studies. 2005;35:115–27.

Khalidi L. Settlement, culture-contact and interaction along the Red Sea coastal plain, Yemen: the Tihamah cultural landscape in the late prehistoric period, 3000–900 BC. Ph.D. dissertation, University of Cambridge Faculty of Archaeology, Cambridge; 2006a.

Khalidi L. Megalithic landscapes: the development of the late prehistoric cultural landscape along the Tihama coastal plain. In: Sholan AM, Antonini S, Arbach M, editors. Sabaean studies: archaeological, epigraphical and historical studies in honor of Yusuf M. Abdallah, Alessandro de Maigret and Christian J. Robin on the occasion of their 60th birthdays. I.U.O. Naples, Sanaa; 2006b. p. 359–75.

Khalidi L. The formation of a southern Red Seascape in the late prehistoric period: tracing cross-Red Sea culture-contact, interaction, and maritime communities along the Tihamah coastal plain, Yemen in the third to first millennium BC. In: Starkey J, Starkey P, Wilkinson T, editors. Natural resources and cultural connections of the Red Sea: Proceedings of Red Sea Project III. Society for Arabian studies monographs No. 5 and BAR International Series 1661. Archaeopress, Oxford; 2007, p. 35–43.

Lewis K, Khalidi L. From prehistoric landscapes to urban sprawl: the Masna'at Marya Region of Highland Yemen. Proceedings of the Seminar for Arabian Studies 2008; 38

Lipparini T. Contributi alla conoscenza geologica del Yemen (SW Arabia). Bolletino del Servizio Geologico 1954; LXXVI:95–118.

Mogessie A, Krenn K, Schaflechner J, Koch U, Egger T, Goritchnig B, Kosednar B, Pichler H, Ofner L, Bauernfeind D, Tadesse S, Hailu K, Demessie M. A geological excursion to the Mesozoic sediments of the Abay Basin (Blue Nile), recent volcanics of the Ethiopian main Rift and basement rocks of the Adola Rea, Ethiopia. Mitt Österr Miner Ges 2002;147:43–74.

Muir ID, Hivernel F. Obsidians from the Melka-Konturé prehistoric site, Ethiopia. Journal of Archaeological Sciences. 1976;3:211–7.

Paribeni R. Ricerche nel Luogo dell'Antica Adulis (Colonia Eritrea). Monumenti Antichi. 1907;18:445–51.

Peate IU, Baker JA, Al-Kadasi M, Al-Subbary A, Knight KB, Riisager P, et al. Volcanic stratigraphy of large-volume silicic pyroclastic eruptions during Oligocene Afro-Arabian flood volcanism in Yemen. Bulletin of Volcanology. 2005;68(2):135–56.

Petraglia MD. The Lower Paleolithic of the Arabian peninsula: occupations, adaptations, and dispersals. Journal of World Prehistory. 2003;17(2):141–79.

Phillipson DW. African archaeology. Cambridge: Cambridge University Press; 1993.

Piel-Desruisseaux J-L. Outils Préhistoriques: Du galet taillé au bistouri d'obsidienne. Paris: Dunod; 2004.

Poisblaud B. Les sites du Ghoubbet à Djibouti dans le cadre de la préhistoire récente de l'Afrique de l'Est: Tome I, Tome II. Ph.D. dissertation, Préhistoire africaine – Anthropologie, Université de Paris – Pantheon –Sorbonne, Paris; 1999.

Rahimi D. Geometric microliths of Yemen: Arabian precursors, African connections. Conference paper, society for American archaeology, parting the Red Sea: Holocene interactions between Northeastern Africa and Arabia, New Orleans; 2001.

Renfrew C. Forword. In: Cauvin MC, Gourgaud A, Gratuze B, Arnaud N, Poupeau G, Poidevin J-L, Chataigner C, editors. L'obsidienne au Proche et Moyen Orient: du volcan à l'outil. BAR International Series 738 and Maison de l'Orient Méditerranéen. Oxford: Archaeopress; 1998, p. 5–6.

Renfrew C, Dixon J. Obsidian in western Asia: a review. In: Sieveking GdG, Longworth IH, Wilson KE, editors. Problems in economic and social archaeology. London: Duckworth; 1976. p. 137–50.

Renfrew C, Dixon J, Cann JR. Further analysis of near eastern obsidian. Proceedings of the Prehistoric Society. 1968;34:319–31.

Sanlaville P. Le Moyen-Orient arabe: Le milieu et l'homme. Paris: Armand Colin; 2000.

Teilhard de Chardin P, Lamare P. Le cañon de l'Aouache et le volcan Fantalé. Etudes géologiques en Ethiopie, Somalie et Arabie Méridionale. Mémoires de la Société géologique de France 1930a;6(II):13–20.

Teilhard de Chardin P, Lamare P. Observations géologiques en Somalie Française et au Harar. Etudes géologiques en Ethiopie, Somalie et Arabie Méridionale. Mémoires de la Société géologique de France 1930b;6(I):5–12.

Tosi M. Archaeological activities in the Yemen Arab Republic, 1986: survey and excavations on the coastal plain (Tihamah). East and West. 1986;36:400–15.

Wiart PAM, Oppenheimer C. Largest known historic eruption in Africa: Dubbi volcano, Eritrea, 1861. Geology. 2000;28:291–4.

Wiart P, Oppenheimer C. Large magnitude silicic volcanism in north Afar: The Nabro Volcanic Range and Ma'alalta volcano. Bulletin of Volcanology. 2005;67:99–115.

Wilkinson TJ. Archaeological landscapes of the Near East. Tucson, AZ: The University of Arizona Press; 2003.

Wilkinson TJ, Edens C. Survey and excavation in the central highlands of Yemen: results of the Dhamar Survey Project, 1996 and 1998. Arabian Archaeology and Epigraphy. 1999;10:1–33.

Wilkinson T, Edens C, Gibson M. The Archaeology of the Yemen high plains: a preliminary chronology. Arabian Archaeology and Epigraphy. 1997;8:99–142.

Zarins J. Ancient Egypt and the Red Sea trade: the case for obsidian in the predynastic and archaic periods. In: Leonard Jr A, William BB, editors. Essays in ancient civilization presented to Helene J. Kantor. SAOC; 1989, p. 339–68.

Zarins J. Obsidian and the Red Sea Trade: prehistoric aspects. In: Taddei M, Callieri P, editors. South Asian archaeology 1987. Rome: Istituto Italiano per il Medio ed Estremo Oriente; 1990. p. 507–41.

Zarins J, Al-Badr H. Archaeological investigation in the Southern Tihama Plain II (Including Sihi, 217-107 and Sharja, 217-172) 1405/1985. Atlal. 1986;10(Part I):36–57.

Part V
Synthesis and Discussion

Chapter 20
The Paleolithic of Arabia in an Inter-regional Context

Anthony E. Marks

Keywords Acheulean • Early Stone Age • Middle Stone Age • Lower Paleolithic • Middle Paleolithic • Post-Acheulean • Upper Paleolithic

Introduction

Very little is known about the Paleolithic of Arabia. In spite of surveys undertaken immediately after the initial exploration of this environmentally marginal region (e.g., Philby, 1933; Caton-Thompson, 1939) and a small but continuous trickle of prehistorians into Arabia over the past 60 years, knowledge of both Arabian Pleistocene occupations and paleoenvironments is woefully poor, compared to what is known about adjacent regions. The reasons for this are myriad, ranging from the absence of extant, large karstic caves with deeply stratified sediments (the highly preferred Paleolithic site type of the twentieth century), to truly difficult logistics, and, until recently, a lack of encouragement from local authorities. Still, prehistorians did try and virtually all found some materials they could attribute to the Paleolithic (e.g., Caton-Thompson, 1954; Van Beek et al., 1963; Gramly, 1971; Pullar, 1974; Inizan and Ortlieb, 1987; Whalen and Pease, 1990; McBrearty, 1993).

With the exception of Amirkhanov (1991, 1994), who excavated some shelters in Yemen that had only poor archaeological deposits, almost all Paleolithic materials were found in either deflated or eroded surface scatters, without geological context and without the possibilities for absolute dating. The discovery of such sites usually led to small, cursory collections, mainly of what were perceived to be "diagnostic" artifacts, since the contemporaneity of the artifacts in such scatters could not be assumed. In addition, there was the problem of the ubiquitous Neolithic arrowheads that cover the surface of Arabia, such that one or two might be found in

proximity to almost any scatter of artifacts and, so, making it possible to view almost all scatters as Neolithic (Edens, 1982; Charpentier, 1999).

The cumulative result of all these collections was to clearly document the presence of Paleolithic materials in Arabia but little more. It was possible, based on the typology and technology of many artifacts, to recognize a conceivable but certainly unproven (Petraglia and Alsharekh, 2003) pre-Acheulean (Amirkhanov, 1994; Whalen and Fritz, 2004; Whalen and Schatte, 1997), an Acheulean in some areas (e.g., Amirkhanov, 1987, 1995; Whalen et al., 1983; Zarins et al., 1981), and what was called Middle Paleolithic, based on recognizable Levallois flakes and cores, as well as "general" Middle Paleolithic type tools (e.g., Caton-Thompson, 1939; de Bayle, 1976). With a single exception (Edens, 2001) beyond the Levantine border area, no evidence had been found for anything either typologically or technologically "Upper Paleolithic" or "Epipaleolithic" in the traditional Levantine sense, with fine blade technology, tools on blades and bladelets, and in later stages, with geometric and backed microlithic tools.

At best, only two periods of occupation were soundly established, a Lower Paleolithic of Acheulean type and a generic Middle Paleolithic and, even then, diagnostic materials were not abundant for the latter. The general sense of these finds was that Arabia was never densely occupied during the Paleolithic and, perhaps, was completely unoccupied during some rather long periods. This understanding fit well with the generally prevailing view of Arabian paleoclimates. While there were pluvial periods when the Indian Ocean monsoons pushed north of the Dhofar Mountains, they were interspersed with long inter-pluvials during which Arabia was thought to have been hyperarid and unsuitable for human occupation.

Given this understanding, the part that adjacent regions played in the Arabian Paleolithic, by necessity, was major since some adjacent region or regions had to have been the source of the populations found in Arabia during pluvial periods, as there would have been no people left in Arabia after inter-pluvial conditions took hold. Even this, however, was not as consistent as might have been expected, since no Upper Paleolithic had been found that, in the Levant, at least, was essentially coeval with the pluvial of MIS 3 (60–24 ka)

A.E. Marks (✉)
Department of Anthropology, Southern Methodist University, 3225 Daniel Avenue, Heroy Building 408, Dallas, TX, 75275, USA
e-mail: amarks@mail.smu.edu

and the number of sites attributable to the Middle Paleolithic that could have fallen into the pluvials of MIS 7 (200–180 ka), MIS 5e (128–120 ka), and MIS 5a (82–72 ka) were, in fact, very few and rather far between. This paucity of sites is in marked contrast with the very high Middle Paleolithic site density in the arid southern Levant (on the northern edge of Arabia), mostly datable to the MIS 7 and MIS 5a pluvials (Bar-Yosef, 2007). A similar contrast exists with East and North East Africa, where mid-MSA/MP sites are abundant (e.g., Marks, 1968; Clark, 1988; Rafalski et al., 1978; Yellen et al., 2005).

The very paucity of known true Paleolithic sites in Arabia, as opposed to isolated findspots, tended to re-enforce the idea of only intermittent occupations during the Pleistocene and, therefore, for their immediate origins in adjacent regions. Yet, perhaps, because of the small size of most collections, little effort had been made to link specific assemblages with any in adjacent regions. Most statements tended to be very general, such as that made by McBrearty (1993) that the material she saw was not inconsistent with MSA material in East Africa, or when Pullar (1974) saw possible African connections in the bifacial foliates from some Arabian surface sites. Others (Whalen and Pease, 1990; Whalen and Schatte, 1997) appeared to think some connections lay to the North, since they used terminology, such as Middle Acheulean and Upper Acheulean, normally applied to materials from the Levant. It is likely, however, that the choice of terminology used by those doing survey in Arabia was less dependant upon the specific attributes of the artifacts found and rather more to do with the backgrounds of the archaeologists involved.

In the past 7 years or so, a series of new in situ Paleolithic sites have been found in southern Arabia and, while results are still mainly preliminary, the assemblages involved and the absolute dates very recently recovered, suggest that the Paleolithic of Arabia may have been quite complex. In addition, recent paleoclimatic studies indicate that southern Arabia may not have been climatically homogeneous and that the view of extreme hyperarid conditions over the whole area during all inter-pluvials is most probably overstated (Parker and Rose, 2008; Rose and Usik, 2009). If, in fact, there were environmental refugia in Arabia during inter-pluvial events, then the model of all Paleolithic materials being associated solely with pluvial conditions may be wrong and the possibility for long-term local cultural development becomes viable.

From this perspective, an Arabian assemblage may fall into one of two different inter-regional contexts:

1. An assemblage's immediate origin is to be found in an adjacent region. This would show in clear technological and typological patterning that closely parallels patterning from a near contemporaneous industry in an adjacent region. Some variability might be expected, owing to potential differences in raw material type and availability between the adjacent region and Arabia, but these should not mask the otherwise strong parallels. Such assemblages would reflect either initial expansions of people from adjacent regions into Arabia or longer term local occupations where interaction with the source area was maintained over time. If such existed, it is most likely that these movements took place during pluvials but, if some inter-pluvials were not as hyperarid as once thought, it is also possible that they could date to beginning stages of dry periods. If such movements were into areas of refugia, then they might well have formed the base for long term, local developments that, over time, would lose their "foreign" aspects. Long-term interactions with adjacent regions would be most likely in areas that were both geographic and environmental extensions of some adjacent region.

2. An assemblage's immediate origin lies in Arabia. Its technological and typological patterning most closely parallels earlier local assemblages that show no direct, specific parallels with any industry from an adjacent region. Of course, the ultimate source of such a developmental sequence would lie in an adjacent region, as described above. The length of time, the continuity of local adaptations, and a low level in inter-regional interactions, however, would have brought about sufficient technological and typological modifications that all but the most general technological patterns would seem out of place in any adjacent region.

It is important to recognize that both contexts may well be valid and not just at the beginning of a local long-term development. Even if there had been a continuous occupation of southern Arabia from, say, MIS 7 until the Neolithic (some 200,000 years), given the probable low population density at any one time and the limited geographic distribution during inter-pluvials (limited to refugia), it is also quite possible that people in adjacent regions spread into the more arid, uninhabited peripheral areas of Arabia, as they became habitable during the onset of later pluvial conditions. Such movements could have been extremely rapid, as was the case from the Nile Valley into the Western Desert with the onset of the Neolithic sub-pluvial (Kuper and Kröpelin, 2006). These may represent brief incursions only associated with temporary wetter conditions or, in fact, may represent long-term expansions of populations. In addition, shifts to pluvial conditions would also have permitted the expansion of Arabian populations out of local refugia and into areas that were uninhabitable during inter-pluvial conditions, such as the Rub' al Khali and its margins. Thus, during pluvial conditions, the potential for inter-regional interactions would have increased, with its possible effects on adaptations and lithic technology.

All of this suggests that any grand generalizations about extra-regional origins, local long-term developments, or inter-regional interactions certainly would be premature. While at this nascent stage of Arabian prehistoric studies, it might be best not to focus on adjacent regions (Tosi, 1986), though it is already clear from a number of papers in this volume (e.g., Crassard, 2009; Khalidi, 2009; Maher, 2009) and others (e.g., Marks, 2008) that inter-regional comparisons will inevitably be addressed. After all, archaeology is a comparative science and, at this point, there is little to compare within Arabia itself. What is important, however, is that the prehistoric sequences from adjacent areas do not become the models for what should be found in Arabia (Marks, 2008). Newly excavated assemblages can be viewed in inter-regional perspectives, with the caveat that no one assemblage or even a few assemblages can give a realistic picture of the spectrum of the Arabian Paleolithic and how it related to cultural developments in adjacent areas.

The geographic position of Arabia, with the Levant to the north, East Africa to the west, and Iran to the east, makes numerous movements, both into and out of Arabia, quite possible and highly probable. In fact, coming down from the north, there are no serious geographic impediments to the expansion of people until they encounter the hyper arid Rub' al Khali, which effectively buffers southern Arabia from the rest of the subcontinent. While this certainly would have deterred movements through it, or even into it, during inter-pluvials, there is abundant evidence that during the Neolithic pluvial people inhabited the area (Zeuner, 1954; Field, 1955; Edens, 1988; McClure, 1994). Farther to the West, the Asir Mountains of Yemen might be thought of as the southern terminus of a north/south line of relatively well watered high ground beginning in the Levant with the Jordanian Plateau that, combined with its adjacent coastal plain, forms a favorable environmental zone running all the way from the southern Levant to the southwestern tip of Arabia. Not only would this area have provided an environmentally friendly route southward during pluvials but, quite possibly, the highest grounds of the Asir Mountains in Yemen and the coastal plain also formed a refugium during inter-pluvial times (Bailey, 2009).

The situation to the east of Arabia during the Pleistocene was quite different than it is today. Since the Persian Gulf did not exist in its present form until 6 ka, there was no major barrier between what is now the western (Arabian) and eastern (Iranian Makran) shores of the basin. Rather than a wide, shallow gulf, there was a large floodplain dubbed the Ur-Schatt River Valley (Seibold and Vollbrecht, 1969), with fresh water springs, estuaries and peat bogs (Godwin et al., 1958; Lambeck, 1996; Alsharhan and Kendall, 2003). This valley must have been environmentally favorable during pluvial conditions and even more so during inter-pluvials, perhaps among the most favorable environments in the region,

when it might have paralleled the relative environmental advantages of the central and southern Nile Valley, compared with the surrounding deserts. It is still quite impossible to judge the potential importance of this now submerged valley as a conduit for and even as a source of Paleolithic developments now being found in southern Arabia (Rose, 2008). It would be a mistake, however, to ignore its potential.

To the west, across the Red Sea, lies East Africa. While during the Pleistocene, the Red Sea never fell sufficiently to create a land bridge between East Africa and Arabia, distances were minor and the movements of a number of animal species from East Africa to Arabia are well documented (cited in Rose, 2006). Thus, movements of people, archaic and modern, would have been possible and most likely. Since contemporaneous environmental conditions on both side of the Red Sea around the Bab al Mandab and to the north were comparable, movements from one side to the other would not have required any new adaptations, being merely range expansions (Rose, 2007).

In summary, access to Arabia was possible and rather simple from the north and east, and only slightly more difficult from the west. While Rub' al Khali, with its adjacent areas, would certainly have been a barrier to local habitation and, perhaps, even to movements through it during inter-pluvials, this would have effected intra-regional, rather than inter-regional, contact and exchange.

Given the likelihood of movements into and out of Arabia throughout the Pleistocene from East Africa, the Levant and Iran, the question arises whether or not there is a reasonable chance of actually recognizing from which of the adjacent regions movements came, and with which of the adjacent regions local Arabian populations might have interacted.

Recent studies proposing a southern coastal route out of Africa for early moderns based on DNA evidence (Macaulay et al., 2005) certainly points to one specific region as a population source at some, as yet, unconfirmed time. Unfortunately, DNA evidence is rarely available from archaeological sites and, thus, evidence for origins and or connections must rely on less certain criteria, such as patterns of lithic technology and typology. The degree to which these criteria may be effective depends upon contemporary comparative patterns in the adjacent regions. If, for instance, during some period the prevailing patterns of lithic technology and typology in the adjacent regions are very similar, then it is highly unlikely that materials in Arabia with similar patterns will be traceable to one specific adjacent region, rather than another. While this situation, in fact, does pertain over a very long period, by the Middle Pleistocene there are significant differences in inter-regional lithic techno-typological patterns, at least between East Africa and the Levant. So little is known of southern Iranian Pleistocene prehistory that, for the moment, such questions cannot be adequately addressed in this area.

Techno-Typological Patterns of Adjacent Regions

Lower Paleolithic/Early Stone Age

If claims for a pre-Acheulean in southern Arabia (Amirkhanov, 1994; Whalen and Pease, 1990; Whalen and Schatte, 1997; Whalen and Fritz, 2004) are confirmed, its origin must lie in Africa, since that is where it evolved. The presently available data, however, are not sufficient to accept these claims. Pre-Acheulean technology is so simple and its typology so limited and temporally ubiquitous that without good geological context, the lithics are simply not diagnostic.

The Acheulean, found in both East Africa and the Levant and, most probably in southern Iran, since it is also found in India (Petraglia et al., 2005) lasts for somewhat more than 1 million years, from about 1.7 Ma to about 500 ka in Africa and for only slightly less in the Levant (Bar-Yosef, 1998). The huge area over which it is found, combined with its impressive temporal span, resulted in considerable technological and typological diversity, although some tool classes, such as handaxes, cleavers, choppers, spheroids and polyhedrons, tend to be found in almost all Early Acheulean sites, regardless of location. The variability noted in Acheulean assemblages prior to the Late Acheulean has been interpreted as functional (Kleindienst, 1961), temporal (Gilead, 1970, 1975), even stochastic (Isaac, 1969), but not cultural. This is reflected in the similarities between the East African Acheulean and that found in the Levant, where for most of the million years no clear regional distinctions can be made. Such Levantine Acheulean sites as Ubeidiya (Bar-Yosef and Goren-Inbar, 1993), Evron Quarry (Ronen, 1991), Gesher Benot Ya'aqov (Goren-Inbar et al., 2000) all have African traits, both in their typologies and, even, in their associated fauna (Tchernov, 1992). Thus, it is unlikely that any pre-Late Acheulean site found in Arabia can be linked more strongly with one adjacent region, as opposed to another.

Different regional patterns arose in the Late Acheulean: for the first time, a distinction can be made between African and Levantine assemblages (Clark, 1975). The Levantine Late Acheulean sites have both symmetric bifaces and Levallois flake production but, often, the typical African cleaver is missing. Even in the Levant, however, sites are often surface scatters and artifact associations are difficult to assess (Bar-Yosef, 1998). In the East African later Acheulean there is little intra-tool class morphological variability, although the proportional occurrences of the major tool classes may vary significantly from site to site (Kleindienst, 1961). Unfortunately, the typological approach to later Acheulean assemblages in East Africa has been very generalized (e.g., heavy duty tools vs. light duty tools), so that meaningful comparisons between East

African and Arabian assemblages are not promising. An exception was the "Upper Acheulean" defined for Nubia, based on a major presence of well made Micoquian and Lanceolate handaxes, a paucity of Levallois reduction, and an absence of cleavers (Guichard and Guichard, 1968), should such an assemblage be found in Arabia, it might well point to an East African connection. With sufficient systematic surface collections and enough patterned assemblage redundancy, it might be possible to differentiate between a Late Acheulean originating from the Levant, as opposed to one coming from Nubia.

Somewhat after 400 ka, there appeared a marked bifurcation in the developmental patterns of lithic reduction and tool production in East Africa and the Levant. Through time, this difference became even more pronounced and there is no problem differentiating the East African patterns from those contemporary Levantine ones.

In some parts of East Africa, among them Nubia, the later Acheulean seems to have evolved into or, at least, was supplanted by the Sangoan (McBrearty and Tryon, 2006), which is a later Acheulean with the addition of large core axes, a few hard hammer blades, and a flake production mainly based on discoidal core reduction but with a Levallois element, as well. By the end of the Sangoan (i.e., beginning of Lumpemban), handaxes have disappeared and in their place are found bifacially produced, often elongated foliates.

Middle Paleolithic/Middle Stone Age

In other parts of East Africa, the later Acheulean may have transformed directly into an early Middle Stone Age, without the larger tools (large handaxes, cleavers and heavy duty tools), but with lighter flake tools and smaller bifacial or partly bifacially retouched tools, such as foliates, points, and ovates. In this Middle Stone Age (MSA), a large range of flake tools became common, including many forms of sidescrapers, endscrapers, perforators, burins, denticulates, etc. In this sense, the range of retouched tools in the East African MSA combined what, in the Levant and Europe, largely would have been temporally separate groups. With this shift to flake tools, Levallois reduction methods became common, but were mainly limited to the production of ovoid to rectangular flakes from preferential or recurrent modes of reduction (Tryon et al., 2005). There is a little evidence for unidirectional converging Levallois reduction but classic Levallois points are very rare. Few elongated blanks of any kind occurred and, although found in most assemblages, they almost never account for over 10% of the debitage.

It is during the early MSA that, for the first time, regional variability can be seen in Africa (Clark, 1988). The East African data are too limited to fully define different sub-regional industries but there is some evidence that the mid-MSA of the Horn and Ethiopia (Yellen et al., 2005) might be somewhat different from the contemporaneous MSA in Tanzania (Mehlman, 1989). Even this variability, however, is trivial when compared to the striking differences between all published East African MSA sites and temporally comparable Levantine assemblages.

Sometime around 400 ka in the central Levant, the Upper Acheulean gives way to the Mugharan Tradition (Jelinek, 1982). While its dating might suggest it belongs in the Lower Paleolithic, technologically and typologically it can be considered Middle Paleolithic. Compared to the Upper Acheulean, the Mugharan Tradition is highly complex, consisting of what have been interpreted as three different facies: the Yabrudian, the Acheuleo-Yabrudian, and the Amudian (Jelinek, 1982, 1990). The Yabrudian and the Acheuleo-Yabrudian are both characterized by a lack of Levallois reduction, many sidescrapers made on thick, wide flakes with overlapping retouch (Quina or demi-Quina) and by the presence of some asymmetric handaxes. These two facies differ only in the proportional occurrence of the handaxes; they are rare in the Yabrudian but more common in the Acheuleo-Yabrudian. Neither the asymmetric handaxes nor the "Quina" sidescrapers have been reported from the southern Levant, much less from East Africa and it appears that this tradition is to be found only as far south as Mt. Carmel.

The third facies of the Mugharan Tradition, the Amudian, is radically different from the other two. Aside from an occasional asymmetric handaxe and a "Quina" scraper here and there, it has little in common with the other two facies except for its stratigraphic contexts in a few sites (e.g., Tabun, Yabrud) where it is interstratified between Yabrudian and Acheuleo-Yabrudian levels (Bordes, 1955; Jelinek, 1990). Technologically, the production of blanks in the Amudian is based almost wholly on hard hammer, unidirectional blade production of Upper Paleolithic aspect (Gopher et al., 2004). While similar reduction strategies are known in pre-Upper Paleolithic contexts to the north in Eastern Europe, usually the elongated blanks were used to make "Middle Paleolithic" tools, such as sidescrapers, denticulates, and retouched points (e.g., Chabai, 2000), while in the Amudian the tools are almost wholly of Upper Paleolithic type: backed knives, endscrapers, burins, virtually all on blades (Gopher et al., 2004).

Around 250 ka, the Mugharan Tradition evolves into the Levantine Mousterian, which lasts in various forms until ca. 50–48 ka (Bar-Yosef, 2007). Unlike the Mugharan Tradition, the Levantine Mousterian has a strong Levallois component and, as in the Amudian, shows a marked tendency toward the production of elongated blanks, both Levallois and from uni-

directional, hard hammer volumetric core reduction (Monigal, 2003). The Levallois method in the Levant includes a number of different operational chains, from preferential centripetal through recurrent centripetal, to elongate unidirectional converging. At almost any time and at any site, all these and other specific chains may be found. The overall development of Levallois production, however, from the beginning of the Levantine Mousterian to its end, with a brief exception around 150 ka, is the tendency to produce triangular blanks, most often Levallois points from unidirectional converging reduction (Meignen, 1995; Mustafa and Clark, 2007) and, then, mainly elongated ones (Meignen and Bar-Yosef, 2005). Associated with this tendency toward elongated Levallois products are a good number of hard hammer blades from volumetric cores during the earliest phase. By the end of the Levantine Mousterian, however, almost all elongated blanks were produced by a Levallois method.

In sharp contrast to East Africa, the Levantine Mousterian lacks any evidence for bifacial reduction – both true façonnage and even bifacial retouch of core-produced blanks. Unlike in East Africa, the tool assemblages of the Levantine Mousterian are either dominated by Upper Paleolithic types or by Middle Paleolithic types; rarely, if ever, are these groups in proportional balance. The initial Levantine Mousterian (Levantine Mousterian of Tabun D type) exhibits high percentages of Upper Paleolithic tool types (rather like the preceding Amudian), particularly burins and backed knives, but by the end of its development (Levantine Mousterian of Tabun B type), most tools are typically Middle Paleolithic (sidescrapers, denticulates, and retouched Levallois points).

Upper Paleolithic/Mid to Late MSA

While significant differences existed between concurrent lithic traditions in East Africa and the Levant after 400 ka, around 50 ka these differences became even greater, particularly in basic technological patterns. In East Africa, nothing really changes after 50 ka. Those technological and typological patterns that were well established by 150 ka merely continue. While there may be some evidence for a minor decrease in overall artifact size, the prevailing reduction strategies (Levallois, discoidal, and a minor hard hammer blade strategy) continue unabated and unchanged until at least 30 ka and in some areas, until much later (Mehlman, 1989). Perhaps, the one change that can be seen is an increase in bipolar reduction (totally absent in the Levant), which is a very minor component at first but slowly gains in popularity, although its prominence is not manifest until the Holocene (Mehlman, 1989). Tool assemblages continue to be dominated by various side and endscrapers, unifacially and bifacially retouched points, denticulates, and poorly retouched pieces, almost all made on flakes, rather than blades. It is possible that by 50 ka a few backed tools appear in Tanzania but it might be late as 40 ka or even later and they never have the proportional dominance seen in the Howiesons Poort of South Africa some 20–15,000 years earlier (Miller et al., 1999; Mellars, 2006). In fact, they are always a very minor component of what is clearly regional developmental continuity.

In the Levant, the tendency toward the production of elongated blanks accelerates. At about 50 ka in the Initial Upper Paleolithic (Emiran) there is a transition from the dominant unidirectional converging Levallois strategy for the production of points and blades to a brief period of bidirectional reduction, also blade and point producing, and then to a single platform hard hammer volumetric production of blades (Monigal, 2002). By 40 ka, in the Early Ahmarian (Marks, 1983), there is a shift to a combined hard and soft hammer reduction, with the initial core shaping being done with a hard hammer but then shifting to a soft hammer for the removal of long, thin, symmetric blades and bladelets with small platforms and naturally pointed tips (Monigal, 2003). Tool assemblages are mainly made up of semi-steep retouched pointed blades and bladelets (El Wad points), burins on blades, endscrapers, simple retouched blades, and only an occasional backed blade or bladelet. By 30 ka, backing becomes more common and the reduction strategies shifted, so that bladelet cores were made on large flakes, while blade cores still utilized flint cobbles (Ferring, 1988). These Ahmarian assemblages are highly laminar and the only flakes produced appear to be by-products of blade core formation and maintenance. Throughout the Ahmarian blade and bladelet tools with semi-steep retouch dominate, until it passes into the Epipaleolithic at about 20 ka, when backing became the preferred retouch (Coinman, 2003).

Actually, the Upper Paleolithic of the Levant is more complex. While the Ahmarian appears to be the indigenous cultural tradition, at ca. 34 ka an Aurignacian moves southward, as far as the central Levant around Mt. Carmel. South of Mt. Carmel a number of flake-based assemblages are found that are distinct from the Ahmarian but broadly similar to the more northern Aurignacian in that they, too, may be characterized by carinated reduction that produces small twisted bladelets and "carinated" tools (Williams, 2006). Only a single site of this type has been dated and it falls at the very end of the southern Levantine Upper Paleolithic (Marks, 1976). Other such assemblages may well be earlier and their lack of close affinities with either the Ahmarian or the true Aurignacian might suggest either origins or, at least, links to Arabia.

The Arabian Paleolithic

Acheulean

As shown by Petraglia et al. (2009), most reported Acheulean surface sites contained relatively few handaxes and where the artifact concentrations were large, such as at Dawadmi, they were associated with quarry/workshop activities, heavily dominated by flaking debris. Even there, the effects of landscape deflation, slopewash, and erosion, have widely scattered the artifacts. At the richest locality, 206-76, for instance, the systematic collection from 900 m² recovered only 3.61 artifacts per square meter. The described collections with their roughly shaped handaxes, picks, cleavers, polyhedrons, spheroids, certainly indicate the presence of Early Acheulean in African and Levantine terms and the few "Levallois flakes" may or may not indicate a real co-association with the other materials, as noted by Petraglia and colleagues. Their discomfort with Whalen's attribution of these sites as being "Middle Acheulean" is fully justified. The sites are simply too spread by deflation and slope wash and the collection areas were too large to have any particular confidence that any two specific artifacts are contemporaneous, beyond the limits of comparable technological patterning. Of course, in this Acheulean sense "contemporaneity" may well encompass a half million years.

The presence of a single locality with "thin symmetric handaxes" does suggest a later Acheulean and, therefore, the possibility of some mixture from quite different temporal periods. Still, Whalen's work (Whalen et al., 1981, 1983, 1984, 1988) and its reappraisal by Petraglia (2003) and Petraglia et al. (2009) are among the strongest and clearest evidence to date for an Early Acheulean and, just possibly, a later Acheulean in Arabia. In spite of a large total artifact sample, neither Whalen nor Petraglia and colleagues were able to link this material to one geographic source, Africa as opposed to the Levant, beyond noting that, ultimately, it had to derive from Africa.

Post-Acheulean

Until recently, knowledge of the post-Acheulean but pre-Neolithic occupation of Arabia has been based on a few poorly reported excavations and a number of unsystematic surface collections. While still poorly known, excavations and systematic surface collections in the past few years finally have provided a sound and tantalizing glimpse of Arabian late Middle and Upper Pleistocene prehistory. Several recent surveys are reported in this volume (Crassard, 2009;

Jagher, 2009; Rose and Usik, 2009; Scott-Jackson et al., 2009; Wahida et al., 2009): while some provide only very preliminary data, others have very useful information and document the potential for further work.

As in Africa (Clark, 1988), it is during the post-Acheulean that sub-regional differences become apparent in the archaeology of Arabia. Although highly preliminary and, perhaps, more intuitive than data driven, three areas in Arabia can be seen where somewhat different influences and developments may have taken place. The first (The West) is the western littoral and the Asir Mountains, stretching from southern Jordan to the Indian Ocean bordering southwestern Yemen. Included are the eastern slopes of the Asir Mountains, leading into and including the western edge of the Rub' al Khali. The second (The South) is the southern littoral of Yemen and the high ground of the Hadramawt, the southern slopes of the Dhofar Mountains, extending eastward toward northern Muscat. The third area (The East), which might well be argued to be merely the northeastern edge of the Rub' al Khali and its eastern sand sheets, is the southwestern shore and hinterlands of the present Persian Gulf. While much smaller than the other two and, certainly, environmentally less friendly, it is the southwestern margin of the Pleistocene Ur-Schatt River Valley and, thus, might well contain cultural materials originating in that valley.

Parts of the first two areas were as environmentally favorable as any in Arabia, even during inter-pluvials. As such, their core areas could well have served as refugia and might have constituted a single large, refugium, extending down the western coast of Arabia and then eastward along the southern coast. In this sense, the two areas share the southwest-most portion of Yemen. The third area would have been inhabitable mainly during pluvials and might well have been abandoned during inter-pluvials, as would have been the case for the all the borderlands of the central Arabian hyperarid zone (Fig. 1).

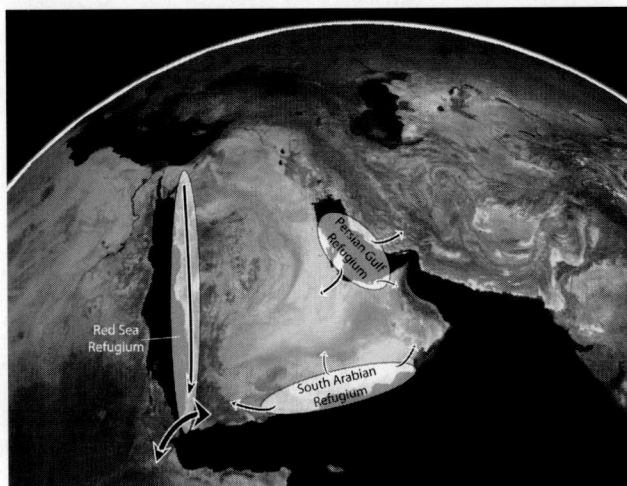

Fig. 1 Map of the Arabian peninsula depicting core zones and predicted population movements out of these refugia

The West

The work of Delagnes et al. (2008a, b) has finally provided important descriptions of what must be called Middle Paleolithic (as opposed to Middle Stone Age) in western Yemen. In addition, extensive surveys of western Saudi Arabia by the CASS (Zarins et al., 1980, 1981) have shown that possible post-Acheulean, Middle Paleolithic, materials may be common, unlike in the rest of Arabia (Petraglia and Alsharekh, 2003). Unfortunately, none of these survey finds has been described in detail and the general attribution of them to the Middle Paleolithic was only based on some "Levallois" radial and tortoise cores and the presence of flake tools (Zarins et al., 1980, 1981). While these might signify Middle Paleolithic, they might be later (Crassard, 2009), or they might also be earlier, since both flake tools and the Levallois method were present in the later Acheulean of adjacent regions and some of these Arabian sites had handaxes, as well. Those with handaxes classified as "Mousterian of Acheulean Tradition" may well be Acheulean, if the handaxes are contemporaneous with the other materials. If so, then these collections with handaxes and "Middle Paleolithic" elements, might suggest local continuity between the well documented Acheulean and less well known "Middle Paleolithic."

Delagnes' (2008a, b) approach permits reasonable, if preliminary, inter-regional comparisons. The site of Shi'bat Dihya 1, in western Yemen, has been dated by OSL to a dry period at 80–70 ka and the materials technologically show marked similarities with the Levantine Mousterian, with a heavy emphasis on Levallois elongated point and blade production (none of which was described or, perhaps, recognized by Zarins). Since the Asir Mountains and the eastern littoral of the Red Sea may have been environmentally more favorable during inter-pluvials than the highlands of the Central Negev and the southern Jordanian Plateau, it could have permitted continuous occupation by Levantine groups, when the Negev and southern Jordan became arid during OIS 4 and occupational densities there dropped sharply.

Not only is this described Yemeni Middle Paleolithic pattern clearly Levantine related, it might well indicate a continuation of the technological proclivities of the Early Levantine Mousterian into Late Levantine Mousterian times, adding to the evidence for long term continuity of that technological tradition (Marks, 1983, 1990; Monigal, 2002; Mustafa and Clark, 2007).

Given the seemingly unbroken environmental richness of the western Arabian littoral, from the southern Jordanian Plateau south to the southern Asir Mountains of Yemen, even under inter-pluvial conditions (Bailey, 2009), a comparable geographic continuity of prehistoric industries from the Levant to the southern Asir Mountains should come as no surprise.

While Shi'bat Dihya 1 points to Levantine connections, other sites reported to be Middle Paleolithic are not so informative. The sites called Middle Paleolithic by Zarins et al. (1980, 1981), however, seem to lack the unidirectional convergent Levallois cores and Levallois points found at Shi'bat Dihya 1. Thus, if these sites are post-Acheulean they indicate that two quite different patterns of Levallois technology were present in this western area and African connections are possible, since the Levallois method dominated during virtually all of the East African MSA. Certainly, larger samples and more detailed studies are needed before detailed comparisons can be made.

One of the more peculiar reported occurrences is that of an "Aterian" site on the southwestern edge of the Rub' al Khali, just beyond the eastern edge of the Asir Mountains (McClure, 1994). Cited by Petraglia and Alsharekh (2003: 678) as being significant in terms of "inter-regional trends," McClure (1994: 5), in spite of his unfortunate choice of article title, concluded "all considered, the assemblage for the present seems best to be designated as of unknown or undetermined affinity, with perhaps only a tentative suggestion of Aterian relationship." Even that suggestion is, at best, unlikely. Aside from the absence of all Aterian technological and typological attributes except a series of tanged points and scrapers (McClure, 1994), the Aterian has never been reported east of the Nile Valley and never confirmed even in the Nile Valley (*contra* Carlson and Sigstad, 1967–1968).

To date, there are only hints of possible occupation in western Arabia during the MIS 3 pluvial (Maher, 2009). This is strange, indeed, given the consistent evidence for marked pluvial conditions even in inland Arabia from about 35 to 20 ka (e.g., McClure, 1976, 1978; Anton, 1984; discussed in detail by Parker, 2009).

A single cluster of two surface scatters, the Fall Well site (Edens, 2001), located east of the Asir Mountains in southern Saudi Arabia, is the only convincing evidence beyond the northern border zone for an Levantine Upper Paleolithic presence in Arabia. While the collections were relatively small, both technologically and typologically they fall comfortably within the later Ahmarian of the southern Levant, between ca. 24 and 20 ka (Coinman, 2003).

As discussed by Maher (2009), there are no other sites that are technologically or typologically "Upper Paleolithic" in the Levantine sense. Of course, occupations during MIS 3, should they have originated or been influenced from East Africa, would not look "Upper Paleolithic" in the Levantine sense, with its fine blade/bladelet technology and its semi-steep, marginally retouched or backed blades/bladelets. In addition, should sites of this period reflect long term, local technological and typological adaptations, they may not look either Levantine or East African, reflecting their local origins (Zarins et al., 1982).

Given the proximity between the western Yemeni littoral and the eastern coast of East Africa, it is strange that no clearly African or even African related assemblages have been found. Granted that Amirkhanov (1991, 1994) saw connections to Africa, but they were not to East Africa, rather, to North and Northeast Africa, which are technologically more Levantine than East African. To date, the only possible East African connections might be the "Middle Paleolithic" reported by Garbini (1970), Zarins et al. (1980, 1981) and de Maigret (1985) but it is not described sufficiently to make any inter-regional comparisons.

The South

There is no sharp geographic or environmental break between the southernmost part of western Arabia and what is referred to here as The South. It is likely that these formed a single macro-environmental zone. There would be no reason why movements of peoples southward from the Levant could not have spread eastward, once they reached the southern terminus of the Asir Mountains. In fact, there is evidence that Levantine Mousterian Levallois reduction patterns did reach the Hadramawt of eastern Yemen (Crassard, 2009). While Crassard's study is on a relatively small sample of undated surface cores from the Wadis Sana and Wa'sha, his detailed technological approach is important to making convincing inter-regional comparisons and also provides a sound methodology for recognizing and organizing Arabian Levallois core assemblages. While his sample seems to lack the tendency toward marked elongation seen at Shi'bat Dihya 1 in that only two of 51 cores were twice as long as wide (Crassard, 2009), the specific Levallois reduction strategies, with their emphasis on unidirectional converging removals, again, show clear and strong relationships to the Levant, rather than to East Africa.

In spite of literally hundreds of recorded artifact findspots and sites, the surveys by Rose (2006), Rose and Usik (2009), Jagher (2009), as well as earlier work by others (referenced in Rose, 2006), have failed to find evidence farther east in Oman for the Levantine related Levallois technology now known from eastern and western Yemen. This suggests a possible eastern terminus for the Levantine related Middle Paleolithic material but may also just reflect the large areas of Oman that remain to be surveyed.

In the surveyed areas, a number of technological patterns have been recorded that are most certainly post-Acheulean and pre-Neolithic, even if most are undated at present. One group of surface sites and a large number of surface findspots (Jagher, 2009), tentatively named the Sibakhan Industry (Rose, 2006), is characterized by unidirectional hard hammer, unfaceted large blade production from cores mainly with broad, flat flaking surfaces, as well as by bifacial façonnage reduction that produced large biconvex bifacial pieces. These latter may be foliate preforms or may be finished tools but they cannot be reasonably characterized as handaxes (Jagher, 2009). In addition, there is a minor component of what could be unidirectional converging Levallois reduction. If so, it might relate to the Levallois strategies seen to the west but the possible variability within the typical flat linear blade cores calls for caution. The large hard hammer blades associated with bifacial tools and a scattering of sidescrapers suggests a late Middle Pleistocene or early Upper Pleistocene date (Biagi, 1994; Rose, 2006; Jagher, 2009).

A second group of sites has been called the Nejd Leptolithic (Rose, 2006). Again, these sites and findspots are characterized by blade production mainly from flat flaking surfaces, although volumetric reduction is present, as well (Rose, 2006; Rose and Usik, 2009). In addition, there are indications of bifacial, façonnage reduction and a small series of sidescrapers and endscrapers. Samples are small, making conclusions difficult, but the blade technology seems to have involved some limited use soft hammer percussion. The sites are undated but seemingly more recent than those of the Sibakhan, based on artifact weathering and geomorphic position on the landscape (Rose, 2006).

A third group of sites has been subsumed under the Khasfian Industry (Rose, 2006; Rose and Usik, 2009). The Khasfian is complex, consisting of specialized workshop sites, campsites, and sites that might fall somewhere in between. All, however, are related through one or both of the following technological traits; the use of façonnage to produce percussion flaked foliates and the production of blades by mainly hard hammer percussion from volumetric cores. The specialized workshop cluster at Bir Khasfa focused on foliate production, while at Ras Aïn Noor and Dhanaqr blade production was emphasized, with only hints of foliate production or, perhaps, foliate resharpening, At the known campsite, al-Hatab, there was an emphasis on blade and retouched tool production but with some elements of foliate production, as well. The striking difference in emphasis can be seen in the low percentages of bifacial retouch/shaping flakes at al-Hatab's two levels (10.2% and 4.8%) and at the blade producing sites of Ras Aïn Noor and Dhanagr (4.8% and 4.5%), compared with the 40.4% found at the specialized foliate workshop of Bir Khasfa (Rose and Usik, 2009). Whether all these sites are contemporaneous or whether this variability represents temporal developmental change is presently unknown. The OSL dates for al-Hatab indicate occupation sometime between ca. 14–12 ka, falling into the later and more humid half of MIS 2.

A possibly similar group of artifacts was briefly described by Zarins et al. (1979) from the Wadi Dawasir in the Rub' al Khali and, if confirmed, would certainly suggest that the Khasfian existed, at least, during pluvial conditions. On the other hand, its full distribution is unknown and if found further

southeast on the southern slopes of the Dhofar Mountains or the coastal plain, then it may well have existed during both pluvial and inter-pluvial conditions.

While the Sibakhan, the Nejd Leptolithic, and the Khasfian, combined, might well span most of the Upper Pleistocene, if not the end of the Middle Pleistocene, as well, all share some technological traits: mainly hard hammer blade production, bifacial façonnage reduction, little or maybe no true Levallois production, and little tendency toward platform faceting. Typologically, this group is hard to characterize, since almost all of the flake tools come from what would seem to be the temporally youngest site, al-Hatab. Still, sidescrapers seems to occur in all sites and at al-Hatab, the presence of retouched tools (burins and scrapers) made on large blades and primary flakes, is striking. While there are bifacial tools, mainly foliates and heavy-duty "bifacials," there is an absence of both true handaxes and small bifacially retouched points. In spite of the foliates, which some have seen as indicating possible connections to East Africa (Pullar, 1974; Rose, 2004), the other technological and typological traits, both present and absent, lead to a firm conclusion that none of these industries was directly related to or influenced by East African developments. By the same token, there is nothing technologically that would suggest direct connection to the Levant. The large hard hammer blades of the Sibakhan remind some of the Amudian (Rose, 2006; Jagher, 2009) but differences in blade core types and the absence of the numerous Amudian Upper Paleolithic type tools in the Sibakhan make any such connection unlikely. By al-Hatab times, the burins, both in number, type, and blank form, are certainly more Levantine than East African, as are the few carinated pieces. Yet, the blade technology is quite different from contemporary Levantine blade technology and the absence of the ubiquitous Levantine el Wad points, microgravettes, and finely retouched blades and bladelets at al-Hatab and the other Khasfian sites, makes any direct connections doubtful. While additional sites and absolute dates are desperately needed for all of these industries, at the moment the most parsimonious view is that they are parts of a long term, local southern Arabian developmental sequence.

The East

It might be argued that this area is not really comparable to the other two used here, since it is merely the eastern edge of the much larger arid region of the Rub' al Khali or, perhaps, merely a thin strip of the western hinterlands of the now submerged Ur-Schatt River Valley (Fig. 1). While either position is certainly reasonable, this area differs from the other two in that it does not include any possible environmental refugia and was, without question, consistently more arid, although there is good evidence for favorable environmental conditions during pluvials (Parker, 2009). Of all the areas, this one has the fewest known Paleolithic sites but also contains one spectacular site, Jebel Faya 1, that has produced not only three sizable in situ, stratified assemblages but also absolute dates associated with them.

A few post-Acheulean but pre-Neolithic sites have been reported from the eastern border of the northern Rub' al Khali, at Harad, in Saudi Arabia (Adams et al., 1977), through Abu Dhabi (McBrearty, 1993, 1999; Wahida et al., 2009), to eastern Sharjah (Uerpmann et al., 2007, 2008; Scott-Jackson et al., 2009). Along with the relative paucity of the reported surface sites, the nature of the reports, the collection techniques, and the small samples collected, tend to be regrettably typical of the traditional surface survey results from most other parts of Arabia.

During the first year of the Comprehensive Archaeological Survey of Saudi Arabia in the Eastern Province of the country, seven surface scatters of pre-Neolithic artifacts were reported near Harad that included Levallois "tortoise" cores and large blades (then referred to as flake-blades). There were no bifacial tools, so the sites were designated as Middle Paleolithic (Adams et al., 1977). Since there were large cleavers found, as well, it is probable that at least some of these scatters also contained Acheulean materials.

Recently, a series of surface sites have been found in the Western Region of Abu Dhabi at Jebel Barakah, along the Gulf Coast (Wahida et al., 2009). Originally reported by McBrearty (1993), recent work has increased the number of concentrations and resulted in small randomly collected artifact samples from three of them (Wahida et al., 2009). While little can be gleaned from these collections, it appears that the cores are consistently of "centripetal radial strategy" and there is no evidence for blade production. These traits suggested an Early Middle Paleolithic status (Wahida et al., 2009) and this may be reasonable. The absence of blade production is in marked contrast to virtually all Pleistocene sites in the southern area, as well as to the Levantine related sites in the West. Perhaps, when systematic collections are made and the cores studied in more detail (e.g., using Crassard's system detailed in this volume), more will be learned.

Farther to the northeast, in Sharjah, a series of outcrops of "red chert" occur on the western side of the Hajar Mountains, on the eastern edge of the Al-Madam Plain. Seven of these outcrops were found to have artifacts around them, consisting of 14 separate concentrations (Scott-Jackson et al., 2009). Small selective collections were made, so that the value of these artifacts resides solely on the individual type level, since no two artifacts can be assumed to be even broadly contemporaneous. Artifacts of this raw material are found in small amounts at Neolithic and Paleolithic sites some 20 km to the east (H.-P. Uerpmann et al., 2008), so it is clear that very different groups exploited these outcrops over a very long time.

Based on the selective collections, it is possible to say that the following technological strategies were used to reduce the raw material: façonnage that can be seen in some bifacial thinning flakes and in foliate preforms; hard hammer production of wide blades from unfaceted unidirectional cores (both blades and cores are present); the centripetal Levallois method for the production of flakes (cores and flakes present): Kombewa reduction, as seen by two flakes and one core; and, simple flake production from 90°, multiple platform, and discoidal cores (Scott-Jackson et al., 2009).

Assuming that all these technological traits are Paleolithic, then these lithic scatters do suggest some similarities (foliates, Kombewa technique, and large unfaceted blades) with materials from the southern area, particularly from Oman. On the other hand, the Levallois cores might relate to the Abu Dhabi materials, while the 90° and multiple platform cores could be a link with the materials from Jebel Faya 1, Assemblages A and B, discussed below. The degree to which additional information may be gleaned from these localities will depend upon the study of systematic collections to determine if one or more of these various technological strategies dominate a particular scatter. It seems unlikely that any one lithic scatter will be attributable to a single short period, but with sufficient collections, associations of technological strategies may be seen. As it now stands, whatever the degree of mixing, the collections confirm technological attributes that occur in the southern and northeastern areas at sites with better integrity.

The final site in the northeastern area is among the most important Pleistocene sites in Arabia, while its later occupations are also of considerable significance. Its potential is still being discovered, since excavations are ongoing. The site, Jebel Faya 1, is a collapsed rockshelter at the western base of the Jebel Faya, some 20 km west of the Hajar Mountains in Sharjah, just at the northeastern edge of the Rub' al Khali. To date, excavations have exposed over 3 m of sediments, containing archaeological materials ranging from Bronze Age, through Neolithic and Late Pleistocene occupations, to early Upper Pleistocene occupational levels (Uerpmann et al., 2008).

There are, at the moment, five occupational levels dating to the Pleistocene. The lowest two were found at the end of the 2008 field season and samples are much too small to permit comment beyond that they date to MIS 5 or earlier. On the other hand, the three top Paleolithic levels, field designated Assemblages A, B, and C, in stratigraphic order, are producing reasonable artifact samples and, most importantly, they are presently being dated by OSL. The stratigraphically lowest, Assemblage C, is dated to >85 ka, while Assemblages A and B fall securely into mid MIS 3 (Uerpmann et al., 2008). Given the position of Jebel Faya 1, on the edge of the Rub' al Khali and west of the Hajar Mountains, it is not surprising that these three occupations date to pluvial times, since during inter-pluvials the area would have been hyper-arid and probably uninhabitable.

Jebel Faya is rich in flint sources and the assemblages all reflect the abundance of this locally available raw material, not only in the quantity of artifacts but also in the significant workshop component in each assemblage. Assemblage C also has a small component of "red chert" artifacts, the source of which was likely to be one of the sites collected by Scott-Jackson et al. (2009).

Assemblage C exhibits a number of different reduction strategies: façonnage to produce both small handaxes and foliates; a little hard hammer blade production from uni-directional volumetric cores; and, flake production from rather crude Levallois and discoidal cores. While there is some platform faceting, particularly on the Levallois flakes, most flakes are unfaceted, as are the blades. Typologically, the tools include small cordiform handaxes, foliates, foliate preforms, endscrapers, sidescrapers, and denticulates, as well as a questionable burin or two.

Assemblage B technologically is characterized by the production of mainly flakes from both 90° cores and from flat cores with converging to 90° removals from the same flaking surface. There are a few multiple platform cores that are merely 90° cores with one additional flaking surface, as well as a few truncated faceted pieces. Blade production exists but is far from dominant. Cores suggest some volumetric reduction but some blades were produced on the flat cores, as well. Tools consist of sidescrapers, endscrapers, denticulates, perforators, and a good number of simple retouched pieces.

Assemblage A is technologically characterized by the production of small rectangular flakes from multiple platform cores. Large numbers of débordant flakes attest to the reworking and changing of flaking surfaces. There is no apparent purposeful blade production, although a few short, wide blades were found. Typologically, there are retouched pieces, some burins, and a number of poor endscrapers, sidescrapers, and denticulates. The percentage of tools is very low and the assemblage appears to mainly reflect workshop activities.

It is difficult to generalize about the post-Acheulean of this northeastern area, since there are so few sites known and reported. Yet, what is known suggests that this area was, at least, partly distinct from the other two areas. The reported Middle Paleolithic (Adams et al., 1977; Wahida et al., 2009) shows not a hint of Levantine relationships, but descriptions are too generalized to go beyond that.

Assemblage C at Jebel Faya 1 certainly dates within what is normally considered "Middle Paleolithic" in Levantine terms, but it is neither technologically nor typologically related to the Levant, nor to the few Levantine related Middle Paleolithic sites in western and southern Arabia. Not only does it lack Levantine Middle Paleolithic tendencies (elongated blanks, Levallois points, unidirectional converging core strategies, etc.) but has tendencies totally unknown in the Levant in Middle Paleolithic contexts (reduction by façonnage for the production of small handaxes and foliates, as well as a balance between "Middle Paleolithic" and "Upper Paleolithic" tool types). It also seems quite distinct from the undated but old Sibakhan, since it does not have the large blades struck from wide flat flaking surfaces or the large "bifacials," although both shared façonnage as a reduction technique. Unique assemblages, by definition, are hard to place into broad constructs. While necessarily tentative, the apparent associated technological patterns in Jebel Faya 1, Assemblage C, show greater similarities with East and Northeast Africa, particularly the late Sangoan, than does any other site known in Arabia. Still, the similarities are at a general technological level. If such similarities, however, do reflect African origins, then the present models for the timing and nature of the movement of moderns humans out of Africa into Arabia need reevaluation. The model proposed by Mellars (2006) calls for movement only after there is strong evidence in South Africa for clear modern behavior; that is, no earlier than 65 ka or, at least, 20,000 years later than the Jebel Faya 1, Assemblage C, occupation. It also calls for the specific technological patterning of the Howiesons Poort, with its exclusive blade technology, its absence of Levallois reduction techniques, and its extensive use of backing to make geometrics. None of these traits is consistent with the Jebel Faya 1, Assemblage C. Might it be that Assemblage C represents an earlier movement out of Africa by not-quite-so-moderns and that the really important movement came later? While this could be the case, the total absence to date of any indication of a non-Levantine, exclusive blade technology associated with backed geometric tools in Arabia makes Mellars' model little more than speculation.

The Jebel Faya 1 assemblages A and B, again, show no obvious technological relations to anything in the Levant, Africa, or in the rest of Arabia, for that matter. They could not be progenitors of the al-Hatab technology with its hard hammer blade technology from volumetric cores and its use of façonnage, both of which are totally missing from Jebel Faya 1, Assemblages A and B. What might they represent? Since they date to MIS 3, a pluvial, they may represent an expansion of peoples of the Ur-Schatt River Valley into its southern hinterlands. Such is mere speculation, since virtually nothing is known of Late Pleistocene industries in that area. Again, though, these assemblages do not indicate any technological influences from either the Levant or Africa.

Conclusions

When inter-regional contexts for the Pleistocene prehistory of Arabia are considered, the admittedly very preliminary results are counter-intuitive. While few sites have absolute dates, both Shi'bat Dihya and al-Hatab fall into inter-pluvial periods of MIS 4 and MIS 2, respectively. This opens the question of whether hyperaridity was universal across Arabia during traditional inter-pluvials. Only the occupations at Jebel Faya 1 fall into pluvials periods, MIS 3 and MIS 5, which is to be expected, given its geographic position.

The DNA derived models for the movement of modern peoples out of Africa and into Arabia seem well based. Yet, those few Arabian sites near the Red Sea, just across from East Africa and to the East in Yemen that probably fall into the general period postulated for the movement, clearly show connections to the Levant and not to Africa. The only occupation described so far that seems to show general East and Northeast African patterns comes from far Eastern Arabia, as close to south Asia as one can get. Not only is it geographically far from Africa, it dates at least 20,000 years earlier than the projected timing of the first moderns who crossed the Bab al Mandab.

With the exception of the Fall Well site, which should date to late MIS 3, based on its close parallels with the mid to late Ahmarian of the Levant, most undated sites throughout Arabia that have been described reasonably, show no close affinities to either the Levant or to East Africa. Combined with the presence of sites dated to inter-pluvials, these sites argue for a robust, local development distinct from both the Levant and Africa. While its ultimate origin may well lie in some adjacent region, more likely Africa than the Levant, it cannot be seen as merely some Upper Pleistocene pseudopodia of an adjacent region. Granted, little hard data are yet available and much may change as more sites are found, systematically collected or excavated and, hopefully, fully published. Only time will tell.

References

Adams R, Parr P, Ibrahim M, al-Mughannum SA. Preliminary report on the first phase of the Comprehensive Survey Program. Atlal. 1977;1:12–40.

Alsharhan A, Kendall CGStC. Holocene coastal carbonates and evaporates of the southern Arabian Gulf and their ancient analogues. Earth Science Reviews. 2003;61:191–243.

Amirkhanov H. The Acheulean of southern Arabia. Sovetskaia Arkheologiia. 1987;4:11–23.

Amirkhanov H. The Paleolithic of South Arabia. Moscow: Academy of Sciences; 1991.

Amirkhanov H. Research on the Palaeolithic and Neolithic of Hadramaut and Mahra. Arabian Archaeology and Epigraphy. 1994;5:217–28.

Amirkhanov H. Acheulean sites of southern Arabia. In: Griaznevich P, editor. The Hadramaut. Moscow: Eastern Literatures Board; 1995.

Anton D. Aspects of geomorphological evolution: paleosols and dunes in Saudi Arabia. In: Jado AR, Zötl JG, editors. Quaternary period in Saudi Arabia. Vol. 2: Sedimentological, hydrogeological, hydrochemical, geomorphological, and climatological investigations of Western Saudi Arabia. Wien: Springer; 1984.

Bailey G. The Red Sea, coastal landscapes, and hominin dispersals. In: Petraglia MD, Rose JI, editors. The evolution of human populations in Arabia: paleoenvironments, prehistory and genetics. The Netherlands: Springer; 2009. p. 15–37.

Bar-Yosef O. Early colonizations and cultural continuities in the Lower Paleolithic of Western Asia. In: Petraglia MD, Korisettar R, editors. Early human behavior in global context: the rise and diversity of the Lower Paleolithic Record. London: Routledge; 1998. p. 221–79.

Bar-Yosef O. The game of dates: another look at the Levantine Middle Paleolithic chronology. In: Bicho N, editor. From the Mediterranean Basin to the Portuguese Atlantic shore: papers in honor of Anthony Marks. Acts of the IV Congress of Peninsular Archaeology. Faro: University of Algarve; 2007. p. 83–100.

Bar-Yosef O, Goren-Inbar N. The Lithic assemblages of Ubeidiya: a Lower Paleolithic site in the Jordan Valley. Qedem 34. Jerusalem: Hebrew University; 1993.

Biagi P. An Early Palaeolithic site near Saiwan (Sultanate of Oman). Arabian Archaeology and Epigraphy. 1994;5:81–8.

Bordes F. Le Paléolithique Inférieur et Moyen de Jabrud (Syrie) et la Question du Pré-Aurignacien. L'Anthropologie. 1955;59:486–507.

Carlson R, Sigstad J. Paleolithic and Late Neolithic sites excavated by the Fourth Colorado Expedition. Kush. 1967–1968;15:51–8.

Caton-Thompson G. Climate, irregation, and early man in the Hadhramaut. The Geographical Journal. 1939;93(1):18–35.

Caton-Thompson G. Some Paleoliths from Arabia. Proceedings of the Prehistoric Society. 1954;29:189–218.

Chabai V. The evolution of the Western Crimean Mousterian industry. In: Orschiedt J, Weniger G-C, editors. Neanderthals and modern humans – discussing the transition: Central and Eastern Europe from 50,000–30,000 B.P. Wissenshaftliche Schriften des Neanderthal Museums, Bd. 2. Mettmann: Neanderthal Museum; 2000. p. 196–211.

Charpentier V. Industries bifacials holocènes d'Arabie orientale, un exemple: Ra's al-Jinz. Proceeding for the Seminar for Arabian Studies. 1999;29:29–44.

Clark DJG. A comparison of the Late Acheulian industries of Africa and the Middle East. In: Butzer K, Isaac G, editors. After the australopithecines. The Hague: Mouton; 1975. p. 605–59.

Clark DJG. The Middle Stone Age of East Africa and the beginnings of regional identity. Journal of World Prehistory. 1988;2:235–305.

Coinman N. The Upper Paleolithic of Jordan: new data from the Wadi al-Hasa. In: Goring-Morris AN, Belfer-Cohen A, editors. More than meets the eye. Oxford: Oxbow Books; 2003. p. 151–70.

Crassard R. The Middle Paleolithic of Arabia: the view from the Hadramawt Region, Yemen. In: Petraglia MD, Rose JI, editors. The evolution of human populations in Arabia: paleoenvironments, prehistory and genetics. The Netherlands: Springer; 2009. p. 151–68.

de Bayle DH. Première mission de recherches préhistoriques en République Arabe du Yemen. L'Anthropologie. 1976;80:5–37.

de Maigret A. Archaeological activities in the Yemen Arab Republic, 1985. East and West. 1985;35:337–97.

Delagnes A, Macchiarelli R, Jaubert J, Peigné S, Tournepiche J-F, Crassard R, et al. A new Middle Paleolithic complex of sites in southern Arabia: preliminary results and interpretations. Paper presented in The Lower and Middle Paleolithic in the Middle East and neighbouring regions. University of Basel, May 8–10, 2008a.

Delagnes A, Macchiarelli R, Jaubert J, Peigné S, Tournepiche J-F, Crassard R, et al. Middle Paleolithic settlement in Arabia: first evidence from a stratified archaeological site in western Yemen. Paper presented at the Meeting of the Paleoanthropology Society, Vancouver, British Colombia, 25–26 March 2008b.

Edens C. Towards a definition of the Rub al Khali "Neolithic". Atlal. 1982;6:109–24.

Edens C. The Rub' al-Khali 'Neolithic' revisited: the view from Nadqan. In: Potts D, editor. Araby the Blest: studies in Arabian

Archaeology. Copenhagen: Carsten Niebuhr Institute Publications 7, Museum Tusculanum; 1988. p. 15–43.

Edens C. A bladelet industry in southwestern Saudi Arabia. Arabian Archaeology and Epigraphy. 2001;12(2):137–42.

Ferring R. Technological change in the Upper Paleolithic of the Negev. In: Dibble H, Montet-White A, editors. Upper Pleistocene prehistory of Western Asia. Philadelphia: The University Museum; 1988. p. 333–48.

Field H. New Stone Age sites in the Arabian peninsula. Man. 1955;55:136–8.

Garbini G. Antichità Yemenite. Annali Instituto Orientale di Napoli. 1970;30:537–48.

Gilead D. Handaxe industries in Israel and the Near East. World Archaeology. 1970;2:1–11.

Gilead D. Lower and Middle Paleolithic settlement patterns in the Levant. In: Wendorf W, Marks AE, editors. Problems in prehistory: North Africa and the Levant. Dallas: Southern Methodist University Press; 1975. p. 273–84.

Godwin FRS, Suggate RP, Willis EH. Radiocarbon dating of the eustatic rise in ocean-level. Nature. 1958;181:1518–9.

Gopher A, Barkai R, Shimelmitz R, Khalily M, Lemorini C, Heshkovitz I, et al. Qesem Cave: an Amudian site in central Israel. Journal of the Israel Prehistoric Society. 2004;35:69–92.

Goren-Inbar N, Feibel C, Verosub K, Melamed Y, Koslev M, Tchernov E, et al. Pleistocene milestones on the Out-of-Africa corridor at Gesher Benot Ya'aqov, Israel. Science. 2000;289:944–7.

Gramly RM. Neolithic flint implement assemblages from Saudi Arabia. Journal of Near Eastern Studies. 1971;30(3):177–85.

Guichard J, Guichard G. Contributions to the study of the Early and Middle Paleolithic of Nubia. In: Wendorf F, editor. The prehistory of Nubia, Volume I. Dallas: Southern Methodist University Press; 1968. p. 148–93.

Inizan ML, Ortlieb L. Prehistoire dans la region de Shabwa au Yemen du sud. Paleorient. 1987;13:5–22.

Isaac G. Studies of early culture in East Africa. World Archaeology. 1969;1:1–28.

Jagher R. The Central Oman Palaeolithic Survey: recent research in Southern Arabia and reflection on the prehistoric evidence. In: Petraglia MD, Rose JI, editors. The evolution of human populations in Arabia: paleoenvironments, prehistory and genetics. The Netherlands: Springer; 2009. p. 139–50.

Jelinek AJ. The Middle Paleolithic in the southern Levant, with comments on the appearance of modern *Homo sapiens*. In: Ronen A, editor. The transition from lower to Middle Paleolithic and the origin of modern man. Oxford: British Archaeological Reports International Series S151; 1982, p. 57–104.

Jelinek A. The Amudian in the context of the Mugharan Tradition at the Tabun cave (Mount Carmel), Israel. In: Mellars P, editor. The emergence of modern humans. Edinburgh: Edinburgh University Press; 1990. p. 81–90.

Khalidi L. Holocene obsidian exchange in the Red Sea region. In: Petraglia MD, Rose JI, editors. The evolution of human populations in Arabia: paleoenvironments, prehistory and genetics. The Netherlands: Springer; 2009. p. 279–291.

Kleindienst M. Variability within the Late Acheulian assemblages in eastern Africa. South African Archaeological Bulletin. 1961;16(62):35–52.

Kuper R, Kröpelin S. Climate-controlled Holocene occupation in the Sahara: motor for Africa's evolution. Science. 2006;313:803–7.

Lambeck K. Shoreline reconstructions for the Persian Gulf since the Last Glacial Maximum. Earth and Planetary Science Letters. 1996;142:43–57.

Macaulay V, Hill C, Achilli A, Rengo C, Clarke D, Meehan W, et al. Single, rapid coastal settlement of Asia revealed by analysis of complete mitochondrial genomes. Science. 2005;308:1034–6.

Maher L. The Late Pleistocene of Arabia in relation to the Levant. In: Petraglia MD, Rose JI, editors. The evolution of human populations in Arabia: paleoenvironments, prehistory and genetics. The Netherlands: Springer; 2009. p. 187–202.

Marks A. The Mousterian industries of Nubia. In: Wendorf F, editor. The prehistory of Nubia, Volume I. Dallas: Southern Methodist University Press; 1968. p. 194–314.

Marks A. Ein Aqev: a Late Levantine Aurignacian site in the Nahal Aqev. In: Marks A, editor. Prehistory and paleoenvironments in the Central Negev, Israel, Volume I. Dallas: Department of Anthropology, Southern Methodist University; 1976. p. 227–91.

Marks A. The sites of Boker Tachtit and Boker: a brief introduction. In: Marks A, editor. Prehistory and paleoenvironments in the Central Negev, Israel, Volume III. Dallas: Department of anthropology, Southern Methodist University; 1983. p. 15–38.

Marks A. The Middle and Upper Paleolithic of the Near East and the Nile Valley: the problem of cultural transformations. In: Mellars P, editor. The emergence of modern humans. Edinburgh: Edinburgh University Press; 1990. p. 56–80.

Marks A. Into Arabia, perhaps, but if so, from where? Proceedings of the Seminar for Arabian Studies. 2008;38:15–24.

McBrearty S, Tryon C. From Acheulean to Middle Stone Age in the Kapthurin Formation, Kenya. In: Hovers E, Kuhn S, editors. Transitions before the transition. New York: Springer; 2006. p. 257–78.

McBrearty S. Lithic artifacts from Abu Dhabi's western region. Tribulus: Bulletin of the Emirates Natural History Group. 1993;3:13–4.

McBrearty S. Earliest stone tools from the Emirate of Abu Dhabi, United Arab Emirates. In: Whybrow PJ, Hill A, editors. Fossil vertebrates of Arabia. New Haven: Yale University Press; 1999. p. 373–88.

McClure HA. Radiocarbonchronology of Late Quaternary lakes in the Arabian Desert. Nature. 1976;263:755–6.

McClure HA. Ar Rub' al Khal. In: Al-Sayari SS, Zötl JG, editors. Quaternary period in Saudi Arabia. Vol 1: Sedimentological, hydrogeological, hydrochemical, geomorphological, and climatological investigations in Central and Eastern Saudi Arabia. Wien: Springer; 1978.

McClure HA. A new Arabian stone tool assemblage and notes on the Aterian industry of North Africa. Arabian Archaeology and Epigraphy. 1994;5:1–16.

Mehlman M. Later Quaternary archaeological sequences in northern Tanzania. Ph.D. dissertation, University of Illinois, Urbana-Champaign; 1989.

Meignen L. Levallois production systems in the Middle Paleolithic of the Near East: the case of the unidirectional method. In: Dibble H, Bar-Yosef O, editors. The definition and interpretation of Levallois technology. Madison: Prehistory Press; 1995. p. 361–80.

Meignen L, Bar-Yosef O. The lithic industries of the Middle and Upper Paleolithic of the Levant: continuity or break? In: Derevianko A, editor. The Middle to Upper Paleolithic transition in Eurasia: hypotheses and facts. Archaeology, ethnology and anthropology of Eurasia. Novosibirsk: Institute of Archaeology and Ethnology; 2005. p. 166–75.

Mellars P. Why did modern human populations disperse from Africa ca. 60000 years ago? A new model. Proceedings of the National Academy of Sciences USA. 2006;103(25):9381–6.

Miller G, Beaumont P, Deacon H, Brooks A, Hare P, Jull A. Earliest modern humans in southern Africa dated by isoleucine epimerization in ostrich egg shell. Quaternary Science Reviews. 1999;18:1537–48.

Monigal K. The Levantine Leptolithic: blade production from the Lower Paleolithic to the dawn of the Upper Paleolithic. Ph.D. dissertation, Southern Methodist University, Dallas, TX; 2002.

Monigal K. Technology, economy, and mobility at the beginning of the Levantine Upper Paleolithic. In: Goring-Morris AN, Belfer-Cohen A, editors. More than meets the eye. Oxford: Oxbow; 2003. p. 118–33.

Mustafa M, Clark GA. Quantifying diachronic variability: the ´Ain Difla rockshelter (Jordan) and the evolution of Levantine Mousterian technology. Eurasian Prehistory. 2007;5(1):47–84.

Parker AG. Pleistocene climate change from Arabia: developing a framework for hominin dispersal over the last 350 ka. In: Petraglia MD, Rose JI, editors. The evolution of human populations in Arabia: paleoenvironments, prehistory and genetics. The Netherlands: Springer; 2009. p. 39–49.

Parker AG, Rose JI. Climate change and human origins in southern Arabia. Proceedings of the Seminar for Arabian Studies. 2008;38:25–42.

Petraglia M. The Lower Paleolithic of the Arabian peninsula: occupations, adaptations, and dispersals. Journal of World Archaeology. 2003;17(2):141–79.

Petraglia M, Alsharekh A. The Middle Palaeolithic of Arabia: implications for modern human origins, behaviour and dispersals. Antiquity. 2003;77(298):671–84.

Petraglia MD, Shipton C, Paddayya K. Life and mind in the Acheulean: a case study from India. In: Gamble C, Porr M, editors. The hominid individual in context: archaeological investigations of Lower and Middle Palaeolithic landscapes, locales and artefacts. London: Routledge; 2005. p. 197–219.

Petraglia MD, Drake N, Alsharekh A. Acheulean landscapes and large cutting tool assemblages in the Arabian peninsula. In: Petraglia MD, Rose JI, editors. The evolution of human populations in Arabia: paleoenvironments, prehistory and genetics. The Netherlands: Springer; 2009. p. 103–16.

Philby H. Rub' Al Khali: an account of exploration in the Great South Desert of Arabia under the auspices and patronage of His Majesty 'Abdul 'Aziz ibu Sa'ud, King of the Hejaz and Nejd and its dependencies. The Geographical Journal. 1933;81(1):1–21.

Pullar J. Harvard Archaeological survey in Oman, 1973: flint sites in Oman. Arabian Seminar. 1974;4:33–48.

Rafalski S, Schröter P, Wagner E. Die funde am Eyasi-Nordtostufer. In: Muller-Beck H, editor. Die Archäologischen und Anthropologischen Ergebnisse der Kohl-Larsen – Expedition in nord – Tanzania, 1933–1939. T bingen: T binger Monographien sur Urgeschichte; 1978, 4(2)

Ronen A. The Lower Paleolithic site Evron-Quarry in western Galilee, Israel. Sonderveroffenlichungen, geolisches Institut der Universität zu Köln. 1991;82:187–212.

Rose JI. The question of Upper Pleistocene connections between East Africa and south Arabia. Current Anthropology. 2004;45:551–5.

Rose JI. Among Arabian sands: defining the Palaeolithic of southern Arabia. Ph.D. dissertation, Southern Methodist University, Dallas, TX; 2006.

Rose JI. The Arabian Corridor Migration Model: archaeological evidence for hominin dispersals into Oman during the Middle and Upper Pleistocene. Proceedings of the Seminar for Arabian Studies. 2007;37:1–19.

Rose JI. Modern human origins: the 'Out of Arabia' hypothesis. In: Cleuziou S, Tosi M, editors. In the shadow of the ancestors. 2nd ed. Muscat: Ministry of Heritage and Culture; 2008. (in Arabic).

Rose JI, Usik VI. The "Upper Paleolithic" of South Arabia. In: Petraglia MD, Rose JI, editors. The evolution of human populations in Arabia: paleoenvironments, prehistory and genetics. The Netherlands: Springer; 2009. p. 169–185.

Scott-Jackson J, Scott-Jackson W, Rose JI. Paleolithic stone tool assemblages from Sharjah and Ras al Khaimah in the United Arab Emirates. In: Petraglia MD, Rose JI, editors. The evolution of human populations in Arabia: paleoenvironments, prehistory and genetics. The Netherlands: Springer; 2009. p. 125–38.

Seibold E, Vollbrecht K. Die bodengestalt des Persischen Golfs. In: Seibold E, Closs H, editors. Meteor Forschungsergebnisse: Herausgegeben von der Deutschen Forschungsgesellschaft. Berlin: Gebrüder Borntraeger; 1969. p. 31–56.

Tchernov A. Eurasian-African biotic exchanges through the Levantine corridor during the Neogene and Quaternary. Courier Forschunsintitut Senckenberg. 1992;153:103–23.

Tosi M. The emerging picture of prehistoric Arabia. Annual Review of Anthropology. 1986;15:461–90.

Tryon C, McBrearty S, Texier P-J. Levallois lithic technology from the Kapthurin Formation, Kenya: Acheulian origin and Middle Paleolithic diversity. African Archaeological Review. 2005;22(4): 199–229.

Uerpmann H-P, Uerpmann M, Kutterer J, Händel M, Jasim SA, Marks A. The Stone Age sequence of Jebel Faya in the Emirate of Sharjah (UAE). Paper presented at the Seminar for Arabian Studies, London; 2007.

Uerpmann H-P, Uerpmann M, Jasin SA, Marks A. Preliminary results of the excavations at Jebel Faya (Sharjah): 2003–2008. Paper presented in Al Ain, April; 2008.

Van Beek G, Cole G, Jamme A. An archaeological reconnaissance in Hadramaut, south Arabia: a preliminary report. Annual Report of the Smithsonian Institution, Washington, DC; 1963. p. 521–45.

Wahida G, Al-Tikriti Y, Beech MJ, Al Meqbali A. A Middle Paleolithic assemblage from Jebel Barakah, Coastal Abu Dhabi Emirate. In: Petraglia MD, Rose JI, editors. The evolution of human populations in Arabia: paleoenvironments, prehistory and genetics. The Netherlands: Springer; 2009. p. 117–24.

Whalen N, Fritz GA. The Oldowan in Arabia. Adumatu 2004;9:7–18.

Whalen N, Pease D. Variability in developed Oldowan and Acheulean bifaces of Saudi Arabia. Atlal. 1990;13:43–8.

Whalen N, Schatte K. Pleistocene sites in southern Yemen. Arabian Archaeology and Epigraphy. 1997;8:1–10.

Whalen N, Killick A, James N, Morsi G, Kamal M. Saudi Arabian archaeological reconnaissance 1980: B. preliminary report on the Western Province survey. Atlal. 1981;5:43–58.

Whalen N, Sindi H, Wahida G, Siraj-ali JS. Excavation of Acheulean sites near Saffaqah in ad-Daw dmi 1402–1982. Atlal. 1983;7:9–21.

Whalen N, Siraj-Ali JS, Davis W. 1–Excavation of Acheulean sites near Saffaqah, Saudi Arabia, 1403 AH 1983. Atlal. 1984;8:9–24.

Whalen N, Siraj-Ali J, Sindi HO, Pease DW, Badein MA. A complex of sites in the Jeddah–Wadi Fatimah area. Atlal. 1988;11:77–85.

Williams J. The Levantine Aurignacian: a closer look. In: Bar-Yosef O, Zilão J, editors. Towards a definition of the Aurignacian. Proceedings of the Symposium held in Lisbon, Portugal, 25–30 June 2002. Portugal: Instituto Português de Arqueologia. Trabalhos de Arqueologia; 2006, p. 317–52.

Yellen J, Brooks A, Helgren D, Tappen M, Ambrose S, Bonnefille R, et al. Archaeology of Aduma: Middle Stone Age sites in the Awash Valley, Ethiopia. Paleoanthropology. 2005;10:25–100.

Zarins J, Ibrahim M, Potts D, Edens C. The preliminary report on the third phase of the Comprehensive Archaeological Survey Program – the Central Province. Atlal. 1979;3:9–42.

Zarins J, Whalen N, Ibrahim M, Mursi A, Khan M. Comprehensive Archaeological Survey Program: preliminary report on the central and Southwestern Province survey: 1979. Atlal. 1980;4:9–36.

Zarins J, al-Jawad Murad A, Al-Yish KS. The Comprehensive Archaeological Survey Program. A. The second preliminary report on the Southwestern Province. Atlal. 1981;5:9–42.

Zarins J, Rahbini A, Kamal M. Preliminary report on the archaeological survey of the Riyadh area. Atlal. 1982;6:25–38.

Zeuner FE. 'Neolithic' sites from the Rub' al-Khali, southern Arabia. Man. 1954;54:133–6.

Index